Study and Solutions Guide

College Algebra:
A Graphing Approach

Fifth Edition

Larson/Hostetler/Edwards

Bruce H. Edwards

University of Florida
Gainesville, Florida

Houghton Mifflin Company   Boston   New York

Publisher: Richard Stratton
Sponsoring Editor: Cathy Cantin
Senior Marketing Manager: Jennifer Jones
Editorial Associate: Jeannine Lawless
Editorial Assistant: Joanna Carter-O'Connell
Marketing Associate: Mary Legere
New Title Project Manager: Susan Peltier

Printed in the U.S.A.

ISBN-10: 0-618-85191-7

ISBN-13: 0-978-0-618-85191-1

123456789-CRS-11 10 09 08 07

Preface

This *Study and Solutions Guide* is a supplement to *College Algebra: A Graphing Approach*, Fifth Edition, by Ron Larson, Robert Hostetler, and Bruce H. Edwards.

Solutions to the exercises in the text are given in two parts. Part I contains solutions to odd-numbered Section and Review Exercises; summaries of the chapters; and Practice Tests with solutions. Part II contains solutions to the Chapter and Cumulative Tests from the textbook.

This *Study and Solutions Guide* is the result of the efforts of Larson Texts, Inc. If you have any corrections or suggestions for improving this guide, we would appreciate hearing from you.

Bruce H. Edwards
358 Little Hall
University of Florida
Gainesville, FL 32611
be@math.ufl.edu

Contents

PART I

CHAPTER P
Prerequisites

CHAPTER P
Prerequisites

Section P.1   Real Numbers

■ You should know the following sets.

(a) The set of real numbers includes the rational numbers and the irrational numbers.

(b) The set of rational numbers includes all real numbers that can be written as the ratio p/q of two integers, where $q \neq 0$.

(c) The set of irrational numbers includes all real numbers which are not rational.

(d) The set of integers: $\{. \ . \ ., -3, -2, -1, 0, 1, 2, 3, . \ . \ .\}$

(e) The set of whole numbers: $\{0, 1, 2, 3, 4, . \ . \ .\}$

(f) The set of natural numbers: $\{1, 2, 3, 4, . \ . \ .\}$

■ The real number line is used to represent the real numbers.

■ Know the inequality symbols.

(a) $a < b$ means a is less than b.   (b) $a \leq b$ means a is less than or equal to b.

(c) $a > b$ means a is greater than b.   (d) $a \geq b$ means a is greater than or equal to b.

■ You should know that
$$|a| = \begin{cases} a, & \text{if } a \geq 0 \\ -a, & \text{if } a < 0. \end{cases}$$

■ Know the properties of absolute value.

(a) $|a| \geq 0$ (b) $|-a| = |a|$ (c) $|ab| = |a|\,|b|$ (d) $\left|\dfrac{a}{b}\right| = \dfrac{|a|}{|b|}$

■ The distance between a and b on the real line is $|b - a| = |a - b|$.

■ You should be able to identify the terms in an algebraic expression.

■ You should know and be able to use the basic rules of algebra.

■ Commutative Property

(a) Addition: $a + b = b + a$ (b) Multiplication: $a \cdot b = b \cdot a$

■ Associative Property

(a) Addition: $(a + b) + c = a + (b + c)$ (b) Multiplication: $(ab)c = a(bc)$

■ Identity Property

(a) Addition: 0 is the identity; $a + 0 = 0 + a = a$

(b) Multiplication: 1 is the identity; $a \cdot 1 = 1 \cdot a = a$

■ Inverse Property

(a) Addition: $-a$ is the inverse of a; $a + (-a) = -a + a = 0$

(b) Multiplication: $1/a$ is the inverse of a, $a \neq 0$; $a(1/a) = (1/a)a = 1$

■ Distributive Property

(a) Left: $a(b + c) = ab + ac$ (b) Right: $(a + b)c = ac + bc$

■ Properties of Negatives

(a) $(-1)a = -a$ (b) $-(-a) = a$

(c) $(-a)b = a(-b) = -ab$ (d) $(-a)(-b) = ab$

(e) $-(a + b) = (-a) + (-b) = -a - b$ —CONTINUED—

2

Section P.1 —CONTINUED—

■ Properties of Zero

(a) $a \pm 0 = a$

(b) $a \cdot 0 = 0$

(c) $0 \div a = 0/a = 0, a \neq 0$

(d) If $ab = 0$, then $a = 0$ or $b = 0$.

(e) $a/0$ is undefined.

■ Properties of Fractions ($b \neq 0, d \neq 0$)

(a) Equivalent Fractions: $a/b = c/d$ if and only if $ad = bc$.

(b) Rule of Signs: $-a/b = a/(-b) = -(a/b)$ and $-a/(-b) = a/b$

(c) Equivalent Fractions: $a/b = ac/bc, c \neq 0$

(d) Addition and Subtraction

1. Like Denominators: $(a/b) \pm (c/b) = (a \pm c)/b$

2. Unlike Denominators: $(a/b) \pm (c/d) = (ad \pm bc)/bd$

(e) Multiplication: $(a/b) \cdot (c/d) = ac/bd$

(f) Division: $(a/b)/(c/d) = (a/b) \cdot (d/c) = ad/bc$, if $c \neq 0$.

■ Properties of Equality

(a) If $a = b$, then $a + c = b + c$.

(b) If $a = b$, then $ac = bc$.

(c) If $a + c = b + c$, then $a = b$.

(d) If $ac = bc$ and $c \neq 0$, then $a = b$.

Vocabulary Check

1. rational

2. Irrational

3. absolute value

4. composite

5. prime

6. variables, constants

7. terms

8. coefficient

9. Zero-Factor Property

---

1. $-9, -\frac{7}{2}, 5, \frac{2}{3}, \sqrt{2}, 0, 1, -4, -1$

(a) Natural numbers: 5, 1

(b) Whole numbers: 5, 1, 0

(c) Integers: $-9, 5, 0, 1, -4, -1$

(d) Rational numbers: $-9, -\frac{7}{2}, 5, \frac{2}{3}, 0, 1, -4, -1$

(e) Irrational numbers: $\sqrt{2}$

3. $2.01, 0.666\ldots, -13, 0.010110111\ldots, 1, -10, 20$

(a) Natural numbers: 1, 20

(b) Whole numbers: 1, 20

(c) Integers: $-13, 1, -10, 20$

(d) Rational numbers:

$2.01, 0.666\ldots, -13, 1, -10, 20$

(e) Irrational numbers: $0.010110111\ldots$

5. $-\pi, -\frac{1}{3}, \frac{6}{3}, \frac{1}{2}\sqrt{2}, -7.5, -2, 3, -3$

(a) Natural numbers: $\frac{6}{3}$ (since it equals 2), 3

(b) Whole numbers: $\frac{6}{3}$, 3

(c) Integers: $\frac{6}{3}, -2, 3, -3$

(d) Rational numbers: $-\frac{1}{3}, \frac{6}{3}, -7.5, -2, 3, -3$

(e) Irrational numbers: $-\pi, \frac{1}{2}\sqrt{2}$

7. $\frac{5}{16} = 0.3125$

9. $\frac{41}{333} = 0.\overline{123}$

11. $-\frac{100}{11} = -9.\overline{09}$

13. $4.6 = 4\frac{6}{10} = \frac{46}{10} = \frac{23}{5}$

15. $-6.5 = -6\frac{5}{10} = -6\frac{1}{2} = -\frac{13}{2}$

17. $-1 < 2.5$

19. $-4 > -8$

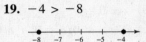

21. $\frac{3}{2} < 7$

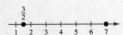

23. $\frac{5}{6} > \frac{2}{3}$

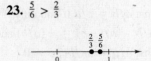

25. (a) The inequality $x \leq 5$ is the set of all real numbers less than or equal to 5.

(b) ![number line graph]
 0 1 2 3 4 5 6 → x

(c) The interval is unbounded.

27. (a) The inequality $x < 0$ is the set of all negative real numbers.

(b) ![number line graph]
 -2 -1 0 1 2 → x

(c) The interval is unbounded.

29. (a) The inequality $-2 < x < 2$ is the set of all real numbers greater than -2 and less than 2.

(b) ![number line graph]
 -2 -1 0 1 2 → x

(c) The interval is bounded.

31. (a) The inequality $-1 \leq x < 0$ is the set of all negative real numbers greater than or equal to -1.

(b) ![number line graph]
 -1 0 → x

(c) The interval is bounded.

33. $x < 0; (-\infty, 0)$

35. $y \geq 0; [0, \infty)$

37. $-1 \leq p < 9; [-1, 9)$

39. $-1 \leq x \leq 5$ or $[-1, 5]$

41. $-3 \leq x < 10$ or $[-3, 10)$

43. $-a \leq x \leq a + 4, \ (a \geq 0)$ or $[-a, a + 4]$

45. The interval $(-6, \infty)$ consists of all real numbers greater than -6.

47. The interval $(-\infty, 2]$ consists of all real numbers less than or equal to 2.

49. $|-10| = -(-10) = 10$

51. $-3|-3| = -3[-(-3)] = -9$

53. If $x + 2 > 0$, then $\dfrac{|x + 2|}{x + 2} = \dfrac{x + 2}{x + 2} = 1$.

If $x + 2 < 0$, then $\dfrac{|x + 2|}{x + 2} = \dfrac{-(x + 2)}{x + 2} = -1$.

Therefore, $\dfrac{|x + 2|}{x + 2} = \begin{cases} 1 \text{ if } x > -2 \\ -1 \text{ if } x < -2 \end{cases}$

The expression is undefined if $x = -2$.

55. $y_1 = \frac{1}{4}(x^2 - 8)$

$y_2 = \frac{1}{4}x^2 - 2$

The graphs are identical. The Distributive Property is illustrated.

57. $|x| - 2|y| = |2| - 2|-1|$
$\qquad = 2 - 2(1) = 2 - 2 = 0$

59. $\left|\dfrac{3x + 2y}{|x|}\right| = \left|\dfrac{3(4) + 2(1)}{|4|}\right|$

$\qquad = \left|\dfrac{12 + 2}{4}\right| = \dfrac{14}{4} = \dfrac{7}{2}$

61. $|-3| > -|-3|$ since $3 > -3$.

63. $-5 = -|5|$ since $-5 = -5$.

65. $-|-2| = -|2|$ since $-2 = -2$.

67. $d(126, 75) = |75 - 126| = 51$

69. $d\left(-\frac{5}{2}, 0\right) = \left|0 - \left(-\frac{5}{2}\right)\right| = \frac{5}{2}$

71. $d\left(\frac{16}{5}, \frac{112}{75}\right) = \left|\frac{112}{75} - \frac{16}{5}\right| = \frac{128}{75}$

73. $d(x, 5) = |x - 5|$ and $d(x, 5) \leq 3$
Thus, $|x - 5| \leq 3$.

75. $d(y, 0) = |y - 0| = |y|$ and $d(y, 0) \geq 6$
Thus, $|y| \geq 6$.

77. $d(57, 236) = |236 - 57|$
$\qquad = 179$ miles

79.

Budgeted Expense, b	Actual Expense, a	$|a - b|$	0.05b
$112,700	$113,356	$656	$5635

The actual expense difference is greater than $500 (but is less than 5% of the budget) so it does not pass the test.

81.

Budgeted Expense, b	Actual Expense, a	$|a - b|$	0.05b
$37,640	$37,335	$305	$1882

Since $305 < $500 and $305 < $1882, it passes the "budget variance test."

83. Receipts: $1351.8 billion
$|1351.8 - 1515.8| = 164 billion budget deficit for 1995

85. Receipts: $1827.5 billion
$|1827.5 - 1701.9| = 125.6 billion budget surplus for 1999

87. Receipts: $1782.3 billion
$|1782.3 - 2157.6| = 375.3 billion budget deficit for 2003

89. Terms: $7x, 4$
Coefficient of $7x$ is 7

91. Terms: $\sqrt{3}x^2, -8x, -11$
Coefficient of $\sqrt{3}x^2$ is $\sqrt{3}$
Coefficient of $-8x$ is -8

93. Terms: $4x^3, \frac{x}{2}, -5$
Coefficient of $4x^3$ is 4
Coefficient of $\frac{x}{2}$ is $\frac{1}{2}$

95. $2x - 5$
(a) $2(3) - 5 = 6 - 5 = 1$
(b) $2\left(-\frac{1}{2}\right) - 5 = -1 - 5 = -6$

97. $x^2 - 4$
(a) $(2)^2 - 4 = 4 - 4 = 0$
(b) $(-2)^2 - 4 = 4 - 4 = 0$

99. $x + 9 = 9 + x$
Commutative (addition)

101. $\dfrac{1}{(h + 6)}(h + 6) = 1, h \neq -6$
Inverse (multiplication)

103. $2(x + 3) = 2x + 6$
Distributive Property

105. $x + (y + 10) = (x + y) + 10$

Associative Property of Addition

107. $\frac{3}{16} + \frac{5}{16} = \frac{8}{16} = \frac{1}{2}$

109. $\frac{5}{8} - \frac{5}{12} + \frac{1}{6} = \frac{15}{24} - \frac{10}{24} + \frac{4}{24} = \frac{9}{24} = \frac{3}{8}$

111. $\frac{x}{6} + \frac{3x}{4} = \frac{2x}{12} + \frac{9x}{12} = \frac{11x}{12}$

113. $\frac{12}{x} \div \frac{1}{8} = \frac{12}{x} \cdot \frac{8}{1} = \frac{96}{x}$

115. $\left(\frac{2}{5} \div 4\right) - \left(4 \cdot \frac{3}{8}\right) = \left(\frac{2}{5} \cdot \frac{1}{4}\right) - \frac{12}{8} = \frac{1}{10} - \frac{3}{2}$

$$= \frac{1}{10} - \frac{15}{10} = -\frac{14}{10} = -\frac{7}{5}$$

117. $14\left(-3 + \frac{3}{7}\right) = -36$

119. $\frac{11.46 - 5.37}{3.91} \approx 1.56$

121. $\frac{\frac{2}{3}(-2 - 6)}{-\frac{2}{5}} \approx 13.33$

123. (a)

n	1	0.5	0.01	0.0001	0.000001
$5/n$	5	10	500	50,000	5,000,000

(b) As n approaches 0, $5/n$ approaches infinity (∞). That is, $5/n$ increases without bound.

125. This is false. For example, $3 > 2$, but $\frac{1}{3} < \frac{1}{2}$.

127. (a) $-A$ is negative, $-A < 0$, because $A > 0$.

(b) $B - A$ is negative, $B - A < 0$, because $B < 0$ and $-A < 0$.

129. Consider $|u + v|$ and $|u| + |v|$

(a) No, the values of the expressions are not always equal. For example, if $u = 5$ and $v = -2$, then $|u + v| = |5 - 2| = 3$, whereas $|u| + |v| = |5| + |-2| = 5 + 2 = 7$. In general, the expressions are unequal if u is positive and v is negative, or if u is negative and v is positive.

(b) $|u + v|$ is always less than or equal to $|u| + |v|$: $|u + v| \le |u| + |v|$. The expressions are equal if u and v have the same sign or one or both of u and v is 0. If they differ in sign (one positive and the other negative), then $|u + v| < |u| + |v|$.

131. Answers will vary. The set of natural numbers is given by $\{1, 2, 3, \ldots\}$.

The whole numbers are the natural numbers together with 0. The integers include the natural numbers, their negatives, and 0: $\{\ldots -3, -2, -1, 0, 1, 2, 3, \ldots\}$. The rational numbers include the integers. The rational numbers consist of all ratios of the form p/q, where p and q are integers, $q \ne 0$. The irrational numbers consist of all real numbers that are not rational numbers.

Section P.2   Exponents and Radicals

■ You should know the properties of exponents.

(a) $a^1 = a$

(b) $a^0 = 1, a \neq 0$

(c) $a^m a^n = a^{m+n}$

(d) $a^m/a^n = a^{m-n}, a \neq 0$

(e) $a^{-n} = 1/a^n, a \neq 0$

(f) $(a^m)^n = a^{mn}$

(g) $(ab)^n = a^n b^n$

(h) $(a/b)^n = a^n/b^n, b \neq 0$

(i) $(a/b)^{-n} = (b/a)^n, a \neq 0, b \neq 0$

(j) $|a^2| = |a|^2 = a^2$

■ You should be able to write numbers in scientific notation, $\pm c \times 10^n$, where $1 \leq c < 10$ and n is an integer.

■ You should be able to use your calculator to evaluate expressions involving exponents.

■ You should know the properties of radicals.

(a) $\sqrt[n]{a^m} = \left(\sqrt[n]{a}\right)^m = a^{m/n}$

(b) $\sqrt[n]{a} \cdot \sqrt[n]{b} = \sqrt[n]{ab}$

(c) $\dfrac{\sqrt[n]{a}}{\sqrt[n]{b}} = \sqrt[n]{\dfrac{a}{b}}$

(d) $\sqrt[m]{\sqrt[n]{a}} = \sqrt[mn]{a}$

(e) $\left(\sqrt[n]{a}\right)^n = a$

(f) For n even, $\sqrt[n]{a^n} = |a|$

For n odd, $\sqrt[n]{a^n} = a$

(g) $a^{1/n} = \sqrt[n]{a}$

■ You should be able to simplify radicals.

(a) All possible factors have been removed from the radical sign.

(b) All fractions have radical-free denominators.

(c) The index for the radical has been reduced as far as possible.

■ You should be able to use your calculator to evaluate radicals.

Vocabulary Check

1. exponent, base

2. scientific notation

3. square root

4. principal nth root

5. index, radicand

6. simplest form

7. conjugates

8. rationalizing

9. power, index

1. (a) $4^2 \cdot 3 = 16 \cdot 3 = 48$

(b) $3 \cdot 3^3 = 3^4 = 81$

3. (a) $(3^3)^2 = 3^6 = 729$

(b) $-3^2 = -9$

5. (a) $\dfrac{3}{3^{-4}} = 3^{1+4} = 3^5 = 243$

(b) $24(-2)^{-5} = \dfrac{24}{(-2)^5} = \dfrac{24}{-32} = -\dfrac{3}{4}$

7. (a) $2^{-1} + 3^{-1} = \dfrac{1}{2} + \dfrac{1}{3} = \dfrac{5}{6}$

(b) $(2^{-1})^{-2} = 2^2 = 4$

9. When $x = 2$, $7x^{-2} = 7(2^{-2}) = 7\left(\dfrac{1}{2^2}\right) = \dfrac{7}{4}$.

11. When $x = -3$,

$2x^3 = 2(-3)^3 = 2(-27) = -54$.

13. When $x = -\frac{1}{2}$, $4x^2 = 4\left(-\frac{1}{2}\right)^2 = 4\left(\frac{1}{4}\right) = 1$.

15. (a) $(-5z)^3 = (-5)^3 z^3 = -125z^3$

(b) $5x^4(x^2) = 5x^{4+2} = 5x^6$

17. (a) $\dfrac{7x^2}{x^3} = 7x^{2-3} = 7x^{-1} = \dfrac{7}{x}$

  (b) $\dfrac{12(x+y)^3}{9(x+y)} = \dfrac{4}{3}(x+y)^{3-1}$

  $\qquad\qquad = \dfrac{4}{3}(x+y)^2, \ x+y \neq 0$

19. (a) $[(x^2y^{-2})^{-1}]^{-1} = x^2y^{-2} = \dfrac{x^2}{y^2}, \ x \neq 0$

  (b) $\left(\dfrac{a^{-2}}{b^{-2}}\right)\left(\dfrac{b}{a}\right)^3 = \dfrac{b^2}{a^2} \cdot \dfrac{b^3}{a^3} = \dfrac{b^5}{a^5}, \ b \neq 0$

21. $(-4)^3(5^2) = (-64)(25)$

  $\qquad\qquad = -1600$

23. $\dfrac{3^6}{7^3} = \dfrac{729}{343} \approx 2.125$

25. $852.25 = 8.5225 \times 10^2$

27. $10{,}252.484 = 1.0252484 \times 10^4$

29. $-1110.25 = -1.11025 \times 10^3$

31. $0.0002485 = 2.485 \times 10^{-4}$

33. $-0.0000025 = -2.5 \times 10^{-6}$

35. $1.25 \times 10^5 = 125{,}000$

37. $-4.816 = -481{,}600{,}000$

39. $3.25 \times 10^{-8} = 0.0000000325$

41. $-9.001 \times 10^{-3} = -0.009001$

43. $57{,}300{,}000 = 5.73 \times 10^7$ square miles

45. $0.0000899 = 8.99 \times 10^{-5}$ gram per cm^3

47. $5.71 \times 10^8 = 571{,}000{,}000$ servings

49. $1.6022 \times 10^{-19} = 0.00000000000000000016022$ coulombs

51. $\sqrt{25 \times 10^8} = \sqrt{5^2 \times (10^4)^2} = 5 \times 10^4 = 50{,}000$

53. (a) $(9.3 \times 10^6)^3(6.1 \times 10^{-4}) \approx 4.907 \times 10^{17}$

  (b) $\dfrac{(2.414 \times 10^4)^6}{(1.68 \times 10^5)^5} \approx 1.479$

55. (a) $\sqrt{4.5 \times 10^9} \approx 67{,}082.039$

  (b) $\sqrt[3]{6.3 \times 10^4} \approx 39.791$

57. $\sqrt{121} = \sqrt{11^2} = 11$

59. $-\sqrt[3]{-27} = -(-3) = 3$

61. $\left(\sqrt[3]{-125}\right)^3 = -125$

63. $32^{-3/5} = \dfrac{1}{32^{3/5}} = \dfrac{1}{\left(\sqrt[5]{32}\right)^3} = \dfrac{1}{(2)^3} = \dfrac{1}{8}$

65. $\left(-\dfrac{1}{64}\right)^{-1/3} = (-64)^{1/3} = \sqrt[3]{-64} = -4$

67. $\sqrt[5]{-27^3} = (-27)^{3/5} \approx -7.225$

69. $(3.4)^{2.5} \approx 21.316$

71. $(1.2^{-2})\sqrt{75} + 3\sqrt{8} \approx 14.499$

73. $\sqrt{\pi + 1} \approx 2.035$

75. $\dfrac{3.14}{\pi} + \sqrt[3]{5} \approx 2.709$

77. $(2.8)^{-2} + 1.01 \times 10^6 \approx 1{,}010{,}000.128$

79. (a) $\left(\sqrt[4]{3}\right)^4 = (3^{1/4})^4 = 3$

  (b) $\sqrt[5]{96x^5} = (2^5 \cdot 3 \cdot x^5)^{1/5} = 2x\sqrt[5]{3}$

81. (a) $\sqrt{54xy^4} = \sqrt{3^2 \cdot 6 \cdot x(y^2)^2} = 3y^2\sqrt{6x}$

  (b) $\sqrt[3]{\dfrac{32a^2}{b^2}} = \left(\dfrac{2^3 \cdot 2^2 a^2}{b^2}\right)^{1/3} = 2\sqrt[3]{\dfrac{4a^2}{b^2}}$

83. (a) $2\sqrt{50} + 12\sqrt{8} = 2\sqrt{25 \cdot 2} + 12\sqrt{4 \cdot 2} = 2\left(5\sqrt{2}\right) + 12\left(2\sqrt{2}\right) = 10\sqrt{2} + 24\sqrt{2} = 34\sqrt{2}$

  (b) $10\sqrt{32} - 6\sqrt{18} = 10\sqrt{16 \cdot 2} - 6\sqrt{9 \cdot 2} = 10\left(4\sqrt{2}\right) - 6\left(3\sqrt{2}\right) = 40\sqrt{2} - 18\sqrt{2} = 22\sqrt{2}$

85. (a) $3\sqrt{x+1} + 10\sqrt{x+1} = 13\sqrt{x+1}$

(b) $7\sqrt{80x} - 2\sqrt{125x} = 7\sqrt{16 \cdot 5x} - 2\sqrt{25 \cdot 5x}$

$$= 7(4)\sqrt{5x} - 2(5)\sqrt{5x}$$

$$= 28\sqrt{5x} - 10\sqrt{5x}$$

$$= 18\sqrt{5x}$$

87. $\sqrt{5} + \sqrt{3} \approx 3.968$ and $\sqrt{5+3} = \sqrt{8} \approx 2.828$

Thus, $\sqrt{5} + \sqrt{3} > \sqrt{5+3}$.

89. $\sqrt{3^2 + 2^2} = \sqrt{9+4} = \sqrt{13} \approx 3.606$

Thus, $5 > \sqrt{3^2 + 2^2}$.

91. $\dfrac{1}{\sqrt{3}} = \dfrac{1}{\sqrt{3}} \cdot \dfrac{\sqrt{3}}{\sqrt{3}} = \dfrac{\sqrt{3}}{3}$

93. $\dfrac{5}{\sqrt{14}-2} = \dfrac{5}{\sqrt{14}-2} \cdot \dfrac{\sqrt{14}+2}{\sqrt{14}+2}$

$$= \dfrac{5(\sqrt{14}+2)}{14-4}$$

$$= \dfrac{5(\sqrt{14}+2)}{10}$$

$$= \dfrac{\sqrt{14}+2}{2}$$

95. $\dfrac{\sqrt{8}}{2} = \dfrac{\sqrt{4\cdot 2}}{2} = \dfrac{2\sqrt{2}}{2} = \dfrac{\sqrt{2}}{1} \cdot \dfrac{\sqrt{2}}{\sqrt{2}} = \dfrac{2}{\sqrt{2}}$

97. $\dfrac{\sqrt{5}+\sqrt{3}}{3} = \dfrac{\sqrt{5}+\sqrt{3}}{3} \cdot \dfrac{\sqrt{5}-\sqrt{3}}{\sqrt{5}-\sqrt{3}}$

$$= \dfrac{5-3}{3(\sqrt{5}-\sqrt{3})}$$

$$= \dfrac{2}{3(\sqrt{5}-\sqrt{3})}$$

	Radical Form		Rational Exponent Form	
99.	$\sqrt[3]{64}$	Given	$64^{1/3}$	Answer
101.	$\sqrt[5]{32}$	Answer	$32^{1/5}$	Given
103.	$\sqrt[3]{-216}$	Given	$(-216)^{1/3}$	Answer
105.	$\sqrt[4]{81^3}$	Given	$81^{3/4}$	Answer

107. $\dfrac{(2x^2)^{3/2}}{2^{1/2}x^4} = \dfrac{2^{3/2}(x^2)^{3/2}}{2^{1/2}x^4} = \dfrac{2^{3/2}x^3}{2^{1/2}x^4} = 2^{3/2 - 1/2}x^{3-4} = 2^1 x^{-1} = \dfrac{2}{x}$

109. $\dfrac{x^{-3} \cdot x^{1/2}}{x^{3/2} \cdot x^{-1}} = \dfrac{x^{1/2} \cdot x^1}{x^{3/2} \cdot x^3} = x^{(1/2)+1-(3/2)-3} = x^{-3} = \dfrac{1}{x^3}, x > 0$

111. (a) $\sqrt[4]{3^2} = 3^{2/4} = 3^{1/2} = \sqrt{3}$

(b) $\sqrt[6]{(x+1)^4} = (x+1)^{4/6}$

$$= (x+1)^{2/3} = \sqrt[3]{(x+1)^2}$$

113. (a) $\sqrt{\sqrt{32}} = (32^{1/2})^{1/2}$

$$= 32^{1/4} = \sqrt[4]{32} = \sqrt[4]{16 \cdot 2} = 2\sqrt[4]{2}$$

(b) $\sqrt{\sqrt[4]{2x}} = ((2x)^{1/4})^{1/2} = (2x)^{1/8} = \sqrt[8]{2x}$

115. For $v = \dfrac{3}{4}$, size $= 0.03\sqrt{\dfrac{3}{4}} = 0.03\dfrac{\sqrt{3}}{\sqrt{4}} = 0.03\dfrac{\sqrt{3}}{2} \approx 0.026$ inches.

117. The sum of all the storms from 1995 to 2005 is

$$19 + 13 + 8 + 14 + 12 + 15 + 15 + 12 + 16 + 15 + 27 = 166.$$

The average per year is $\dfrac{166}{11} \approx 15.09$, a rational number.

119. For $x \neq 0$, this is true because $\dfrac{x^{k+1}}{x} = \dfrac{x^{k+1}}{x^1} = x^k$.

121. For $a \neq 0$, $1 = \dfrac{a}{a} = \dfrac{a^1}{a^1} = a^{1-1} = a^0$.

Thus, $a^0 = 1$.

123. Consider $x^2 = n$, x a positive integer.

Unit digit of x	Unit digit of $n = x^2$
1	1
2	4
3	9
4	6
5	5
6	6
7	9
8	4
9	1
0	0

Therefore, the possible digits are 0, 1, 4, 5, 6, and 9. Thus, $\sqrt{5233}$ is *not* an integer because its unit digit is 3.

Section P.3   Polynomials and Factoring

- Given a polynomial in x, $a_n x^n + a_{n-1} x^{n-1} + \cdots + a_1 x + a_0$, where $a_n \neq 0$, and n is a nonnegative integer, you should be able to identify the following:

  (a) Degree: n

  (b) Terms: $a_n x^n, a_{n-1} x^{n-1}, \ldots, a_1 x, a_0$

  (c) Coefficients: $a_n, a_{n-1}, \ldots, a_1, a_0$

  (d) Leading coefficient: a_n

  (e) Constant term: a_0

- You should be able to add and subtract polynomials.

- You should be able to multiply polynomials by either

  (a) The Distributive Properties

  (b) The Vertical Method

- You should know the special binomial products.

  (a) $(ax + b)(cx + d) = acx^2 + adx + bcx + bd$ FOIL

  $\qquad\qquad\qquad\;\; = acx^2 + (ad + bc)x + bd$

  (b) $(u \pm v)^2 = u^2 \pm 2uv + v^2$

  (c) $(u + v)(u - v) = u^2 - v^2$

  (d) $(u \pm v)^3 = u^3 \pm 3u^2 v + 3uv^2 \pm v^3$

- You should be able to factor out all common factors, the first step in factoring.

- You should be able to factor the following special polynomial forms.

  (a) $u^2 - v^2 = (u + v)(u - v)$

  (b) $u^2 \pm 2uv + v^2 = (u \pm v)^2$

  (c) $mx^2 + nx + r = (ax + b)(cx + d)$, where $m = ac$; $r = bd$, $n = ad + bc$

  Note: Not all trinomials can be factored (using real coefficients).

  (d) $u^3 \pm v^3 = (u \pm v)(u^2 \mp uv + v^2)$

- You should be able to factor by grouping.

Vocabulary Check

1. n, a_n

2. zero polynomial

3. monomial

4. First, Outer, Inner, Last

5. prime

6. perfect square trinomial

---

1. 7 is a polynomial of degree zero. Matches (d).

3. $-4x^3 + 1$ is a binomial with leading coefficient -4. Matches (b).

5. $\frac{3}{4}x^4 + x^2 + 14$ is a trinomial with leading coefficient $\frac{3}{4}$. Matches (f).

7. $-2x^3 + 4x$ is one possible answer.

9. $-4x^4 + 3$ is one possible answer.

11. $3x + 4x^2 + 2 = 4x^2 + 3x + 2$ Standard form
Degree: 2
Leading coefficient: 4

13. $5 - x^6 = -x^6 + 5$ Standard form
Degree: 6
Leading coefficient: -1

15. $1 - x + 6x^4 - 2x^5 = -2x^5 + 6x^4 - x + 1$ Standard form
Degree: 5
Leading coefficient: -2

17. This is a polynomial: $4x^3 + 3x - 5$

19. $\sqrt{x^2 - x^4}$ is not a polynomial.

21. $(6x + 5) - (8x + 15) = 6x + 5 - 8x - 15 = (6x - 8x) + (5 - 15) = -2x - 10$

23. $-(t^3 - 1) + (6t^3 - 5t) = -t^3 + 1 + 6t^3 - 5t = 5t^3 - 5t + 1$

25. $(15x^2 - 6) - (-8.1x^3 - 14.7x^2 - 17) = 15x^2 - 6 + 8.1x^3 + 14.7x^2 + 17 = 8.1x^3 + 29.7x^2 + 11$

27. $3x(x^2 - 2x + 1) = 3x(x^2) + 3x(-2x) + 3x(1) = 3x^3 - 6x^2 + 3x$

29. $-5z(3z - 1) = -5z(3z) + (-5z)(-1)$
$\qquad\qquad\quad = -15z^2 + 5z$

31. $(1 - x^3)(4x) = 1(4x) - x^3(4x) = 4x - 4x^4 = -4x^4 + 4x$

33. $(2.5x^2 + 5)(-3x) = (2.5x^2)(-3x) + 5(-3x) = -7.5x^3 - 15x$

35. $-2x\left(\frac{1}{8}x + 3\right) = -2x\left(\frac{1}{8}x\right) - 2x(3) = -\frac{1}{4}x^2 - 6x$

37. $(x + 3)(x + 4) = x^2 + 4x + 3x + 12$ FOIL
$\qquad\qquad\qquad\ = x^2 + 7x + 12$

39. $(3x - 5)(2x + 1) = 6x^2 + 3x - 10x - 5$ FOIL

$\qquad = 6x^2 - 7x - 5$

41. $(2x - 5y)^2 = 4x^2 - 2(5y)(2x) + 25y^2$

$\qquad = 4x^2 - 20xy + 25y^2$

43. $(x + 10)(x - 10) = x^2 - 100$

45. $(x + 2y)(x - 2y) = x^2 - (2y)^2 = x^2 - 4y^2$

47. $(2r^2 - 5)(2r^2 + 5) = (2r^2)^2 - 5^2 = 4r^4 - 25$

49. $(x + 1)^3 = x^3 + 3x^2(1) + 3x(1^2) + 1^3$

$\qquad = x^3 + 3x^2 + 3x + 1$

51. $(2x - y)^3 = (2x)^3 - 3(2x)^2y + 3(2x)y^2 - y^3$

$\qquad = 8x^3 - 12x^2y + 6xy^2 - y^3$

53. $\left(\frac{1}{2}x - 5\right)^2 = \left(\frac{1}{2}x\right)^2 - 2(5)\left(\frac{1}{2}x\right) + 5^2$

$\qquad = \frac{1}{4}x^2 - 5x + 25$

55. $\left(\frac{1}{4}x - 3\right)\left(\frac{1}{4}x + 3\right) = \left(\frac{1}{4}x\right)^2 - 3^2 = \frac{1}{16}x^2 - 9$

57. $(2.4x + 3)^2 = (2.4x)^2 + 2(3)(2.4x) + 3^2 = 5.76x^2 + 14.4x + 9$

59.

$$
\begin{array}{r}
-x^2 + x - 5 \\
3x^2 + 4x + 1 \\
\hline
-x^2 + x - 5 \\
-4x^3 + 4x^2 - 20x \\
-3x^4 + 3x^3 - 15x^2 \\
\hline
-3x^4 - x^3 - 12x^2 - 19x - 5
\end{array}
$$

Answer: $-3x^4 - x^3 - 12x^2 - 19x - 5$

61. $[(x + 2z) + 5][(x + 2z) - 5] = (x + 2z)^2 - 25 = x^2 + 4xz + 4z^2 - 25$

63. $[(x - 3) + y]^2 = (x - 3)^2 + 2y(x - 3) + y^2$

$\qquad = x^2 - 6x + 9 + 2xy - 6y + y^2$

$\qquad = x^2 + 2xy + y^2 - 6x - 6y + 9$

65. $5x(x + 1) - 3x(x + 1) = (5x - 3x)(x + 1) = 2x(x + 1) = 2x^2 + 2x$

67. $(u + 2)(u - 2)(u^2 + 4) = (u^2 - 4)(u^2 + 4)$

$\qquad = u^4 - 16$

69. $4x + 16 = 4(x + 4)$

71. $2x^3 - 6x = 2x(x^2 - 3)$

73. $3x(x - 5) + 8(x - 5) = (3x + 8)(x - 5)$

75. $x^2 - 64 = x^2 - 8^2 = (x + 8)(x - 8)$

77. $48y^2 - 27 = 3(16y^2 - 9) = 3((4y)^2 - 3^2)$

$\qquad = 3(4y + 3)(4y - 3)$

79. $4x^2 - \frac{1}{9} = (2x)^2 - \left(\frac{1}{3}\right)^2 = \left(2x + \frac{1}{3}\right)\left(2x - \frac{1}{3}\right)$

81. $(x - 1)^2 - 4 = [(x - 1) + 2][(x - 1) - 2]$

$\qquad = (x + 1)(x - 3)$

83. $x^2 - 4x + 4 = x^2 - 2(2)x + 2^2 = (x - 2)^2$

85. $x^2 + x + \frac{1}{4} = x^2 + 2\left(\frac{1}{2}\right)x + \left(\frac{1}{2}\right)^2 = \left(x + \frac{1}{2}\right)^2$

87. $4x^2 - 12x + 9 = (2x)^2 - 2(2x)(3) + 3^2$
$$= (2x - 3)^2$$

89. $4x^2 - \frac{4}{3}x + \frac{1}{9} = (2x)^2 - 2(2x)\left(\frac{1}{3}\right) + \left(\frac{1}{3}\right)^2$
$$= \left(2x - \frac{1}{3}\right)^2$$

91. $x^3 + 64 = x^3 + 4^3 = (x + 4)(x^2 - 4x + 16)$

93. $y^3 + 216 = y^3 + 6^3 = (y + 6)(y^2 - 6y + 36)$

95. $x^3 - \frac{8}{27} = x^3 - \left(\frac{2}{3}\right)^3 = \left(x - \frac{2}{3}\right)\left(x^2 + \frac{2}{3}x + \frac{4}{9}\right)$

97. $8x^3 - 1 = (2x)^3 - 1 = (2x - 1)(4x^2 + 2x + 1)$

99. $(x + 2)^3 - y^3 = [(x + 2) - y][(x + 2)^2 + (x + 2)y + y^2] = (x + 2 - y)(x^2 + 4x + 4 + xy + 2y + y^2)$

101. $x^2 + x - 2 = (x + 2)(x - 1)$

103. $s^2 - 5s + 6 = (s - 3)(s - 2)$

105. $20 - y - y^2 = (5 + y)(4 - y)$
or $-(y + 5)(y - 4)$

107. $3x^2 - 5x + 2 = (3x - 2)(x - 1)$

109. $2x^2 - x - 1 = (2x + 1)(x - 1)$

111. $5x^2 + 26x + 5 = (5x + 1)(x + 5)$

113. $-5u^2 - 13u + 6 = -(5u^2 + 13u - 6)$
$$= -(5u - 2)(u + 3)$$
or $(2 - 5u)(u + 3)$

115. $x^3 - x^2 + 2x - 2 = x^2(x - 1) + 2(x - 1)$
$$= (x - 1)(x^2 + 2)$$

117. $6x^2 + x - 2$

$a = 6, c = -2, ac = -12 = 4(-3)$ where
$4 - 3 = 1 = b.$

Thus, $6x^2 + x - 2 = 6x^2 + 4x - 3x - 2$
$$= 2x(3x + 2) - (3x + 2)$$
$$= (2x - 1)(3x + 2).$$

119. $x^3 - 5x^2 + x - 5 = x^2(x - 5) + (x - 5)$
$$= (x^2 + 1)(x - 5)$$

121. $x^3 - 16x = x(x^2 - 16) = x(x + 4)(x - 4)$

123. $x^3 - x^2 = x^2(x - 1)$

125. $x^2 - 2x + 1 = (x - 1)^2$

127. $1 - 4x + 4x^2 = (1 - 2x)^2 = (2x - 1)^2$

129. $2x^2 + 4x - 2x^3 = -2x(-x - 2 + x^2)$
$$= -2x(x^2 - x - 2)$$
$$= -2x(x + 1)(x - 2)$$

131. $9x^2 + 10x + 1 = (9x + 1)(x + 1)$

133. $\frac{1}{8}x^2 - \frac{1}{96}x - \frac{1}{16} = \frac{1}{8}\left(x^2 - \frac{1}{12}x - \frac{1}{2}\right)$
$$= \frac{1}{8}\left(x - \frac{3}{4}\right)\left(x + \frac{2}{3}\right)$$
$$= \frac{1}{96}(4x - 3)(3x + 2)$$

135. $3x^3 + x^2 + 15x + 5 = x^2(3x + 1) + 5(3x + 1)$
$$= (3x + 1)(x^2 + 5)$$

137. $3u - 2u^2 + 6 - u^3 = -u^3 - 2u^2 + 3u + 6$
$$= -u^2(u + 2) + 3(u + 2)$$
$$= (3 - u^2)(u + 2)$$

139. $2x^3 + x^2 - 8x - 4 = x^2(2x + 1) - 4(2x + 1)$
$$= (x^2 - 4)(2x + 1)$$
$$= (x + 2)(x - 2)(2x + 1)$$

141. $(x^2 + 1)^2 - 4x^2 = [(x^2 + 1) + 2x][(x^2 + 1) - 2x]$ **143.** $2t^3 - 16 = 2(t^3 - 8) = 2(t - 2)(t^2 + 2t + 4)$

$$= (x^2 + 2x + 1)(x^2 - 2x + 1)$$

$$= (x + 1)^2(x - 1)^2$$

145. $4x(2x - 1) + 2(2x - 1)^2 = 2(2x - 1)(2x + (2x - 1))$

$$= 2(2x - 1)(4x - 1)$$

147. $2(x + 1)(x - 3)^2 - 3(x + 1)^2(x - 3) = (x + 1)(x - 3)[2(x - 3) - 3(x + 1)]$

$$= (x + 1)(x - 3)[2x - 6 - 3x - 3]$$

$$= (x + 1)(x - 3)(-x - 9)$$

$$= -(x + 1)(x - 3)(x + 9)$$

149. $(2x + 1)^4(2)(3x - 1)(3) + (3x - 1)^2(4)(2x + 1)^3(2) = 2(2x + 1)^3(3x - 1)[3(2x + 1) + 4(3x - 1)]$

$$= 2(2x + 1)^3(3x - 1)[6x + 3 + 12x - 4]$$

$$= 2(2x + 1)^3(3x - 1)(18x - 1)$$

151. $(x^2 + 5)^4(2)(3x - 1)(3) + (3x - 1)^2(4)(x^2 + 5)^3(2x) = 2(x^2 + 5)^3(3x - 1)[3(x^2 + 5) + 4x(3x - 1)]$

$$= 2(x^2 + 5)^3(3x - 1)[3x^2 + 15 + 12x^2 - 4x]$$

$$= 2(x^2 + 5)^3(3x - 1)(15x^2 - 4x + 15)$$

153. (a) $500(1 + r)^2 = 500(r + 1)^2 = 500(r^2 + 2r + 1) = 500r^2 + 1000r + 500$

(b)

r	$2\frac{1}{2}\%$	3%	4%	$4\frac{1}{2}\%$	5%
$500(1 + r)^2$	\$525.31	\$530.45	\$540.80	\$546.01	\$551.25

Remember to write the interest rate in decimal form: $2\frac{1}{2}\% = 0.025$

(c) As r increases, the amount increases.

155. Volume = (width)(length)(height)

$$= (15 - 2x)\left(\frac{45 - 3x}{2}\right)(x) = \frac{3}{2}x(x - 15)(2x - 15)$$

If $x = 3$, volume = 486 cubic centimeters.

If $x = 5$, volume = 375 cubic centimeters.

If $x = 7$, volume = 84 cubic centimeters.

157. (a) $T = R + B = 1.1x + (0.0475x^2 - 0.001x + 0.23)$

$$= 0.0475x^2 + 1.099x + 0.23$$

(b)

x mi/hr	30	40	55
T feet	75.95	120.19	204.36

(c) As the speed x increases, the total stopping distance increases.

159. $a^2 - b^2 = (a + b)(a - b)$

Matches model (b).

161. $a^2 + 2a + 1 = (a + 1)^2$

Matches model (a).

163. $3x^2 + 7x + 2 = (3x + 1)(x + 2)$

165. $2x^2 + 7x + 3 = (2x + 1)(x + 3)$

167. $A = \pi(r + 2)^2 - \pi r^2 = \pi[(r + 2)^2 - r^2]$

$\qquad = \pi[r^2 + 4r + 4 - r^2] = \pi(4r + 4)$

$\qquad = 4\pi(r + 1)$

169. $A = 8(18) - 4x^2$

$\qquad = 4(36 - x^2)$

$\qquad = 4(6 - x)(6 + x)$

171. $x^4(4)(2x + 1)^3(2x) + (2x + 1)^4(4x^3) = 4x^3(2x + 1)^3[2x^2 + (2x + 1)]$

$\qquad\qquad\qquad\qquad\qquad\qquad\qquad = 4x^3(2x + 1)^3(2x^2 + 2x + 1)$

173. $(2x - 5)^4(3)(5x - 4)^2(5) + (5x - 4)^3(4)(2x - 5)^3(2) = (2x - 5)^3(5x - 4)^2[15(2x - 5) + 8(5x - 4)]$

$\qquad\qquad\qquad\qquad\qquad\qquad\qquad\qquad\qquad\qquad = (2x - 5)^3(5x - 4)^2[30x - 75 + 40x - 32]$

$\qquad\qquad\qquad\qquad\qquad\qquad\qquad\qquad\qquad\qquad = (2x - 5)^3(5x - 4)^2(70x - 107)$

175. $\dfrac{(5x - 1)(3) - (3x + 1)(5)}{(5x - 1)^2} = \dfrac{15x - 3 - 15x - 5}{(5x - 1)^2}$

$\qquad\qquad\qquad\qquad\qquad\quad = \dfrac{-8}{(5x - 1)^2}$

177. For $x^2 + bx - 15 = (x + m)(x + n)$ to be factorable, b must equal $m + n$ where $mn = -15$.

Factors of -15	Sum of factors
$(15)(-1)$	$15 + (-1) = 14$
$(-15)(1)$	$-15 + 1 = -14$
$(3)(-5)$	$3 + (-5) = -2$
$(-3)(5)$	$-3 + 5 = 2$

The possible b-values are $14, -14, -2,$ or 2.

179. For $x^2 + bx + 50 = (x + m)(x + n)$ to be factorable, b must equal $m + n$ where $mn = 50$.

Factors of 50	Sum of factors
$(50)(1)$	51
$(-50)(-1)$	-51
$(25)(2)$	27
$(-25)(-2)$	-27
$(10)(5)$	15
$(-10)(-5)$	-15

The possible b-values are $51, -51, 27, -27, 15,$ or -15.

181. For $2x^2 + 5x + c$ to be factorable, the factors of $2c$ must add up to 5.

Possible c-values	$2c$	Factors of $2c$ that add up to 5
2	4	$(1)(4) = 4$ and $1 + 4 = 5$
3	6	$(2)(3) = 6$ and $2 + 3 = 5$
-3	-6	$(6)(-1) = -6$ and $6 + (-1) = 5$
-7	-14	$(7)(-2) = -14$ and $7 + (-2) = 5$
-12	-24	$(8)(-3) = -24$ and $8 + (-3) = 5$

These are a few possible c-values. There are many correct answers.

If $c = 2$: $2x^2 + 5x + 2 = (2x + 1)(x + 2)$ If $c = -7$: $2x^2 + 5x - 7 = (2x + 7)(x - 1)$

If $c = 3$: $2x^2 + 5x + 3 = (2x + 3)(x + 1)$ If $c = -12$: $2x^2 + 5x - 12 = (2x - 3)(x + 4)$

If $c = -3$: $2x^2 + 5x - 3 = (2x - 1)(x + 3)$

183. For $3x^2 - 10x + c$ to be factorable, the factors of $3c$ must add up to -10.

Possible c-values	$3c$	Factors of $3c$ that must add up to -10
3	9	$(-9)(-1) = 9$ and $-9 - 1 = -10$
7	21	$(-3)(-7) = 21$ and $-3 - 7 = -10$
8	24	$(-4)(-6) = 24$ and $-4 - 6 = -10$
-8	-24	$(2)(-12) = -24$ and $-12 + 2 = -10$

Other c-values are possible. The above values yield the following factorizations. There are many correct answers.

If $c = 3$: $3x^2 - 10x + 3 = (3x - 1)(x - 3)$

If $c = 7$: $3x^2 - 10x + 7 = (3x - 7)(x - 1)$

If $c = 8$: $3x^2 - 10x + 8 = (3x - 4)(x - 2)$

If $c = -8$: $3x^2 - 10x - 8 = (3x + 2)(x - 4)$

185. $V = \pi R^2 h - \pi r^2 h$

(a) $V = \pi h[R^2 - r^2] = \pi h(R - r)(R + r)$

(b) The average radius is $\dfrac{R + r}{2}$. The thickness of the shell is $R - r$. Therefore,

$$V = \pi h(R + r)(R - r) = 2\pi\left(\frac{R + r}{2}\right)(R - r)h = 2\pi(\text{average radius})(\text{thickness})h.$$

187. False. The product of the two binomials is not always a second-degree polynomial. For instance, $(x^2 + 2)(x^2 - 3) = x^4 - x^2 - 6$ is a fourth-degree polynomial.

189. False. For example, $3^2 + 4^2 \neq (3 + 4)^2$.

191. If $m < n$, then the sum of two polynomials of degree m and n is n, the larger degree.

193. $(x - y)^2 \neq x^2 - y^2$ because you have to use the FOIL Method.

$$(x - y)^2 = x^2 - 2xy + y^2 \neq x^2 - y^2$$

195. To cube a binomial difference, cube the first term. Next subtract 3 times the product of the square of the first term times the second term. Next add 3 times the product of the first term times the square of the second term. Finally, subtract the cube of the second term.

$$(x - y)^3 = x^3 - 3x^2y + 3xy^2 - y^3$$

197. No, $(3x - 6)(x + 1)$ is not completely factored because $(3x - 6) = 3(x - 2)$. Completely factored form is $3(x - 2)(x + 1)$.

199. (a) Yes, the sum will always be a fourth-degree polynomial. For example,

$$(x^3 + 2x + 1) + (3x^4 - 3x^3 + x) = 3x^4 - 2x^3 + 3x + 1.$$

(b) No, The sum cannot be a second-degree polynomial. The coefficient of x^4 is not zero.

(c) No, the sum will be a fourth-degree polynomial.

Section P.4   Rational Expressions

- You should be able to find the domain of a fractional expression.
- You should know that a rational expression is the quotient of two polynomials.
- You should be able to simplify rational expressions by reducing them to lowest terms. This may involve factoring both the numerator and the denominator.
- You should be able to add, subtract, multiply, and divide rational expressions.
- You should be able to simplify compound fractions.

Vocabulary Check

1. domain
2. rational expression
3. complex fractions
4. smaller
5. equivalent

1. The domain of the polynomial $3x^2 - 4x + 7$ is the set of all real numbers.

3. The domain of the polynomial $4x^3 + 3$, $x \geq 0$ is the set of non-negative real numbers, since the polynomial is restricted to that set.

5. The domain of $\dfrac{1}{3 - x}$ is the set of all real numbers x such that $x \neq 3$.

7. The denominator $x^2 - 2x + 1 = (x - 1)^2$ cannot equal 0. Hence, the domain is the set of all real numbers x such that $x \neq 1$.

9. The denominator $x^2 - 6x + 9 = (x - 3)^2$ cannot equal 0. Hence, the domain is the set of all real numbers x such that $x \neq 3$.

11. The domain of $\sqrt{x + 7}$ is the set of all real numbers x such that $x \geq -7$.

13. The domain of $\sqrt{2x - 5}$ is the set of all real numbers x such that $2x - 5 \geq 0$, which implies that $x \geq \frac{5}{2}$.

15. The domain of $1/\sqrt{x - 3}$ is the set of all real numbers x such that $x - 3 > 0$, which implies that $x > 3$.

17. $\dfrac{5}{2x} = \dfrac{5(3x)}{(2x)(3x)} = \dfrac{5(3x)}{6x^2}, \ x \neq 0$

Missing factor: $3x$

19. $\dfrac{3}{4} = \dfrac{3(x + 1)}{4(x + 1)}$

The missing factor is $(x + 1)$, where $x \neq -1$.

21. $\dfrac{x - 1}{4(x + 2)} = \dfrac{(x - 1)(x + 2)}{4(x + 2)(x + 2)}$

$= \dfrac{4(x - 1)(x + 2)}{4(x + 2)^2}, \ x \neq -2$

Missing factor: $x + 2$

23. $\dfrac{15x^2}{10x} = \dfrac{5x(3x)}{5x(2)} = \dfrac{3x}{2}, \ x \neq 0$

25. $\dfrac{3xy}{xy + x} = \dfrac{x(3y)}{x(y + 1)} = \dfrac{3y}{y + 1}, \ x \neq 0$

27. $\dfrac{4y - 8y^2}{10y - 5} = \dfrac{4y(1 - 2y)}{5(2y - 1)}$

$= \dfrac{-4y(2y - 1)}{5(2y - 1)} = \dfrac{-4y}{5}, \ y \neq \dfrac{1}{2}$

29. $\dfrac{x - 5}{10 - 2x} = \dfrac{x - 5}{-2(x - 5)} = -\dfrac{1}{2}, \ x \neq 5$

31. $\dfrac{y^2 - 16}{y + 4} = \dfrac{(y + 4)(y - 4)}{y + 4} = y - 4, \ y \neq -4$

33. $\dfrac{x^3 + 5x^2 + 6x}{x^2 - 4} = \dfrac{x(x + 2)(x + 3)}{(x + 2)(x - 2)}$

$= \dfrac{x(x + 3)}{x - 2}, \ x \neq -2$

35. $\dfrac{y^2 - 7y + 12}{y^2 + 3y - 18} = \dfrac{(y - 3)(y - 4)}{(y + 6)(y - 3)} = \dfrac{y - 4}{y + 6}, \ y \neq 3$

37. $\dfrac{2 - x + 2x^2 - x^3}{x - 2} = \dfrac{(2 - x) + x^2(2 - x)}{-(2 - x)} = \dfrac{(2 - x)(1 + x^2)}{-(2 - x)} = -(1 + x^2), \ x \neq 2$

39. $\dfrac{z^3 - 8}{z^2 + 2z + 4} = \dfrac{(z - 2)(z^2 + 2z + 4)}{z^2 + 2z + 4} = z - 2$

41.

x	0	1	2	3	4	5	6
$\dfrac{x^2 - 2x - 3}{x - 3}$	1	2	3	Undef.	5	6	7
$x + 1$	1	2	3	4	5	6	7

The expressions are equivalent except at $x = 3$. In fact,

$\dfrac{x^2 - 2x - 3}{x - 3} = \dfrac{(x - 3)(x + 1)}{x - 3} = x + 1, \ x \neq 3$.

43. $\dfrac{5x^3}{2x^3 + 4} = \dfrac{5x^3}{2(x^3 + 2)}$

There are no common factors so this expression is in reduced form. In this case, factors of terms were incorrectly cancelled.

45. $\dfrac{\pi r^2}{(2r)^2} = \dfrac{\pi r^2}{4r^2} = \dfrac{\pi}{4}$

47. $\dfrac{5}{x-1} \cdot \dfrac{x-1}{25(x-2)} = \dfrac{1}{5(x-2)}, x \neq 1$

49. $\dfrac{r}{r-1} \div \dfrac{r^2}{r^2-1} = \dfrac{r}{r-1} \cdot \dfrac{r^2-1}{r^2}$

$$= \dfrac{r(r+1)(r-1)}{(r-1)r^2} = \dfrac{r+1}{r}, \ r \neq 1, r \neq -1$$

51. $\dfrac{t^2-t-6}{t^2+6t+9} \cdot \dfrac{t+3}{t^2-4} = \dfrac{(t-3)(t+2)(t+3)}{(t+3)^2(t+2)(t-2)} = \dfrac{t-3}{(t+3)(t-2)}, t \neq -2$

53. $\dfrac{3(x+y)}{4} \div \dfrac{x+y}{2} = \dfrac{3(x+y)}{4} \cdot \dfrac{2}{x+y} = \dfrac{3}{2}, x \neq -y$

55. $\dfrac{5}{x-1} + \dfrac{x}{x-1} = \dfrac{5+x}{x-1} = \dfrac{x+5}{x-1}$

57. $\dfrac{6}{2x+1} - \dfrac{x}{x+3} = \dfrac{6(x+3)-x(2x+1)}{(2x+1)(x+3)}$

$$= \dfrac{6x+18-2x^2-x}{(2x+1)(x+3)}$$

$$= \dfrac{-2x^2+5x+18}{(2x+1)(x+3)}$$

$$= -\dfrac{2x^2-5x-18}{(2x+1)(x+3)}$$

59. $\dfrac{3}{x-2} + \dfrac{5}{2-x} = \dfrac{3}{x-2} - \dfrac{5}{x-2} = -\dfrac{2}{x-2}$

61. $\dfrac{1}{x^2-x-2} - \dfrac{x}{x^2-5x+6} = \dfrac{1}{(x-2)(x+1)} - \dfrac{x}{(x-2)(x-3)}$

$$= \dfrac{(x-3)-x(x+1)}{(x+1)(x-2)(x-3)} = \dfrac{-x^2-3}{(x+1)(x-2)(x-3)} = -\dfrac{x^2+3}{(x+1)(x-2)(x-3)}$$

63. $-\dfrac{1}{x} + \dfrac{2}{x^2+1} - \dfrac{1}{x^3+x} = \dfrac{-(x^2+1)}{x(x^2+1)} + \dfrac{2x}{x(x^2+1)} - \dfrac{1}{x(x^2+1)}$

$$= \dfrac{-x^2-1+2x-1}{x(x^2+1)} = -\dfrac{x^2-2x+2}{x(x^2+1)}$$

65. $\dfrac{\left(\dfrac{x}{2}-1\right)}{(x-2)} = \dfrac{\left(\dfrac{x}{2}-\dfrac{2}{2}\right)}{\left(\dfrac{x-2}{1}\right)} = \dfrac{x-2}{2} \cdot \dfrac{1}{x-2} = \dfrac{1}{2}, x \neq 2$

67. $\dfrac{\left[\dfrac{x^2}{(x+1)^2}\right]}{\left[\dfrac{x}{(x+1)^3}\right]} = \dfrac{x^2}{(x+1)^2} \cdot \dfrac{(x+1)^3}{x} = \dfrac{x(x+1)}{1} = x^2+x, \ x \neq 0, -1$

69. $\dfrac{\left[\dfrac{1}{(x+h)^2} - \dfrac{1}{x^2}\right]}{h} = \dfrac{\left[\dfrac{1}{(x+h)^2} - \dfrac{1}{x^2}\right]}{h} \cdot \dfrac{x^2(x+h)^2}{x^2(x+h)^2} = \dfrac{x^2 - (x+h)^2}{hx^2(x+h)^2}$

$$= \dfrac{x^2 - (x^2 + 2xh + h^2)}{hx^2(x+h)^2} = \dfrac{-h(2x+h)}{hx^2(x+h)^2} = -\dfrac{2x+h}{x^2(x+h)^2}, h \neq 0$$

71. $\dfrac{\left(\sqrt{x} - \dfrac{1}{2\sqrt{x}}\right)}{\sqrt{x}} = \dfrac{\left(\sqrt{x} - \dfrac{1}{2\sqrt{x}}\right)}{\sqrt{x}} \cdot \dfrac{2\sqrt{x}}{2\sqrt{x}} = \dfrac{2x-1}{2x}, \ x > 0$

73. $x^5 - 2x^{-2} = x^{-2}(x^7 - 2) = \dfrac{x^7 - 2}{x^2}$

75. $x^2(x^2+1)^{-5} - (x^2+1)^{-4} = (x^2+1)^{-5}[x^2 - (x^2+1)] = -\dfrac{1}{(x^2+1)^5}$

77. $2x^2(x-1)^{1/2} - 5(x-1)^{-1/2} = (x-1)^{-1/2}(2x^2(x-1) - 5) = \dfrac{2x^3 - 2x^2 - 5}{(x-1)^{1/2}}$

79. $\dfrac{2x^{3/2} - x^{-1/2}}{x^2} = \dfrac{x^{-1/2}(2x^2 - 1)}{x^2} = \dfrac{2x^2 - 1}{x^{5/2}}$

81. $\dfrac{-x^2(x^2+1)^{-1/2} + 2x(x^2+1)^{-3/2}}{x^3} = \dfrac{x(x^2+1)^{-3/2}[-x(x^2+1) + 2]}{x^3}$

$$= \dfrac{(x^2+1)^{-3/2}[-x^3 - x + 2]}{x^2} = \dfrac{-x^3 - x + 2}{x^2(x^2+1)^{3/2}}$$

$$= -\dfrac{(x-1)(x^2+x+2)}{x^2(x^2+1)^{3/2}}$$

83. $\dfrac{(x^2+5)\left(\frac{1}{2}\right)(4x+3)^{-1/2}(4) - (4x+3)^{1/2}(2x)}{(x^2+5)^2} = \dfrac{2(4x+3)^{-1/2}[(x^2+5) - x(4x+3)]}{(x^2+5)^2}$

$$= \dfrac{2(-3x^2 - 3x + 5)}{(x^2+5)^2\sqrt{4x+3}}$$

$$= -\dfrac{2(3x^2 + 3x - 5)}{(x^2+5)^2\sqrt{4x+3}}$$

85. $\dfrac{\sqrt{x+2} - \sqrt{x}}{2} = \dfrac{\sqrt{x+2} - \sqrt{x}}{2} \cdot \dfrac{\sqrt{x+2} + \sqrt{x}}{\sqrt{x+2} + \sqrt{x}}$

$$= \dfrac{(x+2) - x}{2(\sqrt{x+2} + \sqrt{x})} = \dfrac{2}{2(\sqrt{x+2} + \sqrt{x})}$$

$$= \dfrac{1}{\sqrt{x+2} + \sqrt{x}}$$

87. $\dfrac{\sqrt{x+2}-\sqrt{2}}{x} = \dfrac{\sqrt{x+2}-\sqrt{2}}{x} \cdot \dfrac{\sqrt{x+2}+\sqrt{2}}{\sqrt{x+2}+\sqrt{2}}$

$$= \dfrac{(x+2)-2}{x(\sqrt{x+2}+\sqrt{2})}$$

$$= \dfrac{x}{x(\sqrt{x+2}+\sqrt{2})}$$

$$= \dfrac{1}{\sqrt{x+2}+\sqrt{2}}, \ x \neq 0$$

89. $\dfrac{\sqrt{x+9}-3}{x} = \dfrac{\sqrt{x+9}-3}{x} \cdot \dfrac{\sqrt{x+9}+3}{\sqrt{x+9}+3}$

$$= \dfrac{(x+9)-9}{x(\sqrt{x+9}+3)}$$

$$= \dfrac{x}{x(\sqrt{x+9}+3)}$$

$$= \dfrac{1}{\sqrt{x+9}+3}, \ x \neq 0$$

91. Probability $= \dfrac{\text{Area shaded rectangle}}{\text{Area large rectangle}} = \dfrac{x(x/2)}{x(2x+1)} = \dfrac{x/2}{2x+1} \cdot \dfrac{2}{2} = \dfrac{x}{2(2x+1)}$

93. (a) $\dfrac{1}{16}$ minute to copy one page

    (b) $x\dfrac{1}{16} = \dfrac{x}{16}$ minutes to copy x pages

    (c) $\dfrac{60}{16} = \dfrac{15}{4}$ minutes to copy 60 pages

95. (a) $R_T = \dfrac{1}{\dfrac{1}{R_1} + \dfrac{1}{R_2} + \dfrac{1}{R_3}}$

$$= \dfrac{1}{\dfrac{R_2 R_3 + R_1 R_3 + R_1 R_2}{R_1 R_2 R_3}}$$

$$= \dfrac{R_1 R_2 R_3}{R_2 R_3 + R_1 R_3 + R_1 R_2}$$

    (b) $R_T = \dfrac{(6)(4)(12)}{(4)(12) + (6)(12) + (6)(4)}$

$$= \dfrac{288}{144} = 2 \text{ ohms}$$

97. (a)

Year	2000	2001	2002	2003	2004	2005
Endangered	565	592	597	598	599	599
Threatened	139	144	146	147	147	147

The models are very close to the data.

    (b) $\dfrac{\text{Threatened}}{\text{Endangered}} = \dfrac{(243.48t^2 + 139)(3.91t^2 + 1)}{(1.65t^2 + 1)(2342.52t^2 + 565)}$

Year	2000	2001	2002	2003	2004	2005
Ratio	0.25	0.24	0.25	0.25	0.25	0.25

99. $\dfrac{(x+h)^2 - x^2}{h} = \dfrac{x^2 + 2xh + h^2 - x^2}{h}$

$$= \dfrac{h(2x + h)}{h}$$

$$= 2x + h, \ h \neq 0$$

101. $\dfrac{\dfrac{1}{(x+h)^2} - \dfrac{1}{x^2}}{h} = \dfrac{x^2 - (x+h)^2}{hx^2(x+h)^2}$

$$= \dfrac{x^2 - (x^2 + 2xh + h^2)}{hx^2(x+h)^2}$$

$$= \dfrac{-2xh - h^2}{hx^2(x+h)^2}$$

$$= \dfrac{-2x - h}{x^2(x+h)^2}, \ h \neq 0$$

103. $\dfrac{\sqrt{2x + h} - \sqrt{2x}}{h} = \dfrac{\sqrt{2x + h} - \sqrt{2x}}{h} \cdot \dfrac{\sqrt{2x + h} + \sqrt{2x}}{\sqrt{2x + h} + \sqrt{2x}}$

$$= \dfrac{(2x + h) - 2x}{h\left(\sqrt{2x + h} + \sqrt{2x}\right)}$$

$$= \dfrac{h}{h\left(\sqrt{2x + h} + \sqrt{2x}\right)}$$

$$= \dfrac{1}{\sqrt{2x + h} + \sqrt{2x}}, \ h \neq 0$$

105. $\dfrac{4}{n}\left(\dfrac{n(n + 1)(2n + 1)}{6}\right) + 2n\left(\dfrac{4}{n}\right) = \dfrac{2(n + 1)(2n + 1)}{3} + 8$

$$= \dfrac{2(2n^2 + 3n + 1) + 24}{3}$$

$$= \dfrac{4n^2 + 6n + 26}{3}, \ n \neq 0$$

107. False. For n odd, the domain of $(x^{2n} - 1)/(x^n - 1)$ is all $x \neq 1$, unlike the domain of the right-hand side.

109. Completely factor the numerator and denominator to determine if they have any common factors.

111. Answers will vary. For example, let $x = y = 1$:

$$\sqrt{1 + 1} = \sqrt{2} \neq \sqrt{1} + \sqrt{1} = 2$$

Section P.5   The Cartesian Plane

- You should be able to plot points.
- You should know that the distance between (x_1, y_1) and (x_2, y_2) in the plane is
  $$d = \sqrt{(x_2 - x_1)^2 + (y_2 - y_1)^2}.$$
- You should know that the midpoint of the line segment joining (x_1, y_1) and (x_2, y_2) is
  $$\left(\dfrac{x_1 + x_2}{2}, \dfrac{y_1 + y_2}{2}\right).$$
- You should know the equation of a circle: $(x - h)^2 + (y - k)^2 = r^2$.
- You should be able to translate points in the plane.

Vocabulary Check

1. (a) iii   (b) vi   (c) i   (d) iv   (e) v   (f) ii   **2.** Cartesian

3. Distance Formula   **4.** Midpoint Formula

5. $(x - h)^2 + (y - k)^2 = r^2$, center, radius

1. $A: (2, 6)$, $B: (-6, -2)$, $C: (4, -4)$, $D: (-3, 2)$

3.

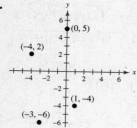

5.

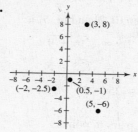

7. $(-5, 4)$

9. $(-6, -6)$

11. $x > 0 \Rightarrow$ The point lies in Quadrant I or in Quadrant IV.

$y < 0 \Rightarrow$ The point lies in Quadrant III or in Quadrant IV.

$x > 0$ and $y < 0 \Rightarrow (x, y)$ lies in Quadrant IV.

13. $x = -4 \Rightarrow$ x is negative $\Rightarrow$ The point lies in Quadrant II or in Quadrant III.

$y > 0 \Rightarrow$ The point lies in Quadrant I or Quadrant II.

$x = -4$ and $y > 0 \Rightarrow (x, y)$ lies in Quadrant II.

15. $y < -5 \Rightarrow y$ is negative $\Rightarrow$ The point lies in either Quadrant III or Quadrant IV.

17. If $-y > 0$, then $y < 0$.

$x < 0 \Rightarrow$ The point lies in Quadrant II or in Quadrant III.

$y < 0 \Rightarrow$ The point lies in Quadrant III or in Quadrant IV.

$x < 0$ and $y < 0 \Rightarrow (x, y)$ lies in Quadrant III.

19. If $xy > 0$, then either x and y are both positive, or both negative. Hence, (x, y) lies in either Quadrant I or Quadrant III.

21.

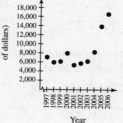

23. $(6, -3)$, $(6, 5)$

$d = \sqrt{(6 - 6)^2 + (5 - (-3))^2} = \sqrt{64} = 8$

25. $(-3, -1)$, $(2, -1)$

$d = \sqrt{(2 - (-3))^2 + (-1 - (-1))^2} = \sqrt{25} = 5$

27. $d = \sqrt{(3 - (-2))^2 + (-6 - 6)^2} = \sqrt{5^2 + (-12)^2}$

$= \sqrt{25 + 144} = \sqrt{169} = 13$

29. $\left(\frac{1}{2}, \frac{4}{3}\right)$, $(2, -1)$

$d = \sqrt{\left(\frac{1}{2} - 2\right)^2 + \left(\frac{4}{3} + 1\right)^2}$

$= \sqrt{\frac{9}{4} + \frac{49}{9}}$

$= \sqrt{\frac{277}{36}} = \frac{\sqrt{277}}{6} \approx 2.77$

31. $(-4.2, 3.1)$, $(-12.5, 4.8)$

$d = \sqrt{(-4.2 + 12.5)^2 + (3.1 - 4.8)^2}$

$= \sqrt{68.89 + 2.89}$

$= \sqrt{71.78} \approx 8.47$

33. (a) The distance between $(0, 2)$ and $(4, 2)$ is 4.

The distance between $(4, 2)$ and $(4, 5)$ is 3.

The distance between $(0, 2)$ and $(4, 5)$ is

$$\sqrt{(4 - 0)^2 + (5 - 2)^2} = \sqrt{16 + 9}$$
$$= \sqrt{25} = 5.$$

(b) $4^2 + 3^2 = 16 + 9 = 25 = 5^2$

35. (a) The distance between $(-1, 1)$ and $(9, 1)$ is 10.

The distance between $(9, 1)$ and $(9, 4)$ is 3.

The distance between $(-1, 1)$ and $(9, 4)$ is

$$\sqrt{(9 - (-1))^2 + (4 - 1)^2} = \sqrt{100 + 9}$$
$$= \sqrt{109}.$$

(b) $10^2 + 3^2 = 109 = \left(\sqrt{109}\right)^2$

37. Find distances between pairs of points.

$$d_1 = \sqrt{(4 - 2)^2 + (0 - 1)^2} = \sqrt{5}$$
$$d_2 = \sqrt{(4 + 1)^2 + (0 + 5)^2} = \sqrt{50}$$
$$d_3 = \sqrt{(2 + 1)^2 + (1 + 5)^2} = \sqrt{45}$$
$$\left(\sqrt{5}\right)^2 + \left(\sqrt{45}\right)^2 = \left(\sqrt{50}\right)^2$$

Because $d_1{}^2 + d_3{}^2 = d_2{}^2$, the triangle is a right triangle.

39. $d_1 = \sqrt{(1 - 3)^2 + (-3 - 2)^2} = \sqrt{4 + 25} = \sqrt{29}$

$d_2 = \sqrt{(3 + 2)^2 + (2 - 4)^2} = \sqrt{25 + 4} = \sqrt{29}$

$d_3 = \sqrt{(1 + 2)^2 + (-3 - 4)^2} = \sqrt{9 + 49} = \sqrt{58}$

$d_1 = d_2$. Triangle is isosceles.

41. Find distances between pairs of points.

$$d_1 = \sqrt{(0 - 2)^2 + (9 - 5)^2} = \sqrt{4 + 16} = \sqrt{20} = 2\sqrt{5}$$
$$d_2 = \sqrt{(-2 - 0)^2 + (0 - 9)^2} = \sqrt{4 + 81} = \sqrt{85}$$
$$d_3 = \sqrt{(0 - (-2))^2 + (-4 - 0)^2} = \sqrt{4 + 16} = \sqrt{20} = 2\sqrt{5}$$
$$d_4 = \sqrt{(0 - 2)^2 + (-4 - 5)^2} = \sqrt{4 + 81} = \sqrt{85}$$

Opposite sides have equal lengths of $2\sqrt{5}$ and $\sqrt{85}$, so the figure is a parallelogram.

43. First show that the diagonals are equal in length.

$$d_1 = \sqrt{(0 - (-3))^2 + (8 - 1)^2} = \sqrt{9 + 49} = \sqrt{58}$$
$$d_2 = \sqrt{(2 - (-5))^2 + (3 - 6)^2} = \sqrt{49 + 9} = \sqrt{58}$$

Now use the Pythagorean Theorem to verify that at least one angle is $90°$ (and, hence, they are all right angles).

$$d_3 = \sqrt{(0 - (-5))^2 + (8 - 6)^2} = \sqrt{25 + 4} = \sqrt{29}$$
$$d_4 = \sqrt{(-3 - (-5))^2 + (1 - 6)^2} = \sqrt{4 + 25} = \sqrt{29}$$

Thus, $d_3{}^2 + d_4{}^2 = d_1{}^2$.

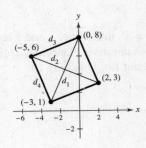

45. (a)

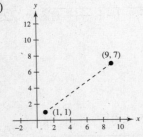

(b) $d = \sqrt{(9 - 1)^2 + (7 - 1)^2}$

$\quad = \sqrt{64 + 36} = 10$

(c) $\left(\dfrac{9 + 1}{2}, \dfrac{7 + 1}{2}\right) = (5, 4)$

47. (a)

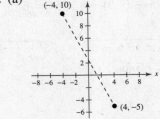

(b) $d = \sqrt{(4 + 4)^2 + (-5 - 10)^2}$

$\quad = \sqrt{64 + 225} = 17$

(c) $\left(\dfrac{4 - 4}{2}, \dfrac{-5 + 10}{2}\right) = \left(0, \dfrac{5}{2}\right)$

49. (a)

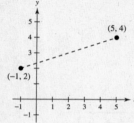

(b) $d = \sqrt{(5 + 1)^2 + (4 - 2)^2}$

$\quad = \sqrt{36 + 4} = \sqrt{40} = 2\sqrt{10}$

(c) $\left(\dfrac{-1 + 5}{2}, \dfrac{2 + 4}{2}\right) = (2, 3)$

51. (a)

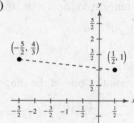

(b) $d = \sqrt{\left(\dfrac{1}{2} + \dfrac{5}{2}\right)^2 + \left(1 - \dfrac{4}{3}\right)^2}$

$\quad d = \sqrt{9 + \dfrac{1}{9}} = \dfrac{\sqrt{82}}{3}$

(c) $\left(\dfrac{-\frac{5}{2} + \frac{1}{2}}{2}, \dfrac{\frac{4}{3} + 1}{2}\right) = \left(-1, \dfrac{7}{6}\right)$

53. (a)

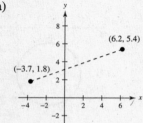

(b) $d = \sqrt{(6.2 + 3.7)^2 + (5.4 - 1.8)^2}$

$\quad = \sqrt{98.01 + 12.96} = \sqrt{110.97}$

(c) $\left(\dfrac{6.2 - 3.7}{2}, \dfrac{5.4 + 1.8}{2}\right) = (1.25, 3.6)$

55. Calculate the midpoint:

$$\left(\dfrac{2000 + 2006}{2}, \dfrac{2237 + 3950}{2}\right) = (2003, 3093.5)$$

The sales in 2003 are $3093.5 million.

57. Since $x_m = \dfrac{x_1 + x_2}{2}$ and $y_m = \dfrac{y_1 + y_2}{2}$ we have:

$\quad 2x_m = x_1 + x_2 \qquad 2y_m = y_1 + y_2$

$2x_m - x_1 = x_2 \qquad 2y_m - y_1 = y_2$

So, $(x_2, y_2) = (2x_m - x_1, 2y_m - y_1)$.

(a) $(x_2, y_2) = (2x_m - x_1, 2y_m - y_1) = (2(4) - 1, 2(-1) - (-2)) = (7, 0)$

(b) $(x_2, y_2) = (2x_m - x_1, 2y_m - y_1) = (2(2) - (-5), 2(4) - 11) = (9, -3)$

59. $(x - 0)^2 + (y - 0)^2 = 3^2$

$\qquad\qquad\quad x^2 + y^2 = 9$

61. $(x - 2)^2 + (y + 1)^2 = 4^2$

$\quad\ (x - 2)^2 + (y + 1)^2 = 16$

63. $(x + 1)^2 + (y - 2)^2 = r^2$

$\quad (0 + 1)^2 + (0 - 2)^2 = r^2 \implies r^2 = 5$

$\quad (x + 1)^2 + (y - 2)^2 = 5$

65. $r = \dfrac{1}{2}\sqrt{(6 - 0)^2 + (8 - 0)^2} = \dfrac{1}{2}\sqrt{100} = 5$

Center: $\left(\dfrac{0 + 6}{2}, \dfrac{0 + 8}{2}\right) = (3, 4)$

$\quad (x - 3)^2 + (y - 4)^2 = 25$

67. Because the circle is tangent to the x-axis, the radius is 1.

$$(x + 2)^2 + (y - 1)^2 = 1$$

69. The center is the midpoint of one of the diagonals of the square.

Center: $\left(\dfrac{7 + (-1)}{2}, \dfrac{-2 + (-10)}{2}\right) = (3, -6)$

The radius is one half the length of a side of the square.

Radius: $\dfrac{1}{2}(7 - (-1)) = 4$

Circle: $(x - 3)^2 + (y + 6)^2 = 16$

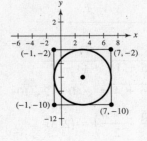

71. $(x - 2)^2 + (y + 1)^2 = 16$

73. $x^2 + y^2 = 25$

Center: $(0, 0)$

Radius: 5

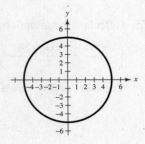

75. Center: $(1, -3)$

Radius: 2

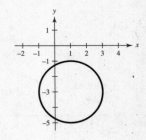

77. Center: $\left(\frac{1}{2}, \frac{1}{2}\right)$

Radius: $\frac{3}{2}$

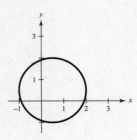

79. The x-coordinates are increased by 2, and the y-coordinates are increased by 5.

Old vertex	Shifted vertex
$(-1, -1)$	$(1, 4)$
$(-2, -4)$	$(0, 1)$
$(2, -3)$	$(4, 2)$

81.

Old vertex	Shifted vertex
$(0, 2)$	$(-1, 5)$
$(3, 5)$	$(2, 8)$
$(5, 2)$	$(4, 5)$
$(2, -1)$	$(1, 2)$

83. The point $(65, 83)$ represents an entrance exam score of 65.

85. (a) Sample answer: The number of artists inducted each year seems to be nearly steady except for the first few years. Estimate: Between 5 and 7 new members

(b) Sample answer: The Rock and Roll Hall of Fame was opened in 1986.

87. $d = \sqrt{(45 - 10)^2 + (40 - 15)^2} = \sqrt{35^2 + 25^2} = \sqrt{1850} = 5\sqrt{74} \approx 43$ yards

89. (a)

(b) Distance at 2 P.M.:

$$\sqrt{(-24 - 0)^2 + (0 - 32)^2} = \sqrt{1600} = 40 \text{ miles}$$

Distance at 4 P.M.:

$$\sqrt{(-48 - 0)^2 + (0 - 64)^2} = \sqrt{6400} = 80 \text{ miles}$$

Yes, the yachts are twice as far from each other.

91. Find the distances between pairs of points.

$$d_1 = \sqrt{\left(2 + 2\sqrt{3} - 2\right)^2 + (0 - 6)^2} = \sqrt{12 + 36} = \sqrt{48} = 4\sqrt{3}$$

$$d_2 = \sqrt{\left((2 + 2\sqrt{3}) - (2 - 2\sqrt{3})\right)^2 + (0 - 0)^2} = 4\sqrt{3}$$

$$d_3 = \sqrt{\left(2 - 2\sqrt{3} - 2\right)^2 + (0 - 6)^2} = \sqrt{12 + 36} = 4\sqrt{3}$$

Because $d_1 = d_2 = d_3$, the triangle is equilateral.

93. False. It would be sufficient to use the midpoint formula 15 times.

95. False. The polygon could be a rhombus. For example, consider the points $(4, 0)$, $(0, 6)$, $(-4, 0)$ and $(0, -6)$.

97. No, the scales can be different. The scales depend on the magnitude of the coordinates. See Figure P.13.

Section P.6   Representing Data Graphically

> ■ You should be able to construct line plots.
> ■ You should be able to construct histograms or frequency distributions.
> ■ You should be able to construct bar graphs.
> ■ You should be able to construct line graphs.

Vocabulary Check

1. Statistics

2. Line plots

3. histogram

4. frequency distribution

5. bar graph

6. Line graphs

1. (a) The price 2.569 occurred with the greatest frequency (6).

(b) The prices range from 2.459 to 2.649. The range is 2.649 − 2.459 = 0.19.

3.

The score of 15 occurred with the greatest frequency.

5. (Answers will depend on intervals selected.)

Interval	Tally
[0, 25)	$\cancel{||||}$ $\cancel{||||}$ $\cancel{||||}$ |||
[25, 50)	$\cancel{||||}$ $\cancel{||||}$ $\cancel{||||}$	
[50, 75)	$\cancel{||||}$	
[75, 100)	$\cancel{||||}$ |||
[100, 125)		
[125, 150)	
[150, 175)	
[175, 200)	
[200, 225)	
[225, 250)		

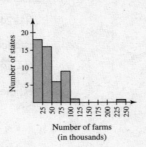

7.

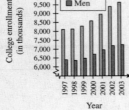

From 1995 to 2006, the number of Wal-Mart stores increases at a fairly constant rate.

9.

1999:	13,428 − 2430 =	$10,998
2000:	14,081 − 2506 =	$11,575
2001:	15,000 − 2562 =	$12,438
2002:	15,742 − 2700 =	$13,042
2003:	16,383 − 2903 =	$13,480
2004:	17,442 − 3313 =	$14,129

11.

College enrollment bar graph, Women and Men, Years 1997–2003.

13. From 1996 to 2005:

$$\frac{2400 - 1100}{1100} \approx 1.18, \text{ or } 118\%$$

15. Highest price was $2.59 in January.

17.

Women in the workforce line graph, Years 1995–2004.

From 1995 to 2004, the total number of women in the work force increases at a fairly constant rate.

19.

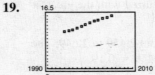

21.

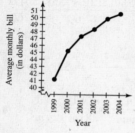

Answers will vary.

23.

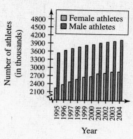

Answers will vary.

25. Answers will vary. Line plots are useful for ordering small sets of data. Histograms or bar graphs can be used to organize larger sets.

Review Exercises for Chapter P

1. $\{11, -14, -\frac{8}{9}, \frac{5}{2}, \sqrt{6}, 0.4\}$

  (a) Natural numbers: 11

  (b) Whole numbers: 11

  (c) Integers: 11, -14

  (d) Rational numbers: $11, -14, -\frac{8}{9}, \frac{5}{2}, 0.4$

  (e) Irrational numbers: $\sqrt{6}$

3. $\frac{5}{6} = 0.8\overline{3}$

  $\frac{7}{8} = 0.875$

  $\frac{5}{6} < \frac{7}{8}$

$$\overset{\underset{\textstyle\frac{17}{24}\quad\frac{3}{4}\quad\frac{19}{24}\quad\bullet\frac{5}{6}\quad\bullet\frac{7}{8}\quad\frac{11}{12}\quad\frac{23}{24}}{}}{\longrightarrow}$$

5. $x \le 7$

The set consists of all real numbers less than or equal to 7.

$$\overset{\underset{\textstyle 5\quad 6\quad 7\quad 8\quad 9}{}}{\longleftarrow\!\!\longrightarrow}\; x$$

7. $-2 \le x < -1,$ or $[-2, -1)$

9. $d(a, b) = |b - a|$
$$= |48 - (-74)|$$
$$= |48 + 74| = 122$$

11. $|x - 7| \ge 6$

13. $d(y, -30) = |y - (-30)| = |y + 30|$ and $d(y, -30) < 5$, so $|y + 30| < 5$.

15. $9x - 2$

  (a) $x = -1$: $9(-1) - 2 = -9 - 2 = -11$

  (b) $x = 3$: $9(3) - 2 = 27 - 2 = 25$

17. $-2x^2 - x + 3$

  (a) $x = 3$: $-2(3)^2 - 3 + 3 = -2(9) = -18$

  (b) $x = -3$:

    $-2(-3)^2 - (-3) + 3 = -18 + 6 = -12$

19. $2x + (3x - 10) = (2x + 3x) - 10$

Associative Property of Addition

21. $(t^2 + 1) + 3 = 3 + (t^2 + 1)$

Commutative Property of Addition

23. $\dfrac{2}{3} + \dfrac{8}{9} = \dfrac{2 \cdot 3}{3 \cdot 3} + \dfrac{8}{9} = \dfrac{6 + 8}{9} = \dfrac{14}{9}$

25. $\dfrac{3}{16} \div \dfrac{9}{2} = \dfrac{3}{16} \cdot \dfrac{2}{9} = \dfrac{3 \cdot 2}{2 \cdot 8 \cdot 3 \cdot 3} = \dfrac{1}{8 \cdot 3} = \dfrac{1}{24}$

27. $\dfrac{x}{5} + \dfrac{7x}{12} = \dfrac{12(x) + 7x(5)}{60} = \dfrac{12x + 35x}{60} = \dfrac{47x}{60}$

29. (a) $(-2z)^3 = (-2)^3 z^3 = -8z^3$

    (b) $(a^2 b^4)(3ab^{-2}) = 3a^{2+1}b^{4-2} = 3a^3 b^2$

31. (a) $\dfrac{6^2 u^3 v^{-3}}{12u^{-2}v} = \dfrac{36u^{3-(-2)}v^{-3-1}}{12} = 3u^5 v^{-4} = \dfrac{3u^5}{v^4}$

    (b) $\dfrac{3^{-4}m^{-1}n^{-3}}{9^{-2}mn^{-3}} = \dfrac{9^2 n^3}{3^4 mmn^3} = \dfrac{81}{81m^2} = \dfrac{1}{m^2} = m^{-2}$

33. $2{,}585{,}000{,}000 = 2.585 \times 10^9$

35. $0.000000125 = 1.25 \times 10^{-7}$

37. $5{,}100{,}000{,}000 = 5.1 \times 10^9$ dollars

39. $1.28 \times 10^5 = 128{,}000$

41. $1.80 \times 10^{-5} = 0.0000180$

43. $4.836 \times 10^8 = 483{,}600{,}000$

45. $\left(\sqrt[4]{78}\right)^4 = (78^{1/4})^4 = 78^1 = 78$

47. $\sqrt{25a^2} = \sqrt{5^2 a^2} = 5|a|$

49. $\sqrt{\dfrac{81}{144}} = \sqrt{\dfrac{9 \cdot 9}{12 \cdot 12}} = \dfrac{9}{12} = \dfrac{3}{4}$

51. $\sqrt[3]{\dfrac{2x^3}{27}} = \sqrt[3]{\dfrac{2x^3}{3^3}} = \dfrac{x}{3}\sqrt[3]{2} = \dfrac{2^{1/3}x}{3}$

53. $\sqrt{48} - \sqrt{27} = \sqrt{3 \cdot 4^2} - \sqrt{3 \cdot 3^2}$

$= 4\sqrt{3} - 3\sqrt{3}$

$= \sqrt{3}$

55. $8\sqrt{3x} - 5\sqrt{3x} = 3\sqrt{3x}$

57. $\sqrt{8x^3} + \sqrt{2x} = \sqrt{2 \cdot 2^2 \cdot x \cdot x^2} + \sqrt{2x}$

$= 2x\sqrt{2x} + \sqrt{2x}$

$= (2x + 1)\sqrt{2x}$

59. $A = wh = \left(12\sqrt{2}\right)\sqrt{24^2 - \left(12\sqrt{2}\right)^2}$

$= 12\sqrt{2}\sqrt{288}$

$= 12\sqrt{2} \cdot 12\sqrt{2}$

$= 288$ square inches

The shape is a square because $w = h = 12\sqrt{2}$.

61. $\dfrac{1}{3 - \sqrt{5}} = \dfrac{1}{3 - \sqrt{5}} \cdot \dfrac{3 + \sqrt{5}}{3 + \sqrt{5}}$

$= \dfrac{3 + \sqrt{5}}{9 - 5} = \dfrac{3 + \sqrt{5}}{4}$

63. $\dfrac{\sqrt{20}}{4} = \dfrac{2\sqrt{5}}{4} = \dfrac{\sqrt{5}}{2} = \dfrac{\sqrt{5}}{2} \cdot \dfrac{\sqrt{5}}{\sqrt{5}} = \dfrac{5}{2\sqrt{5}}$

65. $64^{5/2} = \left(\sqrt{64}\right)^5 = 8^5 = 32{,}768$

67. $(-3x^{2/5})(-2x^{1/2}) = 6x^{2/5 + 1/2} = 6x^{9/10}$

69. $-2x^5 - x^4 + 3x^3 + 15x^2 + 5$

Degree: 5

Leading coefficient: -2

71. $-(3x^2 + 2x) + (1 - 5x) = -3x^2 - 2x + 1 - 5x$

$= -3x^2 - 7x + 1$

73. $(2x^3 - 5x^2 + 10x - 7) + (4x^2 - 7x - 2) = 2x^3 - 5x^2 + 4x^2 + 10x - 7x - 7 - 2$

$= 2x^3 - x^2 + 3x - 9$

75. $(a^2 + a - 3)(a^3 + 2) = a^5 + 2a^2 + a^4 + 2a - 3a^3 - 6 = a^5 + a^4 - 3a^3 + 2a^2 + 2a - 6$

77. $(y^2 - y)(y^2 + 1)(y^2 + y + 1) = (y^4 - y^3 + y^2 - y)(y^2 + y + 1)$

$$= (y^4 - y^3 + y^2 - y)y^2 + (y^4 - y^3 + y^2 - y)y + (y^4 - y^3 + y^2 - y)$$

$$= y^6 - y^5 + y^4 - y^3 + y^5 - y^4 + y^3 - y^2 + y^4 - y^3 + y^2 - y$$

$$= y^6 + y^4 - y^3 - y$$

79. $(x + 8)(x - 8) = x^2 - 8^2 = x^2 - 64$

81. $(x - 4)^3 = x^3 - 12x^2 + 48x - 64$

83. $(m - 4 + n)(m - 4 - n) = [(m - 4) + n][(m - 4) - n]$

$$= (m - 4)^2 - n^2$$

$$= m^2 - 8m + 16 - n^2$$

$$= m^2 - n^2 - 8m + 16$$

85. $(x + 3)(x + 5) = x(x + 5) + 3(x + 5)$

Distributive Property

87. $7x + 35 = 7(x + 5)$

89. $x^3 - x = x(x^2 - 1)$

91. $2x^3 + 18x^2 - 4x = 2x(x^2 + 9x - 2)$

93. (a)

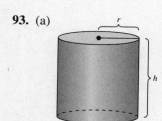

The surface area is the sum of the area of the side, $2\pi rh$, and the areas of the top and bottom which are each πr^2.

$$S = 2\pi rh + \pi r^2 + \pi r^2 = 2\pi rh + 2\pi r^2$$

(b) $S = 2\pi rh + 2\pi r^2 = 2\pi r(r + h)$

95. $x^2 - 169 = (x + 13)(x - 13)$

97. $x^3 + 216 = (x + 6)(x^2 - 6x + 36)$

99. $x^2 - 6x - 27 = (x - 9)(x + 3)$

101. $2x^2 + 21x + 10 = (2x + 1)(x + 10)$

103. $x^3 - 4x^2 - 3x + 12 = x^2(x - 4) - 3(x - 4)$

$$= (x - 4)(x^2 - 3)$$

105. $2x^2 - x - 15$

$a = 2, c = -15, ac = -30 = (-6)5$ and $-6 + 5 = -1 = b$

So, $2x^2 - x - 15 = 2x^2 - 6x + 5x - 15$

$$= 2x(x - 3) + 5(x - 3)$$

$$= (2x + 5)(x - 3).$$

107. Domain: all x

109. Domain: all $x \neq \dfrac{3}{2}$

111. $\dfrac{4x^2}{4x^3 + 28x} = \dfrac{x}{x^2 + 7}$, $x \neq 0$

113. $\dfrac{x^2 - x - 30}{x^2 - 25} = \dfrac{(x - 6)(x + 5)}{(x + 5)(x - 5)} = \dfrac{x - 6}{x - 5}$, $x \neq -5$

115. $\dfrac{x^2 - 4}{x^4 - 2x^2 - 8} \cdot \dfrac{x^2 + 2}{x^2} = \dfrac{(x - 2)(x + 2)}{(x^2 - 4)(x^2 + 2)} \cdot \dfrac{x^2 + 2}{x^2} = \dfrac{(x - 2)(x + 2)}{(x - 2)(x + 2)} \cdot \dfrac{1}{x^2} = \dfrac{1}{x^2}$, $x \neq \pm 2$

117. $\dfrac{x^2(5x - 6)}{2x + 3} \div \dfrac{5x}{2x + 3} = \dfrac{x^2(5x - 6)}{2x + 3} \cdot \dfrac{2x + 3}{5x} = \dfrac{x(5x - 6)}{5}$, $x \neq 0, -\dfrac{3}{2}$

119. $x - 1 + \dfrac{1}{x + 2} + \dfrac{1}{x - 1} = \dfrac{(x - 1)^2(x + 2) + (x - 1) + (x + 2)}{(x + 2)(x - 1)}$

$\qquad = \dfrac{(x^2 - 2x + 1)(x + 2) + 2x + 1}{(x + 2)(x - 1)}$

$\qquad = \dfrac{x^3 - 2x^2 + x + 2x^2 - 4x + 2 + (2x + 1)}{(x + 2)(x - 1)}$

$\qquad = \dfrac{x^3 - x + 3}{(x + 2)(x - 1)}$

121. $\dfrac{1}{x} - \dfrac{x - 1}{x^2 + 1} = \dfrac{1(x^2 + 1) - x(x - 1)}{x(x^2 + 1)} = \dfrac{x^2 + 1 - x^2 + x}{x(x^2 + 1)} = \dfrac{x + 1}{x(x^2 + 1)}$

123. $\dfrac{\dfrac{1}{x} - \dfrac{1}{y}}{(x^2 - y^2)} = \dfrac{y - x}{xy} \cdot \dfrac{1}{(x - y)(x + y)} = \dfrac{-1}{xy(x + y)}$, $x \neq y$

125.

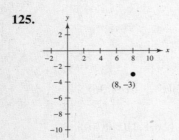

Quadrant IV

127.

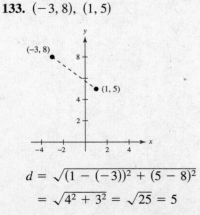

Quadrant II

129. $x > 0, y = -2 \rightarrow$

Quadrant IV

131.

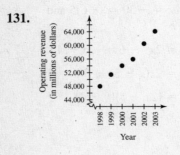

133. $(-3, 8)$, $(1, 5)$

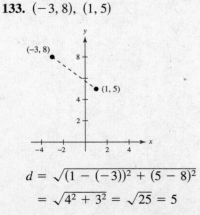

$d = \sqrt{(1 - (-3))^2 + (5 - 8)^2}$

$\quad = \sqrt{4^2 + 3^2} = \sqrt{25} = 5$

135.

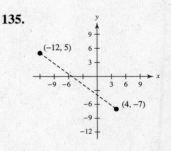

Midpoint:

$\left(\dfrac{-12 + 4}{2}, \dfrac{5 - 7}{2} \right) = (-4, -1)$

137. Radius:

$$\sqrt{(3 - (-5))^2 + (-1 - 1)^2} = \sqrt{64 + 4} = \sqrt{68}$$

$$(x - 3)^2 + (y + 1)^2 = 68$$

139.

Original vertices	Shifted vertices
$(4, 8)$	$(4 - 2, 8 - 3) = (2, 5)$
$(6, 8)$	$(6 - 2, 8 - 3) = (4, 5)$
$(4, 3)$	$(4 - 2, 3 - 3) = (2, 0)$
$(6, 3)$	$(6 - 2, 3 - 3) = (4, 0)$

141.

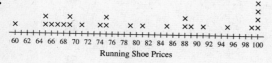

The price of 100 occurs with the greatest frequency (4).

143.

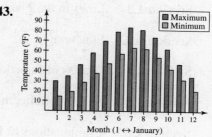

145.

Interval	Tally
$[21.7, 22.8)$	$\cancel{||||}\,|$
$[22.8, 23.9)$	$||$
$[23.9, 25)$	$\cancel{||||}$
$[25, 26.1)$	$|||$
$[26.1, 27.2)$	$|$
$[27.2, 28.3)$	$||$
$[28.3, 29.4)$	
$[29.4, 30.5)$	
$[30.5, 31.6)$	$|$

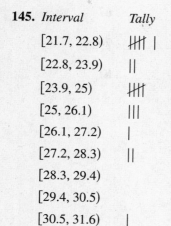

Answers will vary.

147. False. Not true for $x = 1$.

149. $(2x)^4 = 2^4 x^4 = 16x^4$

The exponent is to be applied to the whole quantity inside the parentheses.

151. The expression $\sqrt{5u} + \sqrt{3u}$ does not equal $2\sqrt{2u}$ because radicals cannot be combined unless the index and the radicand are the same. In this case, radicands are not the same.

C H A P T E R 1
Functions and Their Graphs

CHAPTER 1
Functions and Their Graphs

Section 1.1   Graphs of Equations

- ■ You should be able to use the point-plotting method of graphing.
- ■ You should be able to find x- and y-intercepts.
  - (a) To find the x-intercepts, let $y = 0$ and solve for x.
  - (b) To find the y-intercepts, let $x = 0$ and solve for y.
- ■ You should know how to graph an equation with a graphing utility. You should be able to determine an appropriate viewing rectangle.
- ■ You should be able to use the zoom and trace features of a graphing utility.

Vocabulary Check

1. solution point   **2.** graph   **3.** intercepts

---

1. $y = \sqrt{x + 4}$

(a) $(0, 2)$: $2 \stackrel{?}{=} \sqrt{0 + 4}$

$\qquad 2 = 2 \ \checkmark$

Yes, the point *is* on the graph.

(b) $(5, 3)$: $3 \stackrel{?}{=} \sqrt{5 + 4}$

$\qquad 3 = \sqrt{9} \ \checkmark$

Yes, the point *is* on the graph.

3. $y = 4 - |x - 2|$

(a) $(1, 5)$: $5 \stackrel{?}{=} 4 - |1 - 2|$

$\qquad 5 \neq 4 - 1$

No, the point *is not* on the graph.

(b) $(1.2, 3.2)$: $3.2 \stackrel{?}{=} 4 - |1.2 - 2|$

$\qquad 3.2 \stackrel{?}{=} 4 - |-0.8|$

$\qquad 3.2 \stackrel{?}{=} 4 - 0.8$

$\qquad 3.2 \stackrel{?}{=} 3.2 \ \checkmark$

Yes, the point *is* on the graph.

5. $x^2 + y^2 = 20$

(a) $(3, -2)$: $3^2 + (-2)^2 \stackrel{?}{=} 20$

$\qquad\qquad 9 + 4 \stackrel{?}{=} 20$

$\qquad\qquad 13 \neq 20$

No, the point *is not* on the graph.

(b) $(-4, 2)$: $(-4)^2 + 2^2 \stackrel{?}{=} 20$

$\qquad\qquad 16 + 4 \stackrel{?}{=} 20$

$\qquad\qquad 20 = 20$

Yes, the point *is* on the graph.

7. $y = \frac{3}{2}x - 1$

x	-2	0	$\frac{2}{3}$	1	2
y	-4	-1	0	$\frac{1}{2}$	2
Solution point	$(-2, -4)$	$(0, -1)$	$\left(\frac{3}{2}, 0\right)$	$\left(1, \frac{1}{2}\right)$	$(2, 2)$

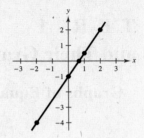

9. (a) $y = \frac{1}{4}x - 3$

x	-2	-1	0	1	2
y	$-\frac{7}{2}$	$-\frac{13}{4}$	-3	$-\frac{11}{4}$	$-\frac{5}{2}$

(b)

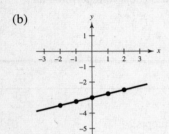

(c) $y = -\frac{1}{4}x - 3$

x	-2	-1	0	1	2
y	$-\frac{5}{2}$	$-\frac{11}{4}$	-3	$-\frac{13}{4}$	$-\frac{7}{2}$

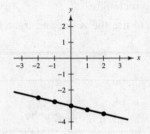

Both graphs are lines. The first graph rises to the right, whereas the second falls. Both pass through $(0, -3)$.

11. $y = 2x + 3$ has intercepts $(0, 3)$ and $\left(-\frac{3}{2}, 0\right)$.

Matches graph (e).

13. $y = x^2 - 2x$ has intercepts $(0, 0)$ and $(2, 0)$.

Matches graph (b).

15. $y = 2\sqrt{x}$ has one intercept $(0, 0)$.

Matches graph (c).

17. $y = -4x + 1$

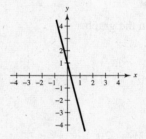

19. $y = 2 - x^2$

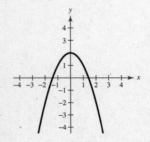

21. $y = x^2 - 3x$

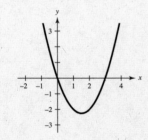

23. $y = x^3 + 2$

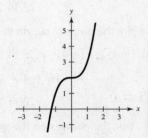

25. $y = \sqrt{x - 3}$

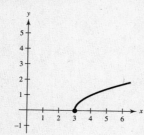

27. $y = |x - 2|$

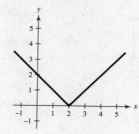

29. $x = y^2 - 1$

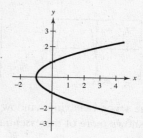

31. $y = x - 7$

Intercepts: $(0, -7), (7, 0)$

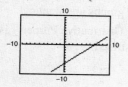

33. $y = 3 - \dfrac{1}{2}x$

Intercepts: $(6, 0), (0, 3)$

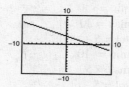

35. $y = \dfrac{2x}{x - 1}$

Intercepts: $(0, 0)$

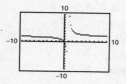

37. $y = x\sqrt{x + 3}$

Intercepts: $(0, 0), (-3, 0)$

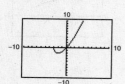

39. $y = \sqrt[3]{x - 8}$

Intercepts: $(8, 0), (0, -2)$

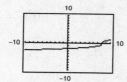

41. $y = x^2 - 4x + 3$

Intercepts: $(3, 0), (1, 0), (0, 3)$

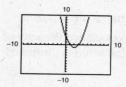

43. $y = x^2(x - 4) + 4x$

$= x^3 - 4x^2 + 4x$

Intercepts: $(0, 0), (2, 0)$

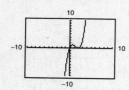

45. $y = \frac{5}{2}x + 5$

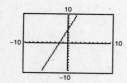

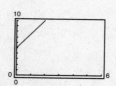

The first setting shows the line and its intercepts.
The first setting is better.

The second setting does not show the x-intercept
$(-2, 0)$.

47. $y = -x^2 + 10x - 5$

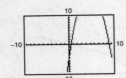 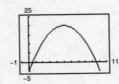

The second viewing window is better because it shows more of the essential features of the function.

49. $y = -10x + 50$

Range/Window

Xmin = -10
Xmax = 10
Xscl = 2
Ymin = -50
Ymax = 100
Yscl = 25

51. $y = \sqrt{x + 2} - 1$

Range/Window

Xmin = -5
Xmax = 1
Xscl = 1
Ymin = -3
Ymax = 1
Yscl = 1

53. $y = |x| + |x - 10|$

Range/Window

Xmin = -30
Xmax = 30
Xscl = 5
Ymin = -10
Ymax = 50
Yscl = 5

55. $y_1 = \frac{1}{4}(x^2 - 8)$

   $y_2 = \frac{1}{4}x^2 - 2$

The graphs are identical.
The Distributive Property is illustrated.

57. $y_1 = \frac{1}{5}[10(x^2 - 1)]$

   $y_2 = 2(x^2 - 1)$

The graphs are identical.

The Associative Property of Multiplication is illustrated.

59. $y = \sqrt{5 - x}$

(a) $(2, y) \approx (2, 1.73)$

(b) $(x, 3) = (-4, 3)$

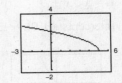

61. $y = x^5 - 5x$

(a) $(-0.5, y) \approx (-0.5, 2.47)$

(b) $(x, -4) = (1, -4)$ or $(x, -4) \approx (-1.65, -4)$

63. $x^2 + y^2 = 16$

   $y^2 = 16 - x^2$

   $y = \pm\sqrt{16 - x^2}$

   Use $y_1 = \sqrt{16 - x^2}$

   $y_2 = -\sqrt{16 - x^2}$.

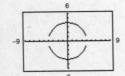

65. $(x - 1)^2 + (y - 2)^2 = 4$

   $(y - 2)^2 = 4 - (x - 1)^2$

   $y - 2 = \pm\sqrt{4 - (x - 1)^2}$

   $y = 2 \pm \sqrt{4 - (x - 1)^2}$

   Use $y_1 = 2 + \sqrt{4 - (x - 1)^2}$

   $y_2 = 2 - \sqrt{4 - (x - 1)^2}$.

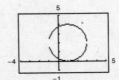

67. The center is in the first quadrant and the circle is tangent to the *x*-axis. Matches (a).

69. $(x - 1)^2 + (y - 2)^2 = 25$

    (a) $(1 - 1)^2 + (2 - 2)^2 = 0 \neq 25$ No

    (b) $(-2 - 1)^2 + (6 - 2)^2 = 9 + 16 = 25$ Yes

    (c) $(5 - 1)^2 + (-1 - 2)^2 = 16 + 9 = 25$ Yes

    (d) $(0 - 1)^2 + \left(2 + 2\sqrt{6} - 2\right)^2 = 1 + 24 = 25$ Yes

71. (a) $y = 225{,}000 - 20{,}000t$, $0 \leq t \leq 8$

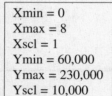

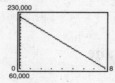

    (b) When $t = 5.8$, $y = 109{,}000$. Algebraically, $225{,}000 - 20{,}000(5.8) = \$109{,}000$.

    (c) When $t = 2.35$, $y = 178{,}000$. Algebraically, $225{,}000 - 20{,}000(2.35) = \$178{,}000$.

73. (a) Model: $y = -0.0049t^3 + 0.443t^2 - 0.75t + 116.7$, $5 \leq t \leq 14$

t	5	6	7	8	9	10	11	12	13	14
Model	123.4	127.1	131.5	136.5	142.3	148.6	155.5	163.0	171.1	179.6

The model is a good fit.

    (b)

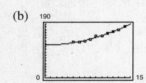

The model is a good fit.

    (c) For 2008, $t = 18$ and $y \approx 218.2$ thousand dollars. For 2010, $t = 20$ and $y \approx 239.7$ thousand dollars. These values seem reasonable.

    (d) For $y = 150$, $t \approx 10.2$, or 2000.

75. (a)

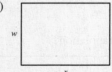

    (b) Perimeter: $12 = 2x + 2w$

$$12 = 2(x + w)$$

$$6 = x + w$$

    Thus, $w = 6 - x$.

    Area: $xw = x(6 - x) \implies A = x(6 - x)$

    (c)

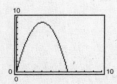

    (d) When $w = 4.9$, $x = 1.1$ and Area $= 5.39$ square meters.

    Algebraically, Area $= xw = (1.1)(4.9) = 5.39$ square meters.

    (e) The maximum area corresponds to the highest point on the graph, which appears to be $(3, 9)$.
Thus, $x = 3$ and $w = 3$, and the rectangle is a square.

77. False. $y = 1 - x^2$ has two x-intercepts, $(1, 0)$ and $(-1, 0)$. Also, $y = x^2 + 1$ has no x-intercepts.

79. Answers will vary.

81. Answers will vary. Sample answer: $y = 250x + 1000$ could represent the amount of money in someone's checking account after x months if they deposited an initial $1000 and added $250 per month.

83. $7\sqrt{72} - 5\sqrt{18} = 7\sqrt{2(6^2)} - 5\sqrt{2(3^2)}$
$$= 42\sqrt{2} - 15\sqrt{2} = 27\sqrt{2}$$

85. $7^{3/2} \cdot 7^{11/2} = 7^{(3/2 + 11/2)} = 7^7 = 823{,}543$

87. $(9x - 4) + (2x^2 - x + 15) = 2x^2 + 8x + 11$

Section 1.2   Lines in the Plane

You should know the following important facts about lines.

■ The graph of $y = mx + b$ is a straight line. It is called a linear equation.

■ The slope of the line through (x_1, y_1) and (x_2, y_2) is
$$m = \frac{y_2 - y_1}{x_2 - x_1}.$$

■ (a) If $m > 0$, the line rises from left to right.   (b) If $m = 0$, the line is horizontal.

(c) If $m < 0$, the line falls from left to right.   (d) If m is undefined, the line is vertical.

■ Equations of Lines

(a) Slope-Intercept: $y = mx + b$ (b) Point-Slope: $y - y_1 = m(x - x_1)$

(c) Two-Point: $y - y_1 = \frac{y_2 - y_1}{x_2 - x_1}(x - x_1)$ (d) General: $Ax + By + c = 0$

(e) Vertical: $x = a$ (f) Horizontal: $y = b$

■ Given two distinct nonvertical lines

$$L_1: y = m_1x + b_1 \quad \text{and} \quad L_2: y = m_2x + b_2$$

(a) L_1 is parallel to L_2 if and only if $m_1 = m_2$ and $b_1 \neq b_2$.

(b) L_1 is perpendicular to L_2 if and only if $m_1 = -1/m_2$.

Vocabulary Check

1. (a) iii   (b) i   (c) v   (d) ii   (e) iv   **2.** slope

3. parallel   **4.** perpendicular   **5.** linear extrapolation

1. (a) $m = \frac{2}{3}$. Since the slope is positive, the line rises. Matches L_2.

(b) m is undefined. The line is vertical. Matches L_3.

(c) $m = -2$. The line falls. Matches L_1.

3.

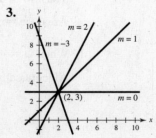

5. Slope $= \dfrac{\text{rise}}{\text{run}} = \dfrac{3}{2}$

7. Slope $= \dfrac{0 - (-10)}{-4 - 0} = \dfrac{10}{-4} = -\dfrac{5}{2}$

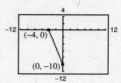

9.

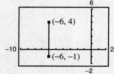

Slope is undefined.

11. Since $m = 0$, y does not change. Three points are $(0, 1)$, $(3, 1)$, and $(-1, 1)$.

13. Since m is undefined, x does not change and the line is vertical. Three points are $(1, 1)$, $(1, 2)$, and $(1, 3)$.

15. Since $m = -2$, y decreases 2 for every unit increase in x. Three points are $(1, -11)$, $(2, -13)$, and $(3, -15)$.

17. Since $m = \frac{1}{2}$, y increases 1 for every increase of 2 in x. Three points are $(9, -1)$, $(11, 0)$, and $(13, 1)$.

19. $5x - y + 3 = 0$

$\qquad y = 5x + 3$

(a) Slope: $m = 5$

$\quad y$-intercept: $(0, 3)$

(b)

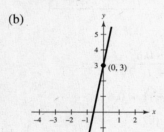

21. $5x - 2 = 0$

$\qquad x = \frac{2}{5}$

(a) Slope: undefined

$\quad$ No y-intercept

(b)

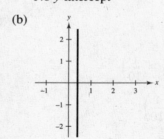

23. $3y + 5 = 0$

$\qquad y = -\frac{5}{3}$

(a) Slope: $m = 0$

$\quad y$-intercept: $\left(0, -\frac{5}{3}\right)$

(b)

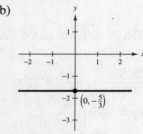

25. $y + 2 = 3(x - 0)$

$\qquad y = 3x - 2 \;\Rightarrow\; 3x - y - 2 = 0$

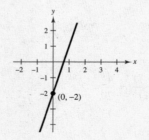

27. $y - (-3) = -\frac{1}{2}(x - 2)$

$\qquad y + 3 = -\frac{1}{2}x + 1$

$\qquad 2y + 4 = -x$

$\qquad x + 2y + 4 = 0$

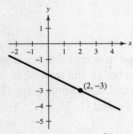

29. $x = 6$

$\qquad x - 6 = 0$

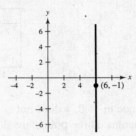

31. $y - \frac{3}{2} = 0\left(x + \frac{1}{2}\right)$

$\qquad y - \frac{3}{2} = 0$, horizontal line

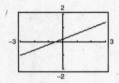

33. $y + 1 = \dfrac{5 + 1}{-5 - 5}(x - 5)$

$\qquad y = -\dfrac{3}{5}(x - 5) - 1$

$\qquad y = -\dfrac{3}{5}x + 2$

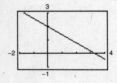

35. Since both points have $x = -8$, the slope is undefined.

$\qquad x = -8$

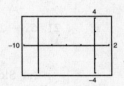

37. $y - \dfrac{1}{2} = \dfrac{\frac{5}{4} - \frac{1}{2}}{\frac{1}{2} - 2}(x - 2)$

$\qquad y = -\dfrac{1}{2}(x - 2) + \dfrac{1}{2}$

$\qquad y = -\dfrac{1}{2}x + \dfrac{3}{2}$

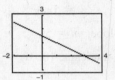

39. $y + \dfrac{3}{5} = \dfrac{-\frac{9}{5} + \frac{3}{5}}{\frac{9}{10} + \frac{1}{10}}\left(x + \dfrac{1}{10}\right)$

$\qquad y + \dfrac{3}{5} = -\dfrac{6}{5}\left(x + \dfrac{1}{10}\right)$

$\qquad y = -\dfrac{6}{5}x - \dfrac{18}{25}$

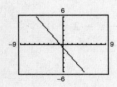

41. $y - 0.6 = \dfrac{-0.6 - 0.6}{-2 - 1}(x - 1)$

$\qquad y = 0.4(x - 1) + 0.6$

$\qquad y = 0.4x + 0.2$

43. The slope is $\dfrac{-3 - (-7)}{1 - (-1)} = \dfrac{4}{2} = 2.$

$\qquad y - (-3) = 2(x - 1)$

$\qquad\quad y + 3 = 2x - 2$

$\qquad\qquad y = 2x - 5$

45. Using the points (2004, 28,500) and (2006, 32,900), you have

$$m = \frac{32,900 - 28,500}{2006 - 2004} = \frac{4400}{2} = 2200$$

$$S - 28,500 = 2200(t - 2004)$$

$$S = 2200t - 4,380,300.$$

When $t = 2008$,

$$S = 2200(2008) - 4,380,300 = \$37,300.$$

47. $x - 2y = 4$

$$-2y = -x + 4$$

$$y = \frac{1}{2}x - 2$$

Slope: $\frac{1}{2}$

y-intercept: $(0, -2)$

The graph passes through $(0, -2)$ and rises 1 unit for each horizontal increase of 2.

49. $x = -6$

Slope is undefined.

No y-intercept

The line is vertical and passes through $(-6, 0)$.

51. $y = 0.5x - 3$

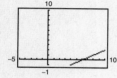 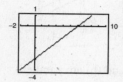

The second setting shows the x- and y-intercepts more clearly.

53. $m_{L_1} = \frac{9 + 1}{5 - 0} = 2$

$$m_{L_2} = \frac{1 - 3}{4 - 0} = -\frac{1}{2} = -\frac{1}{m_{L_1}}$$

L_1 and L_2 are perpendicular.

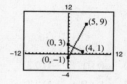

55. $m_{L_1} = \dfrac{0 - 6}{-6 - 3} = \dfrac{2}{3}$

$$m_{L_2} = \frac{\frac{7}{3} + 1}{5 - 0} = \frac{2}{3} = m_{L_1}$$

L_1 and L_2 are parallel.

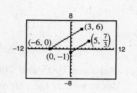

57. $4x - 2y = 3$

$$y = 2x - \frac{3}{2}$$

Slope: $m = 2$

(a) $y - 1 = 2(x - 2)$

$$y = 2x - 3$$

(b) $y - 1 = -\dfrac{1}{2}(x - 2)$

$$y = -\frac{1}{2}x + 2$$

59. $3x + 4y = 7$

$$y = -\tfrac{3}{4}x + \tfrac{7}{4}$$

Slope: $m = -\frac{3}{4}$

(a) $y - \frac{7}{8} = -\frac{3}{4}\left(x + \frac{2}{3}\right)$

$$y = -\tfrac{3}{4}x + \tfrac{3}{8}$$

(b) $y - \frac{7}{8} = \frac{4}{3}\left(x + \frac{2}{3}\right)$

$$y = \tfrac{4}{3}x + \tfrac{127}{72}$$

61. $x - 4 = 0$, vertical line

Slope not defined.

(a) $x - 3 = 0$ passes through $(3, -2)$.

(b) $y = -2$ passes through $(3, -2)$ and is horizontal.

63. The slope is 2 and $(-1, -1)$ lies on the line. Hence,

$$y - (-1) = 2(x - (-1))$$
$$y + 1 = 2(x + 1)$$
$$y = 2x + 1.$$

65. The slope of the given line is 2. Then l has slope $-\frac{1}{2}$. Hence,

$$y - 2 = -\frac{1}{2}(x - (-2))$$
$$y - 2 = -\frac{1}{2}(x + 2)$$
$$y = -\frac{1}{2}x + 1.$$

67. (a) $y = 2x$ (b) $y = -2x$ (c) $y = \frac{1}{2}x$

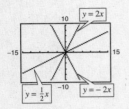

(b) and (c) are perpendicular.

69. (a) $y = -\frac{1}{2}x$ (b) $y = -\frac{1}{2}x + 3$

(c) $y = 2x - 4$

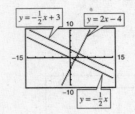

(a) and (b) are parallel.

(c) is perpendicular to (a) and (b).

71. (a)

Years	Slope
1995–1996	$0.69 - 0.91 = -0.22$
1996–1997	$0.57 - 0.69 = -0.12$
1997–1998	$0.74 - 0.57 = 0.17$
1998–1999	$1.60 - 0.74 = 0.86$
1999–2000	$0.82 - 1.60 = -0.78$
2000–2001	$0.92 - 0.82 = 0.10$
2001–2002	$0.20 - 0.92 = -0.72$
2002–2003	$0.00 - 0.20 = -0.20$
2003–2004	$0.31 - 0.00 = 0.31$

Greatest increase: 1998–1999 (0.86)

Greatest decrease: 1999–2000 (−0.78)

(b) $(5, 0.91), (14, 0.31)$:

$$y - 0.91 = \frac{0.31 - 0.91}{14 - 5}(x - 5)$$

$$y = -\frac{1}{15}(x - 5) + \frac{91}{100} = -\frac{1}{15}x + \frac{373}{300}$$

$$y \approx -0.07x + 1.24$$

(c) Between 1995 and 2004, the earnings per share decreased at the rate of 0.07 per year.

(d) For 2010, $x = 20$ and
$y = -0.07(20) + 1.24 = -0.16$, which is reasonable.

73. $\dfrac{\text{rise}}{\text{run}} = \dfrac{3}{4} = \dfrac{x}{\frac{1}{2}(32)}$

$$\frac{3}{4} = \frac{x}{16}$$

$$4x = 48$$

$$x = 12$$

The maximum height in the attic is 12 feet.

75. $(6, 2540), m = 125$

$$V - 2540 = 125(t - 6)$$
$$V = 125t + 1790$$

77. $(6, 20,400), m = -2000$

$$V - 20,400 = -2000(t - 6)$$
$$V = -2000t + 32,400$$

79. The slope is $m = -10$. This represents the decrease in the amount of the loan each week.
Matches graph (b).

81. The slope is $m = 0.35$ This represents the increase in travel cost for each mile driven. Matches graph (a).

83. (a) $(0, 25,000), (10, 2000)$

$$V - 25,000 = \frac{2000 - 25,000}{10 - 0}(t - 0)$$

$$V - 25,000 = -2300t$$

$$V = -2300t + 25,000$$

(b)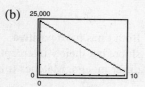

(c) $t = 0$: $V = -2300(0) + 25,000 = 25,000$

$t = 1$: $V = -2300(1) + 25,000 = 22,700$

etc.

t	0	1	2	3	4	5	6	7	8	9	10
V	25,000	22,700	20,400	18,100	15,800	13,500	11,200	8,900	6,600	4,300	2000

85. (a) $C = 36,500 + 5.25t + 11.50t$

$\quad = 16.75t + 36,500$

(c) $P = R - C$

$\quad = 27t - (16.75t + 36,500)$

$\quad = 10.25t - 36,500$

(b) $R = 27t$

(d) $\quad 0 = 10.25t - 36,500$

$\quad 36,500 = 10.25t$

$\quad\quad t \approx 3561$ hours

87. (a) $\dfrac{80,124 - 75,349}{2005 - 1991} = \dfrac{4775}{14} \approx 341$ students per year

(b) 1984: $75,349 - 341(7) \approx 72,962$ students

1997: $75,349 + 341(6) \approx 77,395$ students

2000: $75,349 + 341(9) \approx 78,418$ students

(Answers could vary.)

(c) Let $t = 0$ represent 1990.

$(1, 75,349), (15, 80,124)$

$$y - 75,349 = \frac{80,124 - 75,349}{15 - 1}(t - 1)$$

$$y = \frac{4775}{14}(t - 1) + 75,349$$

$$y \approx 341t + 75,008$$

The slope 341 represents the annual increase in students. It is positive, indicating that Penn State University increased its students from 1991 to 2005.

89. False. The slopes are different:

$$\frac{4 - 2}{-1 + 8} = \frac{2}{7}$$

$$\frac{7 + 4}{-7 - 0} = -\frac{11}{7}$$

91.

$$\frac{x}{5} + \frac{y}{-3} = 1$$

$$-3x + 5y + 15 = 0$$

$a = 5$ and $b = -3$ are the x- and y-intercepts.

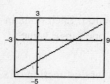

93. $\dfrac{x}{4} + \dfrac{y}{-2/3} = 1$

$-\dfrac{2}{3}x + 4y = \dfrac{-8}{3}$

$-2x + 12y = -8$

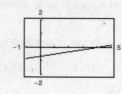

Intercepts: $(4, 0)$, $\left(0, -\dfrac{2}{3}\right)$

95. $\dfrac{x}{2} + \dfrac{y}{3} = 1$

$3x + 2y - 6 = 0$

97. $\dfrac{x}{-1/6} + \dfrac{y}{-2/3} = 1$

$-6x - \dfrac{3}{2}y = 1$

$12x + 3y + 2 = 0$

99. The slope is positive and the y-intercept is positive. Matches (a).

101. Both lines have positive slope, but their y-intercepts differ in sign. Matches (c).

103. No. The line $y = 2$ does not have an x-intercept.

105. Yes. Answers will vary.

107. Yes. $x + 20$

109. No. The term $x^{-1} = \dfrac{1}{x}$ causes the expression to not be a polynomial.

111. No. This expression is not defined for $x = \pm 3$.

113. $x^2 - 6x - 27 = (x - 9)(x + 3)$

115. $2x^2 + 11x - 40 = (2x - 5)(x + 8)$

117. Answers will vary.

Section 1.3   Functions

- Given a set or an equation, you should be able to determine if it represents a function.
- Given a function, you should be able to do the following.
  - (a) Find the domain.
  - (b) Evaluate it at specific values.

Vocabulary Check

1. domain, range, function

2. independent, dependent

3. piecewise-defined

4. implied domain

5. difference quotient

1. Yes, it does represent a function. Each domain value is matched with only one range value.

3. No, it does not represent a function. The domain values are each matched with three range values.

5. Yes, the relation represents y as a function of x. Each domain value is matched with only one range value.

7. No, it does not represent a function. The input values of 10 and 7 are each matched with two output values.

9. (a) Each element of A is matched with exactly one element of B, so it does represent a function.

(b) The element 1 in A is matched with two elements, -2 and 1 of B, so it does not represent a function.

(c) Each element of A is matched with exactly one element of B, so it does represent a function.

(d) The element 2 of A is not matched to any element of B, so it does not represent a function.

11. Each are functions. For each year there corresponds one and only one circulation.

13. $x^2 + y^2 = 4 \implies y = \pm\sqrt{4 - x^2}$

Thus, y *is not* a function of x. For instance, the values $y = 2$ and -2 both correspond to $x = 0$.

15. $y = \sqrt{x^2 - 1}$

This *is* a function of x.

17. $2x + 3y = 4 \implies y = \frac{1}{3}(4 - 2x)$

Thus, y *is* a function of x.

19. $y^2 = x^2 - 1 \implies y = \pm\sqrt{x^2 - 1}$

Thus, y *is not* a function of x. For instance, the values $y = \sqrt{3}$ and $-\sqrt{3}$ both correspond to $x = 2$.

21. $y = |4 - x|$

This *is* a function of x.

23. $x = -7$ does not represent y as a function of x. All values of y correspond to $x = -7$.

25. $f(x) = \dfrac{1}{x + 1}$

(a) $f(4) = \dfrac{1}{(4) + 1} = \dfrac{1}{5}$

(c) $f(4t) = \dfrac{1}{(4t) + 1} = \dfrac{1}{4t + 1}$

(b) $f(0) = \dfrac{1}{(0) + 1} = 1$

(d) $f(x + c) = \dfrac{1}{(x + c) + 1} = \dfrac{1}{x + c + 1}$

27. $f(t) = 3t + 1$

(a) $f(2) = 3(2) + 1 = 7$

(b) $f(-4) = 3(-4) + 1 = -11$

(c) $f(t + 2) = 3(t + 2) + 1 = 3t + 7$

29. $h(t) = t^2 - 2t$

(a) $h(2) = 2^2 - 2(2) = 0$

(b) $h(1.5) = (1.5)^2 - 2(1.5) = -0.75$

(c) $h(x + 2) = (x + 2)^2 - 2(x + 2) = x^2 + 2x$

31. $f(y) = 3 - \sqrt{y}$

(a) $f(4) = 3 - \sqrt{4} = 1$

(b) $f(0.25) = 3 - \sqrt{0.25} = 2.5$

(c) $f(4x^2) = 3 - \sqrt{4x^2} = 3 - 2|x|$

33. $q(x) = \dfrac{1}{x^2 - 9}$

(a) $q(0) = \dfrac{1}{0^2 - 9} = -\dfrac{1}{9}$

(b) $q(3) = \dfrac{1}{3^2 - 9}$ is undefined.

(c) $q(y + 3) = \dfrac{1}{(y + 3)^2 - 9} = \dfrac{1}{y^2 + 6y}$

35. $f(x) = \dfrac{|x|}{x}$

  (a) $f(3) = \dfrac{|3|}{3} = 1$

  (b) $f(-3) = \dfrac{|-3|}{-3} = -1$

  (c) $f(t) = \dfrac{|t|}{t} = \begin{cases} 1, & \text{if } t > 0 \\ -1, & \text{if } t < 0 \end{cases}$

  $f(0)$ is undefined.

37. $f(x) = \begin{cases} 2x + 1, & x < 0 \\ 2x + 2, & x \ge 0 \end{cases}$

  (a) $f(-1) = 2(-1) + 1 = -1$

  (b) $f(0) = 2(0) + 2 = 2$

  (c) $f(2) = 2(2) + 2 = 6$

39. $f(x) = \begin{cases} x^2 + 2, & x \le 1 \\ 2x^2 + 2, & x > 1 \end{cases}$

  (a) $f(-2) = (-2)^2 + 2 = 6$

  (b) $f(1) = (1)^2 + 2 = 3$

  (c) $f(2) = 2(2)^2 + 2 = 10$

41. $f(x) = \begin{cases} x + 2, & x < 0 \\ 4, & 0 \le x < 2 \\ x^2 + 1, & x \ge 2 \end{cases}$

  (a) $f(-2) = (-2) + 2 = 0$

  (b) $f(1) = 4$

  (c) $f(4) = 4^2 + 1 = 17$

43. $h(t) = \frac{1}{2}|t + 3|$

t	-5	-4	-3	-2	-1
$h(t)$	1	$\frac{1}{2}$	0	$\frac{1}{2}$	1

45. $f(x) = \begin{cases} -\frac{1}{2}x + 4, & x \le 0 \\ (x - 2)^2, & x > 0 \end{cases}$

x	-2	-1	0	1	2
$f(x)$	5	$\frac{9}{2}$	4	1	0

47. $f(x) = 15 - 3x = 0$
$$3x = 15$$
$$x = 5$$

49. $f(x) = \dfrac{3x - 4}{5} = 0$
$$3x - 4 = 0$$
$$3x = 4$$
$$x = \frac{4}{3}$$

51. $\qquad f(x) = g(x)$
$$x^2 = x + 2$$
$$x^2 - x - 2 = 0$$
$$(x + 1)(x - 2) = 0$$
$$x = -1 \ \text{ or } \ x = 2$$

53. $f(x) = 5x^2 + 2x - 1$

  Since $f(x)$ is a polynomial, the domain is all real numbers x.

55. $h(t) = \dfrac{4}{t}$

  Domain: All real numbers except $t = 0$

57. $f(x) = \sqrt[3]{x - 4}$

  Domain: All real numbers

59. $g(x) = \dfrac{1}{x} - \dfrac{3}{x + 2}$

  Domain: All real numbers except $x = 0, \ x = -2$

61. $g(y) = \dfrac{y + 2}{\sqrt{y - 10}}$

$y - 10 > 0$

$y > 10$

Domain: All $y > 10$

63. $f(x) = \sqrt{4 - x^2}$

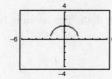

Domain: $[-2, 2]$

Range: $[0, 2]$

65. $g(x) = |2x + 3|$

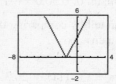

Domain: $(-\infty, \infty)$

Range: $[0, \infty)$

67. $f(x) = x^2$

$\{(-2, 4), (-1, 1), (0, 0), (1, 1), (2, 4)\}$

69. $f(x) = |x| + 2$

$\{(-2, 4), (-1, 3), (0, 2), (1, 3), (2, 4)\}$

71. $A = \pi r^2, \quad C = 2\pi r$

$r = \dfrac{C}{2\pi}$

$A = \pi\left(\dfrac{C}{2\pi}\right)^2 = \dfrac{C^2}{4\pi}$

73. (a) According to the table, the maximum profit is 3375 for $x = 150$.

(b)

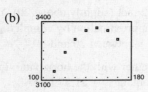

Yes, P is a function of x.

(c) Profit = Revenue − Cost

   = (price per unit)(number of units) − (cost)(number of units)

   = $[90 - (x - 100)(0.15)]x - 60x$

   = $(105 - 0.15x)x - 60x$

   = $45x - 0.15x^2, \quad x > 100$

$P = \begin{cases} 30x, & x \le 100 \\ 45x - 0.15x^2, & x > 100 \end{cases}$

75. $A = \dfrac{1}{2}(\text{base})(\text{height}) = \dfrac{1}{2}xy.$

Since $(0, y)$, $(2, 1)$ and $(x, 0)$ all lie on the same line, the slopes between any pair of points are equal.

$\dfrac{1 - y}{2 - 0} = \dfrac{1 - 0}{2 - x}$

$1 - y = \dfrac{2}{2 - x}$

$y = 1 - \dfrac{2}{2 - x} = \dfrac{x}{x - 2}$

Therefore, $A = \dfrac{1}{2}xy = \dfrac{1}{2}x\left(\dfrac{x}{x - 2}\right) = \dfrac{x^2}{2x - 4}.$

The domain is $x > 2$, since $A > 0$.

77. (a) $V = (\text{length})(\text{width})(\text{height}) = yx^2$

   But, $y + 4x = 108$, or $y = 108 - 4x$.

   Thus, $V = (108 - 4x)x^2$.

   Since $y = 108 - 4x > 0$

   $4x < 108$

   $x < 27$.

   Domain: $0 < x < 27$

(b)

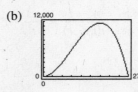

(c) The highest point on the graph occurs at $x = 18$. The dimensions that maximize the volume are $18 \times 18 \times 36$ inches.

79. The domain of $-1.97x + 26.3$ is $7 \leq x \leq 12$.

The domain of $0.505x^2 - 1.47x + 6.3$ is $1 \leq x \leq 6$.

You can tell by comparing the models to the given data. The models fit the data well on the domains above.

81. $f(11) = -1.97(11) + 26.3 = 4.63$

$4,630 in monthly revenue for November

83. $n(t) = \begin{cases} -6.13t^2 + 75.8t + 577, & 0 \leq t \leq 6 \\ 24.9t + 672, & 6 < t \leq 13 \end{cases}$

$t = 0$ corresponds to 1990.

t	0	1	2	3	4	5	6	7	8	9	10	11	12	13
Model	577	647	704	749	782	803	811	846	871	896	921	946	971	996

85. (a) $F(y) = 149.76\sqrt{10}\,y^{5/2}$

y	5	10	20	30	40
F(y)	2.65×10^4	1.50×10^5	8.47×10^5	2.33×10^6	4.79×10^6

(Answers will vary.)

F increases very rapidly as y increases.

(b)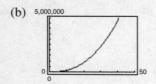

(c) From the table, $y \approx 22$ feet (slightly above 20). You could obtain a better approximation by completing the table for values of y between 20 and 30.

(d) By graphing $F(y)$ together with the horizontal line $y_2 = 1,000,000$, you obtain $y \approx 21.37$ feet.

87. $f(x) = 2x$

$$\frac{f(x + c) - f(x)}{c} = \frac{2(x + c) - 2x}{c}$$

$$= \frac{2c}{c} = 2, \quad c \neq 0$$

89. $f(x) = x^2 - x + 1, \quad f(2) = 3$

$$\frac{f(2 + h) - f(2)}{h} = \frac{(2 + h)^2 - (2 + h) + 1 - 3}{h}$$

$$= \frac{4 + 4h + h^2 - 2 - h + 1 - 3}{h}$$

$$= \frac{h^2 + 3h}{h} = h + 3, \quad h \neq 0$$

91. $f(t) = \dfrac{1}{t}, \quad f(1) = 1$

$$\frac{f(t) - f(1)}{t - 1} = \frac{(1/t) - 1}{t - 1} = \frac{1 - t}{t(t - 1)} = \frac{-1}{t}, \quad t \neq 1$$

93. False. The range of $f(x)$ is $[-1, \infty)$.

95. $f(x) = \begin{cases} x + 4, & x \leq 0 \\ 4 - x^2, & x > 0 \end{cases}$

97. $f(x) = \begin{cases} 2 - x, & x \leq -2 \\ 4, & -2 < x < 3 \\ x + 1, & x \geq 3 \end{cases}$

99. The domain is the set of inputs of the function and the range is the set of corresponding outputs.

101. $12 - \dfrac{4}{x + 2} = \dfrac{12(x + 2) - 4}{x + 2} = \dfrac{12x + 20}{x + 2}$

103. $\dfrac{2x^3 + 11x^2 - 6x}{5x} \cdot \dfrac{x + 10}{2x^2 + 5x - 3} = \dfrac{x(2x^2 + 11x - 6)(x + 10)}{5x(2x - 1)(x + 3)}$

$= \dfrac{(2x - 1)(x + 6)(x + 10)}{5(2x - 1)(x + 3)}$

$= \dfrac{(x + 6)(x + 10)}{5(x + 3)}, x \neq 0, \dfrac{1}{2}$

Section 1.4   Graphs of Functions

- ■ You should be able to determine the domain and range of a function from its graph.
- ■ You should be able to use the vertical line test for functions.
- ■ You should be able to determine when a function is constant, increasing, or decreasing.
- ■ You should be able to find relative maximum and minimum values of a function.
- ■ You should know that f is
  - (a) Odd if $f(-x) = -f(x)$.
  - (b) Even if $f(-x) = f(x)$.

Vocabulary Check

1. ordered pairs   **2.** Vertical Line Test   **3.** decreasing

4. minimum   **5.** greatest integer   **6.** even

---

1. Domain: All real numbers

Range: $(-\infty, 1]$

$f(0) = 1$

3. Domain: $[-4, 4]$

Range: $[0, 4]$

$f(0) = 4$

5. $f(x) = 2x^2 + 3$

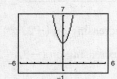

Domain: All real numbers

Range: $[3, \infty)$

7. $f(x) = \sqrt{x - 1}$

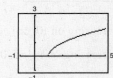

Domain: $x - 1 \geq 0 \implies x \geq 1$
or $[1, \infty)$

Range: $[0, \infty)$

9. $f(x) = |x + 3|$

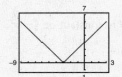

Domain: All real numbers

Range: $[0, \infty)$

11. $f(x) = x^2 - x - 6$

(a) Domain: All real numbers

(b) $f(x) = x^2 - x - 6 = (x - 3)(x + 2)$
$$= 0 \implies x = 3, -2$$

(c) These are the x-intercepts of f.

(d) $f(0) = -6$

(e) This is the y-intercept of f.

(f) $f(1) = 1^2 - 1 - 6 = -6$

The coordinates are $(1, -6)$.

(g) $f(-1) = (-1)^2 - (-1) - 6 = -4$

The coordinates are $(-1, -4)$.

(h) $f(-3) = (-3)^2 - (-3) - 6 = 6$
$$(-3, f(-3)) = (-3, 6)$$

13. $f(x) = |x - 1| - 2$

(a) Domain: All x

(b) $|x - 1| - 2 = 0 \implies |x - 1|$
$$= 2 \implies x = -1, 3$$

(c) x-intercepts

(d) $f(0) = |0 - 1| - 2 = -1$

(e) y-intercept

(f) $f(1) = |1 - 1| - 2 = -2, \quad (1, -2)$

(g) $f(-1) = |-1 - 1| - 2 = 0, \quad (-1, 0)$

(h) $f(-3) = |-3 - 1| - 2 = 2, \quad (-3, 2)$

15. $y = \frac{1}{2}x^2$

A vertical line intersects the graph just once, so y is a function of x. Graph $y_1 = \frac{1}{2}x^2$.

17. $x^2 + y^2 = 25$

A vertical line intersects the graph more than once, so y is not a function of x. Graph the circle as
$$y_1 = \sqrt{25 - x^2}$$
$$y_2 = -\sqrt{25 - x^2}.$$

19. $f(x) = \frac{3}{2}x$

f is increasing on $(-\infty, \infty)$.

21. $f(x) = x^3 - 3x^2 + 2$

f is increasing on $(-\infty, 0)$ and $(2, \infty)$.

f is decreasing on $(0, 2)$.

23. $f(x) = 3$

(a)

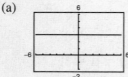

(b) f is constant on $(-\infty, \infty)$.

25. $f(x) = x^{2/3}$

(a)

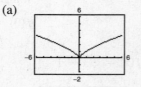

(b) Increasing on $(0, \infty)$

Decreasing on $(-\infty, 0)$

27. $f(x) = x\sqrt{x + 3}$

(a)

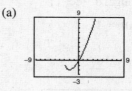

(b) Increasing on $(-2, \infty)$

Decreasing on $(-3, -2)$

29. $f(x) = |x + 1| + |x - 1|$

(a)

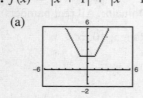

(b) Increasing on $(1, \infty)$, constant on $(-1, 1)$, decreasing on $(-\infty, -1)$

31. $f(x) = x^2 - 6x$

Relative minimum: $(3, -9)$

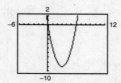

33. $y = 2x^3 + 3x^2 - 12x$

Relative minimum: $(1, -7)$

Relative maximum: $(-2, 20)$

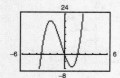

35. $h(x) = (x - 1)\sqrt{x}$

Relative minimum: $(0.33, -0.38)$

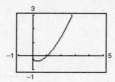

$(0, 0)$ is not a relative maximum because it occurs at the endpoint of the domain $[0, \infty)$.

37. $f(x) = x^2 - 4x - 5$

(a)

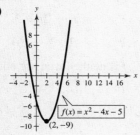

Minimum: $(2, -9)$

(b)

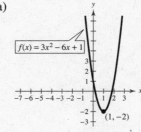

Minimum: $(2, -9)$

(c) Answers are the same.

39. $f(x) = x^3 - 3x$

(a)

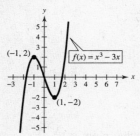

Relative maximum: $(-1, 2)$

Relative minimum: $(1, -2)$

(b) Relative maximum: $(-1, 2)$

Relative minimum: $(1, -2)$

(c) Answers are the same.

41. $f(x) = 3x^2 - 6x + 1$

(a)

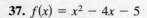

Relative minimum: $(1, -2)$

(b) Relative minimum: $(1, -2)$

(c) Answers are the same.

43. $f(x) = \begin{cases} 2x + 3, & x < 0 \\ 3 - x, & x \geq 0 \end{cases}$

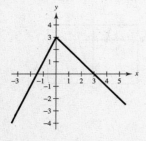

45. $f(x) = \begin{cases} \sqrt{x + 4}, & x < 0 \\ \sqrt{4 - x}, & x \geq 0 \end{cases}$

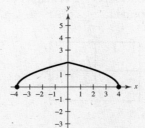

47. $f(x) = \begin{cases} x + 3, & x \leq 0 \\ 3, & 0 < x \leq 2 \\ 2x - 1, & x > 2 \end{cases}$

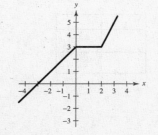

49. $f(x) = \begin{cases} 2x + 1, & x \le -1 \\ x^2 - 2, & x > -1 \end{cases}$

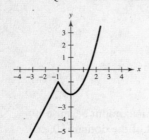

51. $f(x) = [\![x]\!] + 2$

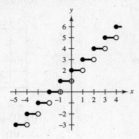

53. $f(x) = [\![x - 1]\!] + 2$

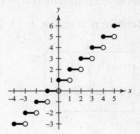

55. $f(x) = [\![2x]\!]$

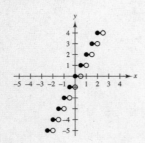

57. $s(x) = 2\left(\frac{1}{4}x - [\![\frac{1}{4}x]\!]\right)$

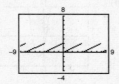

Domain: $(-\infty, \infty)$

Range: $[0, 2)$

Sawtooth pattern

59. $f(-t) = (-t)^2 + 2(-t) - 3$

$\qquad = t^2 - 2t - 3$

$\qquad \ne f(t) \ne -f(t)$

f is neither even nor odd.

61. $g(-x) = (-x)^3 - 5(-x)$

$\qquad = -x^3 + 5x$

$\qquad = -g(x)$

g is odd.

63. $f(-x) = (-x)\sqrt{1 - (-x)^2}$

$\qquad = -x\sqrt{1 - x^2}$

$\qquad = -f(x)$

The function is odd.

65. $g(-s) = 4(-s)^{2/3}$

$\qquad = 4s^{2/3}$

$\qquad = g(s)$

The function is even.

67. $\left(-\frac{3}{2}, 4\right)$

(a) If f is even, another point is $\left(\frac{3}{2}, 4\right)$.

(b) If f is odd, another point is $\left(\frac{3}{2}, -4\right)$.

69. $(4, 9)$

(a) If f is even, another point is $(-4, 9)$.

(b) If f is odd, another point is $(-4, -9)$.

71. $(x, -y)$

(a) If f is even, another point is $(-x, -y)$.

(b) If f is odd, another point is $(-x, y)$.

73. $f(x) = 5$, even

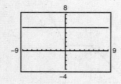

75. $f(x) = 3x - 2$ is neither even nor odd.

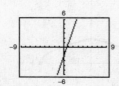

77. $h(x) = x^2 - 4$, even

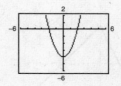

79. $f(x) = \sqrt{1-x}$ is neither even nor odd.

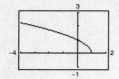

81. $f(x) = |x + 2|$ is neither even nor odd.

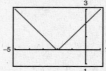

83. $f(x) = 4 - x \geq 0$

$\qquad 4 \geq x$

$\qquad (-\infty, 4]$

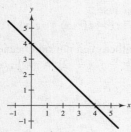

85. $f(x) = x^2 - 9 \geq 0$

$\qquad x^2 \geq 9$

$\qquad x \geq 3 \quad \text{or} \quad x \leq -3$

$\qquad [3, \infty) \text{ or } (-\infty, -3]$

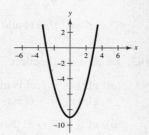

87. (a) The second model is correct. For instance,

$$C_2\left(\tfrac{1}{2}\right) = 1.05 - 0.38\left[\!\left[-\left(\tfrac{1}{2} - 1\right)\right]\!\right]$$

$$= 1.05 - 0.38\left[\!\left[\tfrac{1}{2}\right]\!\right] = 1.05.$$

(b)

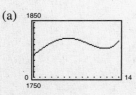

The cost of an 18-minute 45-second call is

$$C_2\left(18\tfrac{45}{60}\right) = C_2(18.75) = 1.05 - 0.38\left[\!\left[-(18.75 - 1)\right]\!\right]$$

$$= 1.05 - 0.38\left[\!\left[-17.75\right]\!\right] = 1.05 - 0.38(-18)$$

$$= 1.05 + 0.38(18) = \$7.89.$$

89. $h = \text{top} - \text{bottom}$

$\qquad = (-x^2 + 4x - 1) - 2$

$\qquad = -x^2 + 4x - 3, \, 1 \leq x \leq 3$

91. $P(t) = 0.0108t^4 - 0.211t^3 + 0.40t^2 + 7.9t + 1791$

$\qquad 0 \leq t \leq 14$

(a)

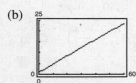

(b) P is increasing from 1990 ($t = 0$) to 1995 ($t \approx 5.7$), and from 2001 ($t \approx 11.8$) to 2004. P is decreasing from 1995 to 2001.

(c) The maximum population was about 1,821,000 in 1995 ($t \approx 5.7$).

93. False. The domain of $f(x) = \sqrt{x^2}$ is the set of all real numbers.

95. c **97.** b **99.** a

101. $f(x) = a_{2n+1}x^{2n+1} + a_{2n-1}x^{2n-1} + \cdots + a_3x^3 + a_1x$

$f(-x) = a_{2n+1}(-x)^{2n+1} + a_{2n-1}(-x)^{2n-1} + \cdots + a_3(-x)^3 + a_1(-x)$

$\quad = -a_{2n+1}x^{2n+1} - a_{2n-1}x^{2n-1} - \cdots - a_3x^3 - a_1x = -f(x)$

Therefore, $f(x)$ is odd.

103. f is an even function.

(a) $g(x) = -f(x)$ is even because
$g(-x) = -f(-x) = -f(x) = g(x).$

(b) $g(x) = f(-x)$ is even because
$g(-x) = f(-(-x)) = f(x) = f(-x) = g(x).$

(c) $g(x) = f(x) - 2$ is even because
$g(-x) = f(-x) - 2 = f(x) - 2 = g(x).$

(d) $g(x) = -f(x - 2)$ is neither even nor odd because
$g(-x) = -f(-x - 2) = -f(x + 2) \neq g(x)$ nor
$-g(x).$

105. No, $x^2 + y^2 = 25$ does not represent x as a function of y. For instance, $(-3, 4)$ and $(3, 4)$ both lie on the graph.

107. $-2x^2 + 8x$

Terms: $-2x^2, 8x$

Coefficients: $-2, 8$

109. $\dfrac{x}{3} - 5x^2 + x^3$

Terms: $\dfrac{x}{3}, -5x^2, x^3$

Coefficients: $\dfrac{1}{3}, -5, 1$

111. (a) $d = \sqrt{(6 - (-2))^2 + (3 - 7)^2}$

$\quad = \sqrt{64 + 16} = \sqrt{80} = 4\sqrt{5}$

(b) midpoint $= \left(\dfrac{-2 + 6}{2}, \dfrac{7 + 3}{2}\right) = (2, 5)$

113. (a) $d = \sqrt{\left(-\dfrac{3}{2} - \dfrac{5}{2}\right)^2 + (4 - (-1))^2} = \sqrt{16 + 25} = \sqrt{410}$

(b) midpoint $= \left(\dfrac{\dfrac{5}{2} - \dfrac{3}{2}}{2}, \dfrac{-1 + 4}{2}\right) = \left(\dfrac{1}{2}, \dfrac{3}{2}\right)$

115. $f(x) = 5x - 1$

(a) $f(6) = 5(6) - 1 = 29$

(b) $f(-1) = 5(-1) - 1 = -6$

(c) $f(x - 3) = 5(x - 3) - 1 = 5x - 16$

117. $f(x) = x\sqrt{x - 3}$

(a) $f(3) = 3\sqrt{3 - 3} = 0$

(b) $f(12) = 12\sqrt{12 - 3}$

$\quad = 12\sqrt{9} = 12(3) = 36$

(c) $f(6) = 6\sqrt{6 - 3} = 6\sqrt{3}$

119. $f(x) = x^2 - 2x + 9$

$f(3 + h) = (3 + h)^2 - 2(3 + h) + 9 = 9 + 6h + h^2 - 6 - 2h + 9$

$\quad = h^2 + 4h + 12$

$f(3) = 3^2 - 2(3) + 9 = 12$

$\dfrac{f(3 + h) - f(3)}{h} = \dfrac{(h^2 + 4h + 12) - 12}{h} = \dfrac{h(h + 4)}{h} = h + 4, \; h \neq 0$

Section 1.5   Shifting, Reflecting, and Stretching Graphs

- You should know the graphs of the most commonly used functions in algebra, and be able to reproduce them on your graphing utility.
  - (a) Constant function: $f(x) = c$
  - (b) Identity function: $f(x) = x$
  - (c) Absolute value function: $f(x) = |x|$
  - (d) Square root function: $f(x) = \sqrt{x}$
  - (e) Squaring function: $f(x) = x^2$
  - (f) Cubing function: $f(x) = x^3$
- You should know how the graph of a function is changed by vertical and horizontal shifts.
- You should know how the graph of a function is changed by reflection.
- You should know how the graph of a function is changed by nonrigid transformations, like stretches and shrinks.
- You should know how the graph of a function is changed by a sequence of transformations.

Vocabulary Check

1. quadratic function

2. absolute value function

3. rigid transformations

4. $-f(x), f(-x)$

5. $c > 1, 0 < c < 1$

6. (a) ii (b) iv (c) iii (d) i

1.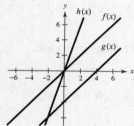

3.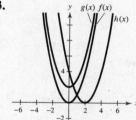

5.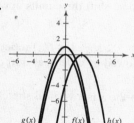

7.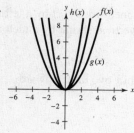

9.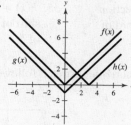

11.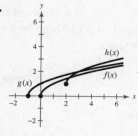

13. (a) $y = f(x) + 2$

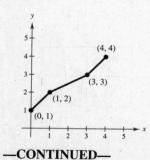

(b) $y = -f(x)$

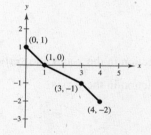

(c) $y = f(x - 2)$

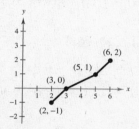

—CONTINUED—

13. —CONTINUED—

(d) $y = f(x + 3)$

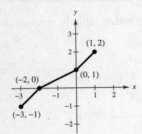

(e) $y = 2f(x)$

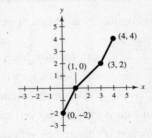

(f) $y = f(-x)$

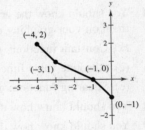

(g) Let $g(x) = f\left(\frac{1}{2}x\right)$. Then from the graph,

$$g(0) = f\left(\frac{1}{2}(0)\right) = f(0) = -1$$

$$g(2) = f\left(\frac{1}{2}(2)\right) = f(1) = 0$$

$$g(6) = f\left(\frac{1}{2}(6)\right) = f(3) = 1$$

$$g(8) = f\left(\frac{1}{2}(8)\right) = f(4) = 2.$$

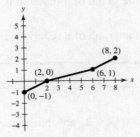

15. Horizontal shift three units to left of $y = x$: $y = x + 3$ (or vertical shift three units upward)

17. Vertical shift one unit downward of $y = x^2$

$$y = x^2 - 1$$

19. Reflection in the x-axis and a vertical shift one unit upward of $y = \sqrt{x}$: $y = 1 - \sqrt{x}$

21. $y = -\sqrt{x} - 1$ is $f(x)$ reflected in the x-axis, followed by a vertical shift one unit downward.

23. $y = \sqrt{x - 2}$ is $f(x)$ shifted right two units.

25. $y = 2\sqrt{x}$ is a vertical stretch of $f(x) = \sqrt{x}$.

27. $y = |x + 5|$ is $f(x)$ shifted left five units.

31. $y = 4|x|$ is a vertical stretch of $f(x)$.

29. $y = -|x|$ is $f(x)$ reflected in the x-axis.

33. $g(x) = 4 - x^3$ is obtained from $f(x)$ by a reflection in the x-axis followed by a vertical shift upward of four units.

35. $h(x) = \frac{1}{4}(x + 2)^3$ is obtained from $f(x)$ by a left shift of two units and a vertical shrink by a factor of $\frac{1}{4}$.

37. $p(x) = \left(\frac{1}{3}x\right)^3 + 2$ is obtained from $f(x)$ by a horizontal stretch followed by a vertical shift two units upward.

39. $f(x) = x^3 - 3x^2$

$g(x) = f(x + 2) = (x + 2)^3 - 3(x + 2)^2$ is a horizontal shift two units to left.

$h(x) = \frac{1}{2}f(x) = \frac{1}{2}(x^3 - 3x^2)$ is a vertical shrink.

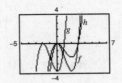

41. $f(x) = x^3 - 3x^2$

$g(x) = -\frac{1}{3}f(x) = -\frac{1}{3}(x^3 - 3x^2)$ reflection in the x-axis and vertical shrink

$h(x) = f(-x) = (-x)^3 - 3(-x)^2$ reflection in the y-axis

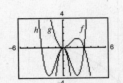

43. (a) $f(x) = x^2$

(b) $g(x) = 2 - (x + 5)^2$ is obtained from f by a horizontal shift to the left five units, a reflection in the x-axis, and a vertical shift upward two units.

(c)

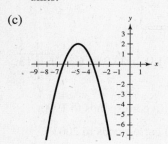

(d) $g(x) = 2 - f(x + 5)$

45. (a) $f(x) = x^2$

(b) $g(x) = 3 + 2(x - 4)^2$ is obtained from f by a horizontal shift four units to the right, a vertical stretch of 2, and a vertical shift upward three units.

(c)

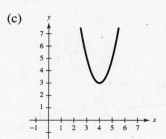

(d) $g(x) = 3 + 2f(x - 4)$

47. (a) $f(x) = x^3$

(b) $g(x) = 3(x - 2)^3$ is obtained from f by a horizontal shift two units to the right followed by a vertical stretch of 3.

(c)

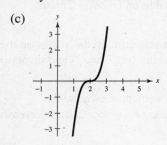

(d) $g(x) = 3f(x - 2)$

49. (a) $f(x) = x^3$

(b) $g(x) = (x - 1)^3 + 2$ is obtained from f by a horizontal shift one unit to the right, and a vertical shift upward two units.

(c)

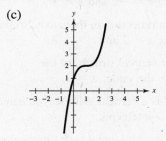

(d) $g(x) = f(x - 1) + 2$

51. (a) $f(x) = |x|$

(b) $g(x) = |x + 4| + 8$ is obtained from f by a horizontal shift four units to the left, followed by a vertical shift eight units upward.

(c)

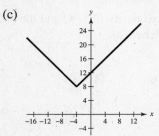

(d) $g(x) = f(x + 4) + 8$

53. (a) $f(x) = |x|$

(b) $g(x) = -2|x - 1| - 4$ is obtained from f by a horizontal shift one unit to the right, a vertical stretch of 2, a reflection in the x-axis, and a vertical shift downward four units.

(c)

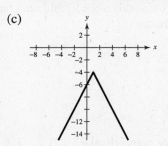

(d) $g(x) = -2f(x - 1) - 4$

55. (a) $f(x) = \sqrt{x}$

   (b) $g(x) = -\frac{1}{2}\sqrt{x + 3} - 1$ is obtained from f by a horizontal shift three units to the left, a vertical shrink, a reflection in the x-axis, and a vertical shift one unit downward.

   (c)

   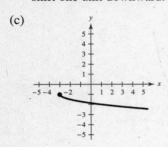

   (d) $g(x) = -\frac{1}{2}f(x + 3) - 1$

59. False. $y = f(-x)$ is a reflection in the y-axis.

61. (a) $y = f(-x)$ is a reflection in the y-axis, so the x-intercepts are $x = -2$ and $x = 3$.

   (b) $y = -f(x)$ is a reflection in the x-axis, so the x-intercepts are $x = 2$ and $x = -3$.

   (c) $y = 2f(x)$ is a vertical stretch, so the x-intercepts are the same: $x = 2, -3$.

   (d) $y = f(x) + 2$ is a vertical shift, so you cannot determine the x-intercepts.

   (e) $y = f(x - 3)$ is a horizontal shift three units to the right, so the x-intercepts are $x = 5$ and $x = 0$.

65. The vertex is approximately at $(2, 1)$ and the graph opens upward. Matches (c).

69. Slope L_1: $\dfrac{10 + 2}{2 + 2} = 3$

   Slope L_2: $\dfrac{9 - 3}{3 + 1} = \dfrac{3}{2}$

   Neither parallel nor perpendicular

73. Domain:
   $100 - x^2 \geq 0 \implies x^2 \leq 100 \implies -10 \leq x \leq 10$

57. (a) $F(t) = 33.0 + 6.2\sqrt{t}$ is a vertical stretch of $f(t) = \sqrt{t}$, followed by a vertical shift of 33.0.

   (b)

   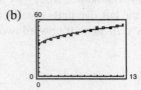

   (c) $G(t) = F(t + 13) = 33.0 + 6.2\sqrt{t + 13}$,

   $-13 \leq t \leq 0$.

   $G(-13) = F(0)$ corresponds to 1990.

   $G(0) = F(13)$ corresponds to 2003.

63. (a) $y = f(-x)$ is a reflection in the y-axis, so the graph is increasing on $(-\infty, -2)$ and decreasing on $(-2, \infty)$.

   (b) $y = -f(x)$ is a reflection in the x-axis, so the graph is decreasing on $(-\infty, 2)$ and increasing on $(2, \infty)$.

   (c) $y = 2f(x)$ is a vertical stretch, so the graph is increasing on $(-\infty, 2)$ and decreasing on $(2, \infty)$.

   (d) $y = f(x) - 3$ is a vertical shift, so the graph is increasing on $(-\infty, 2)$ and decreasing on $(2, \infty)$.

   (e) $y = f(x + 1)$ is a horizontal shift one unit to the left, so the graph is increasing on $(-\infty, 1)$ and decreasing on $(1, \infty)$.

67. The vertex is approximately $(2, -4)$ and the graph opens upward. Matches (c).

71. Domain: All $x \neq 9$

Section 1.6   Combinations of Functions

■ Given two functions, f and g, you should be able to form the following functions (if defined):
   1. Sum: $(f + g)(x) = f(x) + g(x)$
   2. Difference: $(f - g)(x) = f(x) - g(x)$
   3. Product: $(fg)(x) = f(x)g(x)$
   4. Quotient: $(f/g)(x) = f(x)/g(x), g(x) \neq 0$
   5. Composition of f with g: $(f \circ g)(x) = f(g(x))$
   6. Composition of g with f: $(g \circ f)(x) = g(f(x))$

Vocabulary Check

1. addition, subtraction, multiplication, division

2. composition

3. $g(x)$

4. inner, outer

1.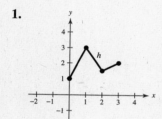

3.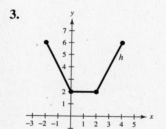

5. $f(x) = x + 3, \quad g(x) = x - 3$

   (a) $(f + g)(x) = f(x) + g(x) = (x + 3) + (x - 3) = 2x$

   (b) $(f - g)(x) = f(x) - g(x) = (x + 3) - (x - 3) = 6$

   (c) $(fg)(x) = f(x)g(x) = (x + 3)(x - 3) = x^2 - 9$

   (d) $\left(\dfrac{f}{g}\right)(x) = \dfrac{f(x)}{g(x)} = \dfrac{x + 3}{x - 3}, \quad x \neq 3$

   Domain: all $x \neq 3$

7. $f(x) = x^2, g(x) = 1 - x$

   (a) $(f + g)(x) = f(x) + g(x) = x^2 + (1 - x) = x^2 - x + 1$

   (b) $(f - g)(x) = f(x) - g(x) = x^2 - (1 - x) = x^2 + x - 1$

   (c) $(fg)(x) = f(x) \cdot g(x) = x^2(1 - x) = x^2 - x^3$

   (d) $\left(\dfrac{f}{g}\right)(x) = \dfrac{f(x)}{g(x)} = \dfrac{x^2}{1 - x}, x \neq 1$

   Domain: all $x \neq 1$

9. $f(x) = x^2 + 5, g(x) = \sqrt{1 - x}$

(a) $(f + g)(x) = x^2 + 5 + \sqrt{1 - x}$

(b) $(f - g)(x) = x^2 + 5 - \sqrt{1 - x}$

(c) $(fg)(x) = (x^2 + 5)\sqrt{1 - x}$

(d) $\left(\dfrac{f}{g}\right)(x) = \dfrac{x^2 + 5}{\sqrt{1 - x}}$

Domain: $x < 1$

11. $f(x) = \dfrac{1}{x}, g(x) = \dfrac{1}{x^2}$

(a) $(f + g)(x) = \dfrac{1}{x} + \dfrac{1}{x^2} = \dfrac{x + 1}{x^2}$

(b) $(f - g)(x) = \dfrac{1}{x} - \dfrac{1}{x^2} = \dfrac{x - 1}{x^2}$

(c) $(fg)(x) = \dfrac{1}{x} \cdot \dfrac{1}{x^2} = \dfrac{1}{x^3}$

(d) $\left(\dfrac{f}{g}\right)(x) = \dfrac{1/x}{1/x^2} = x, x \neq 0$

Domain: $x \neq 0$

13. $(f + g)(3) = f(3) + g(3)$
$= (3^2 - 1) + (3 - 2)$
$= 8 + 1 = 9$

15. $(f - g)(0) = f(0) - g(0)$
$= (0 - 1) - (0 - 2)$
$= 1$

17. $(fg)(4) = f(4)g(4)$
$= (4^2 - 1)(4 - 2)$
$= 15(2)$
$= 30$

19. $\left(\dfrac{f}{g}\right)(-5) = \dfrac{f(-5)}{g(-5)}$
$= \dfrac{(-5)^2 - 1}{-5 - 2}$
$= \dfrac{24}{-7}$
$= -\dfrac{24}{7}$

21. $(f - g)(2t) = f(2t) - g(2t)$
$= ((2t)^2 - 1) - (2t - 2)$
$= 4t^2 - 2t + 1$

23. $(fg)(-5t) = f(-5t)g(-5t)$
$= ((-5t)^2 - 1)(-5t - 2)$
$= (25t^2 - 1)(-5t - 2)$
$= -125t^3 - 50t^2 + 5t + 2$

25. $\left(\dfrac{f}{g}\right)(-t) = \dfrac{f(-t)}{g(-t)}$
$= \dfrac{(-t)^2 - 1}{-t - 2}$
$= \dfrac{t^2 - 1}{-t - 2}$
$= \dfrac{1 - t^2}{t + 2}, \quad t \neq -2$

27.

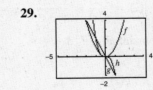

29.

31. $f(x) = 3x, g(x) = -\dfrac{x^3}{10}, (f + g)(x) = 3x - \dfrac{x^3}{10}$

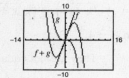

For $0 \le x \le 2, f(x)$ contributes more to the magnitude.

For $x > 6, g(x)$ contributes more to the magnitude.

33. $f(x) = 3x + 2, g(x) = -\sqrt{x + 5},$

$(f + g)(x) = 3x + 2 - \sqrt{x + 5}$

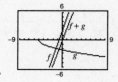

$f(x) = 3x + 2$ contributes more to the magnitude in both intervals.

35. $f(x) = x^2, g(x) = x - 1$

(a) $(f \circ g)(x) = f(g(x)) = f(x - 1) = (x - 1)^2$

(b) $(g \circ f)(x) = g(f(x)) = g(x^2) = x^2 - 1$

(c) $(f \circ g)(0) = (0 - 1)^2 = 1$

37. $f(x) = 3x + 5, g(x) = 5 - x$

(a) $(f \circ g)(x) = f(g(x)) = f(5 - x) = 3(5 - x) + 5 = 20 - 3x$

(b) $(g \circ f)(x) = g(f(x)) = g(3x + 5) = 5 - (3x + 5) = -3x$

(c) $(f \circ g)(0) = 20$

39. (a) The domain of $f(x) = \sqrt{x + 4}$ is $x + 4 \ge 0$ or $x \ge -4$.

(b) The domain of $g(x) = x^2$ is all real numbers.

(c) $(f \circ g)(x) = f(g(x)) = f(x^2) = \sqrt{x^2 + 4}.$

The domain of $(f \circ g)$ is all real numbers.

41. (a) The domain of $f(x) = x^2 + 1$ is all real numbers.

(b) The domain of $g(x) = \sqrt{x}$ is all $x \ge 0$.

(c) $(f \circ g)(x) = f(g(x)) = f(\sqrt{x})$

$= (\sqrt{x})^2 + 1 = x + 1, \quad x \ge 0$

The domain of $f \circ g$ is $x \ge 0$.

43. (a) The domain of $f(x) = \dfrac{1}{x}$ is all $x \ne 0$.

(b) The domain of $g(x) = x + 3$ is all real numbers.

(c) The domain of $(f \circ g)(x) = f(x + 3) = \dfrac{1}{x + 3}$ is all $x \ne -3$.

45. (a) The domain of $f(x) = |x - 4|$ is all real numbers.

(b) The domain of $g(x) = 3 - x$ is all real numbers.

(c) $(f \circ g)(x) = f(g(x)) = f(3 - x) = |(3 - x) - 4| = |-x - 1| = |x + 1|$

Domain: all real numbers

47. (a) The domain of $f(x) = x + 2$ is all real numbers.

(b) The domain of $g(x) = \dfrac{1}{x^2 - 4}$ is all $x \neq \pm 2$

(c) $(f \circ g)(x) = f(g(x)) = f\left(\dfrac{1}{x^2 - 4}\right) = \dfrac{1}{x^2 - 4} + 2$

Domain: $x \neq \pm 2$

49. (a) $(f \circ g)(x) = f(g(x)) = f(x^2) = \sqrt{x^2 + 4}$

Domain: all x

$(g \circ f)(x) = g(f(x)) = g\left(\sqrt{x + 4}\right) = \left(\sqrt{x + 4}\right)^2$

$= x + 4, \; x \geq -4$

(b)   They are not equal.

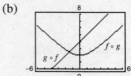

51. (a) $(f \circ g)(x) = f(g(x)) = f(3x + 9)$

$= \tfrac{1}{3}(3x + 9) - 3 = x$

Domain: all x

$(g \circ f)(x) = g(f(x)) = g\left(\tfrac{1}{3}x - 3\right)$

$= 3\left(\tfrac{1}{3}x - 3\right) + 9 = x$

The domain of $f \circ g$ is all real numbers.

(b)   They are equal.

```
         6
   -13 ──────── 7
  f∘g=g∘f
        -6
```

53. (a) $(f \circ g)(x) = f(g(x)) = f(x^6) = (x^6)^{2/3} = x^4$

Domain: all x

$(g \circ f)(x) = g(f(x)) = g(x^{2/3}) = (x^{2/3})^6 = x^4$

(b)   They are equal.

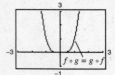

55. (a) $(f \circ g)(x) = f(g(x)) = f(4 - x) = 5(4 - x) + 4 = 24 - 5x$

$(g \circ f)(x) = g(f(x)) = g(5x + 4) = 4 - (5x + 4) = -5x$

(b) No, $(f \circ g)(x) \neq (g \circ f)(x)$ because $24 - 5x \neq -5x$.

(c)

x	$f(g(x))$	$g(f(x))$
0	24	0
1	19	-5
2	14	-10
3	9	-15

57. (a) $(f \circ g)(x) = f(g(x)) = f(x^2 - 5) = \sqrt{(x^2 - 5) + 6} = \sqrt{x^2 + 1}$

$(g \circ f)(x) = g(f(x)) = g\left(\sqrt{x + 6}\right) = \left(\sqrt{x + 6}\right)^2 - 5$

$= (x + 6) - 5 = x + 1, \; x \geq -6$

(b) No, $(f \circ g)(x) \neq (g \circ f)(x)$ because $\sqrt{x^2 + 1} \neq x + 1$.

(c)

x	$f(g(x))$	$g(f(x))$
0	1	1
-2	$\sqrt{5}$	-1
3	$\sqrt{10}$	4

59. (a) $(f \circ g)(x) = f(g(x)) = f(2x - 1) = |(2x - 1) + 3|$

$= |2x + 2| = 2|x + 1|$

$(g \circ f)(x) = g(f(x)) = g(|x + 3|) = 2|x + 3| - 1$

(b) No, $(f \circ g)(x) \neq (g \circ f)(x)$ because $2|x + 1| \neq 2|x + 3| - 1$.

(c)

x	$f(g(x))$	$g(f(x))$
-1	0	3
0	2	5
1	4	7

61. (a) $(f + g)(3) = f(3) + g(3) = 2 + 1 = 3$

    (b) $\left(\dfrac{f}{g}\right)(2) = \dfrac{f(2)}{g(2)} = \dfrac{0}{2} = 0$

63. (a) $(f \circ g)(2) = f(g(2)) = f(2) = 0$

    (b) $(g \circ f)(2) = g(f(2)) = g(0) = 4$

65. Let $f(x) = x^2$ and $g(x) = 2x + 1$, then $(f \circ g)(x) = h(x)$. This is not a unique solution. For example, if $f(x) = (x + 1)^2$ and $g(x) = 2x$, then $(f \circ g)(x) = h(x)$ as well.

67. Let $f(x) = \sqrt[3]{x}$ and $g(x) = x^2 - 4$, then $(f \circ g)(x) = h(x)$. This answer is not unique. Other possibilities may be:

$f(x) = \sqrt[3]{x - 4}$ and $g(x) = x^2$ or

$f(x) = \sqrt[3]{-x}$ and $g(x) = 4 - x^2$ or

$f(x) = \sqrt[9]{x}$ and $g(x) = (x^2 - 4)^3$

69. Let $f(x) = 1/x$ and $g(x) = x + 2$, then $(f \circ g)(x) = h(x)$. Again, this is not a unique solution. Other possibilities may be:

$f(x) = \dfrac{1}{x + 2}$ and $g(x) = x$

or $f(x) = \dfrac{1}{x + 1}$ and $g(x) = x + 1$

71. Let $f(x) = x^2 + 2x$ and $g(x) = x + 4$. Then $(f \circ g)(x) = h(x)$. (Answer is not unique.)

73. (a) $T(x) = R(x) + B(x) = \frac{3}{4}x + \frac{1}{15}x^2$

    (b)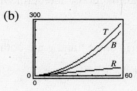

    (c) $B(x)$ contributes more to $T(x)$ at higher speeds.

75. $t = 5$ corresponds to 1995.

Year	1995	1996	1997	1998	1999	2000	2001	2002	2003	2004	2005
y_1	140	151.4	162.8	174.2	185.6	197	208.4	219.8	231.2	242.6	254
y_2	325.8	342.8	364.4	390.6	421.5	457	497.1	541.8	591.2	645.2	703.8
y_3	458.8	475.3	497.9	526.5	561.2	602	648.8	701.7	760.7	825.7	896.8

77. $(A \circ r)(t)$ gives the area of the circle as a function of time.

$(A \circ r)(t) = A(r(t))$

$\qquad = A(0.6t)$

$\qquad = \pi(0.6t)^2 = 0.36\pi t^2$

79. $C(x) = 60x + 750$

$x(t) = 50t$

    (a) $C(x(t)) = C(50t)$

$\qquad\qquad = 60(50t) + 750$

$\qquad\qquad = 3000t + 750$

$\qquad C(x(t))$ represents the cost after t hours.

    (b) $x(4) = 50(4) = 200$ units

    (c)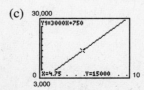

$t = 4.75$, or 4 hours 45 minutes

81. (a) $(N \circ T)(t) = N(T(t))$

$= N(2t + 1)$

$= 10(2t + 1)^2 - 20(2t + 1) + 600$

$= 40t^2 + 590$

$N \circ T$ represents the number of bacteria as a function of time.

(b) $(N \circ T)(6) = 10(13^2) - 20(13) + 600 = 2030$

At time $t = 6$, there are 2030 bacteria.

(c) $N = 800$ when $t \approx 2.3$ hours.

83. $g(f(x)) = g(x - 500{,}000) = 0.03(x - 500{,}000)$ represents 3 percent of the amount over \$500,000.

85. False. $(f \circ g)(x) = f(6x) = 6x + 1$, but $(g \circ f)(x) = g(x + 1) = 6(x + 1)$.

87. Let A, B, and C be the three siblings, in decreasing age. Then $A = 2B$ and $B = \frac{1}{2}C + 6$.

(a) $A = 2B = 2\left(\frac{1}{2}C + 6\right) = C + 12$

(b) If $A = 16$, then $B = 8$ and $C = 4$.

89. Let $f(x)$ and $g(x)$ be odd functions, and define $h(x) = f(x)g(x)$. Then,

$h(-x) = f(-x)g(-x)$

$= [-f(x)][-g(x)]$ since f and g are both odd

$= f(x)g(x) = h(x)$.

Thus, h is even.

Let $f(x)$ and $g(x)$ be even functions, and define $h(x) = f(x)g(x)$. Then,

$h(-x) = f(-x)g(-x)$

$= f(x)g(x)$ since f and g are both even

$= h(x)$.

Thus, h is even.

91. $g(-x) = \frac{1}{2}[f(-x) + f(-(-x))] = \frac{1}{2}[f(-x) + f(x)] = g(x)$,

which shows that g is even.

$h(-x) = \frac{1}{2}[f(-x) - f(-(-x))] = \frac{1}{2}[f(-x) - f(x)]$

$= -\frac{1}{2}[f(x) - f(-x)] = -h(x)$,

which shows that h is odd.

93. $(0, -5), (1, -5), (2, -7)$

(other answers possible)

95. $\left(\sqrt{24}, 0\right), \left(-\sqrt{24}, 0\right), \left(0, \sqrt{24}\right)$

(other answers possible)

97. $y - (-2) = \dfrac{8 - (-2)}{-3 - (-4)}(x - (-4))$

$y + 2 = 10(x + 4)$

$y - 10x - 38 = 0$

99. $y - (-1) = \dfrac{4 - (-1)}{-(1/3) - (3/2)}\left(x - \dfrac{3}{2}\right)$

$y + 1 = \dfrac{5}{-11/6}\left(x - \dfrac{3}{2}\right) = -\dfrac{30}{11}\left(x - \dfrac{3}{2}\right)$

$11y + 11 = -30x + 45$

$30x + 11y - 34 = 0$

Section 1.7   Inverse Functions

- Two functions f and g are inverses of each other if $f(g(x)) = x$ for every x in the domain of g and $g(f(x)) = x$ for every x in the domain of f.
- Be able to find the inverse of a function, if it exists.
  1. Replace $f(x)$ with y.
  2. Interchange x and y.
  3. Solve for y. If this equation represents y as a function of x, then you have found $f^{-1}(x)$. If this equation does not represent y as a function of x, then f does not have an inverse function.
- A function f has an inverse function if and only if no **horizontal** line crosses the graph of f at more than one point.
- A function f has an inverse function if and only if f is one-to-one.

Vocabulary Check

1. inverse, f^{-1} **2.** range, domain   **3.** $y = x$

4. one-to-one   **5.** Horizontal

1. $f(x) = 6x$

$f^{-1}(x) = \frac{1}{6}x$

$f(f^{-1}(x)) = f\left(\frac{1}{6}x\right) = 6\left(\frac{1}{6}x\right) = x$

$f^{-1}(f(x)) = f^{-1}(6x) = \frac{1}{6}(6x) = x$

3. $f(x) = x + 7$

$f^{-1}(x) = x - 7$

$f(f^{-1}(x)) = f(x - 7) = (x - 7) + 7 = x$

$f^{-1}(f(x)) = f^{-1}(x + 7) = (x + 7) - 7 = x$

5. $f^{-1}(x) = \dfrac{x - 1}{2}$

$f(f^{-1}(x)) = f\left(\dfrac{x - 1}{2}\right) = 2\left(\dfrac{x - 1}{2}\right) + 1 = (x - 1) + 1 = x$

$f^{-1}(f(x)) = f^{-1}(2x + 1) = \dfrac{(2x + 1) - 1}{2} = \dfrac{2x}{2} = x$

7. $f^{-1}(x) = x^3$

$f(f^{-1}(x)) = f(x^3) = \sqrt[3]{x^3} = x$

$f^{-1}(f(x)) = f^{-1}(\sqrt[3]{x}) = \left(\sqrt[3]{x}\right)^3 = x$

9. (a) $f(g(x)) = f\left(-\dfrac{2x + 6}{7}\right) = -\dfrac{7}{2}\left(-\dfrac{2x + 6}{7}\right) - 3 = \dfrac{2x + 6}{2} - 3 = (x + 3) - 3 = x$

$g(f(x)) = g\left(-\dfrac{7}{2}x - 3\right) = -\dfrac{2\left(-\frac{7}{2}x - 3\right) + 6}{7} = -\dfrac{-7x - 6 + 6}{7} = \dfrac{7x}{7} = x$

(b)

x	2	0	-2	-4	-6
$f(x)$	-10	-3	4	11	18

x	-10	-3	4	11	18
$g(x)$	2	0	-2	-4	-6

Note that the entries in the tables are the same except that the rows are interchanged.

11. (a) $f(g(x)) = f(\sqrt[3]{x - 5}) = [\sqrt[3]{x - 5}]^3 + 5 = (x - 5) + 5 = x$

$g(f(x)) = g(x^3 + 5) = \sqrt[3]{(x^3 + 5) - 5} = \sqrt[3]{x^3} = x$

(b)

x	-3	-2	-1	0	1
$f(x)$	-22	-3	4	5	6

x	-22	-3	4	5	6
$g(x)$	-3	-2	-1	0	1

Note that the entries in the tables are the same except that the rows are interchanged.

13. (a) $f(g(x)) = f(8 + x^2) = -\sqrt{(8 + x^2) - 8} = -\sqrt{x^2} = -(-x) = x \quad x \le 0$

[Since $x \le 0$, $\sqrt{x^2} = -x$.]

$g(f(x)) = g(-\sqrt{x - 8}) = 8 + [-\sqrt{x - 8}]^2 = 8 + (x - 8) = x$

(b)

x	8	9	12	17	24
$f(x)$	0	-1	-2	-3	-4

x	0	-1	-2	-3	-4
$g(x)$	8	9	12	17	24

Note that the entries in the tables are the same except that the rows are interchanged.

15. $f(g(x)) = f(\sqrt[3]{x}) = (\sqrt[3]{x})^3 = x$

$g(f(x)) = g(x^3) = \sqrt[3]{x^3} = x$

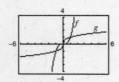

Reflections in the line $y = x$

17. $f(g(x)) = f(x^2 + 4), \quad x \ge 0$

$= \sqrt{(x^2 + 4) - 4} = x$

$g(f(x)) = g(\sqrt{x - 4})$

$= (\sqrt{x - 4})^2 + 4 = x$

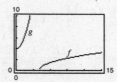

Reflections in the line $y = x$

19. $f(g(x)) = f(\sqrt[3]{1 - x}) = 1 - (\sqrt[3]{1 - x})^3 = 1 - (1 - x) = x$

$g(f(x)) = g(1 - x^3) = \sqrt[3]{1 - (1 - x^3)} = \sqrt[3]{x^3} = x$

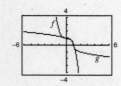

Reflections in the line $y = x$

21. The inverse is a line through $(-1, 0)$.

Matches graph (c).

23. The inverse is half a parabola starting at $(1, 0)$.

Matches graph (a).

25. $f(x) = 2x, \quad g(x) = \dfrac{x}{2}$

(a)

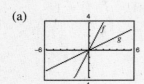

Reflection in the line $y = x$

(b)

x	-2	-1	0	1	2
$f(x)$	-4	-2	0	2	4

x	-4	-2	0	2	4
$g(x)$	-2	-1	0	1	2

The entries in the tables are the same, except that the rows are interchanged.

27. $f(x) = \dfrac{x-1}{x+5}, \quad g(x) = -\dfrac{5x+1}{x-1} = \dfrac{5x+1}{1-x}$

(a)

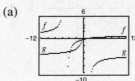

Reflection in the line $y = x$

(b)

x	-2	-1	0	3	5
$f(x)$	-1	$-\frac{1}{2}$	$-\frac{1}{5}$	$\frac{1}{4}$	$\frac{2}{5}$

x	-1	$-\frac{1}{2}$	$-\frac{1}{5}$	$\frac{1}{4}$	$\frac{2}{5}$
$g(x)$	-2	-1	0	3	5

The entries in the tables are the same, except that the rows are interchanged.

29. Not a function

31. It is the graph of a one-to-one function.

33. It is the graph of a one-to-one function.

35. $f(x) = 3 - \dfrac{1}{2}x$

f is one-to-one because a horizontal line will intersect the graph at most once.

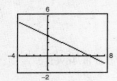

37. $h(x) = \dfrac{x^2}{x^2 + 1}$

h is not one-to-one because some horizontal lines intersect the graph twice.

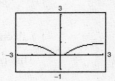

39. $h(x) = \sqrt{16 - x^2}$

h is not one-to-one because some horizontal lines intersect the graph twice.

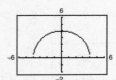

41. $f(x) = 10$

f is not one-to-one because the horizontal line $y = 10$ intersects the graph at every point on the graph.

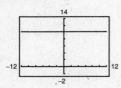

43. $g(x) = (x + 5)^3$

g is one-to-one because a horizontal line will intersect the graph at most once.

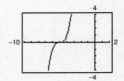

45. $h(x) = |x + 4| - |x - 4|$

h is not one-to-one because some horizontal lines intersect the graph more than once.

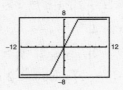

47. $f(x) = x^4$

$y = x^4$

$x = y^4$

$y = \pm\sqrt[4]{x}$

f is not one-to-one. This does not represent y as a function of x. f does not have an inverse.

49. $f(x) = \dfrac{3x + 4}{5}$

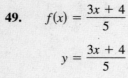

$y = \dfrac{3x + 4}{5}$

$x = \dfrac{3y + 4}{5}$

$5x = 3y + 4$

$5x - 4 = 3y$

$\dfrac{5x - 4}{3} = y$

$f^{-1}(x) = \dfrac{5x - 4}{3}$

f is one-to-one and has an inverse.

53. $f(x) = (x + 3)^2, \ x \geq -3, \ y \geq 0$

$y = (x + 3)^2, \ x \geq -3, \ y \geq 0$

$x = (y + 3)^2, \ y \geq -3, \ x \geq 0$

$\sqrt{x} = y + 3, \ y \geq -3, \ x \geq 0$

$y = \sqrt{x} - 3, \ x \geq 0, \ y \geq -3$

f is one-to-one.

This is a function of x, so f has an inverse.

$f^{-1}(x) = \sqrt{x} - 3, \ x \geq 0$

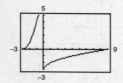

57. $f(x) = |x - 2|, \ x \leq 2, \ y \geq 0$

$y = |x - 2|$

$x = |y - 2|, \quad y \leq 2, \quad x \geq 0$

$x = -(y - 2)$ since $y - 2 \leq 0$.

$x = -y + 2$

$y = -x + 2, \quad x \geq 0, \quad y \leq 2$

$f^{-1}(x) = -x + 2, \quad x \geq 0$

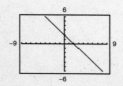

51. $f(x) = \dfrac{1}{x^2}$ is not one-to-one, and does not have an inverse. For example, $f(1) = f(-1) = 1$.

55. $f(x) = \sqrt{2x + 3} \ \Longrightarrow \ x \geq -\dfrac{3}{2}, \ y \geq 0$

$y = \sqrt{2x + 3}, \ x \geq -\dfrac{3}{2}, \ y \geq 0$

$x = \sqrt{2y + 3}, \ y \geq -\dfrac{3}{2}, \ x \geq 0$

$x^2 = 2y + 3, \ x \geq 0, \ y \geq -\dfrac{3}{2}$

$y = \dfrac{x^2 - 3}{2}, \ x \geq 0, \ y \geq -\dfrac{3}{2}$

f is one-to-one.

This is a function of x, so f has an inverse.

$f^{-1}(x) = \dfrac{x^2 - 3}{2}, \ x \geq 0$

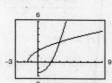

59. $f(x) = 2x - 3$

$y = 2x - 3$

$x = 2y - 3$

$y = \dfrac{x + 3}{2}$

$f^{-1}(x) = \dfrac{x + 3}{2}$

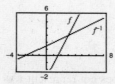

Reflections in the line $y = x$

61. $f(x) = x^5$

$y = x^5$

$x = y^5$

$y = \sqrt[5]{x}$

$f^{-1}(x) = \sqrt[5]{x}$

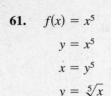

Reflections in the line $y = x$

63. $f(x) = x^{3/5}$

$y = x^{3/5}$

$x = y^{3/5}$

$y = x^{5/3}$

$f^{-1}(x) = x^{5/3}$

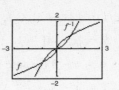

Reflections in the line $y = x$

65. $f(x) = \sqrt{4 - x^2}, 0 \le x \le 2$

$y = \sqrt{4 - x^2}$

$x = \sqrt{4 - y^2}$

$x^2 = 4 - y^2$

$y^2 = 4 - x^2$

$y = \sqrt{4 - x^2}$

$f^{-1}(x) = \sqrt{4 - x^2}, 0 \le x \le 2$

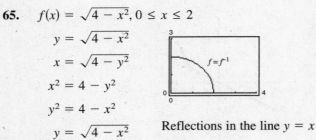

Reflections in the line $y = x$

67. $f(x) = \dfrac{4}{x}$

$y = \dfrac{4}{x}$

$x = \dfrac{4}{y}$

$xy = 4$

$y = \dfrac{4}{x}$

$f^{-1}(x) = \dfrac{4}{x}$

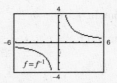

Reflections in the line $y = x$

69. If we let $f(x) = (x - 2)^2$, $x \ge 2$, then f has an inverse. [**Note:** We could also let $x \le 2$.]

$f(x) = (x - 2)^2, \ x \ge 2, \ y \ge 0$

$y = (x - 2)^2, \ x \ge 2, \ y \ge 0$

$x = (y - 2)^2, \ x \ge 0, \ y \ge 2$

$\sqrt{x} = y - 2, \ x \ge 0, \ y \ge 2$

$\sqrt{x} + 2 = y, \ x \ge 0, \ y \ge 2$

Thus, $f^{-1}(x) = \sqrt{x} + 2, \ x \ge 0.$

71. If we let $f(x) = |x + 2|$, $x \ge -2$, then f has an inverse. [**Note:** We could also let $x \le -2$.]

$f(x) = |x + 2|, \ x \ge -2$

$f(x) = x + 2$ when $x - 2$.

$y = x + 2, \ x \ge -2, \ y \ge 0$

$x = y + 2, \ x \ge 0, \ y \ge -2$

$x - 2 = y, \ x \ge 0, \ y \ge -2$

Thus, $f^{-1}(x) = x - 2, \ x \ge 0.$

73. Let $f(x) = (x + 3)^2, \ x \ge -3.$

$y = (x + 3)^2$

$x = (y + 3)^2$

$\sqrt{x} = y + 3$

$y = \sqrt{x} - 3$

$f^{-1}(x) = \sqrt{x} - 3$

Domain f: $x \ge -3$

Range f: $y \ge 0$

Domain f^{-1}: $x \ge 0$

Range f^{-1}: $y \ge -3$

75. Let $f(x) = -2x^2 + 5, \ x \ge 0.$

$y = -2x^2 + 5$

$x = -2y^2 + 5$

$x - 5 = -2y^2$

$y^2 = \dfrac{x - 5}{-2} = \dfrac{5 - x}{2}$

$y = \sqrt{\dfrac{5 - x}{2}}$

$f^{-1}(x) = \sqrt{\dfrac{5 - x}{2}}$

Domain f: $x \ge 0$

Range f: $y \le 5$

Domain f^{-1}: $x \le 5$

Range f^{-1}: $y \ge 0$

77. Let $f(x) = |x - 4| + 1$, $x4$ and $y \geq 1$.

$$y = |x - 4| + 1$$

$$y = x - 3 \text{ because } x \geq 4.$$

$$x = y - 3$$

$$y = x + 3$$

$$f^{-1}(x) = x + 3, \quad x \geq 1$$

Domain f: $x \geq 4$ Range f: $y \geq 1$

Domain f^{-1}: $x \geq 1$ Range f^{-1}: $y \geq 4$

79.

x	$f(x)$
-2	-4
-1	-2
1	2
3	3

x	$f^{-1}(x)$
-4	-2
-2	-1
2	1
3	3

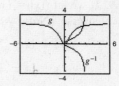

81. $f^{-1}(0) = \frac{1}{2}$ because $f\left(\frac{1}{2}\right) = 0$.

83. $(f \circ g)(2) = f(3) = -2$

85. $f^{-1}(g(0)) = f^{-1}(2) = 0$

87. $(g \circ f^{-1})(2) = g(0) = 2$

89. $f(x) = x^3 + x + 1$

The graph of the inverse relation is an inverse function since it satisfies the Vertical Line Test.

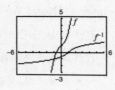

91. $g(x) = \dfrac{3x^2}{x^2 + 1}$

The graph of the inverse relation is not an inverse function since it does not satisfy the Vertical Line Test.

In Exercises 93–97, $f(x) = \frac{1}{8}x - 3$, $f^{-1}(x) = 8(x + 3)$, $g(x) = x^3$, $g^{-1}(x) = \sqrt[3]{x}$.

93. $(f^{-1} \circ g^{-1})(1) = f^{-1}(g^{-1}(1)) = f^{-1}(\sqrt[3]{1}) = 8(\sqrt[3]{1} + 3) = 8(1 + 3) = 32$

95. $(f^{-1} \circ f^{-1})(6) = f^{-1}(f^{-1}(6)) = f^{-1}(8[6 + 3]) = f^{-1}(72) = 8(72 + 3) = 600$

97. $(fg)(x) = f(g(x)) = f(x^3) = \frac{1}{8}x^3 - 3$

Now find the inverse of $(f \circ g)(x) = \frac{1}{8}x^3 - 3$:

$$y = \tfrac{1}{8}x^3 - 3$$

$$x = \tfrac{1}{8}y^3 - 3$$

$$x + 3 = \tfrac{1}{8}y^3$$

$$8(x + 3) = y^3$$

$$\sqrt[3]{8(x + 3)} = y$$

$$(f \circ g)^{-1}(x) = 2\sqrt[3]{x + 3}$$

Note: $(f \circ g)^{-1} = g^{-1} \circ f^{-1}$

In Exercises 99 and 101, $f(x) = x + 4$, $f^{-1}(x) = x - 4$, $g(x) = 2x - 5$, $g^{-1}(x) = \dfrac{x + 5}{2}$.

99. $(g^{-1} \circ f^{-1})(x) = g^{-1}(f^{-1}(x))$

$\qquad = g^{-1}(x - 4)$

$\qquad = \dfrac{(x - 4) + 5}{2}$

$\qquad = \dfrac{x + 1}{2}$

101. $(f \circ g)(x) = f(g(x)) = f(2x - 5)$

$\qquad = (2x - 5) + 4 = 2x - 1$

Now find the inverse of $(f \circ g)(x) = 2x - 1$.

$\qquad y = 2x - 1$

$\qquad x = 2y - 1$

$\qquad x + 1 = 2y$

$\qquad y = \dfrac{x + 1}{2}$

$\qquad (f \circ g)^{-1}(x) = \dfrac{x + 1}{2}$

Note that $(f \circ g)^{-1}(x) = (g^{-1} \circ f^{-1})(x)$;
See Exercise 99.

103. (a) Yes, f is one-to-one. For each European shoe size, there is exactly one U.S. shoe size.

  (b) $f(11) = 45$

  (c) $f^{-1}(43) = 10$ because $f(10) = 43$.

  (d) $f(f^{-1}(41)) = f(8) = 41$

  (e) $f^{-1}(f(13)) = f^{-1}(47) = 13$

105. (a) Yes, f is one-to-one, so f^{-1} exists.

  (b) f^{-1} gives the year corresponding to the 10 values in the second column.

  (c) $f^{-1}(650.3) = 10$ because $f(10) = 650.3$.

  (d) No, because $f(11) = f(15) = 690.4$.

107. False. $f(x) = x^2$ is even, but f^{-1} does not exist.

109. We will show that $(f \circ g)^{-1}(x) = (g^{-1} \circ f^{-1})(x)$ for all x in their domains.
Let $y = (f \circ g)^{-1}(x) \implies (f \circ g)(y) = x$ then $f(g(y)) = x \implies f^{-1}(x) = g(y)$.
Hence, $(g^{-1} \circ f^{-1})(x) = g^{-1}(f^{-1}(x)) = g^{-1}(g(y)) = y = (f \circ g)^{-1}(x)$.
Thus, $g^{-1} \circ f^{-1} = (f \circ g)^{-1}$.

111. No, the graphs are not reflections of each other in the line $y = x$.

113. Yes, the graphs are reflections of each other in the line $y = x$.

115. Yes. The inverse would give the time it took to complete n miles.

117. No. The function oscillates.

119. $\dfrac{27x^3}{3x^2} = 9x,\ x \neq 0$

121. $\dfrac{x^2 - 36}{6 - x} = \dfrac{(x - 6)(x + 6)}{-(x - 6)}$

$\qquad = \dfrac{x + 6}{-1} = -x - 6,\ x \neq 6$

123. $4x - y = 3$

$\qquad y = 4x - 3$

Yes, y is a function of x.

125. $x^2 + y^2 = 9$

$\qquad y = \pm \sqrt{9 - x^2}$

No, y is not a function of x.

127. $y = \sqrt{x + 2}$

Yes, y is a function of x.

Review Exercises for Chapter 1

1. $y = -\frac{1}{2}x + 2$

x	-2	0	2	3	4
y	3	2	1	$\frac{1}{2}$	0
Solution point	$(-2, 3)$	$(0, 2)$	$(2, 1)$	$(3, \frac{1}{2})$	$(4, 0)$

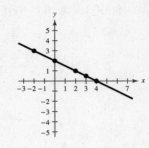

3. $y = 4 - x^2$

x	-2	-1	0	1	2
y	0	3	4	3	0
Solution point	$(-2, 0)$	$(-1, 3)$	$(0, 4)$	$(1, 3)$	$(2, 0)$

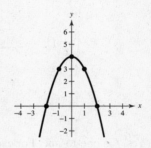

5. $y = \frac{1}{4}(x + 1)^3$

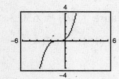

Intercepts: $(-1, 0), \left(0, \frac{1}{4}\right)$

7. $y = \frac{1}{4}x^4 - 2x^2$

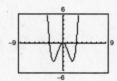

Intercepts:

$(0, 0), \left(\pm 2\sqrt{2}, 0\right) \approx (\pm 2.83, 0)$

9. $y = x\sqrt{9 - x^2}$

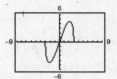

Intercepts: $(0, 0), (\pm 3, 0)$

11. $y = |x - 4| - 4$

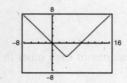

Intercepts: $(0, 0), (8, 0)$

13. The setting is:

Xmin = -20
Xmax = 50
Xscl = 10
Ymin = -2
Ymax = 1
Yscl = 0.5

15. $y = 13{,}500 - 1100t, \quad 0 \le t \le 6$

(a) $0 \le x \le 6, \quad 6900 \le y \le 13{,}500$

(Answers will vary.)

(b)

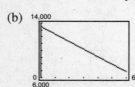

(c) When $y = 9100$, $t = 4$.

17. $m = \dfrac{2 - 2}{8 - (-3)} = \dfrac{0}{11} = 0$

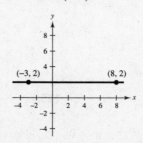

19. $m = \dfrac{(5/2) - 1}{5 - (3/2)} = \dfrac{3/2}{7/2} = \dfrac{3}{7}$

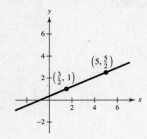

21. $(-4.5, 6), \ (2.1, 3)$

$m = \dfrac{3 - 6}{2.1 - (-4.5)} = \dfrac{-3}{6.6} = -\dfrac{30}{66} = -\dfrac{5}{11}$

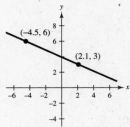

23. (a) $y + 1 = \frac{1}{4}(x - 2)$

$4y + 4 = x - 2$

$-x + 4y + 6 = 0$

(b) Three additional points:

$(2 + 4, -1 + 1) = (6, 0)$

$(6 + 4, 0 + 1) = (10, 1)$

$(10 + 4, 1 + 1) = (14, 2)$

(other answers possible)

25. (a) $y + 5 = \frac{3}{2}(x - 0)$

$2y + 10 = 3x$

$-3x + 2y + 10 = 0$

(b) Three additional points:

$(0 + 2, -5 + 3) = (2, -2)$

$(2 + 2, -2 + 3) = (4, 1)$

$(4 + 2, 1 + 3) = (6, 4)$

(other answers possible)

27. (a) $y + 5 = -1\left(x - \frac{1}{5}\right)$

$y + 5 = -x + \frac{1}{5}$

$5y + 25 = -5x + 1$

$5x + 5y + 24 = 0$

(b) Three additional points:

$\left(\frac{1}{5} + 1, -5 - 1\right) = \left(\frac{6}{5}, -6\right)$

$\left(\frac{6}{5} + 1, -6 - 1\right) = \left(\frac{11}{5}, -7\right)$

$\left(\frac{11}{5} + 1, -7 - 1\right) = \left(\frac{16}{5}, -8\right)$

(other answers possible)

29. (a) $y - 6 = 0(x + 2)$

$y - 6 = 0$

(b) Three additional points:

$(0, 6), (1, 6), (2, 6)$

(other answers possible)

31. (a) m is undefined means that the line is vertical.

$x - 10 = 0$

(b) Three additional points: $(10, 0), (10, 1), (10, 2)$

(other answers possible)

33. $y + 1 = \dfrac{-1 + 1}{4 - 2}(x - 2)$

$= 0(x - 2) = 0 \implies y = -1$

(Slope = 0)

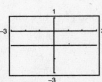

35. $y - 0 = \dfrac{2 - 0}{6 - (-1)}(x + 1)$

$\qquad = \dfrac{2}{7}(x + 1) = \dfrac{2}{7}x + \dfrac{2}{7} \implies y = \dfrac{2}{7}x + \dfrac{2}{7}$

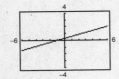

37. $t = 8$ corresponds to 2008.

Point: $(8, 12{,}500)$, slope: 850

$V - 12{,}500 = 850(t - 8)$

$V = 850t + 5700$

39. $(2, 160{,}000)$, $(3, 185{,}000)$

$m = \dfrac{185{,}000 - 160{,}000}{3 - 2} = 25{,}000$

$S - 160{,}000 = 25{,}000(t - 2)$

$S = 25{,}000t + 110{,}000$

For the fourth quarter let $t = 4$. Then we have $S = 25{,}000(4) + 110{,}000 = \$210{,}000$.

41. $5x - 4y = 8 \implies y = \dfrac{5}{4}x - 2$ and $m = \dfrac{5}{4}$

(a) Parallel slope: $m = \dfrac{5}{4}$

$\quad y - (-2) = \dfrac{5}{4}(x - 3)$

$\quad 4y + 8 = 5x - 15$

$\quad 0 = 5x - 4y - 23$

$\quad y = \dfrac{5}{4}x - \dfrac{23}{4}$

(b) Perpendicular slope: $m = -\dfrac{4}{5}$

$\quad y - (-2) = -\dfrac{4}{5}(x - 3)$

$\quad 5y + 10 = -4x + 12$

$\quad 4x + 5y - 2 = 0$

$\quad y = -\dfrac{4}{5}x + \dfrac{2}{5}$

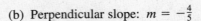

43. $x = 4$ is a vertical line; the slope is not defined.

(a) Parallel line: $x = -6$

(b) Perpendicular slope: $m = 0$

$\quad$ Perpendicular line: $y - 2 = 0(x + 6) = 0 \implies y = 2$

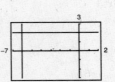

45. (a) Not a function. 20 is assigned two different values.

(b) Function

(c) Function

(d) Not a function. No value is assigned to 30.

47. No, y is not a function of x. Some x-values correspond to two y-values. For example, $x = 1$ corresponds to $y = 4$ and $y = -4$.

49. $y = \sqrt{1 - x}$

Each x value, $x \le 1$, corresponds to only one y-value so y is a function of x.

51. $f(x) = x^2 + 1$

(a) $f(1) = 1^2 + 1 = 2$

(b) $f(-3) = (-3)^2 + 1 = 10$

(c) $f(b^3) = (b^3)^2 + 1 = b^6 + 1$

(d) $f(x - 1) = (x - 1)^2 + 1 = x^2 - 2x + 2$

53. $h(x) = \begin{cases} 2x + 1, & x \le -1 \\ x^2 + 2, & x > -1 \end{cases}$

    (a) $h(-2) = 2(-2) + 1 = -3$

    (b) $h(-1) = 2(-1) + 1 = -1$

    (c) $h(0) = 0^2 + 2 = 2$

    (d) $h(2) = 2^2 + 2 = 6$

55. The domain of $f(x) = \dfrac{x - 1}{x + 2}$

    is all real numbers $x \ne -2$.

57. $f(x) = \sqrt{25 - x^2}$

    Domain: $25 - x^2 \ge 0$

                 $(5 + x)(5 - x) \ge 0$

    Domain: $[-5, 5]$

59. The domain of $g(5) = \dfrac{5s + 5}{3s - 9}$

    is all real numbers $s \ne 3$.

61. (a) $C(x) = 16{,}000 + 5.35x$

    (b) $P(x) = R(x) - C(x)$

             $= 8.20x - (16{,}000 + 5.35x)$

             $= 2.85x - 16{,}000$

63. $f(x) = 2x^2 + 3x - 1$

        $f(x + h) = 2(x + h)^2 + 3(x + h) - 1$

                $= 2x^2 + 4xh + 2h^2 + 3x + 3h - 1$

$$\frac{f(x + h) - f(x)}{h} = \frac{(2x^2 + 4xh + 2h^2 + 3x + 3h - 1) - (2x^2 + 3x - 1)}{h}$$

$$= \frac{4xh + 2h^2 + 3h}{h}$$

$$= 4x + 2h + 3, \quad h \ne 0$$

65. Domain: All real numbers

    Range: $y \le 3$

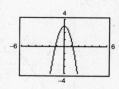

67. Domain: $36 - x^2 \ge 0 \Rightarrow x^2 \le 36 \Rightarrow -6 \le x \le 6$

    Range: $0 \le y \le 6$

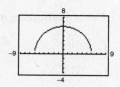

69. (a) $y = \dfrac{x^2 + 3x}{6}$

    (b) y is a function of x.

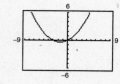

71. (a) $3x + y^2 = 2$

            $y^2 = 2 - 3x$

             $y = \pm\sqrt{2 - 3x}$

    (b) y is not a function of x.

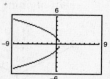

73. $f(x) = x^3 - 3x$

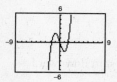

Increasing on $(-\infty, -1)$ and $(1, \infty)$

Decreasing on $(-1, 1)$

75. $f(x) = x\sqrt{x - 6}$

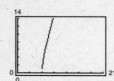

Increasing on $(6, \infty)$

77. $f(x) = (x^2 - 4)^2$

Relative minima: $(-2, 0)$ and $(2, 0)$

Relative maximum: $(0, 16)$

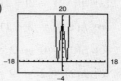

79. $h(x) = 4x^3 - x^4$

Relative maximum $(3, 27)$

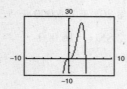

81. $f(x) = \begin{cases} 3x + 5, & x < 0 \\ x - 4, & x \geq 0 \end{cases}$

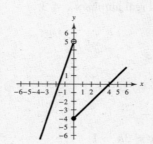

83. $f(x) = [\![x]\!] + 3$

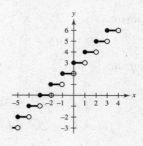

85. $f(-x) = ((-x)^2 - 8)^2$

$= (x^2 - 8)^2$

$= f(x)$

f is even.

87. $f(x) = -2$ is a constant function.

89. $g(x) = -x^3 - 2$ is obtained from $f(x) = x^3$ by a reflection in the x-axis, followed by a vertical shift two units downward.
$g(x) = -f(x) - 2$

91.

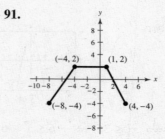

$y = f(-x)$ is a reflection in the y-axis.

93.

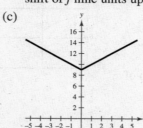

$y = f(x) - 2$ is a vertical shift downward two units.

95. $h(x) = |x| + 9$

(a) $f(x) = |x|$

(b) The graph of h is a vertical shift of f nine units upward.

(c)

(d) $h(x) = |x| + 9$

$= f(x) + 9$

97. $h(x) = -\sqrt{x} + 5$

(a) $f(x) = \sqrt{x}$

(b) The graph of h is a reflection of f in the x-axis, followed by a vertical shift five units upward.

(c)

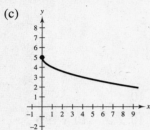

(d) $h(x) = -\sqrt{x} + 5$
$\qquad = -f(x) + 5$

99. $h(x) = \frac{1}{2}(x - 3)^2 - 6$

(a) $f(x) = x^2$

(b) The graph of h is a horizontal shift of f three units to the right, followed by a vertical shrink of $\frac{1}{2}$, followed by a vertical shift six units downward.

(c)

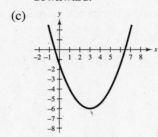

(d) $h(x) = \frac{1}{2}(x - 3)^2 - 6$
$\qquad = \frac{1}{2}f(x - 3) - 6$

101. $f(x) = x^2, g(x) = x + 3$

$(f \circ g)(x) = f(x + 3) = (x + 3)^2 = h(x)$

103. $f(x) = \sqrt{x}, g(x) = 4x + 2$

$(f \circ g)(x) = f(4x + 2) = \sqrt{4x + 2} = h(x)$

105. $f(x) = \dfrac{4}{x}, g(x) = x + 2$

$(f \circ g)(x) = f(x + 2) = \dfrac{4}{x + 2} = h(x)$

107.

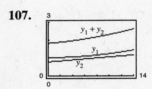

109. $f(x) = \frac{1}{2}x + 3 \implies f^{-1}(x) = 2(x - 3) = 2x - 6$

$f(f^{-1}(x)) = f(2(x - 3)) = \frac{1}{2}(2(x - 3)) + 3 = x - 3 + 3 = x$

$f^{-1}(f(x)) = f^{-1}(\frac{1}{2}x + 3) = 2(\frac{1}{2}x + 3 - 3) = 2(\frac{1}{2}x) = x$

111. (a)

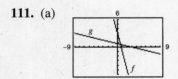

Reflection in the line $y = x$

(b)

x	-5	-1	0	1	3
$f(x)$	23	7	3	-1	-9

x	23	7	3	-1	-9
$g(x)$	-5	-1	0	1	3

The entries in the table are the same except that their rows are interchanged.

113.

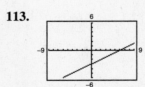

$f(x) = \frac{1}{2}x - 3$ passes the Horizontal Line Test, and hence is one-to-one and has an inverse $(f^{-1}(x) = 2(x + 3))$.

115.

$h(t) = \dfrac{2}{t-3}$ passes the Horizontal Line Test, and hence is one-to-one.

117.
$$y = \frac{1}{2}x - 5$$
$$x = \frac{1}{2}y - 5$$
$$x + 5 = \frac{1}{2}y$$
$$y = 2(x + 5)$$
$$f^{-1}(x) = 2x + 10$$

119.
$$f(x) = 4x^3 - 3$$
$$y = 4x^3 - 3$$
$$x = 4y^3 - 3$$
$$x + 3 = 4y^3$$
$$\frac{x + 3}{4} = y^3$$
$$f^{-1}(x) = \sqrt[3]{\frac{x + 3}{4}}$$

121.
$$f(x) = \sqrt{x + 10}$$
$$y = \sqrt{x + 10},\, x \geq -10,\, y \geq 0$$
$$x = \sqrt{y + 10},\, y \geq -10,\, x \geq 0$$
$$x^2 = y + 10$$
$$x^2 - 10 = y$$
$$f^{-1}(x) = x^2 - 10,\, x \geq 0$$

123. False. $g(x) = -[(x - 6)^2 + 3] = -(x - 6)^2 - 3$ and $g(-1) = -52 \neq 28$

125. False. $f(x) = \dfrac{1}{x}$ or $f(x) = x$ satisfy $f = f^{-1}$.

Chapter 1   Practice Test

1. Use a graphing utility to graph the equation $y = 4/x^2 - 5$. Approximate any x-intercepts of the graph.

2. Use a graphing utility to graph the equation $y = |x - 3| + 2$. Approximate any x-intercepts of the graph.

3. Graph $3x - 5y = 15$ by hand.   **4.** Graph $y = \sqrt{9 - x}$ by hand.

5. Solve $5x + 4 = 7x - 8$.   **6.** Solve $\dfrac{x}{3} - 5 = \dfrac{x}{5} + 1$.

7. Solve $\dfrac{3x + 1}{6x - 7} = \dfrac{2}{5}$ graphically and analytically.

8. Solve $(x - 3)^2 + 4 = (x + 1)^2$ graphically and analytically.

9. Find an equation for the line passing through the points $(3, -2)$ and $(4, -5)$. Use a graphing utility to sketch a graph of the line.

10. Find an equation of the line that passes through the point $(-1, 5)$ and has slope -3. Use a graphing utility to sketch a graph of the line.

11. Does the equation $x^4 + y^4 = 16$ represent y as a function of x?

12. Evaluate the function $f(x) = |x - 2|/(x - 2)$ at the points $x = 0$, $x = 2$, and $x = 4$.

13. Find the domain of the function $f(x) = 5/(x^2 - 16)$.   **14.** Find the domain of the function $g(t) = \sqrt{4 - t}$.

15. Use a graphing utility to sketch the graph of the function $f(x) = 3 - x^6$ and determine if the function is even, odd, or neither.

16. Use a graphing utility to approximate any relative minimum or maximum values of the function $y = 4 - x + x^3$.

17. Compare the graph of $f(x) = x^3 - 3$ with the graph of $y = x^3$.

18. Compare the graph of $f(x) = \sqrt{x - 6}$ with the graph of $y = \sqrt{x}$.

19. Find $g \circ f$ if $f(x) = \sqrt{x}$ and $g(x) = x^2 - 2$. What is the domain of $g \circ f$?

20. Find f/g if $f(x) = 3x^2$ and $g(x) = 16 - x^4$. What is the domain of f/g?

21. Show that $f(x) = 3x + 1$ and $g(x) = \dfrac{x - 1}{3}$ are inverse functions algebraically and graphically.

22. Find the inverse of $f(x) = \sqrt{9 - x^2}$, $0 \le x \le 3$. Graph f and f^{-1} in the same viewing rectangle.

C H A P T E R   2
Solving Equations and Inequalities

C H A P T E R 2
Solving Equations and Inequalities

Section 2.1   Linear Equations and Problem Solving

- ■ You should know how to solve linear equations: $ax + b = 0$.
- ■ An identity is an equation whose solution consists of every real number in its domain.
- ■ To solve an equation you can:
  - (a) Add or subtract the same quantity from both sides.
  - (b) Multiply or divide both sides by the same nonzero quantity.
- ■ To solve an equation that can be simplified to a linear equation:
  - (a) Remove all symbols of grouping and all fractions.   (b) Combine like terms.
  - (c) Solve by algebra.   (d) Check the answer.
- ■ A "solution" that does not satisfy the original equation is called an extraneous solution.
- ■ You should be able to set up mathematical models to solve problems.
- ■ You should be able to translate key words and phrases.
  - (a) Equality:

    Equals, Equal to, is, are, was, will be, represents

  - (b) Addition:

    Sum, plus, greater, increased by, more than, exceeds, total of

  - (c) Subtraction:

    Difference, minus, less than, decreased by, subtracted from, reduced by, the remainder

  - (d) Multiplication:

    Product, multiplied by, twice, times, percent of

  - (e) Division:

    Quotient, divided by, ratio, per

  - (f) Consecutive:

    Next, subsequent

- ■ You should know the following formulas:
  - (a) Perimeter:
    1. Square: $P = 4s$
    2. Rectangle: $P = 2L + 2W$
    3. Circle: $C = 2\pi r$

  - (b) Area:
    1. Square: $A = s^2$ 2. Rectangle: $A = LW$
    3. Circle: $A = \pi r^2$
    4. Triangle: $A = \left(\dfrac{1}{2}\right)bh$

  - (c) Volume
    1. Cube: $V = s^3$
    2. Rectangular solid: $V = LWH$
    3. Cylinder: $V = \pi r^2 h$
    4. Sphere: $V = \left(\dfrac{4}{3}\right)\pi r^3$

  - (d) Simple Interest: $I = Prt$

  - (e) Compound Interest: $A = P\left(1 + \dfrac{r}{n}\right)^{nt}$

  - (f) Distance: $D = r \cdot t$

  - (g) Temperature: $F = \dfrac{9}{5}C + 32$

- ■ You should be able to solve word problems. Study the examples in the text carefully.

Vocabulary Check

1. equation
2. solve
3. identities, conditional equations
4. $ax + b = 0$
5. extraneous
6. Mathematical modeling
7. formulas

1. $\dfrac{5}{2x} - \dfrac{4}{x} = 3$

   (a) $\dfrac{5}{2(-1/2)} - \dfrac{4}{(-1/2)} \overset{?}{=} 3$

   $\qquad\qquad\qquad 3 = 3$

   $x = -\frac{1}{2}$ *is* a solution.

   (b) $\dfrac{5}{2(4)} - \dfrac{4}{4} \overset{?}{=} 3$

   $\qquad\qquad -\dfrac{3}{8} \neq 3$

   $x = 4$ *is not* a solution.

   (c) $\dfrac{5}{2(0)} - \dfrac{4}{0}$ is undefined.

   $x = 0$ *is not* a solution.

   (d) $\dfrac{5}{2(1/4)} - \dfrac{4}{1/4} \overset{?}{=} 3$

   $\qquad\qquad -6 \neq 3$

   $x = \frac{1}{4}$ *is not* a solution.

3. $3 + \dfrac{1}{x + 2} = 4$

   (a) $3 + \dfrac{1}{(-1) + 2} \overset{?}{=} 4$

   $\qquad\qquad\quad 4 = 4$

   $x = -1$ *is* a solution.

   (b) $3 + \dfrac{1}{(-2) + 2} = 3 + \dfrac{1}{0}$ is undefined.

   $x = -2$ *is not* a solution.

   (c) $3 + \dfrac{1}{0 + 2} \overset{?}{=} 4$

   $\qquad\quad \dfrac{7}{2} \neq 4$

   $x = 0$ *is not* a solution.

   (d) $3 + \dfrac{1}{5 + 2} \overset{?}{=} 4$

   $\qquad\quad \dfrac{22}{7} = 4$

   $x = 5$ *is not* a solution.

5. $\dfrac{\sqrt{x + 4}}{6} + 3 = 4$

   (a) $\dfrac{\sqrt{-3 + 4}}{6} + 3 \overset{?}{=} 4$

   $\qquad\quad \dfrac{19}{6} \neq 4$

   $x = -3$ *is not* a solution.

   (b) $\dfrac{\sqrt{0 + 4}}{6} + 3 \overset{?}{=} 4$

   $\qquad\quad \dfrac{10}{3} \neq 4$

   $x = 0$ *is not* a solution.

   (c) $\dfrac{\sqrt{21 + 4}}{6} + 3 \overset{?}{=} 4$

   $\qquad\quad \dfrac{23}{6} \neq 4$

   $x = 21$ *is not* a solution.

   (d) $\dfrac{\sqrt{32 + 4}}{6} + 3 \overset{?}{=} 4$

   $\qquad\qquad 4 = 4$

   $x = 32$ *is* a solution.

7. $2(x - 1) = 2x - 2$ is an *identity* by the Distributive Property. It is true for all real values of x.

9. $x^2 - 8x + 5 = (x - 4)^2 - 11$ is an *identity* since $(x - 4)^2 - 11 = x^2 - 8x + 16 - 11 = x^2 - 8x + 5$.

11. $3 + \dfrac{1}{x + 1} = \dfrac{4x}{x + 1}$ is *conditional*. There are real values of x for which the equation is not true.

13. Method 1: $\dfrac{3x}{8} - \dfrac{4x}{3} = 4$

$$\dfrac{9x - 32x}{24} = 4$$

$$-23x = 96$$

$$x = -\dfrac{96}{23}$$

Method 2: Graph $y_1 = \dfrac{3x}{8} - \dfrac{4x}{3}$ and $y_2 = 4$ in the same viewing window. These lines intersect at $x \approx -4.1739 \approx -\dfrac{96}{23}$.

15. Method 1: $\dfrac{2x}{5} + 5x = \dfrac{4}{3}$

$$\dfrac{2x + 25x}{5} = \dfrac{4}{3}$$

$$27x = \dfrac{20}{3}$$

$$x = \dfrac{20}{3(27)} = \dfrac{20}{81}$$

Method 2: Graph $y_1 = \dfrac{2x}{5} + 5x$ and $y_2 = \dfrac{4}{3}$ in the same viewing window. These lines intersect at $x \approx 0.2469 \approx \dfrac{20}{81}$.

17. $3x - 5 = 2x + 7$

$3x - 2x = 7 + 5$

$x = 12$

19. $4y + 2 - 5y = 7 - 6y$

$-y + 2 = 7 - 6y$

$6y - y = 7 - 2$

$5y = 5$

$y = 1$

21. $3(y - 5) = 3 + 5y$

$3y - 15 = 3 + 5y$

$-18 = 2y$

$y = -9$

23. $\dfrac{x}{5} - \dfrac{x}{2} = 3$

$\dfrac{2x - 5x}{10} = 3$

$-3x = 30$

$x = -10$

25. $\dfrac{3}{2}(z + 5) - \dfrac{1}{4}(z + 24) = 0$

$4\left(\dfrac{3}{2}\right)(z + 5) - 4\left(\dfrac{1}{4}\right)(z + 24) = 4(0)$

$6(z + 5) - (z + 24) = 0$

$6z + 30 - z - 24 = 0$

$5z = -6$

$z = -\dfrac{6}{5}$

27. $\dfrac{2(z - 4)}{5} + 5 = 10z$

$\dfrac{(2z - 8) + 25}{5} = 10z$

$2z + 17 = 50z$

$17 = 48z$

$z = \dfrac{17}{48}$

29. $\dfrac{100 - 4u}{3} = \dfrac{5u + 6}{4} + 6$

$12\left(\dfrac{100 - 4u}{3}\right) = 12\left(\dfrac{5u + 6}{4}\right) + 12(6)$

$4(100 - 4u) = 3(5u + 6) + 72$

$400 - 16u = 15u + 18 + 72$

$-31u = -310$

$u = 10$

31. $\dfrac{5x - 4}{5x + 4} = \dfrac{2}{3}$

$3(5x - 4) = 2(5x + 4)$

$15x - 12 = 10x + 8$

$5x = 20$

$x = 4$

33. $\dfrac{1}{x - 3} + \dfrac{1}{x + 3} = \dfrac{10}{x^2 - 9}$

$\dfrac{(x + 3) + (x - 3)}{x^2 - 9} = \dfrac{10}{x^2 - 9}$

$2x = 10$

$x = 5$

35. $\dfrac{7}{2x + 1} - \dfrac{8x}{2x - 1} = -4$

$7(2x - 1) - 8x(2x + 1) = -4(2x + 1)(2x - 1)$

$14x - 7 - 16x^2 - 8x = -16x^2 + 4$

$6x = 11$

$x = \dfrac{11}{6}$

37. $\dfrac{1}{x} + \dfrac{2}{x - 5} = 0$

$1(x - 5) + 2x = 0$

$3x - 5 = 0$

$3x = 5$

$x = \dfrac{5}{3}$

39. $\dfrac{3}{x(x - 3)} + \dfrac{4}{x} = \dfrac{1}{x - 3}$

$3 + 4(x - 3) = x$

$3 + 4x - 12 = x$

$3x = 9$

$x = 3$

A check reveals that $x = 3$ is an extraneous solution, so there is no solution.

41. $A = \dfrac{1}{2}bh$

$2A = bh$

$\dfrac{2A}{b} = h$

43. $A = P\left(1 + \dfrac{r}{n}\right)^{nt}$

$P = \dfrac{A}{\left(1 + \dfrac{r}{n}\right)^{nt}}$

$P = A\left(1 + \dfrac{r}{n}\right)^{-nt}$

45. $S = \dfrac{rL - a}{r - 1}$

$Sr - S = rL - a$

$Sr - rL = S - a$

$r(S - L) = S - a$

$r = \dfrac{S - a}{S - L}$

47. $V = \dfrac{4}{3}\pi a^2 b$

$\dfrac{3V}{4\pi a^2} = b$

49. $P = 2l + 2w$

$2w = P - 2l$

$w = \dfrac{P - 2l}{2} = \dfrac{P}{2} - l$

51. $V = \pi r^2 h$

$h = \dfrac{V}{\pi r^2}$

53. $S = 2\pi rh$

$r = \dfrac{S}{2\pi h}$

55. $PV = nRT$

$R = \dfrac{PV}{nT}$

57. $16 = 0.432x - 10.44$

$26.44 = 0.432x$

$\dfrac{26.44}{0.432} = x \approx 61.2 \text{ inches}$

59. (a)

(b) $l = 1.5w$

$$P = 2l + 2w = 2(1.5w) + 2w = 5w$$

(c) $25 = 5w \implies w = 5$ meters and $l = (1.5)(5) = 7.5$ meters

61. (a) Test average $= \dfrac{\text{test 1} + \text{test 2} + \text{test 3} + \text{test 4}}{4}$

(b) Four tests at 100 points each $\implies$ 400 points. $(0.90)(400) = 360$ points needed for an A.
So far, $87 + 92 + 84 = 263$ points. $360 - 263 = 97$ points needed on fourth test.

63. Rate $= \dfrac{\text{Distance}}{\text{Time}} = \dfrac{50 \text{ kilometers}}{\frac{1}{2} \text{ hours}} = 100\ 8$ kilometers/hour

Total time $= \dfrac{\text{Total distance}}{\text{Rate}} = \dfrac{300 \text{ kilometers}}{100 \text{ kilometers/hour}} = 3$ hours

65. *Model:*   (Distance) $=$ (rate)(time$_1$ + time$_2$)

Labels: Distance $= 2 \cdot 200 = 400$ miles, rate $= r$,

$$\text{time}_1 = \frac{\text{distance}}{\text{rate}_1} = \frac{200}{55} \text{ hours},\ \text{time}_2 = \frac{\text{distance}}{\text{rate}_2} = \frac{200}{40} \text{ hours}$$

Equation: $400 = r\left(\dfrac{200}{55} + \dfrac{200}{40}\right)$

$$400 = r\left(\frac{1600}{440} + \frac{2200}{440}\right) = \frac{3800}{440}r$$

$$46.3 \approx r$$

The average speed for the round trip was approximately 46.3 miles per hour.

67. Let $h =$ height of the building in feet.

(a)

80 ft $3\frac{1}{2}$ ft   4 ft

Not drawn to scale

(b) $\dfrac{h \text{ feet}}{80 \text{ feet}} = \dfrac{4 \text{ feet}}{3.5 \text{ feet}}$

$$\frac{h}{80} = \frac{4}{3.5}$$

$$3.5h = 320$$

$$h \approx 91.4 \text{ feet}$$

69. $I = Prt$

$= 5000(0.065)6$

$= \$1950$

71. Let x = amount in $4\frac{1}{2}\%$ fund. Then $12{,}000 - x$ = amount in 5% fund.

$$560 = 0.045x + 0.05(12{,}000 - x)$$

$$560 = 0.045x + 600 - 0.05x$$

$$0.005x = 40$$

$$x = 8000$$

You must invest \$8000 in the $4\frac{1}{2}\%$ fund, and $12{,}000 - 8000 = 4000$ in the 5% fund.

73. Let x = amount invested in DVD players, and y = amount invested in VCRs.

$$x + y = 50{,}000 \implies y = 50{,}000 - x$$

$$0.3x + 0.25y = 0.29(50{,}000) = 14{,}500$$

$$0.3x + 0.25(50{,}000 - x) = 14{,}500$$

$$0.05x = 2000$$

$$x = \$40{,}000$$

$$y = \$10{,}000$$

75. Let x = number of pounds of \$2.49 nuts. Then $100 - x$ = number of pounds of \$3.89 nuts.

$$2.49x + 3.89(100 - x) = 3.19(100)$$

$$2.49x + 389 - 3.89x = 319$$

$$-1.40x = -70$$

$$x = \frac{-70}{-1.40} = 50$$

$$x = 50 \text{ lbs of \$2.49 nuts}$$

$$100 - x = 50 \text{ lbs of \$3.89 nuts}$$

Use 50 pounds of each kind.

77. $A = \dfrac{1}{2}bh$

$$h = \frac{2A}{b} = \frac{2(182.25)}{13.5} = 27 \text{ feet}$$

79. (a)

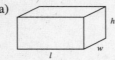

(b) $l = 3w$, $h = \left(1\frac{1}{2}\right)w$

$$V = lwh = (3w)(w)\left(\tfrac{3}{2}w\right) = 2304$$

$$\tfrac{9}{2}w^3 = 2304$$

$$w^3 = 512$$

$$w = 8 \text{ inches}$$

Dimensions $24 \times 8 \times 12$ inches

81.

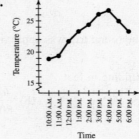

83. $W_1x = W_2(L - x)$

$$50x = 75(10 - x)$$

$$50x = 750 - 75x$$

$$125x = 750$$

$$x = 6 \text{ feet from the 50-pound child}$$

85. False. $x(3 - x) = 10$ is a quadratic equation.

$$-x^2 + 3x = 10 \text{ or } x^2 - 3x + 10 = 0$$

87. You need $a(-3) + b = c(-3)$ or $b = (c - a)(-3) = 3(a - c)$. One answer is $a = 2$, $c = 1$ and $b = 3$ $(2x + 3 = x)$ and another is $a = 6$, $c = 1$, and $b = 15$ $(6x + 15 = x)$.

89. Equivalent equations are derived from the substitution principle and simplification techniques. They have the same solution(s).

$2x + 3 = 8$ and $2x = 5$ are equivalent equations.

91.
$$2x - 5c = 10 + 3c - 3x, \quad x = 3$$
$$2(3) - 5c = 10 + 3c - 3(3)$$
$$6 - 5c = 1 + 3c$$
$$5 = 8c$$
$$c = \tfrac{5}{8}$$

93. $y = \frac{5}{8}x - 2$

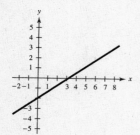

95. $y = (x - 3)^2 + 7$

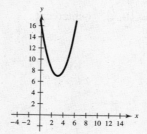

97. $y = -\frac{1}{2}|x + 4| - 1$

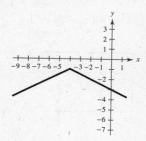

99. $(f + g)(-3) = f(-3) + g(-3) = [-(-3)^2 + 4] + [6(-3) - 5] = -5 - 23 = -28$

101. $(fg)(8) = f(8)g(8)$
$$= [-64 + 4][48 - 5] = -2580$$

103. $(f \circ g)(4) = f(g(4)) = f(19) = -357$

Section 2.2   Solving Equations Graphically

- ■ You should be able to find the intercepts of the graph of an equation.
- ■ You should be able to find the zeros of a function $y = f(x)$ by solving the equation $f(x) = 0$.
- ■ You should be able to find the solutions of an equation graphically using a graphing utility.
- ■ You should be able to use the zoom and trace features to find solutions to any desired accuracy.
- ■ You should be able to find the points of intersection of two graphs.

Vocabulary Check

1. x-intercept, y-intercept   **2.** zero   **3.** point of intersection

1. $y = x - 5$

Let $y = 0$: $0 = x - 5 \implies x = 5 \implies (5, 0)$ x-intercept

Let $x = 0$: $y = 0 - 5 \implies y = -5 \implies (0, -5)$ y-intercept

3. $y = x^2 + x - 2$

Let $y = 0$: $(x^2 + x - 2) = (x + 2)(x - 1) = 0 \implies x = -2, 1 \implies (-2, 0), (1, 0)$ x-intercepts

Let $x = 0$: $y = 0^2 + 0 - 2 = -2 \implies (0, -2)$ y-intercept

5. $y = x\sqrt{x + 2}$

Let $y = 0$: $0 = x\sqrt{x + 2} \implies x = 0, -2 \implies (0, 0), (-2, 0)$ *x*-intercepts

Let $x = 0$: $y = 0\sqrt{0 + 2} = 0 \implies (0, 0)$ *y*-intercept

7. $xy = 4$

If $x = 0$, then $0y = 0 = 4$, which is impossible. Similarly, $y = 0$ is impossible.
Hence there are no intercepts.

9. $xy - 2y - x + 1 = 0$

Let $y = 0$: $-x + 1 = 0 \implies x = 1 \implies (1, 0)$ *x*-intercept

Let $x = 0$: $-2y + 1 = 0 \implies y = \frac{1}{2} \implies \left(0, \frac{1}{2}\right)$ *y*-intercept

11.

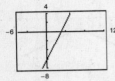 $(3, 0), (0, -6)$

$y = 0 = 2(x - 1) - 4 = 2x - 2 - 4 = 2x - 6 \implies 2x = 6 \implies x = 3$

$x = 0 \implies y = -6$

13.

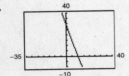 $(10, 0), (0, 30)$

$y = 0 = 20 - (3x - 10) = 20 - 3x + 10 = 30 - 3x \implies 3x = 30 \implies x = 10$

$x = 0 \implies y = 30$

15. $f(x) = 5(4 - x)$

$5(4 - x) = 0$

$4 - x = 0$

$x = 4$

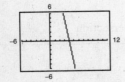

$x = 4$

17. $f(x) = x^3 - 6x^2 + 5x$

$x^3 - 6x^2 + 5x = 0$

$x(x^2 - 6x + 5) = 0$

$x(x - 5)(x - 1) = 0$

$x = 0, 5, 1$

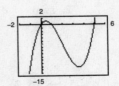

$x = 0, 1, 5$

19.
$$f(x) = \frac{x+2}{3} - \frac{x-1}{5} - 1$$

$$\frac{x+2}{3} - \frac{x-1}{5} - 1 = 0$$

$$5(x+2) - 3(x-1) - 15 = 0$$

$$2x = 2$$

$$x = 1$$

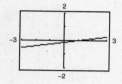

21. $2.7x - 0.4x = 1.2$

$$2.3x = 1.2$$

$$x = \frac{1.2}{2.3} \approx 0.522$$

$$f(x) = 2.7x - 0.4x - 1.2 = 0$$

$$x \approx 0.522$$

23. $25(x-3) = 12(x+2) - 10$

$$25x - 75 = 12x + 24 - 10$$

$$13x - 89 = 0$$

$$x = \frac{89}{13}$$

$$f(x) = 25(x-3) - 12(x+2) + 10 = 0$$

$$x = 6.846$$

25. $\frac{3x}{2} + \frac{1}{4}(x-2) = 10$

$$\frac{6x}{4} + \frac{x}{4} = 10 + \frac{1}{2}$$

$$\frac{7x}{4} = \frac{21}{2}$$

$$x = 6$$

$$f(x) = \frac{3x}{2} + \frac{1}{4}(x-2) - 10 = 0$$

$$x = 6.0$$

27. $0.60x + 0.40(100 - x) = 1.2$

$$0.60x + 40 - 0.40x = 1.2$$

$$0.20x = -38.8$$

$$x = -194$$

$$f(x) = 0.60x + 0.40(100 - x) - 1.2 = 0$$

$$x = -194$$

29. $\frac{x-3}{3} = \frac{3x-5}{2}$

$$2(x-3) = 3(3x-5)$$

$$2x - 6 = 9x - 15$$

$$9 = 7x$$

$$x = \frac{9}{7}$$

$$f(x) = \frac{x-3}{3} - \frac{3x-5}{2}$$

$$x \approx 1.286 \approx \frac{9}{7}$$

31. $\frac{x-5}{4} + \frac{x}{2} = 10$

$$(x-5) + 2x = 40$$

$$3x = 45$$

$$x = 15$$

$$f(x) = \frac{x-5}{4} + \frac{x}{2} - 10$$

33. $(x+2)^2 = x^2 - 6x + 1$

$$x^2 + 4x + 4 = x^2 - 6x + 1$$

$$10x = -3$$

$$x = -\frac{3}{10}$$

$$f(x) = (x+2)^2 - x^2 + 6x - 1$$

$$x = -\frac{3}{10}$$

35. $\frac{1}{4}(x^2 - 10x + 17) = 0$

$x = 2.172, 7.828$

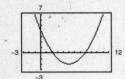

37. $x^3 + x + 4 = 0$

$x = -1.379$

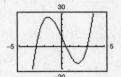

39. $2x^3 - x^2 - 18x + 9 = 0$

$x = -3.0, \ 0.5, \ 3.0$

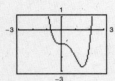

41. $x^4 = 2x^3 + 1$

$x^4 - 2x^3 - 1 = 0$

$x \approx -0.717, \ 2.107$

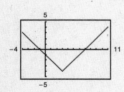

43. $\frac{2}{x + 2} = 3$

$\frac{2}{x + 2} - 3 = 0$

$x = -\frac{4}{3}$

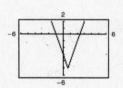

45. $\frac{5}{x} = 1 + \frac{3}{x + 2}$

$\frac{5}{x} - 1 - \frac{3}{x + 2} = 0$

$x = -3.162, 3.162$

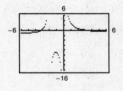

47. $|x - 3| = 4$

$|x - 3| - 4 = 0$

$x = -1, 7$

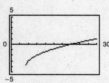

49. $|3x - 2| - 1 = 4$

$|3x - 2| - 5 = 0$

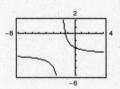

$x \approx -1.0, 2.333$

51. $\sqrt{x - 2} = 3$

$\sqrt{x - 2} - 3 = 0$

$x = 11$

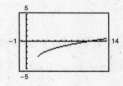

53. $2 + \sqrt{x - 5} = 6$

$\sqrt{x - 5} - 4 = 0$

$x = 21$

55. (a)

x	-1	0	1	2	3	4
$3.2x - 5.8$	-9	-5.8	-2.6	0.6	3.8	7.0

Because of the sign change, $1 < x < 2$.

—CONTINUED—

55. —CONTINUED—

(b)

x	1.5	1.6	1.7	1.8	1.9	2.0
$3.2x - 5.8$	-1	-0.68	-0.36	-0.04	0.28	0.6

Because of the sign change, $1.8 < x < 1.9$. To improve accuracy, evaluate the expression for values in this interval and determine where the sign changes.

(c) Let $y_1 = 3.2x - 5.8$. The graph of y_1 crosses the x-axis at $x = 1.8125$.

57. $y = 2 - x$

$y = 2x - 1$

$2 - x = 2x - 1$

$3 = 3x$

$x = 1, y = 2 - 1 = 1$

$(x, y) = (1, 1)$

59. $2x + y = 6 \implies y = 6 - 2x$

$-x + y = 0 \implies y = x$

$6 - 2x = x$

$6 = 3x$

$x = 2, y = x = 2$

$(x, y) = (2, 2)$

61. $x - y = 10 \implies y = x - 10$

$x + 2y = 4 \implies y = -\frac{1}{2}x + 2$

$x - 10 = -\frac{1}{2}x + 2$

$2x - 20 = -x + 4$

$3x = 24$

$x = 8, y = 8 - 10 = -2$

$(x, y) = (8, -2)$

63. $y = x^2 - x + 1$

$y = x^2 + 2x + 4$

$x^2 - x + 1 = x^2 + 2x + 4$

$-3 = 3x$

$x = -1$

$y = (-1)^2 - (-1) + 1 = 3$

$(x, y) = (-1, 3)$

65. $y = 9 - 2x$

$y = x - 3$

$(4, 1)$

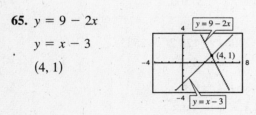

67. $y = 4 - x^2$

$y = 2x - 1$

$(x, y) = (1.449, 1.898), (-3.449, -7.899)$

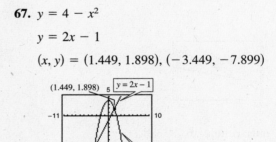

69. $y = 2x^2$

$y = x^4 - 2x^2$

$(x, y) = (0, 0), (2, 8), (-2, 8)$

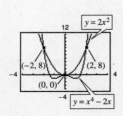

71. (a) $\dfrac{1 + 0.73205}{1 - 0.73205} = \dfrac{1.73205}{0.26795}$

$= 6.464079 = 6.46$

(b) $\dfrac{1 + 0.73205}{1 - 0.73205} = \dfrac{1.73205}{0.26795}$

$= \dfrac{1.73}{0.27}$

$= 6.407407 = 6.41$

The second method decreases the accuracy.

73. (a) $t = \dfrac{x}{63} + \dfrac{(280 - x)}{54}$

(b) Domain: $0 \le x \le 280$

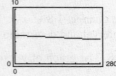

(c) If the time was 4 hours and 45 minutes, then $t = 4\frac{3}{4}$ and $x = 164.5$ miles.

75. (a) $A = x + 0.33(55 - x)$

(b) Domain: $0 \le x \le 55$

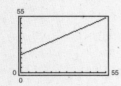

(c) If the final mixture is 60% concentrate, then $A = 0.6(55) = 33$ and $x = 22.2$ gallons.

77. (a) Area $= A(x) = 4x + 8x = 12x$

(b)

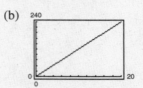

(c) $A(x) = 12x = 200 \implies x \approx 16.67$ units

79. (a) $T = I + S = x + 10{,}000 - \frac{1}{2}x = 10{,}000 + \frac{1}{2}x$

(b) If $S = 6600 = 10{,}000 - \frac{1}{2}x \implies \frac{1}{2}x = 3400 \implies x = \$6800$.

(c) If $T = 13{,}800 = 10{,}000 + \frac{1}{2}x \implies 3800 = \frac{1}{2}x \implies x = \$7600$.

(d) If $T = 12{,}500 = 10{,}000 + \frac{1}{2}x$ then $x = 5000$. Thus, $S = 10{,}000 - \frac{1}{2}x = \$7500$.

81. (a)

Intersection: $(6.7, 3388.7)$

(c) The slopes indicate the change in population per year. Arizona's population is growing faster.

(b) $45.2t + 3087 = 128.2t + 2533$

$83t = 554$

$t \approx 6.7$

$A = S \approx 3388.7$

The point $(6.7, 3388.7)$ indicates the year, 1986, in which the two populations were the same, about 3388.7 thousand.

(d) For 2010, $t = 30$ and $S \approx 4443$ thousand and $A \approx 6379$ thousand. Answers will vary.

83. True

85. False. Two linear equations could have an infinite number of points of intersection.
For example, $x + y = 1$ and $2x + 2y = 2$.

87. From the table, $f(x) = 0$ for $x = 3$.

89. From the table, $g(x) = -f(x)$ for $x = 1$. In this case, $f(x) = -6$ and $g(x) = 6$.

91. $\dfrac{12}{5\sqrt{3}} \cdot \dfrac{\sqrt{3}}{\sqrt{3}} = \dfrac{12\sqrt{3}}{5(3)} = \dfrac{4\sqrt{3}}{5}$

93. $\dfrac{3}{8+\sqrt{11}} \cdot \dfrac{8-\sqrt{11}}{8-\sqrt{11}} = \dfrac{3\left(8-\sqrt{11}\right)}{64-11} = \dfrac{3\left(8-\sqrt{11}\right)}{53}$ **95.** $(x+6)(3x-5) = 3x^2 + 13x - 30$

97. $(2x-9)(2x+9) = 4x^2 - 81$

Section 2.3   Complex Numbers

- You should know how to work with complex numbers.
- Operations on complex numbers
  - (a) Addition: $(a+bi) + (c+di) = (a+c) + (b+d)i$
  - (b) Subtraction: $(a+bi) - (c+di) = (a-c) + (b-d)i$
  - (c) Multiplication: $(a+bi)(c+di) = (ac-bd) + (ad+bc)i$
  - (d) Division: $\dfrac{a+bi}{c+di} = \dfrac{a+bi}{c+di} \cdot \dfrac{c-di}{c-di} = \dfrac{ac+bd}{c^2+d^2} + \dfrac{bc-ad}{c^2+d^2}i$
- The complex conjugate of $a + bi$ is $a - bi$:
  $$(a+bi)(a-bi) = a^2 + b^2$$
- The additive inverse of $a + bi$ is $-a - bi$.
- The multiplicative inverse of $a + bi$ is
  $$\dfrac{a-bi}{a^2+b^2}.$$
- $\sqrt{-a} = \sqrt{a}\,i$ for $a > 0$.

Vocabulary Check

1. (a) ii   (b) iii   (c) i   **2.** $\sqrt{-1}, -1$ **3.** complex, $a + bi$

4. real, imaginary   **5.** Mandelbrot Set

---

1. $a + bi = -9 + 4i$ **3.** $(a-1) + (b+3)i = 5 + 8i$ **5.** $5 + \sqrt{-16} = 5 + \sqrt{16(-1)}$

$\quad\ a = -9$ $\qquad a - 1 = 5 \implies a = 6$ $\qquad\qquad\qquad = 5 + 4i$

$\quad\ b = 4$ $\qquad b + 3 = 8 \implies b = 5$

7. $-6 = -6 + 0i$ **9.** $-5i + i^2 = -5i - 1 = -1 - 5i$ **11.** $\left(\sqrt{-75}\right)^2 = -75$

13. $\sqrt{-0.09} = \sqrt{0.09}\,i = 0.3i$ **15.** $(4+i) - (7-2i) = (4-7) + (1+2)i$

$\qquad\qquad\qquad\qquad\qquad\qquad\qquad\qquad\qquad = -3 + 3i$

17. $\left(-1 + \sqrt{-8}\right) + \left(8 - \sqrt{-50}\right) = 7 + 2\sqrt{2}i - 5\sqrt{2}i = 7 - 3\sqrt{2}i$

19. $13i - (14 - 7i) = 13i - 14 + 7i = -14 + 20i$ **21.** $\left(\dfrac{3}{2} + \dfrac{5}{2}i\right) + \left(\dfrac{5}{3} + \dfrac{11}{3}i\right) = \left(\dfrac{3}{2} + \dfrac{5}{3}\right) + \left(\dfrac{5}{2} + \dfrac{11}{3}\right)i$

$\qquad\qquad\qquad\qquad\qquad\qquad\qquad\qquad\qquad\qquad\qquad\quad = \dfrac{9+10}{6} + \dfrac{15+22}{6}i$

$\qquad\qquad\qquad\qquad\qquad\qquad\qquad\qquad\qquad\qquad\qquad\quad = \dfrac{19}{6} + \dfrac{37}{6}i$

23. $(1.6 + 3.2i) + (-5.8 + 4.3i) = -4.2 + 7.5i$

25. $\sqrt{-6} \cdot \sqrt{-2} = (\sqrt{6}i)(\sqrt{2}i) = \sqrt{12}i^2 = (2\sqrt{3})(-1) = -2\sqrt{3}$

27. $(\sqrt{-10})^2 = (\sqrt{10}i)^2 = 10i^2 = -10$

29. $(1 + i)(3 - 2i) = 3 - 2i + 3i - 2i^2$
$$= 3 + i + 2$$
$$= 5 + i$$

31. $4i(8 + 5i) = 32i + 20i^2$
$$= 32i + 20(-1)$$
$$= -20 + 32i$$

33. $(\sqrt{14} + \sqrt{10}\,i)(\sqrt{14} - \sqrt{10}\,i) = 14 - 10i^2$
$$= 14 + 10 = 24$$

35. $(4 + 5i)^2 - (4 - 5i)^2 = [(4 + 5i) + (4 - 5i)][(4 + 5i) - (4 - 5i)] = 8(10i) = 80i$

37. $4 - 3i$ is the complex conjugate of $4 + 3i$.
$(4 + 3i)(4 - 3i) = 16 + 9 = 25$

39. $-6 + \sqrt{5}i$ is the complex conjugate of $-6 - \sqrt{5}i$.
$(-6 - \sqrt{5}i)(-6 + \sqrt{5}i) = 36 + 5 = 41$

41. $-\sqrt{20}i$ is the complex conjugate of
$\sqrt{-20} = \sqrt{20}i$.
$(\sqrt{20}i)(-\sqrt{20}i) = 20$

43. $3 + \sqrt{2}i$ is the complex conjugate of
$3 - \sqrt{-2} = 3 - \sqrt{2}i$.
$(3 - \sqrt{2}i)(3 + \sqrt{2}i) = 9 + 2 = 11$

45. $\dfrac{6}{i} = \dfrac{6}{i} \cdot \dfrac{-i}{-i} = \dfrac{-6i}{-i^2} = \dfrac{-6i}{1} = -6i$

47. $\dfrac{2}{4 - 5i} = \dfrac{2}{4 - 5i} \cdot \dfrac{4 + 5i}{4 + 5i} = \dfrac{8 + 10i}{16 + 25} = \dfrac{8}{41} + \dfrac{10}{41}i$

49. $\dfrac{2 + i}{2 - i} = \dfrac{2 + i}{2 - i} \cdot \dfrac{2 + i}{2 + i}$
$$= \dfrac{4 + 4i + i^2}{4 + 1}$$
$$= \dfrac{3 + 4i}{5} = \dfrac{3}{5} + \dfrac{4}{5}i$$

51. $\dfrac{i}{(4 - 5i)^2} = \dfrac{i}{16 - 25 - 40i}$
$$= \dfrac{i}{-9 - 40i} \cdot \dfrac{-9 + 40i}{-9 + 40i}$$
$$= \dfrac{-40 - 9i}{81 + 40^2}$$
$$= \dfrac{-40}{1681} - \dfrac{9}{1681}i$$

53. $\dfrac{2}{1 + i} - \dfrac{3}{1 - i} = \dfrac{2(1 - i) - 3(1 + i)}{(1 + i)(1 - i)}$
$$= \dfrac{2 - 2i - 3 - 3i}{1 + 1}$$
$$= \dfrac{-1 - 5i}{2} = -\dfrac{1}{2} - \dfrac{5}{2}i$$

55. $\dfrac{i}{3 - 2i} + \dfrac{2i}{3 + 8i} = \dfrac{3i + 8i^2 + 6i - 4i^2}{(3 - 2i)(3 + 8i)}$
$$= \dfrac{-4 + 9i}{9 + 18i + 16}$$
$$= \dfrac{-4 + 9i}{25 + 18i} \cdot \dfrac{25 - 18i}{25 - 18i}$$
$$= \dfrac{-100 + 72i + 225i + 162}{25^2 + 18^2}$$
$$= \dfrac{62 + 297i}{949}$$
$$= \dfrac{62}{949} + \dfrac{297}{949}i$$

57. $-6i^3 + i^2 = -6i^2i + i^2$

$$= -6(-1)i + (-1)$$

$$= 6i - 1$$

$$= -1 + 6i$$

59. $\left(\sqrt{-75}\right)^3 = \left(5\sqrt{3}i\right)^3 = 5^3\left(\sqrt{3}\right)^3 i^3$

$$= 125\left(3\sqrt{3}\right)(-i)$$

$$= -375\sqrt{3}i$$

61. $\dfrac{1}{i^3} = \dfrac{1}{i^3} \cdot \dfrac{i}{i} = \dfrac{i}{i^4} = \dfrac{i}{1} = i$

63. $(2)^3 = 8$

$$\left(-1 + \sqrt{3}i\right)^3 = (-1)^3 + 3(-1)^2\left(\sqrt{3}i\right) + 3(-1)\left(\sqrt{3}i\right)^2 + \left(\sqrt{3}i\right)^3$$

$$= -1 + 3\sqrt{3}i - 9i^2 + 3\sqrt{3}i^3$$

$$= -1 + 3\sqrt{3}i + 9 - 3\sqrt{3}i$$

$$= 8$$

$$\left(-1 - \sqrt{3}i\right)^3 = (-1)^3 + 3(-1)^2\left(-\sqrt{3}i\right) + 3(-1)\left(-\sqrt{3}i\right)^2 + \left(-\sqrt{3}i\right)^3$$

$$= -1 - 3\sqrt{3}i - 9i^2 - 3\sqrt{3}i^3$$

$$= -1 - 3\sqrt{3}i + 9 + 3\sqrt{3}i$$

$$= 8$$

The three numbers are cube roots of 8.

65. $4 + 3i$

67. $5i$

69. 2

71. $4 - 5i$

73. $3i$

75. 1

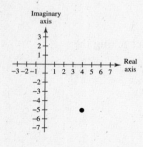

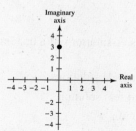

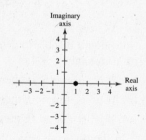

77. The complex number $\frac{1}{2}i$, is in the Mandelbrot Set since for $c = \frac{1}{2}i$, the corresponding Mandelbrot sequence is

$$\frac{1}{2}i, \; -\frac{1}{4} + \frac{1}{2}i, \; -\frac{3}{16} + \frac{1}{4}i, \; -\frac{7}{256} + \frac{13}{32}i, \; -\frac{10,767}{65,536} + \frac{1957}{4096}i, \; -\frac{864,513,055}{4,294,967,296} + \frac{46,037,845}{134,217,728}i$$

which is bounded. Or in decimal form

$$0.5i, \; -0.25 + 0.5i, \; -0.1875 + 0.25i, \; -0.02734 + 0.40625i,$$

$$-0.164291 + 0.477783i, \; -0.201285 + 0.343009i.$$

79. $z_1 = 5 + 2i$

$z_2 = 3 - 4i$

$$\frac{1}{z} = \frac{1}{z_1} + \frac{1}{z_2} = \frac{1}{5 + 2i} + \frac{1}{3 - 4i} = \frac{(3 - 4i) + (5 + 2i)}{(5 + 2i)(3 - 4i)} = \frac{8 - 2i}{23 - 14i}$$

$$z = \frac{23 - 14i}{8 - 2i}\left(\frac{8 + 2i}{8 + 2i}\right) = \frac{212 - 66i}{68} \approx 3.118 - 0.971i$$

81. False. A real number $a + 0i = a$ is equal to its conjugate.

83. False. For example, $(1 + 2i) + (1 - 2i) = 2$, which is not an imaginary number.

85. True. Let $z_1 = a_1 + b_1 i$ and $z_2 = a_2 + b_2 i$. Then

$$\overline{z_1 z_2} = \overline{(a_1 + b_1 i)(a_2 + b_2 i)}$$
$$= \overline{(a_1 a_2 - b_1 b_2) + (a_1 b_2 + b_1 a_2)i}$$
$$= (a_1 a_2 - b_1 b_2) - (a_1 b_2 + b_1 a_2)i$$
$$= (a_1 - b_1 i)(a_2 - b_2 i)$$
$$= \overline{a_1 + b_1 i} \; \overline{a_2 + b_2 i}$$
$$= \overline{z_1} \; \overline{z_2}.$$

87. $(4x - 5)(4x + 5) = 16x^2 - 20x + 20x - 25$
$$= 16x^2 - 25$$

89. $\left(3x - \frac{1}{2}\right)(x + 4) = 3x^2 - \frac{1}{2}x + 12x - 2$
$$= 3x^2 + \frac{23}{2}x - 2$$

Section 2.4   Solving Quadratic Equations Algebraically

- ■ You should be able to solve a quadratic equation by factoring, if possible.
- ■ You should be able to solve a quadratic equation of the form $u^2 = d$ by extracting square roots.
- ■ You should be able to solve a quadratic equation by completing the square.
- ■ You should know and be able to use the Quadratic Formula: For $ax^2 + bx + c = 0$, $a \neq 0$,

$$x = \frac{-b \pm \sqrt{b^2 - 4ac}}{2a}.$$

- ■ You should be able to determine the types of solutions of a quadratic equation by checking the discriminant $b^2 - 4ac$.
  - (a) If $b^2 - 4ac > 0$, there are two distinct real solutions.
  - (b) If $b^2 - 4ac = 0$, there is one repeated real solution.
  - (c) If $b^2 - 4ac < 0$, there is no real solution.
- ■ You should be able to solve certain types of nonlinear or nonquadratic equations.
- ■ For equations involving radicals or fractional powers, raise both sides to the same power.
- ■ For equations that are of the quadratic type, $au^2 + bu + c = 0$, $a \neq 0$, use either factoring or the quadratic equation.
- ■ For equations with fractions, multiply both sides by the least common denominator to clear the fractions.
- ■ For equations involving absolute value, remember that the expression inside the absolute value can be positive or negative.
- ■ Always check for extraneous solutions.

Vocabulary Check

1. quadratic equation

2. factoring, extracting square roots, completing the square, Quadratic Formula

3. discriminant

4. position, $-16t^2 + v_0 t + s_0$, initial velocity, initial height

1. $2x^2 = 3 - 5x$

Standard form: $2x^2 + 5x - 3 = 0$

3. $\frac{1}{5}(3x^2 - 10) = 12x$

$3x^2 - 10 = 60x$

Standard form: $3x^2 - 60x - 10 = 0$

5. $6x^2 + 3x = 0$

$3x(2x + 1) = 0$

$3x = 0$ or $2x + 1 = 0$

$x = 0$ or $\qquad x = -\frac{1}{2}$

7. $x^2 - 2x - 8 = 0$

$(x - 4)(x + 2) = 0$

$x - 4 = 0$ or $x + 2 = 0$

$x = 4$ or $\qquad x = -2$

9. $3 + 5x - 2x^2 = 0$

$(3 - x)(1 + 2x) = 0$

$3 - x = 0$ or $1 + 2x = 0$

$x = 3$ or $\qquad x = -\frac{1}{2}$

11. $\qquad x^2 + 4x = 12$

$x^2 + 4x - 12 = 0$

$(x + 6)(x - 2) = 0$

$x + 6 = 0$ or $x - 2 = 0$

$x = -6$ or $\quad x = 2$

13. $\qquad (x + a)^2 - b^2 = 0$

$[(x + a) + b][(x + a) - b] = 0$

$x + a + b = 0 \implies x = -a - b$

$x + a - b = 0 \implies x = -a + b$

15. $x^2 = 49$

$x = \pm\sqrt{49} = \pm 7$

17. $(x - 12)^2 = 16$

$x - 12 = \pm\sqrt{16} = \pm 4$

$x = 12 \pm 4$

$x = 16, 8$

19. $(3x - 1)^2 + 6 = 0$

$(3x - 1)^2 = -6$

$3x - 1 = \pm\sqrt{-6} = \pm\sqrt{6}i$

$x = \frac{1}{3} \pm \frac{\sqrt{6}}{3}i \approx 0.33 \pm 0.82i$

21. $(x - 7)^2 = (x + 3)^2$

$x - 7 = \pm(x + 3)$

$x - 7 = x + 3, \quad$ impossible

$x - 7 = -(x + 3) \implies 2x = 4$

$\implies \quad x = 2$

23. $\qquad x^2 + 4x = 32$

$x^2 + 4x + 4 = 32 + 4$

$(x + 2)^2 = 36$

$x + 2 = \pm 6$

$x = -2 \pm 6$

$x = -8, 4$

25. $x^2 + 6x + 2 = 0$

$x^2 + 6x = -2$

$x^2 + 6x + 3^2 = -2 + 3^2$

$(x + 3)^2 = 7$

$x + 3 = \pm\sqrt{7}$

$x = -3 \pm \sqrt{7}$

27. $9x^2 - 18x + 3 = 0$

$x^2 - 2x + \frac{1}{3} = 0$

$x^2 - 2x = -\frac{1}{3}$

$x^2 - 2x + 1^2 = -\frac{1}{3} + 1^2$

$(x - 1)^2 = \frac{2}{3}$

$x - 1 = \pm\sqrt{\frac{2}{3}}$

$x = 1 \pm \sqrt{\frac{2}{3}}$

$x = 1 \pm \frac{\sqrt{6}}{3}$

29. $-6 + 2x - x^2 = 0$

$(x^2 - 2x + 1) = -6 + 1$

$(x - 1)^2 = -5$

$x - 1 = \pm\sqrt{-5}$

$= \pm\sqrt{5}i$

$x = 1 \pm \sqrt{5}i$

31. $2x^2 + 5x - 8 = 0$

$x^2 + \dfrac{5}{2}x - 4 = 0$

$x^2 + \dfrac{5}{2}x + \dfrac{25}{16} = 4 + \dfrac{25}{16}$

$\left(x + \dfrac{5}{4}\right)^2 = \dfrac{89}{16}$

$x + \dfrac{5}{4} = \pm\dfrac{\sqrt{89}}{4}$

$x = \dfrac{-5}{4} \pm \dfrac{\sqrt{89}}{4}$

33. $y = (x + 3)^2 - 4$

(a)

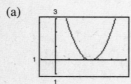

(b) The x-intercepts are $(-1, 0)$ and $(-5, 0)$.

(c) $0 = (x + 3)^2 - 4$

$4 = (x + 3)^2$

$\pm\sqrt{4} = x + 3$

$-3 \pm 2 = x$

$x = -1$ or $x = -5$

35. $y = -4x^2 + 4x + 3$

(a)

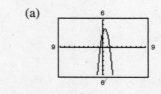

(b) The x-intercepts are $\left(-\frac{1}{2}, 0\right)$ and $\left(\frac{3}{2}, 0\right)$.

(c) $0 = -4x^2 + 4x + 3$

$4x^2 - 4x = 3$

$4(x^2 - x) = 3$

$x^2 - x = \dfrac{3}{4}$

$x^2 - x + \left(\dfrac{1}{2}\right)^2 = \dfrac{3}{4} + \left(\dfrac{1}{2}\right)^2$

$\left(x - \dfrac{1}{2}\right)^2 = 1$

$x - \dfrac{1}{2} = \pm\sqrt{1}$

$x = \dfrac{1}{2} \pm 1$

$x = \dfrac{3}{2}$ or $x = -\dfrac{1}{2}$

37. $y = \frac{1}{4}(4x^2 - 20x + 25)$

(a)

(b) The x-intercept is $\left(\frac{5}{2}, 0\right)$.

(c) $\frac{1}{4}(4x^2 - 20x + 25) = 0$

$4x^2 - 20x + 25 = 0$

$(2x - 5)^2 = 0$

$2x - 5 = 0$

$x = \dfrac{5}{2}$

39. $2x^2 - 5x + 5 = 0$

The graph does not have any x-intercepts and thus the equation has no real solution.

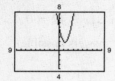

41. $\frac{4}{7}x^2 - 8x + 28 = 0$

The graph has one x-intercept $(7, 0)$ and hence the equation has one real solution.

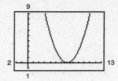

43. $-0.2x^2 + 1.2x - 8 = 0$

The graph does not have any x-intercepts and hence the equation has no real solution.

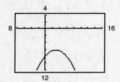

45. $-x^2 + 2x + 2 = 0$

$$x = \frac{-b \pm \sqrt{b^2 - 4ac}}{2a}$$

$$= \frac{-2 \pm \sqrt{2^2 - 4(-1)(2)}}{2(-1)}$$

$$= \frac{-2 \pm 2\sqrt{3}}{-2} = 1 \pm \sqrt{3}$$

47. $2x^2 - 3x + 4 = 0$

$$x = \frac{3 \pm \sqrt{9 - 4(2)(4)}}{2(2)}$$

$$= \frac{3 \pm \sqrt{-23}}{4}$$

$$= \frac{3}{4} \pm \frac{\sqrt{23}}{4}i$$

49. $x^2 + 3x + 8 = 0$

$$x = \frac{-3 \pm \sqrt{9 - 4(8)}}{2}$$

$$= \frac{-3 \pm \sqrt{-23}}{2}$$

$$= -\frac{3}{2} \pm \frac{\sqrt{23}i}{2}$$

51. $4x^2 + 16x + 17 = 0$

$$x = \frac{-b \pm \sqrt{b^2 - 4ac}}{2a}$$

$$= \frac{-16 \pm \sqrt{16^2 - 4(4)(17)}}{2(4)}$$

$$= \frac{-16 \pm \sqrt{-16}}{8}$$

$$= \frac{-16 \pm 4i}{8}$$

$$= -2 \pm \frac{1}{2}i$$

53. $x^2 - 2x - 1 = 0$

$$x^2 - 2x = 1$$

$$x^2 - 2x + 1^2 = 1 + 1^2$$

$$(x - 1)^2 = 2$$

$$x - 1 = \pm\sqrt{2}$$

$$x = 1 \pm \sqrt{2}$$

55. $(x + 3)^2 = 81$

$$x + 3 = \pm 9$$

$$x + 3 = 9 \text{ or } x + 3 = -9$$

$$x = 6 \text{ or } \quad x = -12$$

57. $x^2 - 2x + \frac{13}{4} = 0$

$$x = \frac{2 \pm \sqrt{4 - 4(13/4)}}{2}$$

$$= \frac{2 \pm \sqrt{-9}}{2}$$

$$= 1 \pm \frac{3}{2}i$$

59.

$$(x + 1)^2 = x^2$$

$$(x + 1)^2 - x^2 = 0$$

$$(x + 1 - x)(x + 1 + x) = 0$$

$$2x + 1 = 0$$

$$2x = -1$$

$$x = -\frac{1}{2}$$

61. $2x^2 + 7x - 3 = 0$

$$x = \frac{-b \pm \sqrt{b^2 - 4ac}}{2a}$$

$$= \frac{-7 \pm \sqrt{7^2 - 4(2)(-3)}}{2(2)}$$

$$= \frac{-7 \pm \sqrt{73}}{4}$$

$$= -\frac{7}{4} \pm \frac{\sqrt{73}}{4}$$

63. $-4x^2 + 12x - 9 = 0$

$$4x^2 - 12x + 9 = 0$$

$$(2x - 3)^2 = 0$$

$$x = \frac{3}{2}$$

65. $2x^2 - 5x + 1 = 0$

$$x = \frac{-b \pm \sqrt{b^2 - 4ac}}{2a}$$

$$= \frac{5 \pm \sqrt{(-5)^2 - 4(2)(1)}}{2(2)}$$

$$= \frac{5 \pm \sqrt{17}}{4}$$

$$= \frac{5}{4} \pm \frac{\sqrt{17}}{4}$$

67. $(x + 6)(x - 5) = 0$

$$x^2 + x - 30 = 0$$

(other answers possible)

69. $\left(x + \frac{7}{3}\right)\left(x - \frac{6}{7}\right) = 0$

$$\frac{(3x + 7)(7x - 6)}{21} = 0$$

$$21x^2 + 31x - 42 = 0$$

(other answers possible)

71. $\left(x - 5\sqrt{3}\right)\left(x + 5\sqrt{3}\right) = 0$

$$x^2 - \left(5\sqrt{3}\right)^2 = 0$$

$$x^2 - 75 = 0$$

(other answers possible)

73. $\left(x - 1 - 2\sqrt{3}\right)\left(x - 1 + 2\sqrt{3}\right) = 0$

$$\left((x - 1) - 2\sqrt{3}\right)\left((x - 1) + 2\sqrt{3}\right) = 0$$

$$(x - 1)^2 - \left(2\sqrt{3}\right)^2 = 0$$

$$x^2 - 2x + 1 - 12 = 0$$

$$x^2 - 2x - 11 = 0$$

(other answers possible)

75. $[x - (2 + i)][x - (2 - i)] = 0$

$$[(x - 2) - i][(x - 2) + i] = 0$$

$$(x - 2)^2 + 1 = 0$$

$$x^2 - 4x + 5 = 0$$

(other answers possible)

77. (a)

(b) $w(w + 14) = 1632$

$$w^2 + 14w - 1632 = 0$$

(c) $(w + 48)(w - 34) = 0$

$$w = 34, \text{ length} = w + 14 = 48$$

width 34 feet, length 48 feet

79. Let x be the length of the square base. Then $2x^2 = 200 \Rightarrow x^2 = 100 \Rightarrow x = 10$. Thus, the original piece of material is of length $x + 4 = 14$ cm.

Size: 14×14 centimeters

81. (a) $s_0 = 1815$ and $v_0 = 0$

$$s = -16t^2 + 1815$$

(b)

t	0	2	4	6	8	10	12
s	1815	1751	1559	1239	791	215	-489

(c) The object reaches the ground between 10 and 12 seconds, $[10, 12]$. In fact, $t \approx 10.651$ seconds.

83. (a) $s = -16t^2 + v_0 t + s_0$

$$0 = -16t^2 + 8000$$

$$16t^2 = 8000$$

$$t^2 = 500$$

$$t = 10\sqrt{5} \approx 22.36 \text{ seconds}$$

(b) Distance $= (600 \text{ miles/hour})\left(10\sqrt{5} \text{ seconds}\right)/(3600 \text{ seconds/hour})$

$$= \tfrac{1}{6}\left(10\sqrt{5}\right) \text{ miles} \approx 3.73 \text{ miles} \approx 19{,}677.4 \text{ feet}$$

85. (a) $P = 0.1220t^2 + 1.529t + 18.72 = 40$

$$0.122t^2 + 1.529t - 21.28 = 0$$

$$t = \frac{-1.529 \pm \sqrt{(1.529)^2 - 4(0.122)(-21.28)}}{2(0.122)}$$

$$= \frac{-1.529 \pm \sqrt{12.7225}}{0.244}$$

$t \approx 8.35, -20.89$

Taking the positive root, $t \approx 8.35$ or 1998.
Similarly, $P = 50$ yields 10.93, or 2000.

(b) Answers will vary.

(c)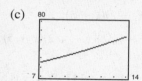

(d) $P = 75$ when $t \approx 16.1$, or 2006.

(e) Answers will vary.

87. (a) $C = 0.45x^2 - 1.65x + 50.75, \ 10 \le x \le 25$

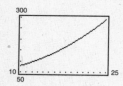

(b) If $C = 150$, then $x = 16.797$ degrees.

(c) If the temperature is increased 10° to 20°, then C increases from 79.25 to 197.75, a factor of 2.5.

89. Let u be the speed of the eastbound plane. Then $u + 50 =$ speed of northbound plane.

$$[u(3)]^2 + [(u + 50)3]^2 = 2440^2$$

$$9u^2 + 9(u + 50)^2 = 2440^2$$

$$18u^2 + 900u + 22{,}500 = 2440^2$$

$$18u^2 + 900u - 5{,}931{,}100 = 0$$

Using a graphing utility, $u \approx 549.57$ mph and $u + 50 = 599.57$ mph.

91. False. The solutions are complex numbers.

93. False. The solutions are either both imaginary or both real.

95. $3(x + 4)^2 + (x + 4) - 2 = 0$

(a) $3u^2 + u - 2 = 0, \quad u = x + 4$

$(3u - 2)(u + 1) = 0$

$3u - 2 = 0 \implies u = \frac{2}{3}$

$u + 1 = 0 \implies u = -1$

$u = \frac{2}{3} = x + 4$

$-\frac{10}{3} = x$

$u = -1 = x + 4$

$-5 = x$

(b) $3x^2 + 24x + 48 + x + 2 = 0$

$3x^2 + 25x + 50 = 0$

$(3x + 10)(x + 5) = 0$

$x = -\frac{10}{3}, x = -5$

(c) Answers will vary.

97. Add the two solutions and the radicals cancel.

$$S = \frac{-b + \sqrt{b^2 - 4ac}}{2a} + \frac{-b - \sqrt{b^2 - 4ac}}{2a}$$

$$= -\frac{b}{a}$$

99. Answers will vary.

101. The parabola opens upward and its vertex is at $(1, 2)$. Matches (e).

103. $x^5 - 27x^2 = x^2(x^3 - 27) = x^2(x - 3)(x^2 + 3x + 9)$

105. $x^3 + 5x^2 - 2x - 10 = x^2(x + 5) - 2(x + 5)$

$= (x^2 - 2)(x + 5)$

$= (x + \sqrt{2})(x - \sqrt{2})(x + 5)$

107. Yes, y is a function of x.

$y = \frac{1}{8}(-1 - 5x)$

109. No, y is not a function of x.

$y = \pm\sqrt{10 - x}$

111. Yes, y is a function of x.

$y = |x - 3|$

113. Answers will vary.

Section 2.5   Solving Other Types of Equations Algebraically

■ You should know the properties of inequalities.

(a) Transitive: $a < b$ and $b < c$ implies $a < c$.

(b) Addition: $a < b$ and $c < d$ implies $a + c < b + d$.

(c) Adding or Subtracting a Constant: $a \pm c < b \pm c$ if $a < b$.

(d) Multiplying or Dividing by a Constant: For $a < b$,

  1. If $c > 0$, then $ac < bc$ and $\frac{a}{c} < \frac{b}{c}$.   2. If $c < 0$, then $ac > bc$ and $\frac{a}{c} > \frac{b}{c}$.

■ You should know that

$$|x| = \begin{cases} x & \text{if } x \geq 0 \\ -x & \text{if } x < 0 \end{cases}.$$

■ You should be able to solve absolute value inequalities.

(a) $|x| < a$ if and only if $-a < x < a$.

(b) $|x| > a$ if and only if $x < -a$ or $x > a$.

■ You should be able to solve polynomial inequalities.

(a) Find the critical numbers.

  1. Values that make the expression zero

  2. Values that make the expression undefined

(b) Test one value in each interval on the real number line resulting from the critical numbers.

(c) Determine the solution intervals.

■ You should be able to solve rational and other types of inequalities.

Vocabulary Check

1. n **2.** extraneous   **3.** quadratic type

---

1. $4x^4 - 16x^2 = 0$

$4x^2(x^2 - 4) = 0$

$4x^2(x - 2)(x + 2) = 0$

$x = 0, \pm 2$

3. $5x^3 + 30x^2 + 45x = 0$

$5x(x^2 + 6x + 9) = 0$

$5x(x + 3)^2 = 0$

$5x = 0 \implies x = 0$

$x + 3 = 0 \implies x = -3$

5. $x^3 - 3x^2 - x + 3 = 0$

$x^2(x - 3) - (x - 3) = 0$

$(x^2 - 1)(x - 3) = 0$

$(x - 1)(x + 1)(x - 3) = 0$

$x = 1, -1, 3$

7. $x^4 - 4x^2 + 3 = 0$

$(x^2 - 3)(x^2 - 1) = 0$

$\left(x + \sqrt{3}\right)\left(x - \sqrt{3}\right)(x + 1)(x - 1) = 0$

$x + \sqrt{3} = 0 \implies x = -\sqrt{3}$

$x - \sqrt{3} = 0 \implies x = \sqrt{3}$

$x + 1 = 0 \implies x = -1$

$x - 1 = 0 \implies x = 1$

9. $4x^4 - 65x^2 + 16 = 0$

$(4x^2 - 1)(x^2 - 16) = 0$

$(2x + 1)(2x - 1)(x + 4)(x - 4) = 0$

$2x + 1 = 0 \implies x = -\dfrac{1}{2}$

$2x - 1 = 0 \implies x = \dfrac{1}{2}$

$x + 4 = 0 \implies x = -4$

$x - 4 = 0 \implies x = 4$

11. $\dfrac{1}{t^2} + \dfrac{8}{t} + 15 = 0$

$1 + 8t + 15t^2 = 0$

$(1 + 3t)(1 + 5t) = 0$

$1 + 3t = 0 \implies t = -\dfrac{1}{3}$

$1 + 5t = 0 \implies t = -\dfrac{1}{5}$

13. $6\left(\dfrac{s}{s + 1}\right)^2 + 5\left(\dfrac{s}{s + 1}\right) - 6 = 0$

Let $u = \dfrac{s}{s + 1}$.

$6u^2 + 5u - 6 = 0$

$(3u - 2)(2u + 3) = 0$

$3u - 2 = 0 \implies u = \dfrac{2}{3}$

$2u + 3 = 0 \implies u = -\dfrac{3}{2}$

$\dfrac{s}{s + 1} = \dfrac{2}{3} \implies s = 2$

$\dfrac{s}{s + 1} = -\dfrac{3}{2} \implies s = -\dfrac{3}{5}$

15. $y = x^3 - 2x^2 - 3x$

(a)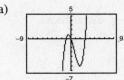

(b) x-intercepts: $(-1, 0), (0, 0), (3, 0)$

(c) $0 = x^3 - 2x^2 - 3x$

$0 = x(x + 1)(x - 3)$

$x = 0$

$x + 1 = 0 \implies x = -1$

$x - 3 = 0 \implies x = 3$

(d)   The x-intercepts are the same as the solutions.

17. $y = x^4 - 10x^2 + 9$

(a)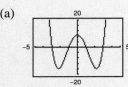

(b) x-intercepts: $(\pm 1, 0), (\pm 3, 0)$

(c) $0 = x^4 - 10x^2 + 9$

$0 = (x^2 - 1)(x^2 - 9)$

$0 = (x + 1)(x - 1)(x + 3)(x - 3)$

$x + 1 = 0 \implies x = -1$

$x - 1 = 0 \implies x = 1$

$x + 3 = 0 \implies x = -3$

$x - 3 = 0 \implies x = 3$

(d)   The x-intercepts are the same as the solutions.

19. $3\sqrt{x} - 10 = 0$

$\qquad 3\sqrt{x} = 10$

$\qquad 9x = 100$

$\qquad x = \frac{100}{9}$

21. $\sqrt{x - 10} - 4 = 0$

$\qquad \sqrt{x - 10} = 4$

$\qquad x - 10 = 16$

$\qquad x = 26$

23. $\sqrt[3]{2x + 5} + 3 = 0$

$\qquad \sqrt[3]{2x + 5} = -3$

$\qquad 2x + 5 = (-3)^3 = -27$

$\qquad 2x = -32$

$\qquad x = -16$

25. $\sqrt[3]{2x + 1} + 8 = 0$

$\qquad \sqrt[3]{2x + 1} = -8$

$\qquad 2x + 1 = -512$

$\qquad 2x = -513$

$\qquad x = -\frac{513}{2}$

$\qquad = -256.5$

27. $\sqrt{5x - 26} + 4 = x$

$\qquad \sqrt{5x - 26} = x - 4$

$\qquad 5x - 26 = x^2 - 8x + 16$

$\qquad x^2 - 13x + 42 = 0$

$\qquad (x - 6)(x - 7) = 0$

$\qquad x = 6, 7$

29. $\sqrt{x + 1} - 3x = 1$

$\qquad \sqrt{x + 1} = 3x + 1$

$\qquad x + 1 = 9x^2 + 6x + 1$

$\qquad 0 = 9x^2 + 5x$

$\qquad 0 = x(9x + 5)$

$\qquad x = 0$

$9x + 5 = 0 \implies x = -\frac{5}{9}$, extraneous

31. $\sqrt{x + 1} = \sqrt{3x + 1}$

$\qquad x + 1 = 3x + 1$

$\qquad 0 = 2x$

$\qquad x = 0$

33. $\qquad 2x + 9\sqrt{x} - 5 = 0$

$\left(2\sqrt{x} - 1\right)\left(\sqrt{x} + 5\right) = 0$

$\sqrt{x} = \frac{1}{2} \implies x = \frac{1}{4}$

$\left(\sqrt{x} = -5 \text{ is not possible.}\right)$

Note: You can see graphically that there is only one solution.

35. $\sqrt{x} - \sqrt{x - 5} = 1$

$\qquad \sqrt{x} = 1 + \sqrt{x - 5}$

$\qquad \left(\sqrt{x}\right)^2 = \left(1 + \sqrt{x - 5}\right)^2$

$\qquad x = 1 + 2\sqrt{x - 5} + x - 5$

$\qquad 4 = 2\sqrt{x - 5}$

$\qquad 2 = \sqrt{x - 5}$

$\qquad 4 = x - 5$

$\qquad 9 = x$

37. $3\sqrt{x - 5} - \sqrt{x - 1} = 0$

$\qquad 3\sqrt{x - 5} = \sqrt{x - 1}$

$\qquad 9(x - 5) = x - 1$

$\qquad 8x = 44$

$\qquad x = \frac{44}{8} = \frac{11}{2}$

39. $\qquad 3x^{1/3} + 2x^{2/3} = 5$

$\qquad 2x^{2/3} + 3x^{1/3} - 5 = 0$

$\qquad (2x^{1/3} + 5)(x^{1/3} - 1) = 0$

$\qquad x^{1/3} = -\frac{5}{2} \implies x = -\frac{125}{8}$

$\qquad x^{1/3} = 1 \implies x = 1$

41. $(x - 5)^{2/3} = 16$

$\qquad x - 5 = \pm 16^{3/2}$

$\qquad x - 5 = \pm 64$

$\qquad\quad x = 69, -59$

43. $(x - 8)^{2/3} = 25$

$\qquad (x - 8) = \pm(25)^{3/2}$

$\qquad\quad x - 8 = \pm 125$

$\qquad\qquad x = -117, 133$

45. $(x^2 - 5x - 2)^{1/3} = -2$

$\qquad x^2 - 5x - 2 = (-2)^3 = -8$

$\qquad x^2 - 5x + 6 = 0$

$\qquad (x - 3)(x - 2) = 0$

$\qquad\qquad\quad x = 2, 3$

47. $3x(x - 1)^{1/2} + 2(x - 1)^{3/2} = 0$

$\qquad (x - 1)^{1/2}[3x + 2(x - 1)] = 0$

$\qquad\quad (x - 1)^{1/2}(5x - 2) = 0$

$\qquad (x - 1)^{1/2} = 0 \implies x - 1 = 0 \implies x = 1$

$\qquad 5x - 2 = 0 \implies x = \frac{2}{5}$ which is extraneous.

49. $y = \sqrt{11x - 30} - x$

(a)

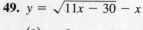
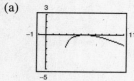

(b) x-intercepts: $(5, 0), (6, 0)$

(c) $\qquad\qquad 0 = \sqrt{11x - 30} - x$

$\qquad\qquad\quad x = \sqrt{11x - 30}$

$\qquad\qquad\quad x^2 = 11x - 30$

$\qquad x^2 - 11x + 30 = 0$

$\qquad (x - 5)(x - 6) = 0$

$\qquad x - 5 = 0 \implies x = 5$

$\qquad x - 6 = 0 \implies x = 6$

(d) The x-intercepts and the solutions are the same.

51. $y = \sqrt{7x + 36} - \sqrt{5x + 16} - 2$

(a)

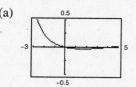

(b) x-intercepts: $(0, 0), (4, 0)$

(c) $\qquad\quad 0 = \sqrt{7x + 36} - \sqrt{5x + 16} - 2$

$\qquad \sqrt{7x + 36} = 2 + \sqrt{5x + 16}$

$\qquad \left(\sqrt{7x + 36}\right)^2 = \left(2 + \sqrt{5x + 16}\right)^2$

$\qquad 7x + 36 = 4 + 4\sqrt{5x + 16} + 5x + 16$

$\qquad 7x + 36 = 5x + 20 + 4\sqrt{5x + 16}$

$\qquad 2x + 16 = 4\sqrt{5x + 16}$

$\qquad\quad x + 8 = 2\sqrt{5x + 16}$

$\qquad x^2 + 16x + 64 = 4(5x + 16)$

$\qquad x^2 + 16x + 64 = 20x + 64$

$\qquad\quad x^2 - 4x = 0$

$\qquad\quad x(x - 4) = 0$

$\qquad\qquad\quad x = 0$

$\qquad x - 4 = 0 \implies x = 4$

(d) The x-intercepts and the solutions are the same.

53.
$$x = \frac{3}{x} + \frac{1}{2}$$

$$(2x)(x) = (2x)\left(\frac{3}{x}\right) + (2x)\left(\frac{1}{2}\right)$$

$$2x^2 = 6 + x$$

$$2x^2 - x - 6 = 0$$

$$(2x + 3)(x - 2) = 0$$

$$2x + 3 = 0 \implies x = -\frac{3}{2}$$

$$x - 2 = 0 \implies x = 2$$

55.
$$\frac{1}{x} - \frac{1}{x + 1} = 3$$

$$x(x + 1)\frac{1}{x} - x(x + 1)\frac{1}{x + 1} = x(x + 1)(3)$$

$$x + 1 - x = 3x(x + 1)$$

$$1 = 3x^2 + 3x$$

$$0 = 3x^2 + 3x - 1$$

$$a = 3, \ b = 3, \ c = -1$$

$$x = \frac{-3 \pm \sqrt{(3)^2 - 4(3)(-1)}}{2(3)} = \frac{-3 \pm \sqrt{21}}{6}$$

57. $\dfrac{20 - x}{x} = x$

$$20 - x = x^2$$

$$0 = x^2 + x - 20$$

$$0 = (x + 5)(x - 4)$$

$$x + 5 = 0 \implies x = -5$$

$$x - 4 = 0 \implies x = 4$$

59.
$$\frac{x}{x^2 - 4} + \frac{1}{x + 2} = 3$$

$$(x + 2)(x - 2)\frac{x}{x^2 - 4} + (x + 2)(x - 2)\frac{1}{x + 2} = 3(x + 2)(x - 2)$$

$$x + x - 2 = 3x^2 - 12$$

$$3x^2 - 2x - 10 = 0$$

$$a = 3, \ b = -2, \ c = -10$$

$$x = \frac{-(-2) \pm \sqrt{(-2)^2 - 4(3)(-10)}}{2(3)} = \frac{2 \pm \sqrt{124}}{6} = \frac{2 \pm 2\sqrt{31}}{6} = \frac{1 \pm \sqrt{31}}{3}$$

61. $|2x - 1| = 5$

$$2x - 1 = 5 \implies x = 3$$

$$-(2x - 1) = 5 \implies x = -2$$

63. $|x| = x^2 + x - 3$

$$x = x^2 + x - 3 \quad \text{OR} \qquad -x = x^2 + x - 3$$

$$x^2 - 3 = 0 \qquad\qquad x^2 + 2x - 3 = 0$$

$$x = \pm\sqrt{3} \qquad\qquad (x - 1)(x + 3) = 0$$

$$x - 1 = 0 \implies x = 1$$

$$x + 3 = 0 \implies x = -3$$

Only $x = \sqrt{3}$, and $x = -3$ are solutions to the original equation. $x = -\sqrt{3}$ and $x = 1$ are extraneous. Note that the graph of $y = x^2 + x - 3 - |x|$ has two x-intercepts.

65.
$$|x + 1| = x^2 - 5$$

$$x + 1 = x^2 - 5 \quad \text{OR} \quad -(x + 1) = x^2 - 5$$

$$x^2 - x - 6 = 0 \qquad\qquad x^2 + x - 4 = 0$$

$$(x - 3)(x + 2) = 0 \qquad x = \frac{-1 \pm \sqrt{1 - 4(-4)}}{2}$$

$$x = 3 \qquad\qquad\qquad x = -\frac{1}{2} - \frac{\sqrt{17}}{2}$$

$(x = -2 \text{ is extraneous.})$

$$\left(x = -\frac{1}{2} + \frac{\sqrt{17}}{2} \text{ is extraneous.}\right)$$

$$x = -\frac{1}{2} - \frac{\sqrt{17}}{2}, 3$$

67. $y = \dfrac{1}{x} - \dfrac{4}{x - 1} - 1$

(a)

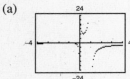

(b) x-intercept: $(-1, 0)$

(c) $0 = \dfrac{1}{x} - \dfrac{4}{x - 1} - 1$

$$0 = (x - 1) - 4x - x(x - 1)$$

$$0 = x - 1 - 4x - x^2 + x$$

$$0 = -x^2 - 2x - 1$$

$$0 = x^2 + 2x + 1$$

$$x + 1 = 0 \implies x = -1$$

(d) The x-intercepts and the solutions are the same.

69. $y = |x + 1| - 2$

(a)

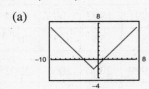

(b) x-intercepts: $(1, 0)$, $(-3, 0)$

(c) $0 = |x + 1| - 2$

$$2 = |x + 1|$$

$$x + 1 = 2 \quad \text{or} \quad -(x + 1) = 2$$

$$x = 1 \quad \text{or} \quad -x - 1 = 2$$

$$-x = 3$$

$$x = -3$$

(d) The x-intercepts and the solutions are the same.

71. Let $x =$ original number of students. The original cost per student is $1700/x$ and the new cost per student is $1700/(x + 6)$. Hence,

$$\frac{1700}{x} - 7.5 = \frac{1700}{x + 6}$$

$$1700(x + 6) - 7.5x(x + 6) = 1700x$$

$$10{,}200 - 7.5x^2 - 45x = 0$$

$$7.5x^2 + 45x - 10{,}200 = 0$$

$$x^2 + 6x - 1360 = 0$$

$$(x - 34)(x + 40) = 0$$

$$x = 34 \text{ students.}$$

73. Let $v =$ the average speed of the plane. The time for the 145-mile trip is $145/v$ hours. By increasing the speed by 40 mph, the time is $145/(v + 40)$ hours. Hence,

$$\frac{145}{v} - \frac{1}{5} = \frac{145}{v + 40}$$

$$145(5)(v + 40) - v(v + 40) = 145(5)(v)$$

$$725v + 29{,}000 - v^2 - 40v = 725v$$

$$v^2 + 40v - 29{,}000 = 0$$

$$v = \frac{-40 \pm \sqrt{40^2 - 4(-29{,}000)}}{2}$$

$$= \frac{-40 \pm \sqrt{117{,}600}}{2}.$$

Taking the positive square root, $v \approx 151.5$ mph, and $v + 40 = 191.5$ mph.

75. $A = P\left(1 + \dfrac{r}{n}\right)^{nt}, n = 12, t = 5$

$$3052.49 = 2500\left(1 + \dfrac{r}{12}\right)^{(12)(5)}$$

$$\left(1 + \dfrac{r}{12}\right)^{60} = 1.220996$$

$$1 + \dfrac{r}{12} = 1.003333$$

$$\dfrac{r}{12} = 0.003333$$

$$r \approx 0.04, \text{ or } 4\%$$

77. (a) $N(t) = 206 + 248.7\sqrt{t}, \ 0 \le t \le 14$

Year	1990	1991	1992	1993	1994	1995	1996	1997
N	206	455	558	637	703	762	815	864

Year	1998	1999	2000	2001	2002	2003	2004
N	909	952	992	1031	1068	1103	1137

(b) 500 in 1991; 1000 in 2000

(c) $206 + 248.7\sqrt{t} = 500$

$$248.7\sqrt{t} = 294$$

$$\sqrt{t} \approx 1.1821$$

$$t \approx 1.4, \text{ or } 1991$$

$$206 + 248.7\sqrt{t} = 1000$$

$$248.7\sqrt{t} = 794$$

$$\sqrt{t} \approx 3.1926$$

$$t \approx 10.2, \text{ or } 2000$$

(d) $t \approx 10.2, 1.4$

(e) For 1500, $t \approx 27.1$, or 2017. For 2000, $t \approx 52.0$, or 2042.

Answers will vary.

79. $\sqrt{0.2x + 1} = C = 2.5$

$$0.2x + 1 = 6.25$$

$$0.2x = 5.25$$

$$x = 26.25, \text{ or } 26{,}250 \text{ passengers}$$

81. $p = 750 = 800 - \sqrt{0.01x + 1}$

$$\sqrt{0.01x + 1} = 50$$

$$0.01x + 1 = 2500$$

$$0.01x = 2499$$

$$x = 249{,}900 \text{ units}$$

83. (a)

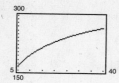

(b) $x = 14.696$

$75.82 - 2.11(14.696) + 43.51\sqrt{14.696} \approx 211.6°F$

(c) $240 = 75.82 - 2.11x + 43.51\sqrt{x}$

$0 = -164.18 - 2.11x + 43.51\sqrt{x}$

$a = -2.11, \ b = 43.51, \ c = -164.18$

$$\sqrt{x} = \frac{-43.51 \pm \sqrt{(43.51)^2 - 4(-2.11)(-164.18)}}{2(-2.11)}$$

$$x = \left[\frac{-43.51 - \sqrt{507.4409}}{-4.22}\right]^2 \approx 244.874$$

$$x = \left[\frac{-43.51 + \sqrt{507.4409}}{-4.22}\right]^2 \approx 24.725$$

Because x is restricted to $5 \le x \le 40$, choose $x = 24.725$ pounds per square inch.

85. False. An equation can have any number of extraneous solutions. For example, see Example 8.

87. $\left(x - \sqrt{2}\right)\left(x + \sqrt{2}\right)(x - 4) = 0$

$(x^2 - 2)(x - 4) = 0$

$x^3 - 4x^2 - 2x + 8 = 0$

(other answers possible)

89. $(x - 2)(x + 1)(2x - 1)(x + 3) = 0$

$(x^2 - x - 2)(2x^2 + 5x - 3) = 0$

$2x^4 + 3x^3 - 12x^2 - 7x + 6 = 0$

(other answers possible)

91. $(x + 2)(x - 2)(x + i)(x - i) = 0$

$(x^2 - 4)(x^2 + 1) = 0$

$x^4 - 3x^2 - 4 = 0$

(other answers possible)

93. The distance between $(1, 2)$ and $(x, -10)$ is 13.

$\sqrt{(x - 1)^2 + (-10 - 2)^2} = 13$

$(x - 1)^2 + (-12)^2 = 13^2$

$x^2 - 2x + 1 + 144 = 169$

$x^2 - 2x - 24 = 0$

$(x + 4)(x - 6) = 0$

$x + 4 = 0 \implies x = -4$

$x - 6 = 0 \implies x = 6$

95. $x + |x - a| = b, \ x = 9$

$9 + |9 - a| = b$

One solution is $a = 9, b = 9$. Another solution is $a = 0, b = 18$.

97. $x + \sqrt{x - a} = b, \ x = 20$

$20 + \sqrt{20 - a} = b$

One solution is $a = 19$ and $b = 21$. Another solution is $a = b = 20$.

99. $\dfrac{8}{3x} + \dfrac{3}{2x} = \dfrac{16}{6x} + \dfrac{9}{6x} = \dfrac{25}{6x}$

101. $\dfrac{2}{z + 2} - \left(3 - \dfrac{2}{z}\right) = \dfrac{2}{z + 2} - 3 + \dfrac{2}{z}$

$$= \dfrac{2z - 3(z + 2)z + 2(z + 2)}{z(z + 2)}$$

$$= \dfrac{2z - 3z^2 - 6z + 2z + 4}{z(z + 2)}$$

$$= \dfrac{-3z^2 - 2z + 4}{z(z + 2)}$$

103. $x^2 - 22x + 121 = 0$

$$(x - 11)^2 = 0$$

$$x = 11$$

Section 2.6   Solving Inequalities Algebraically and Graphically

- ■   You should be able to solve an inequality algebraically using the Properties of Inequalities.
- ■   You should be able to solve inequalities involving absolute values.
- ■   You should be able to solve polynomial inequalities using critical numbers and test intervals.
- ■   You should be able to solve rational inequalities.
- ■   You should be able to solve inequalities using a graphing utility.

Vocabulary Check

1. negative   **2.** double   **3.** $-a \leq x \leq a$

4. $x \leq -a, x \geq a$ **5.** zeros, undefined values

---

1. $x < 3$

   Matches (f).

3. $-3 < x \leq 4$

   Matches (d).

5. $-1 \leq x \leq \frac{5}{2}$

   Matches (e).

7. (a) $x = 3$

     $5(3) - 12 \overset{?}{>} 0$

           $3 > 0$

   Yes, $x = 3$ is a solution.

(b) $x = -3$

     $5(-3) - 12 \overset{?}{>} 0$

        $-27 \not> 0$

   No, $x = -3$ is not a solution.

(c) $x = \frac{5}{2}$

     $5\left(\frac{5}{2}\right) - 12 \overset{?}{>} 0$

        $\frac{1}{2} > 0$

   Yes, $x = \frac{5}{2}$ is a solution.

(d) $x = \frac{3}{2}$

     $5\left(\frac{3}{2}\right) - 12 \overset{?}{>} 0$

        $-\frac{9}{2} \not> 0$

   No, $x = \frac{3}{2}$ is not a solution.

9. $-1 < \dfrac{3-x}{2} \leq 1$

(a) $x = 0$

$$-1 \overset{?}{<} \dfrac{3-0}{2} \overset{?}{\leq} 1$$

$$-1 \overset{?}{<} \dfrac{3}{2} \overset{?}{\leq} 1$$

No, $x = 0$ is not a solution.

(b) $x = \sqrt{5}$

$$-1 \overset{?}{<} \dfrac{3-\sqrt{5}}{2} \overset{?}{\leq} 1$$

$$-1 \overset{?}{<} 0.382 \overset{?}{\leq} 1$$

Yes, $x = \sqrt{5}$ is a solution.

(c) $x = 1$

$$-1 \overset{?}{<} \dfrac{3-1}{2} \overset{?}{\leq} 1$$

$$-1 \overset{?}{<} 1 \overset{?}{\leq} 1$$

Yes, $x = 1$ is a solution.

(d) $x = 5$

$$-1 \overset{?}{<} \dfrac{3-5}{2} \leq 1$$

$$-1 \overset{?}{<} -1 \overset{?}{\leq} 1$$

No, $x = 5$ is not a solution.

11.
$$-10x < 40$$
$$-\tfrac{1}{10}(-10x) > -\tfrac{1}{10}(40)$$
$$x > -4$$

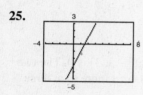

13. $4(x+1) < 2x + 3$

$$4x + 4 < 2x + 3$$
$$2x < -1$$
$$x < -\tfrac{1}{2}$$

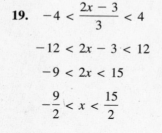

15. $\tfrac{3}{4}x - 6 \leq x - 7$

$$1 \leq \tfrac{1}{4}x$$
$$4 \leq x$$
$$x \geq 4$$

17. $-8 \leq 1 - 3(x-2) < 13$

$$-8 \leq 1 - 3x + 6 < 13$$
$$-8 \leq -3x + 7 < 13$$
$$-15 \leq -3x < 6$$
$$5 \geq x > -2 \implies -2 < x \leq 5$$

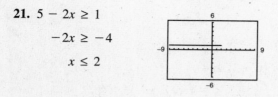

19. $-4 < \dfrac{2x-3}{3} < 4$

$$-12 < 2x - 3 < 12$$
$$-9 < 2x < 15$$
$$-\dfrac{9}{2} < x < \dfrac{15}{2}$$

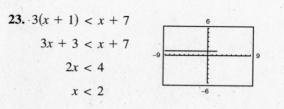

21. $5 - 2x \geq 1$

$$-2x \geq -4$$
$$x \leq 2$$

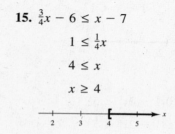

23. $3(x+1) < x + 7$

$$3x + 3 < x + 7$$
$$2x < 4$$
$$x < 2$$

25.

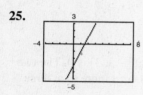

Using the graph, (a) $y \geq 1$ for $x \geq 2$ and (b) $y \leq 0$ for $x \leq \tfrac{3}{2}$.
Algebraically:

(a) $y \geq 1$

$$2x - 3 \geq 1$$
$$2x \geq 4$$
$$x \geq 2$$

(b) $y \leq 0$

$$2x - 3 \leq 0$$
$$2x \leq 3$$
$$x \leq \tfrac{3}{2}$$

27.

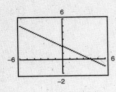

Using the graph, (a) $0 \le y \le 3$ for $-2 \le x \le 4$ and (b) $y \ge 0$ for $x \le 4$. Algebraically:

(a) $0 \le y \le 3$

$0 \le -\frac{1}{2}x + 2 \le 3$

$-2 \le -\frac{1}{2}x \le 1$

$4 \ge x \ge -2$

(b) $y \ge 0$

$-\frac{1}{2}x + 2 \ge 0$

$2 \ge \frac{1}{2}x$

$4 \ge x$

29. $|5x| > 10$

$5x < -10$ or $5x > 10$

$x < -2$ or $x > 2$

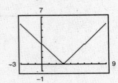

31. $|x - 7| < 6$

$-6 < x - 7 < 6$

$1 < x < 13$

33. $|x + 14| + 3 > 17$

$|x + 14| > 14$

$x + 14 < -14$ or $x + 14 > 14$

$x < -28$ or $x > 0$

35. $10|1 - 2x| < 5$

$|1 - 2x| < \frac{1}{2}$

$-\frac{1}{2} < 1 - 2x < \frac{1}{2}$

$-\frac{3}{2} < -2x < -\frac{1}{2}$

$\frac{3}{4} > x > \frac{1}{4}$

$\frac{1}{4} < x < \frac{3}{4}$

37. $y = |x - 3|$

Graphically, (a) $y \le 2$ for $1 \le x \le 5$ and (b) $y \ge 4$ for $x \le -1$ or $x \ge 7$. Algebraically:

(a) $y \le 2$

$|x - 3| \le 2$

$-2 \le x - 3 \le 2$

$1 \le x \le 5$

(b) $y \ge 4$

$|x - 3| \ge 4$

$x - 3 \le -4$ or $x - 3 \ge 4$

$x \le -1$ or $x \ge 7$

39. The midpoint of the interval $[-3, 3]$ is 0. The interval represents all real numbers x no more than three units from 0.

$|x - 0| \le 3$

$|x| \le 3$

41. The midpoint of the interval $[-3, 3]$ is 0. The two intervals represent all numbers x more than three units from 0.

$|x - 0| > 3$

$|x| > 3$

43. All real numbers within 10 units of 7

$|x - 7| \le 10$

45. All real numbers at least five units from 3

$|x - 3| \ge 5$

47. $x^2 - 4x - 5 > 0$

$(x - 5)(x + 1) > 0$

Critical numbers: $-1, 5$

Testing the intervals $(-\infty, -1), (-1, 5)$ and $(5, \infty)$, we have $x^2 - 4x - 5 > 0$ on $(-\infty, -1)$ and $(5, \infty)$. Similarly, $x^2 - 4x - 5 < 0$ on $(-1, 5)$.

49. $2x^2 - 4x - 3 = 0$

$$x = \frac{4 \pm \sqrt{16 + 24}}{4} = 1 \pm \frac{\sqrt{10}}{2}$$

Entirely negative:

$$\left(1 - \frac{\sqrt{10}}{2}, 1 + \frac{\sqrt{10}}{2}\right) \approx (-0.581, 2.581)$$

Entirely positive:

$$\left(-\infty, 1 - \frac{\sqrt{10}}{2}\right) \cup \left(1 + \frac{\sqrt{10}}{2}, \infty\right)$$

51. $x^2 - 4x + 5 > 0$ for all x. There are no critical numbers because $x^2 - 4x + 5 \ne 0$. The only test interval is $(-\infty, \infty)$.

53. $(x + 2)^2 < 25$

$x^2 + 4x + 4 < 25$

$x^2 + 4x - 21 < 0$

$(x + 7)(x - 3) < 0$

Critical numbers: $x = -7, x = 3$

Test intervals: $(-\infty, -7), (-7, 3), (3, \infty)$

Test: Is $(x + 7)(x - 3) < 0$?

Solution set: $(-7, 3)$

55. $x^2 + 4x + 4 \ge 9$

$x^2 + 4x - 5 \ge 0$

$(x + 5)(x - 1) \ge 0$

Critical numbers: $x = -5, x = 1$

Test intervals: $(-\infty, -5), (-5, 1), (1, \infty)$

Test: Is $(x + 5)(x - 1) \ge 0$?

Solution set: $(-\infty, -5] \cup [1, \infty)$

57. $x^3 - 4x \ge 0$

$x(x + 2)(x - 2) \ge 0$

Critical number: $x = 0, x = \pm 2$

Test intervals: $(-\infty, -2), (-2, 0), (0, 2), (2, \infty)$

Test: Is $x(x + 2)(x - 2) \ge 0$?

Solution set: $[-2, 0] \cup [2, \infty)$

59. $3x^2 - 11x + 16 \le 0$

Since $b^2 - 4ac = -71 < 0$, there are no real solutions to $3x^2 - 11x + 16 = 0$. In fact, $3x^2 - 11x + 16 > 0$ for all x.

No solution

61. $2x^3 + 5x^2 - 6x - 9 > 0$

$(x + 1)(x + 3)(2x - 3) > 0$

Critical numbers: $-3, -1, \frac{3}{2}$

Testing the four intervals, we see that $2x^3 + 5x^2 - 6x - 9 > 0$ on $(-3, -1)$ and $\left(\frac{3}{2}, \infty\right)$.

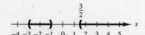

63. (a) $f(x) = g(x)$ when $x = 1$.

(b) $f(x) \ge g(x)$ when $x \ge 1$.

(c) $f(x) > g(x)$ when $x > 1$.

65. $y = -x^2 + 2x + 3$

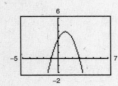

(a) $y \leq 0$ when $x \leq -1$ or $x \geq 3$.

(b) $y \geq 3$ when $0 \leq x \leq 2$.

Algebraically,

$$-x^2 + 2x + 3 \leq 0$$

$$x^2 - 2x - 3 \geq 0$$

$$(x - 3)(x + 1) \geq 0$$

Critical numbers: $x = -1, x = 3$

Testing the intervals $(-\infty, -1)$, $(-1, 3)$, and $(3, \infty)$, you obtain $x \leq -1$ or $x \geq 3$.

$$-x^2 + 2x + 3 \geq 3$$

$$-x^2 + 2x \geq 0$$

$$x^2 - 2x \leq 0$$

$$x(x - 2) \leq 0$$

Critical numbers: $x = 0, x = 2$

Testing the intervals $(-\infty, 0)$, $(0, 2)$, and $(2, \infty)$, you obtain $0 \leq x \leq 2$.

67. $\dfrac{1}{x} - x > 0$

$$\dfrac{1 - x^2}{x} > 0$$

Critical numbers: $x = 0, x = \pm 1$

Test intervals: $(-\infty, -1), (-1, 0), (0, 1), (1, \infty)$

Test: Is $\dfrac{1 - x^2}{x} > 0$?

Solution set: $(-\infty, -1) \cup (0, 1)$

69. $\dfrac{x + 6}{x + 1} - 2 < 0$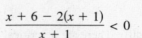

$$\dfrac{x + 6 - 2(x + 1)}{x + 1} < 0$$

$$\dfrac{4 - x}{x + 1} < 0$$

Critical numbers: $x = -1, x = 4$

Test intervals: $(-\infty, -1), (-1, 4), (4, \infty)$

Test: Is $\dfrac{4 - x}{x + 1} < 0$?

Solution set: $(-\infty, -1) \cup (4, \infty)$

71. $y = \dfrac{3x}{x - 2}$

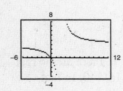

(a) $y \leq 0$ when $0 \leq x < 2$.

(b) $y \geq 6$ when $2 < x \leq 4$.

73. $\sqrt{x - 5}$

Need: $x - 5 \geq 0$

$$x \geq 5$$

Domain: $[5, \infty)$

75. $\sqrt[3]{6 - x}$

Domain: all real x

77. $\sqrt{x^2 - 4}$

Need: $x^2 - 4 \geq 0$

$$(x + 2)(x - 2) \geq 0$$

$$x \leq -2 \text{ or } x \geq 2$$

Domain: $(-\infty, -2] \cup [2, \infty)$

79. (a) $P(t) = 1000$

This occurs at the point of intersection, $t \approx 4$, or 1994.

(b) Less than one million: $P(t) < 1000$
This occurs for $t < 4$, or before 1994.

Greater than one million: $P(t) > 1000$
This occurs for $t > 4$, or after 1994.

81. (a) $s = -16t^2 + v_0t + s_0$

$s = -16t^2 + 160t$

$s = 16t(10 - t)$

$s = 0$ when $t = 10$ seconds.

(b) $s = -16t^2 + 160t > 384$

$16t^2 - 160t + 384 < 0$

$16(t - 6)(t - 4) < 0$

$s > 384$ when $4 < t < 6$.

83. (a)

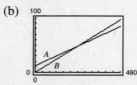

(b) $15 < D < 20$ for $1.28 < t < 10.09$, or between 1991 and 2000

(c) $15 < D < 20$

$15 < -0.0165t^2 + 0.755t + 14.06 < 20$

To solve these inequalities, find the critical numbers.

$0.0165t^2 - 0.755t + 0.94 = 0$

$t = \dfrac{0.755 \pm \sqrt{(-0.755)^2 - 4(0.0165)(0.94)}}{2(0.0165)}$

$= \dfrac{0.755 \pm \sqrt{0.507985}}{0.033}$

Because $0 < t < 13$, select the negative sign, $t \approx 1.28$. Hence, $15 < D$ for $1.28 < t$.
Similarly, $D < 20$ for $t < 10.09$.

(d) No. $D(t) < 30$ for all t.

85. $V(t) \geq 65$

$3.37t + 57.9 \geq 65$

$3.37t \geq 7.1$

$t \geq 2.11$

The number of hours playing video games
exceeded 65 in 2002.

87. $V(t) = N(t)$

$3.37t + 57.9 = -2.51t + 179.6$

$5.88t = 121.7$

$t \approx 20.7$

According to these models, the number of hours
reading daily newspapers and playing video games
will be the same in 2020.

89. When $t = 2$, $v \approx 333$ vibrations per second.

91. When $200 \leq v \leq 400$, $1.2 < t < 2.4$.

93. (a) Option A: $A(t) = 0.15t + 12$

Option B: $B(t) = 0.20t$

(b)

(c) $A(t) = B(t)$ when $t = 240$. $B(t)$ is the better
choice if you use less than 240 minutes.
$A(t)$ is the better choice if you use more than
240 minutes.

(d) Answers will vary.

95. False. If $-10 \leq x \leq 8$, then $10 \geq -x$ and
$-x \geq -8$.

97. The polynomial $f(x) = (x - a)(x - b)$ is zero at $x = a$ and $x = b$.

99. (iv) $a < b$

(ii) $2a < 2b$

(iii) $2a < a + b < 2b$

(i) $a < \dfrac{a + b}{2} < b$

101. $f(x) = -x^2 + 6$

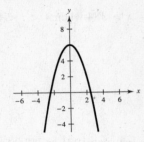

103. $f(x) = -|x + 5| - 6$

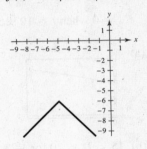

105.
$$y = 12x$$
$$x = 12y$$
$$\frac{x}{12} = y$$
$$f^{-1}(x) = \frac{x}{12}$$

107.
$$y = x^3 + 7$$
$$x = y^3 + 7$$
$$x - 7 = y^3$$
$$\sqrt[3]{x - 7} = y$$
$$f^{-1}(x) = \sqrt[3]{x - 7}$$

109. Answers will vary.

Section 2.7   Linear Models and Scatter Plots

- You should know how to construct a scatter plot for a set of data.
- You should recognize if a set of data has a positive correlation, negative correlation, or neither.
- You should be able to fit a line to data using the point-slope formula.
- You should be able to use the regression feature of a graphing utility to find a linear model for a set of data.
- You should be able to find and interpret the correlation coefficient of a linear model.

Vocabulary Check

1. positive

2. negative

3. fitting a line to data

4. $-1, 1$

1. (a)

(b) Yes, the data appears somewhat linear. The more experience, x, corresponds to higher sales, y.

3. Negative correlation—y decreases as x increases.

5. No correlation

7. (a)

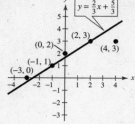

(b) $y = 0.46x + 1.62$

Correlation coefficient: 0.95095

(c)

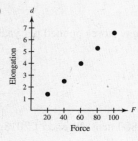

(d) Yes, the model appears valid.

9. (a)

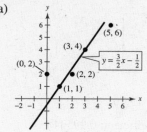

(b) $y = 0.95x + 0.92$

Correlation coefficient: 0.90978

(c)

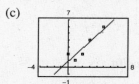

(d) Yes, the model appears valid.

11. (a)

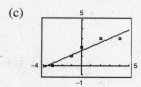

(b) $d = 0.07F - 0.3$

(c) $d = 0.066F$ or $F = 15.13d + 0.096$

(d) If $F = 55$, $d = 0.066(55) \approx 3.63$ cm.

13. (a)

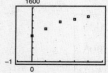

(b) $y = 136.1t + 836$

(c)

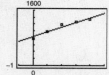

Yes, the model is a good fit.

(d) For 2005, $t = 5$ and $y \approx 1516.5$, or \$1,516,500.

For 2010, $t = 10$ and $y \approx 2197$, or \$2,197,000.

Yes, the answers seem reasonable.

(e) The slope is 136.1. It says that the mean salary increases by \$136,100 per year.

15. (a)
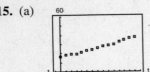

(b) $C = 1.552t + 15.70$

Correlation coefficient: 0.99544

(c)
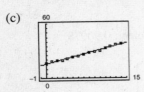

(d) The model is a good fit.

(e) For 2005, $t = 15$, $y_1 \approx \$38.98$.

For 2010, $t = 20$, $y_1 \approx \$46.74$.

(f) Answers will vary.

17. (a)
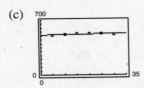

(b) $P = 0.6t + 512$

(c)
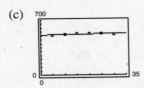

The model is not a good fit.

(d) For 2050, $t = 50$ and $P = 542$, or 542,000 people. Answers will vary.

19. (a) $T = 36.7t + 926$

Correlation coefficient: 0.79495

(b)
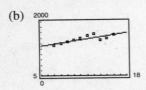

(c) The slope indicated the number of new stores opened per year.

(d) $T = 36.7t + 926 > 1800$

$36.7t > 874$

$t > 23.8$

The number of stores will exceed 1800 near the end of 2013.

(e)

Year	1997	1998	1999	2000	2001	2002	2003	2004	2005	2006
Data	1130	1182	1243	1307	1381	1475	1553	1308	1400	1505
Model	1183	1220	1256	1293	1330	1366	1403	1440	1477	1513

The model is not a good fit, especially around $t = 14$.

21. True. To have positive correlation, the y-values tend to increase as x increases.

23. Answers will vary.

25. $f(x) = 2x^2 - 3x + 5$

(a) $f(-1) = 2 + 3 + 5 = 10$

(b) $f(w + 2) = 2(w + 2)^2 - 3(w + 2) + 5$

$= 2w^2 + 5w + 7$

27. $h(x) = \begin{cases} 1 - x^2, & x \le 0 \\ 2x + 3, & x > 0 \end{cases}$

(a) $h(1) = 2(1) + 3 = 5$

(b) $h(0) = 1 - 0 = 1$

29. $6x + 1 = -9x - 8$

$15x = -9$

$x = -\frac{9}{15} = -\frac{3}{5}$

31. $8x^2 - 10x - 3 = 0$

$(4x + 1)(2x - 3) = 0$

$x = -\frac{1}{4}, \frac{3}{2}$

33. $2x^2 - 7x + 4 = 0$

$x = \dfrac{7 \pm \sqrt{49 - 4(4)(2)}}{4} = \dfrac{7 \pm \sqrt{17}}{4}$

Review Exercises for Chapter 2

1. $6 + \dfrac{3}{x-4} = 5$

  (a) $x = 5$

$$6 + \frac{3}{5-4} \overset{?}{=} 5$$
$$6 + 3 \overset{?}{=} 5$$
$$9 \neq 5$$

No, $x = 5$ is not a solution.

  (c) $x = -2$

$$6 + \frac{3}{-2-4} \overset{?}{=} 5$$
$$6 - \frac{1}{2} \overset{?}{=} 5$$
$$5.5 \neq 5$$

No, $x = -2$ is not a solution.

  (b) $x = 0$

$$6 + \frac{3}{0-4} \overset{?}{=} 5$$
$$6 - \frac{3}{4} \overset{?}{=} 5$$
$$5.25 \neq 5$$

No, $x = 0$ is not a solution.

  (d) $x = 1$

$$6 + \frac{3}{1-4} \overset{?}{=} 5$$
$$6 - 1 \overset{?}{=} 5$$
$$5 = 5$$

Yes, $x = 1$ is a solution.

3.
$$\frac{18}{x} = \frac{10}{x-4}$$
$$18(x-4) = 10x$$
$$18x - 72 = 10x$$
$$8x = 72$$
$$x = 9$$

5.
$$\frac{5}{x-2} = \frac{13}{2x-3}$$
$$10x - 15 = 13x - 26$$
$$11 = 3x$$
$$x = \frac{11}{3}$$

7. $14 + \dfrac{2}{x-1} = 10$

$$\frac{2}{x-1} = -4$$
$$2 = -4(x-1)$$
$$2 = -4x + 4$$
$$4x = 2$$
$$x = \frac{1}{2}$$

9. $6 - \dfrac{11}{x} = 3 + \dfrac{7}{x}$

$$3 = \frac{18}{x}$$
$$3x = 18$$
$$x = 6$$

11.
$$\frac{9x}{3x-1} - \frac{4}{3x+1} = 3$$
$$9x(3x+1) - 4(3x-1) = 3(3x-1)(3x+1)$$
$$27x^2 + 9x - 12x + 4 = 3(9x^2 - 1)$$
$$27x^2 - 3x + 4 = 27x^2 - 3$$
$$-3x = -7$$
$$x = \frac{7}{3}$$

13. September's profit + October's profit = 689,000

Let $x =$ September's profit. Then
$x + 0.12x =$ October's profit.

$$x + (x + 0.12x) = 689{,}000$$
$$2.12x = 689{,}000$$
$$x = 325{,}000$$
$$x + 0.12x = 364{,}000$$

September: \$325,000; October: \$364,000

15. (a)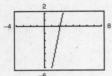

(b) $\dfrac{h}{8} = \dfrac{2}{3/4} = \dfrac{8}{3}$

$h = \dfrac{64}{3} = 21\dfrac{1}{3}$ meters high

17. $F = \dfrac{9}{5}C + 32$

$25.7 = \dfrac{9}{5}C + 32$

$-6.3 = \dfrac{9}{5}C$

$C = -3.5$

$-3.5°$ Celsius

19. $-x + y = 3$

Let $x = 0$: $y = 3$, y-intercept: $(0, 3)$

Let $y = 0$: $x = -3$, x-intercept: $(-3, 0)$

21. $y = x^2 - 9x + 8 = (x - 8)(x - 1)$

Let $x = 0$: $y = 8$, y-intercept: $(0, 8)$

Let $y = 0$: $x = 1, 8$, x-intercepts: $(1, 0), (8, 0)$

23. $5(x - 2) - 1 = 0$

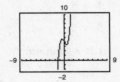

Solution: $x = 2.2$

25. $3x^3 - 2x + 4 = 0$

Solution: $x = -1.301$

27. $x^4 - 3x + 1 = 0$

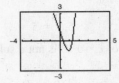

Solutions: $x = 1.307, x = 0.338$

29. $3x + 5y = -7$

$-x - 2y = 3$

From second equation, $x = -2y - 3$. Then:

$3(-2y - 3) + 5y = 7$

$-y - 9 = -7$

$y = -2$ and

$x = -2(-2) - 3 = 1$

Intersection point: $(1, -2)$

31. $x^2 + 2y = 14$

$3x + 4y = 1$

From equation 2, $y = \dfrac{1}{4}(1 - 3x)$. Then:

$x^2 + 2\left(\dfrac{1}{4}\right)(1 - 3x) = 14$

$x^2 + \dfrac{1}{2} - \dfrac{3}{2}x = 14$

$2x^2 - 3x - 27 = 0$

$(2x - 9)(x + 3) = 0$

$x = \dfrac{9}{2} \Rightarrow y = \dfrac{1}{4}\left(1 - 3\left(\dfrac{9}{2}\right)\right) = -\dfrac{25}{8}$

$x = -3 \Rightarrow y = \dfrac{1}{4}\left(1 - 3(-3)\right) = \dfrac{5}{2}$

Intersection points: $\left(-3, \dfrac{5}{2}\right), \left(\dfrac{9}{2}, -\dfrac{25}{8}\right)$

33. $6 + \sqrt{-25} = 6 + 5i$

35. $-2i^2 + 7i = 2 + 7i$

37. $(7 + 5i) + (-4 + 2i) = (7 - 4) + (5i + 2i)$

$= 3 + 7i$

39. $5i(13 - 8i) = 65i - 40i^2 = 40 + 65i$

41. $\left(\sqrt{-16} + 3\right)\left(\sqrt{-25} - 2\right) = (4i + 3)(5i - 2)$

$= -20 - 8i + 15i - 6$

$= -26 + 7i$

43. $\sqrt{-9} + 3 + \sqrt{-36} = 3i + 3 + 6i$

$= 3 + 9i$

45. $(10 - 8i)(2 - 3i) = 20 - 30i - 16i + 24i^2$

$= -4 - 46i$

47. $(3 + 7i)^2 + (3 - 7i)^2 = (9 + 42i - 49) + (9 - 42i - 49)$

$$= -80$$

49. $\dfrac{6 + i}{i} = \dfrac{6 + i}{i} \cdot \dfrac{-i}{-i} = \dfrac{-6i - i^2}{-i^2}$

$$= \dfrac{-6i + 1}{1} = 1 - 6i$$

51. $\dfrac{3 + 2i}{5 + i} \cdot \dfrac{5 - i}{5 - i} = \dfrac{15 + 10i - 3i + 2}{25 + 1}$

$$= \dfrac{17}{26} + \dfrac{7}{26}i$$

53. $-3 - 2i$

55. $2 - 5i$

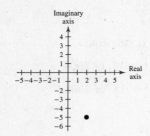

57. $-6i$

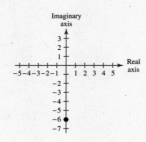

59. 3

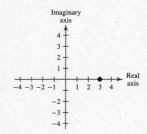

61. $(2x - 1)(x + 3) = 0$

$$x = \tfrac{1}{2}, -3$$

63. $(3x - 2)(x - 5) = 0$

$$x = \tfrac{2}{3}, 5$$

65. $6x = 3x^2$

$$0 = 3x^2 - 6x$$

$$0 = 3x(x - 2)$$

$$3x = 0 \implies x = 0$$

$$x - 2 = 0 \implies x = 2$$

67. $x^2 - 4x = 5$

$$x^2 - 4x - 5 = 0$$

$$(x - 5)(x + 1) = 0$$

$$x = 5, -1$$

69. $x^2 - 3x = 4$

$$x^2 - 3x - 4 = 0$$

$$(x - 4)(x + 1) = 0$$

$$x = 4, -1$$

71. $2x^2 - x - 3 = 0$

$$(2x - 3)(x + 1) = 0$$

$$x = \tfrac{3}{2}, -1$$

73. $15 + x - 2x^2 = 0$

$$(5 + 2x)(3 - x) = 0$$

$$5 + 2x = 0 \implies x = -\tfrac{5}{2}$$

$$3 - x = 0 \implies x = 3$$

75. $(x + 4)^2 = 18$

$$x + 4 = \pm\sqrt{18}$$

$$x = -4 \pm 3\sqrt{2}$$

77. $x^2 - 12x + 30 = 0$

$$x^2 - 12x = -30$$

$$x^2 - 12x + 36 = -30 + 36$$

$$(x - 6)^2 = 6$$

$$x - 6 = \pm\sqrt{6}$$

$$x = 6 \pm \sqrt{6}$$

79. $2x^2 + 9x - 5 = 0$

$$(2x - 1)(x + 5) = 0$$

$$x = \tfrac{1}{2}, -5$$

81. $-x^2 - x + 15 = 0$

$$x^2 + x - 15 = 0$$

$$x = \dfrac{-1 \pm \sqrt{1 - 4(-15)}}{2}$$

$$= \dfrac{-1 \pm \sqrt{61}}{2}$$

83. $x^2 + 4x + 10 = 0$

$$x = \dfrac{-4 \pm \sqrt{16 - 40}}{2}$$

$$= \dfrac{-4 \pm \sqrt{-24}}{2}$$

$$= -2 \pm \sqrt{6}\,i$$

85. $2x^2 - 6x + 21 = 0$

$$x = \frac{6 \pm \sqrt{36 - 168}}{4}$$

$$= \frac{6 \pm \sqrt{-132}}{4}$$

$$= \frac{3}{2} \pm \frac{\sqrt{33}}{2}i$$

87. (a)

(b) $C = 1250$ when $t \approx 11.5$, or 2001.

(c) $6t^2 - 62.9t + 1182 = 1250$

$$6t^2 - 62.9t - 68 = 0$$

$$t = \frac{62.9 \pm \sqrt{(62.9)^2 - 4(6)(-68)}}{2(6)}$$

$$= \frac{62.9 \pm \sqrt{5588.41}}{12}$$

$$\approx 11.5, -0.99 \text{ (extraneous)}$$

(d) $C = 1500$ when $t \approx 14$, or 2004.

$C = 2000$ when $t \approx 18$, or 2008.

(e) Answers will vary.

89. $3x^3 - 26x^2 + 16x = 0$

$$x(3x^2 - 26x + 16) = 0$$

$$x(3x - 2)(x - 8) = 0$$

$$x = 0, \tfrac{2}{3}, 8$$

91. $5x^4 - 12x^3 = 0$

$$x^3(5x - 12) = 0$$

$$x^3 = 0 \quad \text{or} \quad 5x - 12 = 0$$

$$x = 0 \quad \text{or} \quad x = \tfrac{12}{5}$$

93. $x^4 - x^2 - 12 = 0$

$$(x^2 - 4)(x^2 + 3) = 0$$

$$x = \pm 2, \pm \sqrt{3}\,i$$

95. $2x^4 - 22x^2 + 56 = 0$

$$(x^2 - 4)(2x^2 - 14) = 0$$

$$x = \pm 2, \pm \sqrt{7}$$

97. $\sqrt{x + 4} = 3$

$$\left(\sqrt{x + 4}\right)^2 = (3)^2$$

$$x + 4 = 9$$

$$x = 5$$

99. $2\sqrt{x} - 5 = 0$

$$2\sqrt{x} = 5$$

$$4x = 25$$

$$x = \tfrac{25}{4}$$

101. $\sqrt{2x + 3} + \sqrt{x - 2} = 2$

$$\left(\sqrt{2x + 3}\right)^2 = \left(2 - \sqrt{x - 2}\right)^2$$

$$2x + 3 = 4 - 4\sqrt{x - 2} + x - 2$$

$$x + 1 = -4\sqrt{x - 2}$$

$$(x + 1)^2 = \left(-4\sqrt{x - 2}\right)^2$$

$$x^2 + 2x + 1 = 16(x - 2)$$

$$x^2 - 14x + 33 = 0$$

$$(x - 3)(x - 11) = 0$$

$x = 3$, extraneous or $x = 11$, extraneous

No solution. (You can verify that the graph of $y = \sqrt{2x + 3} + \sqrt{x - 2} - 2$ lies above the x-axis.)

103. $(x - 1)^{2/3} - 25 = 0$

$$(x - 1)^{2/3} = 25$$

$$(x - 1)^2 = 25^3$$

$$x - 1 = \pm \sqrt{25^3}$$

$$x = 1 \pm 125$$

$$x = 126 \quad \text{or} \quad x = -124$$

105. $(x + 4)^{1/2} + 5x(x + 4)^{3/2} = 0$

$(x + 4)^{1/2}[1 + 5x(x + 4)] = 0$

$(x + 4)^{1/2}(5x^2 + 20x + 1) = 0$

$(x + 4)^{1/2} = 0 \quad$ or $\quad 5x^2 + 20x + 1 = 0$

$$x = -4 \qquad\qquad x = \frac{-20 \pm \sqrt{400 - 20}}{10}$$

$$x = \frac{-20 \pm 2\sqrt{95}}{10}$$

$$x = -2 \pm \frac{\sqrt{95}}{5}$$

107.
$$\frac{x}{8} + \frac{3}{8} = \frac{1}{2x}$$

$$x + 3 = \frac{4}{x}$$

$$x^2 + 3x - 4 = 0$$

$$(x + 4)(x - 1) = 0$$

$$x = -4, 1$$

109. $3\left(1 - \frac{1}{5t}\right) = 0$

$$1 - \frac{1}{5t} = 0$$

$$1 = \frac{1}{5t}$$

$$5t = 1$$

$$t = \frac{1}{5}$$

111. $\dfrac{4}{(x - 4)^2} = 1$

$$4 = (x - 4)^2$$

$$\pm 2 = x - 4$$

$$4 \pm 2 = x$$

$$x = 6 \text{ or } x = 2$$

113. $|x - 5| = 10$

$x - 5 = -10$ or $x - 5 = 10$

$x = -5 \qquad\quad x = 15$

115. $|x^2 - 3| = 2x$

$x^2 - 3 = 2x \quad$ or $\quad x^2 - 3 = -2x$

$x^2 - 2x - 3 = 0 \qquad x^2 + 2x - 3 = 0$

$(x - 3)(x + 1) = 0 \qquad (x + 3)(x - 1) = 0$

$x = 3 \text{ or } x = -1 \qquad x = -3 \text{ or } x = 1$

The only solutions to the original equation are $x = 3$ or $x = 1$. ($x = -3$ and $x = -1$ are extraneous.)

117. Let $x =$ number of farmers.

$$\frac{48{,}000}{x} = \frac{48{,}000}{x + 2} + 4000$$

$$48{,}000(x + 2) = 48{,}000x + 4000x(x + 2)$$

$$96{,}000 = 4000x^2 + 8000x$$

$$x^2 + 2x - 24 = 0$$

$$(x - 4)(x + 6) = 0$$

$$x = 4 \text{ farmers}$$

119.
$$A = P\left(1 + \frac{r}{n}\right)^{nt}$$

$$1196.95 = 1000\left(1 + \frac{r}{12}\right)^{12(6)}$$

$$1.19695 = \left(1 + \frac{r}{12}\right)^{72}$$

$$1 + \frac{r}{12} = (1.19695)^{1/72}$$

$$r = 0.03, \text{ or } 3\%$$

121. (a)

Year	2000	2001	2002	2003	2004
P (millions)	18.31	18.51	18.59	18.65	18.71

(b)

(c) $P = 18.5$ when $t \approx 0.9$, or late 2000.

(d) $18.31 + 0.1989\sqrt{t} = 18.5$

$0.1989\sqrt{t} = 0.19$

$\sqrt{t} \approx 0.95525$

$t \approx 0.9$

(e) $P = 19$ when $t \approx 12$, or 2012.

(f) Answers will vary.

123. $8x - 3 < 6x + 15$

$2x < 18$

$x < 9$

125. $\frac{1}{2}(3 - x) > \frac{1}{3}(2 - 3x)$

$3(3 - x) > 2(2 - 3x)$

$9 - 3x > 4 - 6x$

$3x > -5$

$x > -\frac{5}{3}, \left(-\frac{5}{3}, \infty\right)$

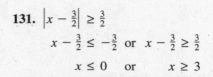

127. $-2 < -x + 7 \le 10$

$-9 < -x \le 3$

$9 > x \ge -3$

$-3 \le x < 9$

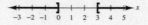

129. $|x - 2| < 1$

$-1 < x - 2 < 1$

$1 < x < 3$

which can be written as $(1, 3)$.

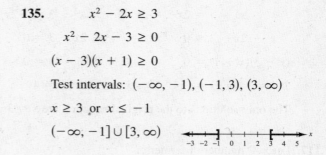

131. $\left|x - \frac{3}{2}\right| \ge \frac{3}{2}$

$x - \frac{3}{2} \le -\frac{3}{2}$ or $x - \frac{3}{2} \ge \frac{3}{2}$

$x \le 0$ or $x \ge 3$

which can be written as $(-\infty, 0] \cup [3, \infty)$.

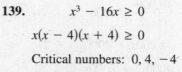

133. $4|3 - 2x| \le 16$

$|3 - 2x| \le 4$

$-4 \le 3 - 2x \le 4$

$-7 \le -2x \le 1$

$\frac{7}{2} \ge x \ge -\frac{1}{2}$

$-\frac{1}{2} \le x \le \frac{7}{2}$

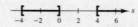

135. $x^2 - 2x \ge 3$

$x^2 - 2x - 3 \ge 0$

$(x - 3)(x + 1) \ge 0$

Test intervals: $(-\infty, -1), (-1, 3), (3, \infty)$

$x \ge 3$ or $x \le -1$

$(-\infty, -1] \cup [3, \infty)$

137. $4x^2 - 23x \le 6$

$4x^2 - 23x - 6 \le 0$

$(x - 6)(4x + 1) \le 0$

Critical numbers: $6, -\frac{1}{4}$

Testing the three intervals, we obtain $-\frac{1}{4} \le x \le 6$.

139. $x^3 - 16x \ge 0$

$x(x - 4)(x + 4) \ge 0$

Critical numbers: $0, 4, -4$

Testing the four intervals, we obtain $-4 \le x \le 0$ or $x \ge 4$.

141. $\dfrac{x-5}{3-x} < 0$

Critical numbers: $x = 5, x = 3$

Test intervals: $(-\infty, 3), (3, 5), (5, \infty)$

Test: Is $\dfrac{x-5}{3-x} < 0$?

Solution set: $(-\infty, 3) \cup (5, \infty)$

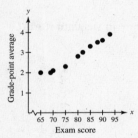

143. $\dfrac{3x+8}{x-3} - 4 \le 0$

$\dfrac{3x + 8 - 4(x-3)}{x-3} \le 0$

$\dfrac{20 - x}{x-3} \le 0$

$\dfrac{x-20}{x-3} \ge 0$

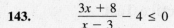

Critical numbers: $x = 3, 20$

Testing the three intervals, we obtain $x \ge 20$ or $x < 3$.

145. $x - 4 \ge 0$

$[4, \infty)$

147. $\sqrt[3]{2 - 3x}$ is defined for all x.

$(-\infty, \infty)$

149. $(0.1)(2.59) = 0.259 \approx \$0.26$

151. (a)

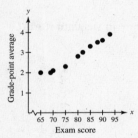

Exam score

(b) Yes, the relationship is approximately linear. Higher entrance exam scores, x, are associated with higher grade-point averages, y.

153. (a)

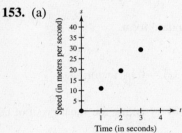

Time (in seconds)

(b) $s \approx 10t$ (Approximations will vary.)

(c) $s = 9.7t + 0.4$; 0.99933

(d) For $t = 2.5$, $S \approx 24.7$ m/sec.

155. False. A function can have only one y-intercept.

(Vertical Line Test)

157. False. The slope can be positive, negative, or 0.

159. They are the same. A point $(a, 0)$ is an x-intercept if it is a solution point of the equation. In other words, a is a zero of the function.

161. $\sqrt{-6}\sqrt{-6} \ne \sqrt{(-6)(-6)}$

In fact, $\sqrt{-6}\sqrt{-6} = \sqrt{6}i\,\sqrt{6}i = -6$.

163. (a) $i^{40} = (i^4)^{10} = 1^{10} = 1$

(b) $i^{25} = i(i^{24}) = i(1) = i$

(c) $i^{50} = i^2(i^{48}) = (-1)(1) = -1$

(d) $i^{67} = i^3(i^{64}) = -i(1) = -i$

Chapter 2   Practice Test

1. Solve the equation $\frac{1}{2}x - \frac{1}{3}(x - 1) = 10$. Verify your answer with a graphing utility.

2. Solve the equation $(x + 1)^2 - 6 = x^2 + 3x$ and verify your answer with a graphing utility.

3. Solve $A = \frac{1}{2}(a + b)h$ for a.

4. 301 is what percent of 4300?

5. Cindy has \$6.05 in quarters and nickels. How many of each coin does she have if there are 53 coins in all?

6. Ed has \$15,000 invested in two funds paying $9\frac{1}{2}\%$ and 11% simple interest, respectively. How much is invested in each if the yearly interest is \$1582.50?

7. Use a graphing utility to approximate any points of intersection of $y = 3x^2 - 4$ and $y = 2 - x$.

8. Use a graphing utility to approximate any points of intersection of $y = 2x^2 + 3$ and $y = 5 + \sqrt{x}$.

9. Write $\dfrac{2}{1 + i}$ in standard form.

10. Write $\dfrac{3 + i}{2} - \dfrac{i + 1}{4}$ in standard form.

11. Solve $28 + 5x - 3x^2 = 0$ by factoring.

12. Solve $(x - 2)^2 = 24$ by taking the square root of both sides.

13. Solve $x^2 - 4x - 9 = 0$ by completing the square.

14. Solve $x^2 + 5x - 1 = 0$ by the Quadratic Formula.

15. Solve $3x^2 - 2x + 4 = 0$ by the Quadratic Formula.

16. The perimeter of a rectangle is 1100 feet. Find the dimension so that the enclosed area will be 60,000 square feet.

17. Find two consecutive even positive integers whose product is 624.

18. Solve $x^3 - 10x^2 + 24x = 0$ by factoring.

19. Solve $\sqrt[3]{6 - x} = 4$.

20. Solve $(x^2 - 8)^{2/5} = 4$.

21. Solve $x^4 - x^2 - 12 = 0$.

22. Solve $4 - 3x > 16$.

23. Solve $\left| \dfrac{x - 3}{2} \right| < 5$.

24. Solve $\dfrac{x + 1}{x - 3} < 2$.

25. Solve $|3x - 4| \geq 9$.

26. Use a graphing utility to find the least squares regression line for the points $(-1, 0)$, $(0, 1)$, $(3, 3)$ and $(4, 5)$. Graph the points and the line.

C H A P T E R   3
Polynomial and Rational Functions

CHAPTER 3
Polynomial and Rational Functions

Section 3.1   Quadratic Functions

You should know the following facts about parabolas.

■ $f(x) = ax^2 + bx + c$, $a \neq 0$, is a quadratic function, and its graph is a parabola.

■ If $a > 0$, the parabola opens upward and the vertex is the minimum point. If $a < 0$, the parabola opens downward and the vertex is the maximum point.

■ The vertex is $(-b/2a, f(-b/2a))$.

■ To find the x-intercepts (if any), solve
$$ax^2 + bx + c = 0.$$

■ The standard form of the equation of a parabola is
$$f(x) = a(x - h)^2 + k$$
where $a \neq 0$.

(a) The vertex is (h, k).

(b) The axis is the vertical line $x = h$.

Vocabulary Check

1. nonnegative integer, real

2. quadratic, parabola

3. axis

4. positive, minimum

5. negative, maximum

---

1. $f(x) = (x - 2)^2$ opens upward and has vertex $(2, 0)$. Matches graph (c).

3. $f(x) = x^2 + 3$ opens upward and has vertex $(0, 3)$. Matches graph (b).

5.

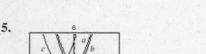 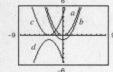

(a) $y = \frac{1}{2}x^2$, vertical shrink

(b) $y = \frac{1}{2}x^2 - 1$, vertical shrink and vertical shift one unit downward

(c) $y = \frac{1}{2}(x + 3)^2$, vertical shrink and horizontal shift three units to the left

(d) $y = -\frac{1}{2}(x + 3)^2 - 1$, horizontal shift three units to the left, vertical shrink, reflection in x-axis, and vertical shift one unit downward

7. $f(x) = 25 - x^2$

Vertex: $(0, 25)$

x-intercepts: $(-5, 0), (5, 0)$

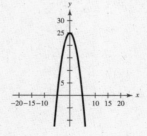

9. $f(x) = \frac{1}{2}x^2 - 4$

Vertex: $(0, -4)$

x-intercepts: $\left(\pm 2\sqrt{2}, 0\right)$

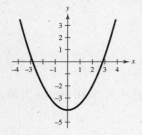

11. $f(x) = (x + 4)^2 - 3$

Vertex: $(-4, -3)$

x-intercepts: $\left(-4 \pm \sqrt{3}, 0\right)$

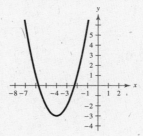

13. $h(x) = x^2 - 8x + 16 = (x - 4)^2$

Vertex: $(4, 0)$

x-intercepts: $(4, 0)$

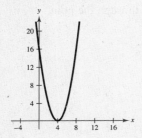

15. $f(x) = x^2 - x + \frac{5}{4} = \left(x - \frac{1}{2}\right)^2 + 1$

Vertex: $\left(\frac{1}{2}, 1\right)$

x-intercepts: None

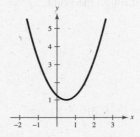

17. $f(x) = -x^2 + 2x + 5 = -(x - 1)^2 + 6$

Vertex: $(1, 6)$

x-intercepts: $\left(1 - \sqrt{6}, 0\right), \left(1 + \sqrt{6}, 0\right)$

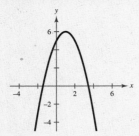

19. $h(x) = 4x^2 - 4x + 21 = 4\left(x - \frac{1}{2}\right)^2 + 20$

Vertex: $\left(\frac{1}{2}, 20\right)$

x-intercept: None

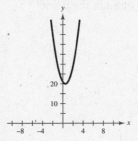

21. $f(x) = -(x^2 + 2x - 3) = -(x + 1)^2 + 4$

Vertex: $(-1, 4)$

x-intercepts: $(-3, 0), (1, 0)$

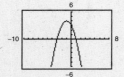

23. $g(x) = x^2 + 8x + 11 = (x + 4)^2 - 5$

Vertex: $(-4, -5)$

x-intercepts: $\left(-4 \pm \sqrt{5}, 0\right)$

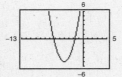

25. $f(x) = -2x^2 + 16x - 31$

$\qquad = -2\left(x^2 - 8x + \frac{31}{2}\right)$

$\qquad = -2\left(x^2 - 8x + 16 - \frac{1}{2}\right)$

$\qquad = -2(x - 4)^2 + 1$

Vertex: $(4, 1)$

x-intercept: $\left(4 \pm \frac{1}{2}\sqrt{2}, 0\right)$

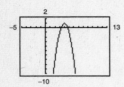

27. $(-1, 4)$ is the vertex.

$f(x) = a(x + 1)^2 + 4$

Since the graph passes through the point $(1, 0)$, we have:

$0 = a(1 + 1)^2 + 4$

$0 = 4a + 4$

$-1 = a$

Thus, $f(x) = -(x + 1)^2 + 4$. Note that $(-3, 0)$ is on the parabola.

29. $(-2, 5)$ is the vertex.

$f(x) = a(x + 2)^2 + 5$

Since the graph passes through the point $(0, 9)$, we have:

$9 = a(0 + 2)^2 + 5$

$4 = 4a$

$1 = a$

$f(x) = 1(x + 2)^2 + 5 = (x + 2)^2 + 5$

31. $(1, -2)$ is the vertex.

$f(x) = a(x - 1)^2 - 2$

Since the graph passes through the point $(-1, 14)$, we have:

$14 = a(-1 - 1)^2 - 2$

$14 = 4a - 2$

$16 = 4a$

$4 = a$

$f(x) = 4(x - 1)^2 - 2$

33. $\left(\frac{1}{2}, 1\right)$ is the vertex.

$f(x) = a\left(x - \frac{1}{2}\right)^2 + 1$

Since the graph passes through the point $\left(-2, -\frac{21}{5}\right)$, we have:

$-\frac{21}{5} = a\left(-2 - \frac{1}{2}\right)^2 + 1$

$-\frac{21}{5} = \frac{25}{4}a + 1$

$-\frac{26}{5} = \frac{25}{4}a$

$-\frac{104}{125} = a$

$f(x) = -\frac{104}{125}\left(x - \frac{1}{2}\right)^2 + 1$

35. $y = x^2 - 4x - 5$

x-intercepts: $(5, 0), (-1, 0)$

$0 = x^2 - 4x - 5$

$0 = (x - 5)(x + 1)$

$x = 5$ or $x = -1$

37. $y = x^2 + 8x + 16$

x-intercept: $(-4, 0)$

$0 = x^2 + 8x + 16$

$0 = (x + 4)^2$

$x = -4$

39. $y = x^2 - 4x$

$0 = x^2 - 4x$

$0 = x(x - 4)$

$x = 0$ or $x = 4$

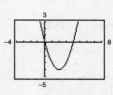

x-intercepts: $(0, 0), (4, 0)$

41. $y = 2x^2 - 7x - 30$

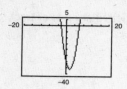

$0 = 2x^2 - 7x - 30$

$0 = (2x + 5)(x - 6)$

$x = -\frac{5}{2}$ or $x = 6$

x-intercepts: $\left(-\frac{5}{2}, 0\right), (6, 0)$

45. $f(x) = [x - (-1)](x - 3)$, opens upward

$\quad = (x + 1)(x - 3)$

$\quad = x^2 - 2x - 3$

$g(x) = -[x - (-1)](x - 3)$, opens downward

$\quad = -(x + 1)(x - 3)$

$\quad = -(x^2 - 2x - 3)$

$\quad = -x^2 + 2x + 3$

Note: $f(x) = a(x + 1)(x - 3)$ has x-intercepts $(-1, 0)$ and $(3, 0)$ for all real numbers $a \neq 0$.

49. Let x = the first number and y = the second number. Then the sum is

$x + y = 110 \implies y = 110 - x$.

The product is

$P(x) = xy = x(110 - x) = 110x - x^2$.

$P(x) = -x^2 + 110x$

$\quad = -(x^2 - 110x + 3025 - 3025)$

$\quad = -[(x - 55)^2 - 3025]$

$\quad = -(x - 55)^2 + 3025$

The maximum value of the product occurs at the vertex of $P(x)$ and is 3025. This happens when $x = y = 55$.

43. $y = -\frac{1}{2}(x^2 - 6x - 7)$

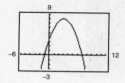

$0 = -\frac{1}{2}(x^2 - 6x - 7)$

$0 = x^2 - 6x - 7$

$0 = (x + 1)(x - 7)$

$x = -1, 7$

x-intercepts: $(-1, 0), (7, 0)$

47. $f(x) = [x - (-3)]\left[x - \left(-\frac{1}{2}\right)\right](2)$, opens upward

$\quad = (x + 3)\left(x + \frac{1}{2}\right)(2)$

$\quad = (x + 3)(2x + 1)$

$\quad = 2x^2 + 7x + 3$

$g(x) = -(2x^2 + 7x + 3)$, opens downward

$\quad = -2x^2 - 7x - 3$

Note: $f(x) = a(x + 3)(2x + 1)$ has x-intercepts $(-3, 0)$ and $\left(-\frac{1}{2}, 0\right)$ for all real numbers $a \neq 0$.

51. Let x be the first number and y be the second number. Then $x + 2y = 24 \implies x = 24 - 2y$. The product is $P = xy = (24 - 2y)y = 24y - 2y^2$.

Completing the square,

$P = -2y^2 + 24y$

$\quad = -2(y^2 - 12y + 36) + 72$

$\quad = -2(y - 6)^2 + 72$.

The maximum value of the product P occurs at the vertex of the parabola and equals 72. This happens when $y = 6$ and $x = 24 - 2(6) = 12$.

53. (a)

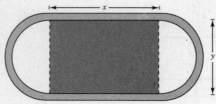

(b) Radius of semicircular ends of track: $r = \frac{1}{2}y$

Distance around two semicircular parts of track:

$$d = 2\pi r = 2\pi\left(\frac{1}{2}y\right) = \pi y$$

(c) Distance traveled around track in one lap:

$$d = \pi y + 2x = 200$$

$$\pi y = 200 - 2x$$

$$y = \frac{200 - 2x}{\pi}$$

(d) Area of rectangular region:

$$A = xy = x\left(\frac{200 - 2x}{\pi}\right)$$

$$= \frac{1}{\pi}(200x - 2x^2)$$

$$= -\frac{2}{\pi}(x^2 - 100x)$$

$$= -\frac{2}{\pi}(x^2 - 100x + 2500 - 2500)$$

$$= -\frac{2}{\pi}(x - 50)^2 + \frac{5000}{\pi}$$

The area is maximum when $x = 50$ and

$$y = \frac{200 - 2(50)}{\pi} = \frac{100}{\pi}.$$

(e)

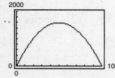

The area is maximum when $x = 50$ and

$$y = \frac{200 - 2(50)}{\pi} = \frac{100}{\pi}.$$

55. (a)

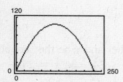

(b) When $x = 0$, $y = \frac{3}{2}$ feet.

(c) The vertex occurs at

$$x = \frac{-b}{2a} = \frac{-9/5}{2(-16/2025)} = \frac{3645}{32} \approx 113.9.$$

The maximum height is

$$y = \frac{-16}{2025}\left(\frac{3645}{32}\right)^2 + \frac{9}{5}\left(\frac{3645}{32}\right) + \frac{3}{2}$$

$$\approx 104.0 \text{ feet.}$$

(d) Using a graphing utility, the zero of y occurs at $x \approx 228.6$, or 228.6 feet from the punter.

57. $C = 800 - 10x + 0.25x^2$

x	10	15	20	25	30
C	725	706.25	700	706.25	725

From the table, the minimum cost seems to be at $x = 20$.

The minimum cost occurs at the vertex.

$$x = \frac{-b}{2a} = -\frac{(-10)}{2(0.25)} = \frac{10}{0.5} = 20$$

$C(20) = 700$ is the minimum cost.

Graphically, you could graph $C = 800 - 10x + 0.25x^2$ in the window $[0, 40] \times [0, 1000]$ and find the vertex $(20, 700)$.

59. (a) $R(20) = -25(20)^2 + 1200(20)$

   $= \$14{,}000$ thousand

   $R(25) = -25(25)^2 + 1200(25)$

   $= \$14{,}375$ thousand

   $R(30) = -25(30)^2 + 1200(30)$

   $= \$13{,}500$ thousand

(b) The vertex occurs at

   $p = \dfrac{-b}{2a} = \dfrac{-1200}{2(-25)} = \$24.$

(c) $R(24) = -25(24)^2 + 1200(24)$

   $= \$14{,}400$ thousand

(d) Answers will vary.

61. $C(t) = 4306 - 3.4t - 1.32t^2,\ 0 \le t \le 44$

   ($t = 0$ corresponds to 1960.)

(a)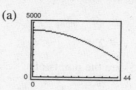

(b) The maximum consumption per year of 4306 cigarettes per person per year occurred in 1960 ($t = 0$). Answers will vary.

(c) For 2000, $C(40) = 2058.$

   $(2058)\dfrac{209{,}117{,}000}{48{,}306{,}000} \approx 8909$ cigarettes per smoker per year,

   $\dfrac{8909}{365} \approx 24$ cigarettes per smoker per day

63. True

   $-12x^2 - 1 = 0$

   $12x^2 = -1,$ impossible

65. The parabola opens downward and the vertex is $(-2, -4)$. Matches (c) and (d).

67. For $a < 0,\ f(x) = a\left(x + \dfrac{b}{2a}\right)^2 + \left(c - \dfrac{b^2}{4a}\right)$ is a maximum when $x = \dfrac{-b}{2a}$. In this case, the maximum value is $c - \dfrac{b^2}{4a}$. Hence,

   $25 = -75 - \dfrac{b^2}{4(-1)}$

   $-100 = 300 - b^2$

   $400 = b^2$

   $b = \pm 20.$

69. For $a > 0,\ f(x) = a\left(x + \dfrac{b}{2a}\right)^2 + \left(c - \dfrac{b^2}{4a}\right)$ is a minimum when $x = \dfrac{-b}{2a}$. In this case, the minimum value is $c - \dfrac{b^2}{4a}$. Hence,

   $10 = 26 - \dfrac{b^2}{4}$

   $40 = 104 - b^2$

   $b^2 = 64$

   $b = \pm 8.$

71. Model (a) is preferable. $a > 0$ means the parabola opens upward and profits are increasing for t to the right of the vertex,

   $t \ge -\dfrac{b}{(2a)}.$

73. $x + y = 8 \implies y = 8 - x$

   Then $-\frac{2}{3}x + y = -\frac{2}{3}x + (8 - x) = 6 \implies -\frac{5}{3}x = -2 \implies x = \frac{6}{5}$ and $y = 8 - \frac{6}{5} = \frac{34}{5}.$

   $(1.2, 6.8)$

75. $y = x + 3 = 9 - x^2$

$x^2 + x - 6 = 0$

$(x + 3)(x - 2) = 0$

$x = -3, x = 2$

Thus, $(-3, 0)$ and $(2, 5)$ are the points of intersection.

77. $(6 - i) - (2i + 11) = 6 - 11 - i - 2i$

$= -5 - 3i$

79. $(3i + 7)(-4i + 1) = -12i^2 + 3i - 28i + 7$

$= 19 - 25i$

81. Answers will vary. (Make a Decision)

Section 3.2   Polynomial Functions of Higher Degree

- You should know the following basic principles about polynomials.
- $f(x) = a_n x^n + a_{n-1} x^{n-1} + \cdots + a_2 x^2 + a_1 x + a_0, a_n \neq 0$, is a polynomial function of degree n.
- If f is of odd degree and
  - (a) $a_n > 0$, then
    1. $f(x) \to \infty$ as $x \to \infty$.
    2. $f(x) \to -\infty$ as $x \to -\infty$.
  - (b) $a_n < 0$, then
    1. $f(x) \to -\infty$ as $x \to \infty$.
    2. $f(x) \to \infty$ as $x \to -\infty$.
- If f is of even degree and
  - (a) $a_n > 0$, then
    1. $f(x) \to \infty$ as $x \to \infty$.
    2. $f(x) \to \infty$ as $x \to -\infty$.
  - (b) $a_n < 0$, then
    1. $f(x) \to -\infty$ as $x \to \infty$.
    2. $f(x) \to -\infty$ as $x \to -\infty$.
- The following are equivalent for a polynomial function.
  - (a) $x = a$ is a zero of a function.
  - (b) $x = a$ is a solution of the polynomial equation $f(x) = 0$.
  - (c) $(x - a)$ is a factor of the polynomial.
  - (d) $(a, 0)$ is an x-intercept of the graph of f.
- A polynomial of degree n has at most n distinct zeros.
- If f is a polynomial function such that $a < b$ and $f(a) \neq f(b)$, then f takes on every value between $f(a)$ and $f(b)$ in the interval $[a, b]$.
- If you can find a value where a polynomial is positive and another value where it is negative, then there is at least one real zero between the values.

Vocabulary Check

1. continuous

2. Leading Coefficient Test

3. $n, n - 1$, relative extrema

4. solution, $(x - a)$, x-intercept

5. touches, crosses

6. Intermediate Value

1. $f(x) = -2x + 3$ is a line with y-intercept $(0, 3)$. Matches graph (f).

3. $f(x) = -2x^2 - 5x$ is a parabola with x-intercepts $(0, 0)$ and $\left(-\frac{5}{2}, 0\right)$ and opens downward. Matches graph (c).

5. $f(x) = -\frac{1}{4}x^4 + 3x^2$ has intercepts $(0, 0)$ and $\left(\pm 2\sqrt{3}, 0\right)$. Matches graph (e).

7. $f(x) = x^4 + 2x^3$ has intercepts $(0, 0)$ and $(-2, 0)$. Matches graph (g).

9. $y = x^3$

    (a) $f(x) = (x - 2)^3$

      Horizontal shift two units to the right

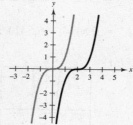

    (b) $f(x) = x^3 - 2$

      Vertical shift two units downward

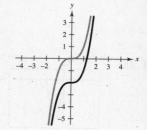

    (c) $f(x) = -\frac{1}{2}x^3$

      Reflection in the x-axis and a vertical shrink

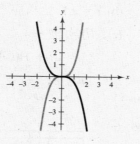

    (d) $f(x) = (x - 2)^3 - 2$

      Horizontal shift two units to the right and a vertical shift two units downward

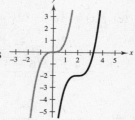

11. $f(x) = 3x^3 - 9x + 1$; $g(x) = 3x^3$

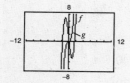

13. $f(x) = -(x^4 - 4x^3 + 16x)$; $g(x) = -x^4$

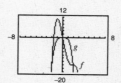

15. $f(x) = 2x^4 - 3x + 1$

Degree: 4

Leading coefficient: 2

The degree is even and the leading coefficient is positive. The graph rises to the left and right.

17. $g(x) = 5 - \frac{7}{2}x - 3x^2$

Degree: 2

Leading coefficient: -3

The degree is even and the leading coefficient is negative. The graph falls to the left and right.

19. Degree: 5 (odd)

Leading coefficient: $\frac{6}{3} = 2 > 0$

Falls to the left and rises to the right

21. $h(t) = -\frac{2}{3}(t^2 - 5t + 3)$

Degree: 2

Leading coefficient: $-\frac{2}{3}$

The degree is even and the leading coefficient is negative. The graph falls to the left and right.

23. $f(x) = x^2 - 25$

  $= (x + 5)(x - 5)$

  $x = \pm 5$

25. $h(t) = t^2 - 6t + 9$

  $= (t - 3)^2$

  $t = 3$ (multiplicity 2)

27. $f(x) = x^2 + x - 2$

  $= (x + 2)(x - 1)$

  $x = -2, 1$

29. $f(t) = t^3 - 4t^2 + 4t$

  $= t(t - 2)^2$

  $t = 0, 2$ (multiplicity 2)

31. $f(x) = \frac{1}{2}x^2 + \frac{5}{2}x - \frac{3}{2}$

  $= \frac{1}{2}(x^2 + 5x - 3)$

  $x = \dfrac{-5 \pm \sqrt{25 - 4(-3)}}{2} = -\dfrac{5}{2} \pm \dfrac{\sqrt{37}}{2}$

  $\approx 0.5414, -5.5414$

33. (a)

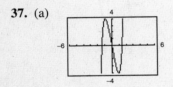

(b) $x \approx 3.732, 0.268$

(c) $f(x) = 3x^2 - 12x + 3$

  $= 3(x^2 - 4x + 1)$

  $x = \dfrac{4 \pm \sqrt{16 - 4}}{2} = 2 \pm \sqrt{3}$

35. (a)

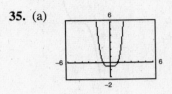

(b) $t = \pm 1$

(c) $g(t) = \frac{1}{2}t^4 - \frac{1}{2}$

  $= \frac{1}{2}(t + 1)(t - 1)(t^2 + 1)$

  $t = \pm 1$

37. (a)

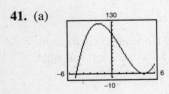

(b) $x = 0, 1.414, -1.414$

(c) $f(x) = x^5 + x^3 - 6x$

  $= x(x^4 + x^2 - 6)$

  $= x(x^2 + 3)(x^2 - 2)$

  $x = 0, \pm\sqrt{2}$

39. (a)

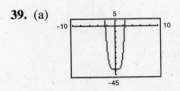

(b) $2.236, -2.236$

(c) $f(x) = 2x^4 - 2x^2 - 40$

  $= 2(x^4 - x^2 - 20)$

  $= 2(x^2 + 4)(x + \sqrt{5})(x - \sqrt{5})$

  $x = \pm\sqrt{5}$

41. (a)

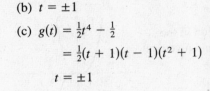

(b) $x = 4, 5, -5$

(c) $f(x) = x^3 - 4x^2 - 25x + 100$

  $= x^2(x - 4) - 25(x - 4)$

  $= (x^2 - 25)(x - 4)$

  $= (x - 5)(x + 5)(x - 4)$

  $x = \pm 5, 4$

43. (a)

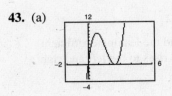

(b) $x = 0, \frac{5}{2}$

(c) $y = 4x^3 - 20x^2 + 25x$

  $0 = 4x^3 - 20x^2 + 25x$

  $0 = x(2x - 5)^2$

  $x = 0$ or $x = \frac{5}{2}$ (multiplicity 2)

45. $f(x) = 2x^4 - 6x^2 + 1$

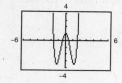

Zeros: $x \approx \pm 0.421, \pm 1.680$

Relative maximum: $(0, 1)$

Relative minimums:
$(1.225, -3.5), (-1.225, -3.5)$

49. $f(x) = (x - 0)(x - 4) = x^2 - 4x$

Note: $f(x) = a(x - 0)(x - 4) = ax(x - 4)$ has zeros 0 and 4 for all nonzero real numbers a.

53. $f(x) = (x - 4)(x + 3)(x - 3)(x - 0)$

$= (x - 4)(x^2 - 9)x$

$= x^4 - 4x^3 - 9x^2 + 36x$

Note: $f(x) = a(x^4 - 4x^3 - 9x^2 + 36x)$ has zeros 4, -3, 3, and 0 for all nonzero real numbers a.

57. $f(x) = (x - 2)\left[x - \left(4 + \sqrt{5}\right)\right]\left[x - \left(4 - \sqrt{5}\right)\right]$

$= (x - 2)\left[(x - 4) - \sqrt{5}\right]\left[(x - 4) + \sqrt{5}\right]$

$= (x - 2)\left[(x - 4)^2 - 5\right]$

$= x^3 - 10x^2 + 27x - 22$

Note: $f(x) = a(x - 2)\left[(x - 4)^2 - 5\right]$ has zeros 2, $4 + \sqrt{5}$, and $4 - \sqrt{5}$ for all nonzero real numbers a.

61. $f(x) = (x + 4)^2(x - 3)^2$

$= x^4 + 2x^3 - 23x^2 - 24x + 144$

Note: $f(x) = a(x + 4)^2(x - 3)^2$ has zeros -4, -4, 3, 3 for all nonzero real numbers a.

65.

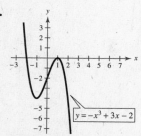

For example,

$f(x) = -(x + 2)(x - 1)^2$

$= -x^3 + 3x - 2.$

47. $f(x) = x^5 + 3x^3 - x + 6$

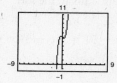

Zeros: $x \approx -1.178$

Relative maximum: $(-0.324, 6.218)$

Relative minimum: $(0.324, 5.782)$

51. $f(x) = (x - 0)(x + 2)(x + 3) = x^3 + 5x^2 + 6x$

Note: $f(x) = ax(x + 2)(x + 3)$ has zeros 0, -2, and -3 for all nonzero real numbers a.

55. $f(x) = \left[x - \left(1 + \sqrt{3}\right)\right]\left[x - \left(1 - \sqrt{3}\right)\right]$

$= \left[(x - 1) - \sqrt{3}\right]\left[(x - 1) + \sqrt{3}\right]$

$= (x - 1)^2 - \left(\sqrt{3}\right)^2$

$= x^2 - 2x + 1 - 3$

$= x^2 - 2x - 2$

Note: $f(x) = a(x^2 - 2x - 2)$ has zeros $1 + \sqrt{3}$ and $1 - \sqrt{3}$ for all nonzero real numbers a.

59. $f(x) = (x + 2)^2(x + 1) = x^3 + 5x^2 + 8x + 4$

Note: $f(x) = a(x + 2)^2(x + 1)$ has zeros -2, -2, and -1 for all nonzero real numbers a.

63. $f(x) = -(x + 1)^2(x + 2)$

$= -x^3 - 4x^2 - 5x - 2$

Note: $f(x) = a(x + 1)^2(x + 2)^2$, $a < 0$, has zeros -1, -1, -2, rises to the left, and falls to the right.

67.

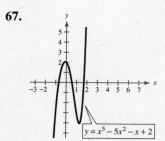

69. (a) The degree of f is odd and the leading coefficient is 1. The graph falls to the left and rises to the right.

(b) $f(x) = x^3 - 9x = x(x^2 - 9) = x(x - 3)(x + 3)$

Zeros: 0, 3, -3

(c) and (d)

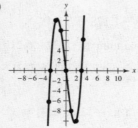

71. (a) The degree of f is odd and the leading coefficient is 1. The graph falls to the left and rises to the right.

(b) $f(x) = x^3 - 3x^2 = x^2(x - 3)$

Zeros: 0, 3

(c) and (d)

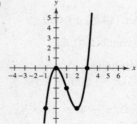

73. (a) The degree of f is even and the leading coefficient is -1. The graph falls to the left and falls to the right.

(b) $f(x) = -x^4 + 9x^2 - 20 = -(x^2 - 4)(x^2 - 5)$

Zeros: $\pm 2, \pm \sqrt{5}$: $(\pm 2, 0), (\pm \sqrt{5}, 0)$

(c) and (d)

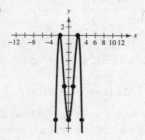

75. (a) The degree is odd and the leading coefficient is 1. The graph falls to the left and rises to the right.

(b) $x^3 + 3x^2 - 9x - 27 = x^2(x + 3) - 9(x + 3)$

$$= (x^2 - 9)(x + 3)$$

$$= (x - 3)(x + 3)^2$$

Zeros: 3, -3: $(3, 0), (-3, 0)$

(c) and (d)

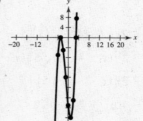

77. $g(t) = -\frac{1}{4}t^4 + 2t^2 - 4$

(a) Falls to left and falls to right; $\left(-\frac{1}{4} < 0\right)$

(b) $g(t) = -\frac{1}{4}(t^4 - 8t^2 + 16) = -\frac{1}{4}(t^2 - 4)^2$

$t = -2, -2, 2, 2 \Rightarrow (-2, 0), (2, 0)$; zeros

(c) and (d)

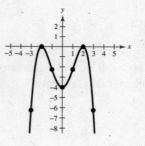

79. $f(x) = x^3 - 3x^2 + 3$

(a)

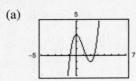

The function has three zeros.
They are in the intervals
$(-1, 0), (1, 2)$ and $(2, 3)$.

(b) Zeros: $-0.879, 1.347, 2.532$

(c)

x	y_1	x	y_1	x	y_1
-0.9	-0.159	1.3	0.127	2.5	-0.125
-0.89	-0.0813	1.31	0.09979	2.51	-0.087
-0.88	-0.0047	1.32	0.07277	2.52	-0.0482
-0.87	0.0708	1.33	0.04594	2.53	-0.0084
-0.86	0.14514	1.34	0.0193	2.54	0.03226
-0.85	0.21838	1.35	-0.0071	2.55	0.07388
-0.84	0.2905	1.36	-0.0333	2.56	0.11642

81. $g(x) = 3x^4 + 4x^3 - 3$

(a)

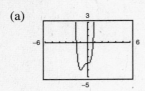

The function has two zeros.
They are in the intervals
$(-2, -1)$ and $(0, 1)$.

(b) Zeros: $-1.585, 0.779$

(c)

x	y_1	x	y_1
-1.6	0.2768	0.75	-0.3633
-1.59	0.09515	0.76	-0.2432
-1.58	-0.0812	0.77	-0.1193
-1.57	-0.2524	0.78	0.00866
-1.56	-0.4184	0.79	0.14066
-1.55	-0.5795	0.80	0.2768
-1.54	-0.7356	0.81	0.41717

83.

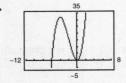

$f(x) = x^2(x + 6)$

No symmetry

Two x-intercepts

85.

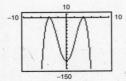

$g(t) = -\frac{1}{2}(t - 4)^2(t + 4)^2$

Symmetric about the y-axis

Two x-intercepts

87.

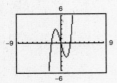

$f(x) = x^3 - 4x$

$\quad = x(x + 2)(x - 2)$

Symmetric to origin

Three x-intercepts

89.

$g(x) = \frac{1}{5}(x + 1)^2(x - 3)(2x - 9)$

Three x-intercepts

No symmetry

91. (a) Volume = length × width × height

Because the box is made from a square, length = width.

Thus:

Volume = (length)² × height = $(36 - 2x)^2 x$

(b) Domain: $0 < 36 - 2x < 36$

$-36 < -2x < 0$

$18 > x > 0$

(c)

Height, x	Length and Width	Volume, V
1	$36 - 2(1)$	$1[36 - 2(1)]^2 = 1156$
2	$36 - 2(2)$	$2[36 - 2(2)]^2 = 2048$
3	$36 - 2(3)$	$3[36 - 2(3)]^2 = 2700$
4	$36 - 2(4)$	$4[36 - 2(4)]^2 = 3136$
5	$36 - 2(5)$	$5[36 - 2(5)]^2 = 3380$
6	$36 - 2(6)$	$6[36 - 2(6)]^2 = 3456$
7	$36 - 2(7)$	$7[36 - 2(7)]^2 = 3388$

Maximum volume 3456 for $x = 6$

(d)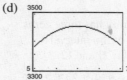

$x = 6$ when $V(x)$ is maximum.

93. The point of diminishing returns (where the graph changes from curving upward to curving downward) occurs when $x = 200$. The point is (200, 160) which corresponds to spending \$2,000,000 on advertising to obtain a revenue of \$160 million.

95.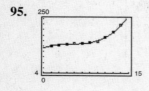

The model is a good fit.

97. For 2010, $t = 20$, and

$y_1 \approx \$730.2$ thousand

$y_2 \approx \$285.0$ thousand.

Answers will vary.

99. True. $f(x) = x^6$ has only one zero, 0.

101. False. The graph touches at $x = 1$, but does not cross the x-axis there.

103. True. The exponent of $(x + 2)$ is odd (3).

105. The zeros are 0, 1, 1, and the graph rises to the right. Matches (b).

107. The zeros are 1, 1, −2, −2, and the graph rises to the right. Matches (a).

109. $(g - f)(3) = g(3) - f(3) = 8(3)^2 - [14(3) - 3]$

$= 72 - 39 = 33$

111. $\left(\dfrac{f}{g}\right)(-1.5) = \dfrac{f(-1.5)}{g(-1.5)} = \dfrac{-24}{18} = -\dfrac{4}{3}$

113. $(g \circ f)(0) = g(f(0)) = g(-3) = 8(-3)^2 = 72$

115. $2x^2 - x \geq 1$

$2x^2 - x - 1 \geq 0$

$(2x + 1)(x - 1) \geq 0$

$[2x + 1 \geq 0 \text{ and } x - 1 \geq 0]$ or $[2x + 1 \leq 0 \text{ and } x - 1 \leq 0]$

$[x \geq -\tfrac{1}{2} \text{ and } x \geq 1]$ or $[x \leq -\tfrac{1}{2} \text{ and } x \leq 1]$

$x \geq 1$ or $x \leq -\tfrac{1}{2}$

117. $|x + 8| - 1 \geq 15$

$$|x + 8| \geq 16$$

$$x + 8 \geq 16 \quad \text{or} \quad x + 8 \leq -16$$

$$x \geq 8 \quad \text{or} \quad x \leq -24$$

Section 3.3   Real Zeros of Polynomial Functions

You should know the following basic techniques and principles of polynomial division.

- The Division Algorithm (Long Division of Polynomials)
- Synthetic Division
- $f(k)$ is equal to the remainder of $f(x)$ divided by $(x - k)$.
- $f(k) = 0$ if and only if $(x - k)$ is a factor of $f(x)$.
- The Rational Zero Test
- The Upper and Lower Bound Rule

Vocabulary Check

1. $f(x)$ is the dividend, $d(x)$ is the divisor, $q(x)$ is the quotient, and $r(x)$ is the remainder.

2. improper, proper   **3.** synthetic division   **4.** Rational Zero

5. Descartes's Rule, Signs   **6.** Remainder Theorem   **7.** upper bound, lower bound

1.

$$\begin{array}{r} 2x + 4 \\ x + 3 \overline{)\, 2x^2 + 10x + 12} \\ \underline{2x^2 + 6x} \\ 4x + 12 \\ \underline{4x + 12} \\ 0 \end{array}$$

$$\frac{2x^2 + 10x + 12}{x + 3} = 2x + 4, \, x \neq -3$$

3.

$$\begin{array}{r} x^3 + 3x^2 \qquad\; - 1 \\ x + 2 \overline{)\, x^4 + 5x^3 + 6x^2 - x - 2} \\ \underline{x^4 + 2x^3} \\ 3x^3 + 6x^2 \\ \underline{3x^3 + 6x^2} \\ -x - 2 \\ \underline{-x - 2} \\ 0 \end{array}$$

$$\frac{x^4 + 5x^3 + 6x^2 - x - 2}{x + 2} = x^3 + 3x^2 - 1, \, x \neq -2$$

5.

$$\begin{array}{r} x^2 - 3x + 1 \\ 4x + 5 \overline{)\, 4x^3 - \quad 7x^2 - 11x + 5} \\ \underline{-(4x^3 + \quad 5x^2)} \\ -12x^2 - 11x \\ \underline{-(-12x^2 - 15x)} \\ 4x + 5 \\ \underline{-(4x + 5)} \\ 0 \end{array}$$

$$\frac{4x^3 - 7x^2 - 11x + 5}{4x + 5} = x^2 - 3x + 1, \quad x \neq -\frac{5}{4}$$

7.

$$\begin{array}{r} 7x^2 - 14x + 28 \\ x + 3 \overline{)\, 7x^3 + 0x^2 + \;\; 0x + \;\; 3} \\ \underline{7x^3 + 14x^2} \\ -14x^2 \\ \underline{-14x^2 - 28x} \\ 28x + \;\; 3 \\ \underline{28x + 56} \\ -53 \end{array}$$

$$\frac{7x^3 + 3}{x + 2} = 7x^2 - 14x + 28 - \frac{53}{x + 2}$$

9.

$$\begin{array}{r} 3x + 5 \\ 2x^2 + 0x + 1 \overline{\smash{\big)}\ 6x^3 + 10x^2 + x + 8} \\ -\underline{(6x^3 + 0x^2 + 3x)} \\ 10x^2 - 2x + 8 \\ -\underline{(10x^2 + 0x + 5)} \\ -2x + 3 \end{array}$$

$$\frac{6x^3 + 10x^2 + x + 8}{2x^2 + 1} = 3x + 5 - \frac{2x - 3}{2x^2 + 1}$$

11.

$$\begin{array}{r} x \\ x^2 + 1 \overline{\smash{\big)}\ x^3 + 0x^2 + 0x - 9} \\ \underline{x^3 \quad\quad + x} \\ -x - 9 \end{array}$$

$$\frac{x^3 - 9}{x^2 + 1} = x - \frac{x + 9}{x^2 + 1}$$

13.

$$\begin{array}{r} 2x \\ x^2 - 2x + 1 \overline{\smash{\big)}\ 2x^3 - 4x^2 - 15x + 5} \\ \underline{2x^3 - 4x^2 + 2x} \\ -17x + 5 \end{array}$$

$$\frac{2x^3 - 4x^2 - 15x + 5}{(x - 1)^2} = 2x - \frac{17x - 5}{(x - 1)^2}$$

15.
$$\begin{array}{r|rrrr} 5 & 3 & -17 & 15 & -25 \\ & & 15 & -10 & 25 \\ \hline & 3 & -2 & 5 & 0 \end{array}$$

$$\frac{3x^3 - 17x^2 + 15x - 25}{x - 5} = 3x^2 - 2x + 5, \ x \neq 5$$

17.
$$\begin{array}{r|rrrr} 3 & 6 & 7 & -1 & 26 \\ & & 18 & 75 & 222 \\ \hline & 6 & 25 & 74 & 248 \end{array}$$

$$\frac{6x^3 + 7x^2 - x + 26}{x - 3} = 6x^2 + 25x + 74 + \frac{248}{x - 3}$$

19.
$$\begin{array}{r|rrrr} 2 & 9 & -18 & -16 & 32 \\ & & 18 & 0 & -32 \\ \hline & 9 & 0 & -16 & 0 \end{array}$$

$$\frac{9x^3 - 18x^2 - 16x + 32}{x - 2} = 9x^2 - 16, \ x \neq 2$$

21.
$$\begin{array}{r|rrrr} -8 & 1 & 0 & 0 & 512 \\ & & -8 & 64 & -512 \\ \hline & 1 & -8 & 64 & 0 \end{array}$$

$$\frac{x^3 + 512}{x + 8} = x^2 - 8x + 64, \ x \neq -8$$

23.
$$\begin{array}{r|rrrr} -\frac{1}{2} & 4 & 16 & -23 & -15 \\ & & -2 & -7 & 15 \\ \hline & 4 & 14 & -30 & 0 \end{array}$$

$$\frac{4x^3 + 16x^2 - 23x - 15}{x + \frac{1}{2}} = 4x^2 + 14x - 30, \ x \neq -\frac{1}{2}$$

25. $y_2 = x - 2 + \dfrac{4}{x + 2}$

$$= \frac{(x - 2)(x + 2) + 4}{x + 2}$$

$$= \frac{x^2 - 4 + 4}{x + 2}$$

$$= \frac{x^2}{x + 2}$$

$$= y_1$$

27. $y_2 = x^2 - 8 + \dfrac{39}{x^2 + 5}$

$$= \frac{(x^2 - 8)(x^2 + 5) + 39}{x^2 + 5}$$

$$= \frac{x^4 - 8x^2 + 5x^2 - 40 + 39}{x^2 + 5}$$

$$= \frac{x^4 - 3x^2 - 1}{x^2 + 5}$$

$$= y_1$$

29. $f(x) = x^3 - x^2 - 14x + 11, \quad k = 4$

$$
\begin{array}{r|rrrr}
4 & 1 & -1 & -14 & 11 \\
  &   & 4 & 12 & -8 \\
\hline
  & 1 & 3 & -2 & 3
\end{array}
$$

$f(x) = (x - 4)(x^2 + 3x - 2) + 3$

$f(4) = (0)(26) + 3 = 3$

31.
$$
\begin{array}{r|rrrr}
\sqrt{2} & 1 & 3 & -2 & -14 \\
 & & \sqrt{2} & 2 + 3\sqrt{2} & 6 \\
\hline
 & 1 & 3 + \sqrt{2} & 3\sqrt{2} & -8
\end{array}
$$

$f(x) = (x - \sqrt{2})(x^2 + (3 + \sqrt{2})x + 3\sqrt{2}) - 8$

$f(\sqrt{2}) = 0(4 + 6\sqrt{2}) - 8 = -8$

33.
$$
\begin{array}{r|rrrr}
1 - \sqrt{3} & 4 & -6 & -12 & -4 \\
 & & 4 - 4\sqrt{3} & 10 - 2\sqrt{3} & 4 \\
\hline
 & 4 & -2 - 4\sqrt{3} & -2 - 2\sqrt{3} & 0
\end{array}
$$

$f(x) = (x - 1 + \sqrt{3})[4x^2 - (2 + 4\sqrt{3})x - (2 + 2\sqrt{3})]$

$f(1 - \sqrt{3}) = 0$

35. $f(x) = 2x^3 - 7x + 3$

(a)
$$
\begin{array}{r|rrrr}
1 & 2 & 0 & -7 & 3 \\
 & & 2 & 2 & -5 \\
\hline
 & 2 & 2 & -5 & -2 \quad = f(1)
\end{array}
$$

(b)
$$
\begin{array}{r|rrrr}
-2 & 2 & 0 & -7 & 3 \\
 & & -4 & 8 & -2 \\
\hline
 & 2 & -4 & 1 & 1 \quad = f(-2)
\end{array}
$$

(c)
$$
\begin{array}{r|rrrr}
\frac{1}{2} & 2 & 0 & -7 & 3 \\
 & & 1 & \frac{1}{2} & -\frac{13}{4} \\
\hline
 & 2 & 1 & -\frac{13}{2} & -\frac{1}{4} \quad = f\left(\frac{1}{2}\right)
\end{array}
$$

(d)
$$
\begin{array}{r|rrrr}
2 & 2 & 0 & -7 & 3 \\
 & & 4 & 8 & 2 \\
\hline
 & 2 & 4 & 1 & 5 \quad = f(2)
\end{array}
$$

37. $h(x) = x^3 - 5x^2 - 7x + 4$

(a)
$$
\begin{array}{r|rrrr}
3 & 1 & -5 & -7 & 4 \\
 & & 3 & -6 & -39 \\
\hline
 & 1 & -2 & -13 & -35 \quad = h(3)
\end{array}
$$

(b)
$$
\begin{array}{r|rrrr}
2 & 1 & -5 & -7 & 4 \\
 & & 2 & -6 & -26 \\
\hline
 & 1 & -3 & -13 & -22 \quad = h(2)
\end{array}
$$

(c)
$$
\begin{array}{r|rrrr}
-2 & 1 & -5 & -7 & 4 \\
 & & -2 & 14 & -14 \\
\hline
 & 1 & -7 & 7 & -10 \quad = h(-2)
\end{array}
$$

(d)
$$
\begin{array}{r|rrrr}
-5 & 1 & -5 & -7 & 4 \\
 & & -5 & 50 & -215 \\
\hline
 & 1 & -10 & 43 & -211 \quad = h(-5)
\end{array}
$$

39.
$$
\begin{array}{r|rrrr}
2 & 1 & 0 & -7 & 6 \\
 & & 2 & 4 & -6 \\
\hline
 & 1 & 2 & -3 & 0
\end{array}
$$

$x^3 - 7x + 6 = (x - 2)(x^2 + 2x - 3)$

$\qquad\qquad = (x - 2)(x + 3)(x - 1)$

Zeros: $2, -3, 1$

41.
$$
\begin{array}{r|rrrr}
\frac{1}{2} & 2 & -15 & 27 & -10 \\
 & & 1 & -7 & 10 \\
\hline
 & 2 & -14 & 20 & 0
\end{array}
$$

$2x^3 - 15x^2 + 27x - 10$

$\qquad = \left(x - \frac{1}{2}\right)(2x^2 - 14x + 20)$

$\qquad = (2x - 1)(x - 2)(x - 5)$

Zeros: $\frac{1}{2}, 2, 5$

43. (a)
$$-2 \; \vert \begin{array}{rrrr} 2 & 1 & -5 & 2 \\ & -4 & 6 & -2 \\ \hline 2 & -3 & 1 & 0 \end{array}$$

(b) $2x^2 - 3x + 1 = (2x - 1)(x - 1)$

Remaining factors: $(2x - 1)$, $(x - 1)$

(c) $f(x) = (x + 2)(2x - 1)(x - 1)$

(d) Real zeros: $-2, \frac{1}{2}, 1$

(e)

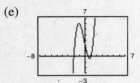

45. (a)
$$5 \; \vert \begin{array}{rrrrr} 1 & -4 & -15 & 58 & -40 \\ & 5 & 5 & -50 & 40 \\ \hline 1 & 1 & -10 & 8 & 0 \end{array}$$

$$-4 \; \vert \begin{array}{rrrr} 1 & 1 & -10 & 8 \\ & -4 & 12 & -8 \\ \hline 1 & -3 & 2 & 0 \end{array}$$

(b) $x^2 - 3x + 2 = (x - 2)(x - 1)$

Remaining factors: $(x - 2)$, $(x - 1)$

(c) $f(x) = (x - 5)(x + 4)(x - 2)(x - 1)$

(d) Real zeros: $5, -4, 2, 1$

(e)

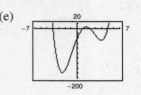

47. (a)
$$-\frac{1}{2} \; \vert \begin{array}{rrrr} 6 & 41 & -9 & -14 \\ & -3 & -19 & 14 \\ \hline 6 & 38 & -28 & 0 \end{array}$$

(b) $6x^2 + 38x - 28 = (3x - 2)(2x + 14)$

Remaining factors: $(3x - 2)$, $(x + 7)$

(c) $f(x) = (2x + 1)(3x - 2)(x + 7)$

(d) Real zeros: $-\frac{1}{2}, \frac{2}{3}, -7$

(e)

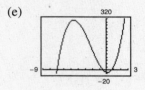

49. $f(x) = x^3 + 3x^2 - x - 3$

p = factor of -3

q = factor of 1

Possible rational zeros: $\pm 1, \pm 3$

$f(x) = x^2(x + 3) - (x + 3) = (x + 3)(x^2 - 1)$

Rational zeros: $\pm 1, -3$

51. $f(x) = 2x^4 - 17x^3 + 35x^2 + 9x - 45$

p = factor of -45

q = factor of 2

Possible rational zeros: $\pm 1, \pm 3, \pm 5, \pm 9, \pm 15, \pm 45,$

$\pm \frac{1}{2}, \pm \frac{3}{2}, \pm \frac{5}{2}, \pm \frac{9}{2}, \pm \frac{15}{2}, \pm \frac{45}{2}$

Using synthetic division, -1, 3, and 5 are zeros.

$f(x) = (x + 1)(x - 3)(x - 5)(2x - 3)$

Rational zeros: $-1, 3, 5, \frac{3}{2}$

53. $z^4 - z^3 - 2z - 4 = 0$

Possible rational zeros: $\pm 1, \pm 2, \pm 4$

$$-1 \; \vert \begin{array}{rrrrr} 1 & -1 & 0 & -2 & -4 \\ & -1 & 2 & -2 & 4 \\ \hline 1 & -2 & 2 & -4 & 0 \end{array}$$

$$2 \; \vert \begin{array}{rrrr} 1 & -2 & 2 & -4 \\ & 2 & 0 & 4 \\ \hline 1 & 0 & 2 & 0 \end{array}$$

$z^4 - z^3 - 2z - 4 = (z + 1)(z - 2)(z^2 + 2) = 0$

The only real zeros are -1 and 2. You can verify this by graphing the function $f(z) = z^4 - z^3 - 2z - 4$.

55. $2y^4 + 7y^3 - 26y^2 + 23y - 6 = 0$

Using a graphing utility and synthetic division, $1/2$, 1, and -6 are rational zeros. Hence,

$$(y + 6)(y - 1)^2(2y - 1) = 0 \implies y = -6, 1, \tfrac{1}{2}.$$

57. $4x^4 - 55x^2 - 45x + 36 = 0$

Using a graphing utility and synthetic division, 4, -3, $\tfrac{1}{2}$, $-\tfrac{3}{2}$ are rational zeros. Hence,
$(x - 4)(x + 3)(2x - 1)(2x + 3) = 0 \implies$
$x = 4, -3, \tfrac{1}{2}, -\tfrac{3}{2}.$

59. $4x^5 + 12x^4 - 11x^3 - 42x^2 + 7x + 30 = 0$

Using a graphing utility and synthetic division, 1, -1, -2, $\tfrac{3}{2}$, and $-\tfrac{5}{2}$ are rational zeros. Hence,
$(x - 1)(x + 1)(x + 2)(2x - 3)(2x + 5) = 0 \implies$
$x = 1, -1, -2, \tfrac{3}{2}, -\tfrac{5}{2}.$

61. $h(t) = t^3 - 2t^2 - 7t + 2$

(a) Zeros: $-2, 3.732, 0.268$

(b)
$$
\begin{array}{r|rrrr}
-2 & 1 & -2 & -7 & 2 \\
   &   & -2 & 8 & -2 \\
\hline
   & 1 & -4 & 1 & 0
\end{array}
$$
$t = -2$ is a zero.

(c) $h(t) = (t + 2)(t^2 - 4t + 1)$

$\qquad = (t + 2)\big[t - (\sqrt{3} + 2)\big]\big[t + (\sqrt{3} - 2)\big]$

63. $h(x) = x^5 - 7x^4 + 10x^3 + 14x^2 - 24x$

(a) $h(x) = x(x^4 - 7x^3 + 10x^2 + 14x - 24)$

From the calculator we have $x = 0, 3, 4$ and $x \approx \pm 1.414$.

(c) $h(x) = x(x - 3)(x - 4)(x^2 - 2)$

$\qquad = x(x - 3)(x - 4)(x - \sqrt{2})(x + \sqrt{2})$

The exact roots are $x = 0, 3, 4, \pm\sqrt{2}.$

(b)
$$
\begin{array}{r|rrrrr}
3 & 1 & -7 & 10 & 14 & -24 \\
  &   & 3   & -12 & -6 & 24 \\
\hline
  & 1 & -4 & -2 & 8 & 0 \\
4 & 1 & -4 & -2 & 8 &   \\
  &   & 4   & 0   & -8 &   \\
\hline
  & 1 & 0   & -2 & 0 &
\end{array}
$$

65. $f(x) = 2x^4 - x^3 + 6x^2 - x + 5$

4 variations in sign $\implies$ 4, 2 or 0 positive real zeros

$f(-x) = 2x^4 + x^3 + 6x^2 + x + 5$

0 variations in sign $\implies$ 0 negative real zeros

67. $g(x) = 4x^3 - 5x + 8$

2 variations in sign $\implies$ 2 or 0 positive real zeros

$g(-x) = -4x^3 + 5x + 8$

1 variation in sign $\implies$ 1 negative real zero

69. $f(x) = x^3 + x^2 - 4x - 4$

(a) $f(x)$ has 1 variation in sign $\implies$ 1 positive real zero.

$f(-x) = -x^3 + x^2 + 4x - 4$ has 2 variations in sign $\implies$ 2 or 0 negative real zeros.

(b) Possible rational zeros: $\pm 1, \pm 2, \pm 4$

(c)
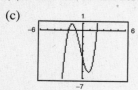

(d) Real zeros: $-2, -1, 2$

71. $f(x) = -2x^4 + 13x^3 - 21x^2 + 2x + 8$

(a) $f(x)$ has 3 variations in sign $\implies$ 3 or 1 positive real zeros.

$f(-x) = -2x^4 - 13x^3 - 21x^2 - 2x + 8$ has 1 variation in sign $\implies$ 1 negative real zero.

(b) Possible rational zeros: $\pm\tfrac{1}{2}, \pm 1, \pm 2, \pm 4, \pm 8$

(c)
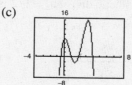

(d) Real zeros: $-\tfrac{1}{2}, 1, 2, 4$

73. $f(x) = 32x^3 - 52x^2 + 17x + 3$

   (a) $f(x)$ has 2 variations in sign $\Rightarrow$ 2 or 0 positive real zeros.

      $f(-x) = -32x^3 - 52x^2 - 17x + 3$ has 1 variation in sign $\Rightarrow$ 1 negative real zero.

   (b) Possible rational zeros: $\pm\frac{1}{32}, \pm\frac{1}{16}, \pm\frac{1}{8}, \pm\frac{1}{4},$

      $\pm\frac{1}{2}, \pm 1, \pm\frac{3}{32}, \pm\frac{3}{16}, \pm\frac{3}{8}, \pm\frac{3}{4}, \pm\frac{3}{2}, \pm 3$

   (c)

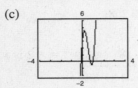

   (d) Real zeros: $1, \dfrac{3}{4}, -\dfrac{1}{8}$

---

75. $f(x) = x^4 - 4x^3 + 15$

$$
\begin{array}{r|rrrrr}
4 & 1 & -4 & 0 & 0 & 15 \\
 &   & 4 & 0 & 0 & 0 \\
\hline
 & 1 & 0 & 0 & 0 & 15
\end{array}
$$

4 is an upper bound.

$$
\begin{array}{r|rrrrr}
-1 & 1 & -4 & 0 & 0 & 15 \\
 &   & -1 & 5 & -5 & 5 \\
\hline
 & 1 & -5 & 5 & -5 & 20
\end{array}
$$

-1 is a lower bound.

Real zeros: 1.937, 3.705

---

77. $f(x) = x^4 - 4x^3 + 16x - 16$

$$
\begin{array}{r|rrrrr}
5 & 1 & -4 & 0 & 16 & -16 \\
 &   & 25 & 105 & 525 & 2705 \\
\hline
 & 5 & 21 & 105 & 541 & 2689
\end{array}
$$

5 is an upper bound.

$$
\begin{array}{r|rrrrr}
-3 & 1 & -4 & 0 & 16 & -16 \\
 &   & -3 & 21 & -63 & 141 \\
\hline
 & 1 & -7 & 21 & -47 & 125
\end{array}
$$

-3 is a lower bound.

Real zeros: $-2, 2$

---

79. $P(x) = x^4 - \frac{25}{4}x^2 + 9$

$\qquad = \frac{1}{4}(4x^4 - 25x^2 + 36)$

$\qquad = \frac{1}{4}(4x^2 - 9)(x^2 - 4)$

$\qquad = \frac{1}{4}(2x + 3)(2x - 3)(x + 2)(x - 2)$

The rational zeros are $\pm\frac{3}{2}$ and ± 2.

---

81. $f(x) = x^3 - \frac{1}{4}x^2 - x + \frac{1}{4}$

$\qquad = \frac{1}{4}(4x^3 - x^2 - 4x + 1)$

$\qquad = \frac{1}{4}[x^2(4x - 1) - 1(4x - 1)]$

$\qquad = \frac{1}{4}(4x - 1)(x^2 - 1)$

$\qquad = \frac{1}{4}(4x - 1)(x + 1)(x - 1)$

The rational zeros are $\frac{1}{4}$ and ± 1.

83. $f(x) = x^3 - 1$

$\qquad = (x - 1)(x^2 + x + 1)$

Rational zeros: 1 ($x = 1$)

Irrational zeros: 0

Matches (d).

85. $f(x) = x^3 - x$

$\qquad = x(x + 1)(x - 1)$

Rational zeros: 3 ($x = 0, \pm 1$)

Irrational zeros: 0

Matches (b).

---

87. $y = 2x^4 - 9x^3 + 5x^2 + 3x - 1$

Using the graph and synthetic division, $-1/2$ is a zero:

$$
\begin{array}{r|rrrrr}
-\frac{1}{2} & 2 & -9 & 5 & 3 & -1 \\
 &   & -1 & 5 & -5 & 1 \\
\hline
 & 2 & -10 & 10 & -2 & 0
\end{array}
$$

$y = \left(x + \frac{1}{2}\right)(2x^3 - 10x^2 + 10x - 2)$

$x = 1$ is a zero of the cubic, so $y = (2x + 1)(x - 1)(x^2 - 4x + 1)$.

For the quadratic term, use the Quadratic Formula. $x = \dfrac{4 \pm \sqrt{16 - 4}}{2} = 2 \pm \sqrt{3}$

The real zeros are $-\frac{1}{2}, 1, 2 \pm \sqrt{3}$.

89. $y = -2x^4 + 17x^3 - 3x^2 - 25x - 3$

Using the graph and synthetic division, -1 and $3/2$ are zeros:

$y = -(x + 1)(2x - 3)(x^2 - 8x - 1)$

Using the Quadratic Formula:

$$x = \frac{8 \pm \sqrt{64 + 4}}{2} = 4 \pm \sqrt{17}$$

The real zeros are $-1, 3/2, 4 \pm \sqrt{17}$.

91. (a) $P(t) = 0.0058t^3 + 0.500t^2 + 1.38t + 4.6$

(b)

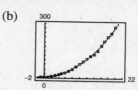

(c) The model fits the data well.

(d) For 2010, $t = 21$ and:

$$\begin{array}{r|rrrr}
21 & 0.0058 & 0.5 & 1.38 & 4.6 \\
& & 0.1218 & 13.0578 & 303.1938 \\
\hline
& 0.0058 & 0.6218 & 14.4378 & 307.7938
\end{array}$$

Hence, the population will be about 307.8 million, which seems reasonable.

93. (a) Combined length and width:

$4x + y = 120 \implies y = 120 - 4x$

Volume $= l \cdot w \cdot h = x^2 y$

$\qquad\qquad = x^2(120 - 4x)$

$\qquad\qquad = 4x^2(30 - x)$

(b)

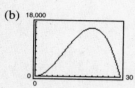

Dimensions with maximum volume:
$20 \times 20 \times 40$

(c) $\qquad\qquad 13{,}500 = 4x^2(30 - x)$

$4x^3 - 120x^2 + 13{,}500 = 0$

$x^3 - 30x^2 + 3375 = 0$

$$\begin{array}{r|rrrr}
15 & 1 & -30 & 0 & 3375 \\
& & 15 & -225 & -3375 \\
\hline
& 1 & -15 & -225 & 0
\end{array}$$

$(x - 15)(x^2 - 15x - 225) = 0$

Using the Quadratic Formula,

$$x = 15 \text{ or } \frac{15 \pm 15\sqrt{5}}{2}.$$

The value of $\dfrac{15 - 15\sqrt{5}}{2}$ is not possible because it is negative.

95. False, $-\frac{4}{7}$ is a zero of f.

97. The zeros are 1, 1, and -2. The graph falls to the right.

$y = a(x - 1)^2(x + 2) \quad a < 0$

Since $f(0) = -4$, $a = -2$.

$y = -2(x - 1)^2(x + 2) = -2x^3 + 6x - 4$

99. $f(x) = -(x + 1)(x - 1)(x + 2)(x - 2)$

101. $\begin{array}{r|rrrr}
4 & 1 & -k & 2k & -8 \\
& & 4 & 16 - 4k & 64 - 8k \\
\hline
& 1 & 4 - k & 16 - 2k & 56 - 8k
\end{array}$

Hence, $56 - 8k = 0 \implies k = 7$.

103. (a) $\dfrac{x^2 - 1}{x - 1} = x + 1, \quad x \neq 1$

(b) $\dfrac{x^3 - 1}{x - 1} = x^2 + x + 1, \quad x \neq 1$

(c) $\dfrac{x^4 - 1}{x - 1} = x^3 + x^2 + x + 1, \quad x \neq 1$

In general,

$\dfrac{x^n - 1}{x - 1} = x^{n-1} + x^{n-2} + \cdots + x + 1, \quad x \neq 1.$

107. $2x^2 + 6x + 3 = 0$

$x = \dfrac{-6 \pm \sqrt{6^2 - 4(2)(3)}}{2(2)}$

$\quad = \dfrac{-6 \pm \sqrt{12}}{4}$

$\quad = \dfrac{-3 \pm \sqrt{3}}{2}$

$x = -\dfrac{3}{2} + \dfrac{\sqrt{3}}{2}, \quad -\dfrac{3}{2} - \dfrac{\sqrt{3}}{2}$

111. $f(x) = (x - 0)(x + 1)(x - 2)(x - 5)$

$\quad = (x^2 + x)(x^2 - 7x + 10)$

$\quad = x^4 - 6x^3 + 3x^2 + 10x$

[Answer not unique]

105. $9x^2 - 25 = 0$

$(3x + 5)(3x - 5) = 0$

$x = -\dfrac{5}{3}, \dfrac{5}{3}$

109. $f(x) = (x - 0)(x + 12) = x^2 + 12x$

[Answer not unique]

Section 3.4   The Fundamental Theorem of Algebra

- ■ You should know that if f is a polynomial of degree $n > 0$, then f has at least one zero in the complex number system. (Fundamental Theorem of Algebra)
- ■ You should know that if $a + bi$ is a complex zero of a polynomial f, with real coefficients, then $a - bi$ is also a complex zero of f.
- ■ You should know the difference between a factor that is irreducible over the rationals (such as $x^2 - 7$) and a factor that is irreducible over the reals (such as $x^2 + 9$).

Vocabulary Check

1. Fundamental Theorem, Algebra

2. Linear Factorization Theorem

3. irreducible, reals

4. complex conjugate

---

1. $f(x) = x^2(x + 3)$

The three zeros are $x = 0$, $x = 0$ and $x = -3$.

3. $f(x) = (x + 9)(x + 4i)(x - 4i)$

Zeros: $-9, \pm 4i$

5. $f(x) = x^3 - 4x^2 + x - 4 = x^2(x - 4) + 1(x - 4)$

$\qquad = (x - 4)(x^2 + 1)$

Zeros: $4, \pm i$

The only real zero of $f(x)$ is $x = 4$. This corresponds to the x-intercept of $(4, 0)$ on the graph.

7. $f(x) = x^4 + 4x^2 + 4 = (x^2 + 2)^2$

Zeros: $\pm \sqrt{2}i, \pm \sqrt{2}i$

$f(x)$ has no real zeros and the graph of $f(x)$ has no x-intercepts.

9. $h(x) = x^2 - 4x + 1$

h has no rational zeros. By the Quadratic Formula, the zeros are

$$x = \frac{4 \pm \sqrt{16 - 4}}{2} = 2 \pm \sqrt{3}.$$

$$h(x) = \left[x - \left(2 + \sqrt{3}\right)\right]\left[x - \left(2 - \sqrt{3}\right)\right]$$

$$\qquad = \left(x - 2 - \sqrt{3}\right)\left(x - 2 + \sqrt{3}\right)$$

11. $f(x) = x^2 - 12x + 26$

f has no rational zeros. By the Quadratic Formula, the zeros are

$$x = \frac{12 \pm \sqrt{(-12)^2 - 4(26)}}{2} = 6 \pm \sqrt{10}.$$

$$f(x) = \left[x - \left(6 + \sqrt{10}\right)\right]\left[x - \left(6 - \sqrt{10}\right)\right]$$

$$\qquad = \left(x - 6 - \sqrt{10}\right)\left(x - 6 + \sqrt{10}\right)$$

13. $f(x) = x^2 + 25$

$\qquad = (x + 5i)(x - 5i)$

The zeros of $f(x)$ are $x = \pm 5i$.

15. $f(x) = 16x^4 - 81$

$\qquad = (4x^2 - 9)(4x^2 + 9)$

$\qquad = (2x - 3)(2x + 3)(2x + 3i)(2x - 3i)$

Zeros: $\pm \frac{3}{2}, \pm \frac{3}{2}i$

17. $f(z) = z^2 - z + 56$

$$z = \frac{1 \pm \sqrt{1 - 4(56)}}{2}$$

$$\quad = \frac{1 \pm \sqrt{-223}}{2}$$

$$\quad = \frac{1}{2} \pm \frac{\sqrt{223}}{2}i$$

$$f(z) = \left(z - \frac{1}{2} + \frac{\sqrt{223}\,i}{2}\right)\left(z - \frac{1}{2} - \frac{\sqrt{223}\,i}{2}\right)$$

19. $f(x) = x^4 + 10x^2 + 9$

$\qquad = (x^2 + 1)(x^2 + 9)$

$\qquad = (x + i)(x - i)(x + 3i)(x - 3i)$

The zeros of $f(x)$ are $x = \pm i$ and $x = \pm 3i$.

21. $f(x) = 3x^3 - 5x^2 + 48x - 80$

Using synthetic division, $\frac{5}{3}$ is a zero:

$$\begin{array}{r|rrrr} \tfrac{5}{3} & 3 & -5 & 48 & -80 \\ & & 5 & 0 & 80 \\ \hline & 3 & 0 & 48 & 0 \end{array}$$

$f(x) = \left(x - \frac{5}{3}\right)(3x^2 + 48)$

$\qquad = (3x - 5)(x^2 + 16)$

$\qquad = (3x - 5)(x + 4i)(x - 4i)$

The zeros are $\frac{5}{3}, 4i, -4i$.

23. $f(t) = t^3 - 3t^2 - 15t + 125$

Possible rational zeros: $\pm 1, \pm 5, \pm 25, \pm 125$

$$\begin{array}{r|rrrr} -5 & 1 & -3 & -15 & 125 \\ & & -5 & 40 & -125 \\ \hline & 1 & -8 & 25 & 0 \end{array}$$

By the Quadratic Formula, the zeros of $t^2 - 8t + 25$ are

$$t = \frac{8 \pm \sqrt{64 - 100}}{2} = 4 \pm 3i.$$

The zeros of $f(t)$ are $t = -5$ and $t = 4 \pm 3i$.

$$f(t) = [t - (-5)][t - (4 + 3i)][t - (4 - 3i)]$$

$$\qquad = (t + 5)(t - 4 - 3i)(t - 4 + 3i)$$

25. $f(x) = 5x^3 - 9x^2 + 28x + 6$

Possible rational zeros:

$$\pm 6, \pm\frac{6}{5}, \pm 3, \pm\frac{3}{5}, \pm 2, \pm\frac{2}{5}, \pm 1, \pm\frac{1}{5}$$

$$
\begin{array}{r|rrrr}
-\frac{1}{5} & 5 & -9 & 28 & 6 \\
 & & -1 & 2 & -6 \\
\hline
 & 5 & -10 & 30 & 0
\end{array}
$$

By the Quadratic Formula, the zeros of
$5x^2 - 10x + 30$ are those of $x^2 - 2x + 6$:

$$x = \frac{2 \pm \sqrt{4 - 4(6)}}{2} = 1 \pm \sqrt{5}\,i$$

Zeros: $-\frac{1}{5}, 1 \pm \sqrt{5}\,i$

$$f(x) = 5\left(x + \frac{1}{5}\right)\left(x - \left(1 + \sqrt{5}\,i\right)\right)\left(x - \left(1 - \sqrt{5}\,i\right)\right)$$

$$= (5x + 1)\left(x - 1 - \sqrt{5}\,i\right)\left(x - 1 + \sqrt{5}\,i\right)$$

27. $g(x) = x^4 - 4x^3 + 8x^2 - 16x + 16$

Possible rational zeros: $\pm 1, \pm 2, \pm 4, \pm 8, \pm 16$

$$
\begin{array}{r|rrrrr}
2 & 1 & -4 & 8 & -16 & 16 \\
 & & 2 & -4 & 8 & -16 \\
\hline
2 & 1 & -2 & 4 & -8 & 0 \\
 & & 2 & 0 & 8 & \\
\hline
 & 1 & 0 & 4 & 0 &
\end{array}
$$

$g(x) = (x - 2)(x - 2)(x^2 + 4)$

$\quad = (x - 2)^2(x + 2i)(x - 2i)$

The zeros of g are 2, 2, and $\pm 2i$.

29. (a) $f(x) = x^2 - 14x + 46$.
By the Quadratic Formula,

$$x = \frac{14 \pm \sqrt{(-14)^2 - 4(46)}}{2} = 7 \pm \sqrt{3}.$$

The zeros are $7 + \sqrt{3}$ and $7 - \sqrt{3}$.

(b) $f(x) = \left[x - \left(7 + \sqrt{3}\right)\right]\left[x - \left(7 - \sqrt{3}\right)\right]$

$\quad = \left(x - 7 - \sqrt{3}\right)\left(x - 7 + \sqrt{3}\right)$

(c) x-intercepts: $\left(7 + \sqrt{3}, 0\right)$ and $\left(7 - \sqrt{3}, 0\right)$

(d)

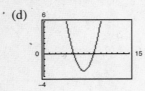

31. (a) $f(x) = 2x^3 - 3x^2 + 8x - 12$

$\quad = (2x - 3)(x^2 + 4)$

The zeros are $\frac{3}{2}, \pm 2i$.

(b) $f(x) = (2x - 3)(x + 2i)(x - 2i)$

(c) x-intercept: $\left(\frac{3}{2}, 0\right)$

(d)

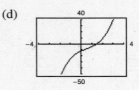

33. (a) $f(x) = x^3 - 11x + 150$

$\quad = (x + 6)(x^2 - 6x + 25)$

Use the Quadratic Formula to find the zeros of
$x^2 - 6x + 25$.

$$x = \frac{6 \pm \sqrt{(-6)^2 - 4(25)}}{2} = 3 \pm 4i.$$

The zeros are $-6, 3 + 4i$, and $3 - 4i$.

(b) $f(x) = (x + 6)(x - 3 + 4i)(x - 3 - 4i)$

(c) x-intercept: $(-6, 0)$

(d)

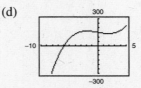

35. (a) $f(x) = x^4 + 25x^2 + 144$

$= (x^2 + 9)(x^2 + 16)$

The zeros are $\pm 3i, \pm 4i$.

(b) $f(x) = (x^2 + 9)(x^2 + 16)$

$= (x + 3i)(x - 3i)(x + 4i)(x - 4i)$

(c) No x-intercepts

(d)

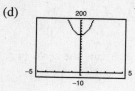

37. $f(x) = (x - 2)(x - i)(x + i)$

$= (x - 2)(x^2 + 1)$

$= (x^3 - 2x^2 + x - 2)$

Note that $f(x) = a(x^3 - 2x^2 + x - 2)$, where a is any nonzero real number, has zeros 2, $\pm i$.

39. $f(x) = (x - 2)^2(x - 4 - i)(x - 4 + i)$

$= (x - 2)^2(x^2 - 8x + 16 + 1)$

$= (x^2 - 4x + 4)(x^2 - 8x + 17)$

$= x^4 - 12x^3 + 53x^2 - 100x + 68$

Note that $f(x) = a(x^4 - 12x^3 + 53x^2 - 100x + 68)$, where a is any nonzero real number, has zeros 2, 2, $4 \pm i$.

41. Because $1 + \sqrt{2}i$ is a zero, so is $1 - \sqrt{2}i$.

$f(x) = (x - 0)(x + 5)(x - 1 - \sqrt{2}i)(x - 1 + \sqrt{2}i)$

$= (x^2 + 5x)(x^2 - 2x + 1 + 2)$

$= (x^2 + 5x)(x^2 - 2x + 3)$

$= x^4 + 3x^3 - 7x^2 + 15x$

Note that $f(x) = a(x^4 + 3x^3 - 7x^2 + 15x)$, where a is any nonzero real number, has zeros 0, -5, $1 \pm \sqrt{2}i$.

43. (a) $f(x) = a(x - 1)(x + 2)(x - 2i)(x + 2i)$

$= a(x - 1)(x + 2)(x^2 + 4)$

$f(-1) = 10 = a(-2)(1)(5) \Rightarrow a = -1$

$f(x) = -(x - 1)(x + 2)(x - 2i)(x + 2i)$

(b) $f(x) = -(x - 1)(x + 2)(x^2 + 4)$

$= -(x^2 + x - 2)(x^2 + 4)$

$= -x^4 - x^3 - 2x^2 - 4x + 8$

45. (a) $f(x) = a(x + 1)(x - 2 - \sqrt{5}i)(x - 2 + \sqrt{5}i)$

$= a(x + 1)(x^2 - 4x + 4 + 5)$

$= a(x + 1)(x^2 - 4x + 9)$

$f(-2) = 42 = a(-1)(4 + 8 + 9) \Rightarrow a = -2$

$f(x) = -2(x + 1)(x - 2 - \sqrt{5}i)(x - 2 + \sqrt{5}i)$

(b) $f(x) = -2(x + 1)(x^2 - 4x + 9)$

$= -2x^3 + 6x^2 - 10x - 18$

47. $f(x) = x^4 - 6x^2 - 7$

(a) $f(x) = (x^2 - 7)(x^2 + 1)$

(b) $f(x) = (x - \sqrt{7})(x + \sqrt{7})(x^2 + 1)$

(c) $f(x) = (x - \sqrt{7})(x + \sqrt{7})(x + i)(x - i)$

49. $f(x) = x^4 - 2x^3 - 3x^2 + 12x - 18$

(a) $f(x) = (x^2 - 6)(x^2 - 2x + 3)$

(b) $f(x) = (x + \sqrt{6})(x - \sqrt{6})(x^2 - 2x + 3)$

(c) $f(x) = (x + \sqrt{6})(x - \sqrt{6})(x - 1 - \sqrt{2}i)(x - 1 + \sqrt{2}i)$

51. $f(x) = 2x^3 + 3x^2 + 50x + 75$

Since $5i$ is a zero, so is $-5i$.

$$
\begin{array}{r|rrrr}
5i & 2 & 3 & 50 & 75 \\
 & & 10i & -50+15i & -75 \\
\hline
 & 2 & 3+10i & 15i & 0
\end{array}
$$

$$
\begin{array}{r|rrr}
-5i & 2 & 3+10i & 15i \\
 & & -10i & -15i \\
\hline
 & 2 & 3 & 0
\end{array}
$$

The zero of $2x + 3$ is $x = -\frac{3}{2}$. The zeros of f are $x = -\frac{3}{2}$ and $x = \pm 5i$.

Alternate Solution

Since $x = \pm 5i$ are zeros of $f(x)$, $(x + 5i)(x - 5i) = x^2 + 25$ is a factor of $f(x)$. By long division we have:

$$
\begin{array}{r}
2x + 3 \\
x^2 + 0x + 25 \overline{)\, 2x^3 + 3x^2 + 50x + 75} \\
\underline{2x^3 + 0x^2 + 50x} \\
3x^2 + 0x + 75 \\
\underline{3x^2 + 0x + 75} \\
0
\end{array}
$$

Thus, $f(x) = (x^2 + 25)(2x + 3)$ and the zeros of f are $x = \pm 5i$ and $x = -\frac{3}{2}$.

53. $g(x) = x^3 - 7x^2 - x + 87$. Since $5 + 2i$ is a zero, so is $5 - 2i$.

$$
\begin{array}{r|rrrr}
5+2i & 1 & -7 & -1 & 87 \\
 & & 5+2i & -14+6i & -87 \\
\hline
 & 1 & -2+2i & -15+6i & 0
\end{array}
$$

$$
\begin{array}{r|rrr}
5-2i & 1 & -2+2i & -15+6i \\
 & & 5-2i & 15-6i \\
\hline
 & 1 & 3 & 0
\end{array}
$$

The zero of $x + 3$ is $x = -3$.

The zeros of f are $-3, 5 \pm 2i$.

55. $h(x) = 3x^3 - 4x^2 + 8x + 8$. Since $1 - \sqrt{3}i$ is a zero, so is $1 + \sqrt{3}i$.

$$
\begin{array}{r|rrrr}
1-\sqrt{3}i & 3 & -4 & 8 & 8 \\
 & & 3-3\sqrt{3}i & -10-2\sqrt{3}i & -8 \\
\hline
 & 3 & -1-3\sqrt{3}i & -2-2\sqrt{3}i & 0
\end{array}
$$

$$
\begin{array}{r|rrr}
1+\sqrt{3}i & 3 & -1-3\sqrt{3}i & -2-2\sqrt{3}i \\
 & & 3+3\sqrt{3}i & 2+2\sqrt{3}i \\
\hline
 & 3 & 2 & 0
\end{array}
$$

The zero of $3x + 2$ is $x = -\frac{2}{3}$. The zeros of h are $x = -\frac{2}{3}, 1 \pm \sqrt{3}i$.

57. $h(x) = 8x^3 - 14x^2 + 18x - 9$. Since $\frac{1}{2}\left(1 - \sqrt{5}i\right)$ is a zero, so is $\frac{1}{2}\left(1 + \sqrt{5}i\right)$.

$$
\begin{array}{r|rrrr}
\frac{1}{2}\left(1-\sqrt{5}i\right) & 8 & -14 & 18 & -9 \\
 & & 4-4\sqrt{5}i & -15+3\sqrt{5}i & 9 \\
\hline
 & 8 & -10-4\sqrt{5}i & 3+3\sqrt{5}i & 0
\end{array}
$$

$$
\begin{array}{r|rrr}
\frac{1}{2}\left(1+\sqrt{5}i\right) & 8 & -10-4\sqrt{5}i & 3+3\sqrt{5}i \\
 & & 4+4\sqrt{5}i & -3-3\sqrt{5}i \\
\hline
 & 8 & -6 & 0
\end{array}
$$

The zero of $8x - 6$ is $x = \frac{3}{4}$. The zeros of h are $x = \frac{3}{4}, \frac{1}{2}\left(1 \pm \sqrt{5}i\right)$.

59. $f(x) = x^4 + 3x^3 - 5x^2 - 21x + 22$

(a) The root feature yields the real roots 1 and 2, and the complex roots $-3 \pm 1.414i$.

(b) By synthetic division:

$$
\begin{array}{r|rrrrr}
1 & 1 & 3 & -5 & -21 & 22 \\
 & & 1 & 4 & -1 & -22 \\
\hline
 & 1 & 4 & -1 & -22 & 0
\end{array}
$$

$$
\begin{array}{r|rrrr}
2 & 1 & 4 & -1 & -22 \\
 & & 2 & 12 & 22 \\
\hline
 & 1 & 6 & 11 & 0
\end{array}
$$

The complex roots of $x^2 + 6x + 11$ are $x = \dfrac{-6 \pm \sqrt{6^2 - 4(11)}}{2} = -3 \pm \sqrt{2}i$.

61. $h(x) = 8x^3 - 14x^2 + 18x - 9$

(a) The root feature yields the real root 0.75, and the complex roots $0.5 \pm 1.118i$.

(b) By synthetic division:

$$
\begin{array}{r|rrrr}
\frac{3}{4} & 8 & -14 & 18 & -9 \\
 & & 6 & -6 & 9 \\
\hline
 & 8 & -8 & 12 & 0 \\
\end{array}
$$

The complex roots of $8x^2 - 8x + 12$ are

$$x = \frac{8 \pm \sqrt{64 - 4(8)(12)}}{2(8)} = \frac{1}{2} \pm \frac{\sqrt{5}}{2}i.$$

63. $-16t^2 + 48t = 64, \quad 0 \le t \le 3$

$-16t^2 + 48t - 64 = 0$

$$t = \frac{-48 \pm \sqrt{1792i}}{-32}$$

Since the roots are imaginary, the ball never will reach a height of 64 feet. You can verify this graphically by observing that $y_1 = -16t^2 + 48t$ and $y_2 = 64$ do not intersect.

65. False, a third degree polynomial must have at least one real zero.

67. $f(x) = x^4 - 4x^2 + k$

(a) f has two real zeros each of multiplicity 2 for $k = 4$: $f(x) = x^4 - 4x^2 + 4 = (x^2 - 2)^2$.

(b) f has two real zeros and two complex zeros if $k < 0$.

69. $f(x) = x^2 - 7x - 8 = \left(x^2 - 7x + \frac{49}{4}\right) - 8 - \frac{49}{4}$

$$= \left(x - \frac{7}{2}\right)^2 - \frac{81}{4}$$

Vertex: $\left(\frac{7}{2}, -\frac{81}{4}\right)$

$f(x) = (x - 8)(x + 1)$

Intercepts: $(8, 0), (-1, 0), (0, -8)$

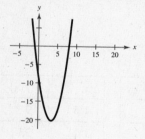

71. $f(x) = 6x^2 + 5x - 6 = (3x - 2)(2x + 3)$

Intercepts: $\left(\frac{2}{3}, 0\right), \left(-\frac{3}{2}, 0\right), (0, -6)$

$f(x) = 6x^2 + 5x - 6$

$$= 6\left(x^2 + \frac{5}{6}x + \frac{25}{144}\right) - 6 - \frac{25}{24}$$

$$= 6\left(x + \frac{5}{12}\right)^2 - \frac{169}{24}$$

Vertex: $\left(-\frac{5}{12}, -\frac{169}{24}\right)$

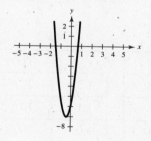

Section 3.5   Rational Functions and Asymptotes

■   You should know the following basic facts about rational functions.

(a) A function of the form $f(x) = P(x)/Q(x)$, $Q(x) \neq 0$, where $P(x)$ and $Q(x)$ are polynomials, is called a rational function.

(b) The domain of a rational function is the set of all real numbers except those which make the denominator zero.

(c) If $f(x) = P(x)/Q(x)$ is in reduced form, and a is a value such that $Q(a) = 0$, then the line $x = a$ is a vertical asymptote of the graph of f. $f(x) \to \infty$ or $f(x) \to -\infty$ as $x \to a$.

(d) The line $y = b$ is a horizontal asymptote of the graph of f if $f(x) \to b$ as $x \to \infty$ or $x \to -\infty$.

(e) Let $f(x) = \dfrac{P(x)}{Q(x)} = \dfrac{a_n x^n + a_{n-1} x^{n-1} + \cdots + a_1 x + a_0}{b_m x^m + b_{m-1} x^{m-1} + \cdots + b_1 x + b_0}$ where $P(x)$ and $Q(x)$ have no common factors.

1. If $n < m$, then the x-axis $(y = 0)$ is a horizontal asymptote.

2. If $n = m$, then $y = \dfrac{a_n}{b_m}$ is a horizontal asymptote.

3. If $n > m$, then there are no horizontal asymptotes.

Vocabulary Check

1. rational functions   2. vertical asymptote   3. horizontal asymptote

---

1. $f(x) = \dfrac{1}{x-1}$

(a) Domain: all $x \neq 1$

(b)

x	$f(x)$
0.5	-2
0.9	-10
0.99	-100
0.999	-1000

x	$f(x)$
1.5	2
1.1	10
1.01	100
1.001	1000

x	$f(x)$
5	0.25
10	$0.\overline{1}$
100	$0.0\overline{1}$
1000	$0.00\overline{1}$

x	$f(x)$
-5	$-0.\overline{16}$
-10	$-0.\overline{09}$
-100	$-0.\overline{0099}$
-1000	$-0.\overline{00099}$

(c) f approaches $-\infty$ from the left of 1 and ∞ from the right of 1.

3. $f(x) = \dfrac{3x}{|x-1|}$

(a) Domain: all $x \neq 1$

(b)

x	$f(x)$
0.5	3
0.9	27
0.99	297
0.999	2997

x	$f(x)$
1.5	9
1.1	33
1.01	303
1.001	3003

x	$f(x)$
5	3.75
10	$3.\overline{33}$
100	$3.\overline{03}$
1000	$3.\overline{003}$

x	$f(x)$
-5	-2.5
-10	-2.727
-100	-2.970
-1000	-2.997

(c) f approaches ∞ from both the left and the right of 1.

5. $f(x) = \dfrac{3x^2}{x^2 - 1}$

(a) Domain: all $x \neq \pm 1$

(b)

x	$f(x)$
0.5	-1
0.9	-12.79
0.99	-147.8
0.999	-1498

x	$f(x)$
1.5	5.4
1.1	17.29
1.01	152.3
1.001	1502.3

x	$f(x)$
5	3.125
10	$3.\overline{03}$
100	$3.\overline{0003}$
1000	3

x	$f(x)$
-5	3.125
-10	$3.\overline{03}$
-100	$3.\overline{0003}$
-1000	3

(c) f approaches $-\infty$ from the left of 1, and ∞ from the right of 1. f approaches ∞ from the left of -1, and $-\infty$ from the right of -1.

7. $f(x) = \dfrac{2}{x + 2}$

Vertical asymptote: $x = -2$

Horizontal asymptote: $y = 0$

Matches graph (a).

9. $f(x) = \dfrac{4x + 1}{x}$

Vertical asymptote: $x = 0$

Horizontal asymptote: $y = 4$

Matches graph (c).

11. $f(x) = \dfrac{x - 2}{x - 4}$

Vertical asymptote: $x = 4$

Horizontal asymptote: $y = 1$

Matches graph (b).

13. $f(x) = \dfrac{1}{x^2}$

(a) Vertical asymptote: $x = 0$

   Horizontal asymptote: $y = 0$

(b) Holes: none

15. $f(x) = \dfrac{x(2 + x)}{2x - x^2} = \dfrac{2 + x}{2 - x}, \; x \neq 0$

(a) Vertical asymptote: $x = 2$

   Horizontal asymptote: $y = -1$

(b) Hole at $x = 0$: $(0, 1)$

17. $f(x) = \dfrac{x^2 - 25}{x^2 + 5x}$

$= \dfrac{(x - 5)(x + 5)}{x(x + 5)}$

$= \dfrac{x - 5}{x}, \; x \neq -5$

(a) Vertical asymptote: $x = 0$

   Horizontal asymptote: $y = 1$

(b) Hole at $x = -5$: $(-5, 2)$

19. $f(x) = \dfrac{3x^2 + x - 5}{x^2 + 1}$

(a) Domain: all real numbers

(b) Vertical asymptote: none

   Horizontal asymptote: $y = 3$

(c)

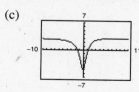

21. $f(x) = \dfrac{x - 3}{|x|}$

(a) Domain: all real numbers except $x = 0$

(b) Vertical asymptote: $x = 0$

Horizontal asymptote:

$y = 1$ to the right

$y = -1$ (to the left)

(c)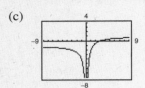

23. $f(x) = \dfrac{x^2 - 16}{x - 4},\ g(x) = x + 4$

(a) Domain of f: all real numbers except 4

Domain of g: all real numbers

(b) $f(x) = \dfrac{(x - 4)(x + 4)}{x - 4} = x + 4,\ x \neq 4$

f has no vertical asymptotes.

(c) Hole at $x = 4$

(d)

x	1	2	3	4	5	6	7
$f(x)$	5	6	7	Undef.	9	10	11
$g(x)$	5	6	7	8	9	20	11

(e) f and g differ at $x = 4$, where f is undefined.

25. $f(x) = \dfrac{x^2 - 1}{x^2 - 2x - 3} = \dfrac{(x - 1)(x + 1)}{(x + 1)(x - 3)},\ g(x) = \dfrac{x - 1}{x - 3}$

(a) Domain of f: all real numbers except $-1, 3$

Domain of g: all real numbers except 3

(b) $f(x) = \dfrac{(x - 1)(x + 1)}{(x + 1)(x - 3)} = \dfrac{x - 1}{x - 3},\ x \neq -1$

f has a vertical asymptote at $x = 3$.

(c) The graph has a hole at $x = -1$.

(d)

x	-2	-1	0	1	2	3	4
$f(x)$	$\frac{3}{5}$	Undef.	$\frac{1}{3}$	0	-1	Undef.	3
$g(x)$	$\frac{3}{5}$	$\frac{1}{2}$	$\frac{1}{3}$	0	-1	Undef.	3

(e) f and g differ at $x = -1$, where f is undefined.

27. $f(x) = 4 - \dfrac{1}{x}$

(a) As $x \rightarrow \pm\infty, f(x) \rightarrow 4$

(b) As $x \rightarrow \infty, f(x) \rightarrow 4$ but is less than 4.

(c) As $x \rightarrow -\infty, f(x) \rightarrow 4$ but is greater than 4.

29. $f(x) = \dfrac{2x - 1}{x - 3}$

(a) As $x \rightarrow \pm\infty, f(x) \rightarrow 2$.

(b) As $x \rightarrow \infty, f(x) \rightarrow 2$ but is greater than 2.

(c) As $x \rightarrow -\infty, f(x) \rightarrow 2$ but is less than 2.

31. $g(x) = \dfrac{x^2 - 4}{x + 3} = \dfrac{(x - 2)(x + 2)}{x + 3}$

The zeros of g are the zeros of the numerator: $x = \pm 2$

33. $f(x) = 1 - \dfrac{2}{x - 5} = \dfrac{x - 7}{x - 5}$

The zero of f corresponds to the zero of the numerator and is $x = 7$.

35. $g(x) = \dfrac{x^2 - 2x - 3}{x^2 + 1} = \dfrac{(x - 3)(x + 1)}{x^2 + 1} = 0$

Zeros: $x = -1, 3$

37. $f(x) = \dfrac{2x^2 - 5x + 2}{2x^2 - 7x + 3} = \dfrac{(2x - 1)(x - 2)}{(2x - 1)(x - 3)} = \dfrac{x - 2}{x - 3}$,

$x \neq \dfrac{1}{2}$

Zero: $x = 2\ \left(x = \dfrac{1}{2} \text{ is not in the domain.}\right)$

39. $C = \dfrac{255p}{100 - p}, \ 0 \le p < 100$

(a) $C(10) = \dfrac{255(10)}{100 - 10} \approx 28.33$ million dollars

(b) $C(40) = \dfrac{255(40)}{100 - 40} = 170$ million dollars

(c) $C(75) = \dfrac{255(75)}{100 - 75} = 765$ million dollars

(d)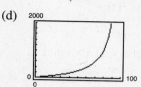

(e) $C \to \infty$ as $x \to 100$. No, it would not be possible to remove 100% of the pollutants.

41. (a) Use data $\left(16, \dfrac{1}{3}\right), \left(32, \dfrac{1}{4.7}\right), \left(44, \dfrac{1}{9.8}\right),$

$\left(50, \dfrac{1}{19.7}\right), \left(60, \dfrac{1}{39.4}\right).$

$\dfrac{1}{y} = -0.007x + 0.445$

$y = \dfrac{1}{0.445 - 0.007x}$

(b)

x	16	32	44	50	60
y	3.0	4.5	7.3	10.5	40

(Answers will vary.)

(c) No, the function is negative for $x = 70$.

43. $N = \dfrac{20(5 + 3t)}{1 + 0.04t}, \ 0 \le t$

(a)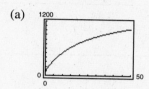

(b) $N(5) \approx 333$ deer

$N(10) = 500$ deer

$N(25) = 800$ deer

(c) The herd is limited by the horizontal asymptote:

$N = \dfrac{60}{0.04} = 1500$ deer

45. False. A rational function can have at most n vertical asymptotes, where n is the degree of the denominator.

47. There are vertical asymptotes at $x = \pm 3$, and zeros at $x = \pm 2$. Matches (b).

49. $f(x) = \dfrac{x - 1}{x^3 - 8}$

51. $f(x) = \dfrac{2(x + 3)(x - 3)}{(x + 2)(x - 1)} = \dfrac{2x^2 - 18}{x^2 + x - 2}$

53. $y - 2 = \dfrac{-1 - 2}{0 - 3}(x - 3) = 1(x - 3)$

$y = x - 1$

$y - x + 1 = 0$

55. $y - 7 = \dfrac{10 - 7}{3 - 2}(x - 2) = 3(x - 2)$

$y = 3x + 1$

$3x - y + 1 = 0$

57.

$$
\begin{array}{r}
x + 9 \\
x - 4 \overline{\smash{)}\, x^2 + 5x + 6} \\
\underline{x^2 - 4x} \\
9x + 6 \\
\underline{9x - 36} \\
42
\end{array}
$$

$\dfrac{x^2 + 5x + 6}{x - 4} = x + 9 + \dfrac{42}{x - 4}$

59.

$$
\begin{array}{r}
2x^2 - 9 \\
x^2 + 5 \overline{\smash{)}\, 2x^4 + 0x^3 + x^2 + 0x - 11} \\
\underline{2x^4 + 10x^2} \\
-9x^2 - 11 \\
\underline{-9x^2 - 45} \\
34
\end{array}
$$

$\dfrac{2x^4 + x^2 - 11}{x^2 + 5} = 2x^2 - 9 + \dfrac{34}{x^2 + 5}$

Section 3.6   Graphs of Rational Functions

■   You should be able to graph $f(x) = \dfrac{p(x)}{q(x)}$.

(a) Find the x- and y-intercepts.

(b) Find any vertical or horizontal asymptotes.

(c) Plot additional points.

(d) If the degree of the numerator is one more than the degree of the denominator, use long division to find the slant asymptote.

Vocabulary Check

1. slant, asymptote

2. vertical

---

1. $g(x) = \dfrac{2}{x} + 1$

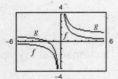

Vertical shift one unit upward

3. $g(x) = -\dfrac{2}{x}$

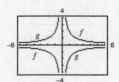

Reflection in the x-axis

5. $g(x) = \dfrac{2}{x^2} - 2$

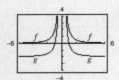

Vertical shift two units downward

7. $g(x) = \dfrac{2}{(x - 2)^2}$

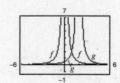

Horizontal shift two units to the right

9. $f(x) = \dfrac{1}{x + 2}$

y-intercept: $\left(0, \dfrac{1}{2}\right)$

Vertical asymptote:
$x = -2$

Horizontal asymptote:
$y = 0$

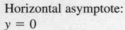

x	-4	-3	-1	0	1
y	$-\frac{1}{2}$	-1	1	$\frac{1}{2}$	$\frac{1}{3}$

11. $C(x) = \dfrac{5 + 2x}{1 + x} = \dfrac{2x + 5}{x + 1}$

x-intercept: $\left(-\dfrac{5}{2}, 0\right)$

y-intercept: $(0, 5)$

Vertical asymptote:
$x = -1$

Horizontal asymptote:
$y = 2$

x	-4	-3	-2	0	1	2
$C(x)$	1	$\frac{1}{2}$	-1	5	$\frac{7}{2}$	3

13. $f(t) = \dfrac{1 - 2t}{t} = -\dfrac{2t - 1}{t}$

t-intercept: $\left(\dfrac{1}{2}, 0\right)$

Vertical asymptote: $t = 0$

Horizontal asymptote: $y = -2$

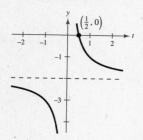

x	-2	-1	$\frac{1}{2}$	1	2
y	$-\frac{5}{2}$	-3	0	-1	$-\frac{3}{2}$

15. $f(x) = \dfrac{x^2}{x^2 - 4}$

Intercept: $(0, 0)$

Vertical asymptotes: $x = 2,\ x = -2$

Horizontal asymptote: $y = 1$

y-axis symmetry

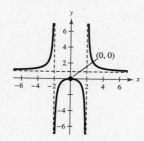

x	-4	-1	0	-1	4
y	$\frac{4}{3}$	$-\frac{1}{3}$	0	$-\frac{1}{3}$	$\frac{4}{3}$

17. $f(x) = \dfrac{x}{x^2 - 1} = \dfrac{x}{(x + 1)(x - 1)}$

Intercept: $(0, 0)$

Vertical asymptotes: $x = 1$ and $x = -1$

Horizontal asymptote: $y = 0$

Origin symmetry

x	-3	-2	$-\frac{1}{2}$	0	$\frac{1}{2}$	2	3	4
y	$-\frac{3}{8}$	$-\frac{2}{3}$	$\frac{2}{3}$	0	$-\frac{2}{3}$	$\frac{2}{3}$	$\frac{3}{8}$	$\frac{4}{15}$

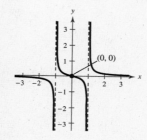

19. $g(x) = \dfrac{4(x + 1)}{x(x - 4)}$

Intercept: $(-1, 0)$

Vertical asymptotes: $x = 0$ and $x = 4$

Horizontal asymptote: $y = 0$

x	-2	-1	1	2	3	5	6
y	$-\frac{1}{3}$	0	$-\frac{8}{3}$	-3	$-\frac{16}{3}$	$\frac{24}{5}$	$\frac{7}{3}$

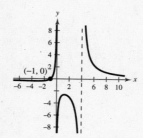

21. $f(x) = \dfrac{3x}{x^2 - x - 2} = \dfrac{3x}{(x + 1)(x - 2)}$

Intercept: $(0, 0)$

Vertical asymptotes: $x = -1, 2$

Horizontal asymptote: $y = 0$

x	-3	0	1	3	4
y	$-\frac{9}{10}$	0	$-\frac{3}{2}$	$\frac{9}{4}$	$\frac{6}{5}$

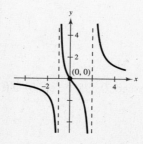

23. $f(x) = \dfrac{x^2 + 3x}{x^2 + x - 6} = \dfrac{x(x + 3)}{(x - 2)(x + 3)} = \dfrac{x}{x - 2}$,

$x \neq -3$

Intercept: $(0, 0)$

Vertical asymptote: $x = 2$

(There is a hole at $x = -3$.)

Horizontal asymptote: $y = 1$

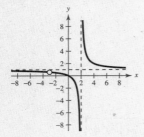

25. $f(x) = \dfrac{x^2 - 1}{x + 1} = \dfrac{(x + 1)(x - 1)}{x + 1} = x - 1$,

$x \neq -1$

The graph is a line, with a hole at $x = -1$.

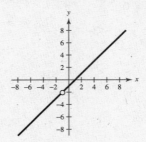

27. $f(x) = \dfrac{2 + x}{1 - x} = -\dfrac{x + 2}{x - 1}$

Vertical asymptote: $x = 1$

Horizontal asymptote: $y = -1$

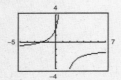

Domain: $x \neq 1$ or $(-\infty, 1) \cup (1, \infty)$

29. $f(t) = \dfrac{3t + 1}{t}$

Vertical asymptote: $t = 0$

Horizontal asymptote: $y = 3$

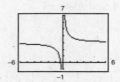

Domain: $t \neq 0$ or $(-\infty, 0) \cup (0, \infty)$

31. $h(t) = \dfrac{4}{t^2 + 1}$

Domain: all real numbers OR $(-\infty, \infty)$

Horizontal asymptote: $y = 0$

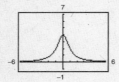

33. $f(x) = \dfrac{x + 1}{x^2 - x - 6} = \dfrac{x + 1}{(x - 3)(x + 2)}$

Domain: all real numbers except $x = 3, -2$

Vertical asymptotes: $x = 3$, $x = -2$

Horizontal asymptote: $y = 0$

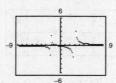

35. $f(x) = \dfrac{20x}{x^2 + 1} - \dfrac{1}{x} = \dfrac{19x^2 - 1}{x(x^2 + 1)}$

Domain: all real numbers except 0, OR $(-\infty, 0) \cup (0, \infty)$

Vertical asymptote: $x = 0$

Horizontal asymptote: $y = 0$

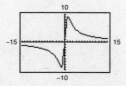

37. $h(x) = \dfrac{6x}{\sqrt{x^2 + 1}}$

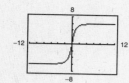

There are two horizontal asymptotes, $y = \pm 6$.

39. $g(x) = \dfrac{4|x - 2|}{x + 1}$

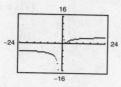

There are two horizontal asymptotes, $y = \pm 4$.

One vertical asymptote: $x = -1$

41. $f(x) = \dfrac{4(x - 1)^2}{x^2 - 4x + 5}$

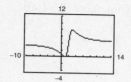

The graph crosses its horizontal asymptote, $y = 4$.

43. $f(x) = \dfrac{2x^2 + 1}{x} = 2x + \dfrac{1}{x}$

Vertical asymptote: $x = 0$

Slant asymptote: $y = 2x$

Origin symmetry

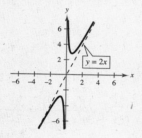

45. $h(x) = \dfrac{x^2}{x - 1} = x + 1 + \dfrac{1}{x - 1}$

Intercept: $(0, 0)$

Vertical asymptote: $x = 1$

Slant asymptote: $y = x + 1$

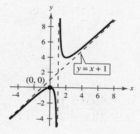

47. $g(x) = \dfrac{x^3}{2x^2 - 8} = \dfrac{1}{2}x + \dfrac{4x}{2x^2 - 8}$

Intercept: $(0, 0)$

Vertical asymptotes: $x = \pm 2$

Slant asymptote: $y = \dfrac{1}{2}x$

Origin symmetry

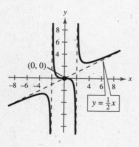

49. $f(x) = \dfrac{x^3 + 2x^2 + 4}{2x^2 + 1} = \dfrac{x}{2} + 1 + \dfrac{3 - \dfrac{x}{2}}{2x^2 + 1}$

Intercepts: $(-2.594, 0)$, $(0, 4)$

Slant asymptote: $y = \dfrac{x}{2} + 1$

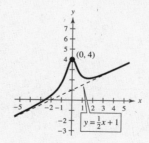

51. $y = \dfrac{x+1}{x-3}$

(a) x-intercept: $(-1, 0)$

(b) $0 = \dfrac{x+1}{x-3}$

$0 = x+1$

$-1 = x$

53. $y = \dfrac{1}{x} - x$

(a) x-intercepts: $(\pm 1, 0)$

(b) $0 = \dfrac{1}{x} - x$

$x = \dfrac{1}{x}$

$x^2 = 1$

$x = \pm 1$

55. $y = \dfrac{2x^2+x}{x+1} = 2x - 1 + \dfrac{1}{x+1}$

Domain: all real numbers except $x = -1$

Vertical asymptote: $x = -1$

Slant asymptote: $y = 2x - 1$

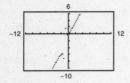

57. $y = \dfrac{1+3x^2-x^3}{x^2} = \dfrac{1}{x^2} + 3 - x = -x + 3 + \dfrac{1}{x^2}$

Domain: all real numbers except 0

or $(-\infty, 0) \cup (0, \infty)$

Vertical asymptote: $x = 0$

Slant asymptote: $y = -x + 3$

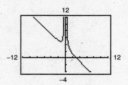

59. $f(x) = \dfrac{x^2-5x+4}{x^2-4} = \dfrac{(x-4)(x-1)}{(x-2)(x+2)}$

Vertical asymptotes: $x = 2, x = -2$

Horizontal asymptote: $y = 1$

No slant asymptotes, no holes

61. $f(x) = \dfrac{2x^2-5x+2}{2x^2-x-6} = \dfrac{(2x-1)(x-2)}{(2x+3)(x-2)} = \dfrac{2x-1}{2x+3}$,

$x \neq 2$

Vertical asymptote: $x = -\dfrac{3}{2}$

Horizontal asymptote: $y = 1$

No slant asymptotes

Hole at $x = 2$, $\left(2, \dfrac{3}{7}\right)$

63. $f(x) = \dfrac{2x^3-x^2-2x+1}{x^2+3x+2}$

$= \dfrac{(x-1)(x+1)(2x-1)}{(x+1)(x+2)}$

$= \dfrac{(x-1)(2x-1)}{x+2}$, $x \neq -1$

Long division gives $f(x) = \dfrac{2x^2-3x+1}{x+2} = 2x - 7 + \dfrac{15}{x+2}$.

Vertical asymptote: $x = -2$

No horizontal asymptote

Slant asymptote: $y = 2x - 7$

Hole at $x = -1$, $(-1, 6)$

65. $y = \dfrac{1}{x+5} + \dfrac{4}{x}$

(a)

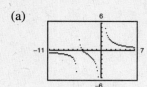

x-intercept: $(-4, 0)$

(b) $\qquad 0 = \dfrac{1}{x+5} + \dfrac{4}{x}$

$$-\dfrac{4}{x} = \dfrac{1}{x+5}$$

$$-4(x+5) = x$$

$$-4x - 20 = x$$

$$-5x = 20$$

$$x = -4$$

67. $y = \dfrac{1}{x+2} + \dfrac{2}{x+4}$

(a)

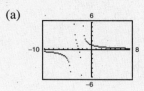

x-intercept: $\left(-\dfrac{8}{3}, 0\right)$

(b) $\dfrac{1}{x+2} + \dfrac{2}{x+4} = 0$

$$\dfrac{1}{x+2} = \dfrac{-2}{x+4}$$

$$x + 4 = -2x - 4$$

$$3x = -8$$

$$x = -\dfrac{8}{3}$$

69. $y = x - \dfrac{6}{x-1}$

(a)

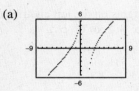

x-intercept: $(-2, 0), (3, 0)$

(b) $\qquad 0 = x - \dfrac{6}{x-1}$

$$\dfrac{6}{x-1} = x$$

$$6 = x(x-1)$$

$$0 = x^2 - x - 6$$

$$0 = (x+2)(x-3)$$

$$x = -2, \quad x = 3$$

71. $y = x + 2 - \dfrac{1}{x+1}$

(a)

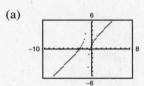

x-intercepts: $(-2.618, 0), (-0.382, 0)$

(b) $\qquad x + 2 = \dfrac{1}{x+2}$

$$x^2 + 3x + 2 = 1$$

$$x^2 + 3x + 1 = 0$$

$$x = \dfrac{-3 \pm \sqrt{9-4}}{2}$$

$$= \dfrac{-3}{2} \pm \dfrac{\sqrt{5}}{2}$$

$$\approx -2.618, -0.382$$

73. $y = x + 1 + \dfrac{2}{x-1}$

(a)

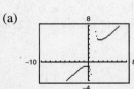

No *x*-intercepts

(b) $x + 1 + \dfrac{2}{x-1} = 0$

$$\dfrac{2}{x-1} = -x - 1$$

$$2 = -x^2 + 1$$

$$x^2 + 1 = 0$$

No real zeros

75. $y = x + 3 - \dfrac{2}{2x - 1}$

(a)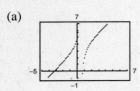

x-intercepts: $(0.766, 0)$, $(-3.266, 0)$

(b) $x + 3 - \dfrac{2}{2x - 1} = 0$

$$x + 3 = \dfrac{2}{2x - 1}$$

$$2x^2 + 5x - 3 = 2$$

$$2x^2 + 5x - 5 = 0$$

$$x = \dfrac{-5 \pm \sqrt{25 - 4(2)(-5)}}{4}$$

$$= \dfrac{-5 \pm \sqrt{65}}{4}$$

$$\approx 0.766, -3.266$$

79. (a) $A = xy$ and

$$(x - 2)(y - 4) = 30$$

$$y - 4 = \dfrac{30}{x - 2}$$

$$y = 4 + \dfrac{30}{x - 2} = \dfrac{4x + 22}{x - 2}$$

Thus, $A = xy = x\left(\dfrac{4x + 22}{x - 2}\right) = \dfrac{2x(2x + 11)}{x - 2}$.

81. $C = 100\left(\dfrac{200}{x^2} + \dfrac{x}{x + 30}\right)$, $1 \le x$

The minimum occurs when $x \approx 40.4 \approx 40$.

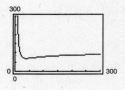

83. $C = \dfrac{3t^2 + t}{t^3 + 50}$, $0 \le t$

(a) The horizontal asymptote is the t-axis, or $C = 0$. This indicates that the chemical eventually dissipates.

(b)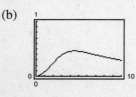

The maximum occurs when $t \approx 4.5$.

77. (a) $0.25(50) + 0.75(x) = C(50 + x)$

$$\dfrac{12.5 + 0.75x}{50 + x} = C$$

$$\dfrac{50 + 3x}{200 + 4x} = C$$

$$C = \dfrac{3x + 50}{4(x + 50)}$$

(b) Domain: $x \ge 0$ and $x \le 1000 - 50 = 950$

Thus, $0 \le x \le 950$.

(c)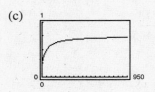

As the tank fills, the rate that the concentration is increasing slows down. It approaches the horizontal asymptote $C = \frac{3}{4} = 0.75$. When the tank is full $(x = 950)$, the concentration is $C = 0.725$.

(b) Domain: Since the margins on the left and right are each 1 inch, $x > 2$, or $(2, \infty)$.

(c)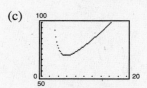

The area is minimum when $x \approx 5.87$ in. and $y \approx 11.75$ in.

(c) Graph C together with $y = 0.345$. The graphs intersect at $t \approx 2.65$ and $t \approx 8.32$. $C < 0.345$ when $0 \le t < 2.65$ hours and when $t > 8.32$ hours.

85. (a) $y_1 = 583.8t + 2414$ ($t = 0$ corresponds to 1990)

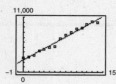

(b) Using the data $\left(t = \dfrac{1}{A}\right)$, we obtain:

$$y_2 = -0.00001855t + 0.0003150$$

$$y_3 = \frac{1}{-0.00001855t + 0.000315} = \frac{1}{A}$$

(c)

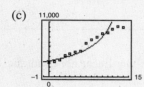

87. False, you will have to lift your pencil to cross the vertical asymptote.

89. $h(x) = \dfrac{6 - 2x}{3 - x} = \dfrac{2(3 - x)}{3 - x} = 2, \quad x \neq 3$

Since $h(x)$ is not reduced and $(3 - x)$ is a factor of both the numerator and the denominator, $x = 3$ is not a horizontal asymptote.

There is a hole in the graph at $x = 3$.

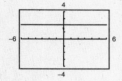

91. $y = x + 1 + \dfrac{a}{x + 2}$ has a slant asymptote $y = x + 1$ and a vertical asymptote $x = -2$.

$$0 = 2 + 1 + \frac{a}{2 + 2}$$

$$0 = 3 + \frac{a}{4}$$

$$\frac{a}{4} = -3$$

$$a = -12$$

Hence, $y = x + 1 - \dfrac{12}{x + 2} = \dfrac{x^2 + 3x - 10}{x + 2}$.

93. $\left(\dfrac{x}{8}\right)^{-3} = \left(\dfrac{8}{x}\right)^3 = \dfrac{512}{x^3}$

95. $\dfrac{3^{7/6}}{3^{1/6}} = 3^{6/6} = 3$

97.

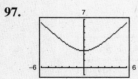

Domain: all x

Range: $y \geq \sqrt{6}$

99.

Domain: all x

Range: $y \leq 0$

101. Answers will vary.
(Make a Decision)

Section 3.7   Quadratic Models

You should know how to
- ■ Construct and classify scatter plots.
- ■ Fit a quadratic model to data.
- ■ Choose an appropriate model given a set of data.

Vocabulary Check

1. linear

2. quadratic

---

1. A quadratic model is better.   **3.** A linear model is better.   **5.** Neither linear nor quadratic

7. (a)

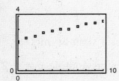

(b) Linear model is better.

(d)

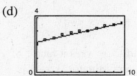

(c) $y = 0.14x + 2.2$, linear

[$y = -0.00478x^2 + 0.1887x + 2.1692$, quadratic]

(e)

x	0	1	2	3	4	5	6	7	8	9	10
y	2.1	2.4	2.5	2.8	2.9	3.0	3.0	3.2	3.4	3.5	3.6
Model	2.2	2.4	2.5	2.7	2.8	2.9	3.1	3.2	3.4	3.5	3.6

9. (a)

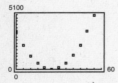

(b) Quadratic model is better.

(d)

(c) $y = 5.55x^2 - 277.5x + 3478$

(e)

x	0	5	10	15	20	25	30	35	40	45	50	55
y	3480	2235	1250	565	150	12	145	575	1275	2225	3500	5010
Model	3478	2229	1258	564	148	9	148	564	1258	2229	3478	5004

11. (a) $y = 2.48x + 1.1$, linear

$y = 0.071x^2 + 1.69x + 2.7$, quadratic

(b) 0.98995 for linear model

0.99519 for quadratic model

(c) Quadratic fits better.

13. (a) $y = -0.89x + 5.3$, linear

$y = 0.001x^2 - 0.90x + 5.3$, quadratic

(b) 0.99982 for the linear model

0.99987 for the quadratic model

(c) The quadratic model is slightly better.

15. (a)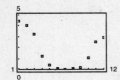

(b) $P = 0.1322t^2 - 1.901t + 6.87$

(c)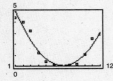

(d) The model's minimum is $H \approx 0.03$ at $t = 7.2$. This corresponds to July.

17. (a)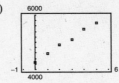

(b) $y = -2.630t^2 + 301.74t + 4270.2$

(c)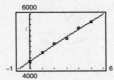

(d) According to the model, $y > 10,000$ when $t \approx 24$, or 2024.

(e) Answers will vary.

19. (a)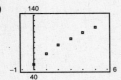

(b)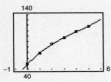

(c) $y = -1.1357t^2 + 18.999t + 50.32$, quadratic model,   (0.99859)

(d)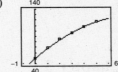

(e) The cubic model is a better fit.

(f)

Year	2006 ($t = 6$)	2007 ($t = 7$)	2008 ($t = 8$)
A^*	127.76	140.15	154.29
Cubic	129.91	145.13	164.96
Quadratic	123.40	127.64	129.60

21. True

23. The model is above all data points.

25. (a) $f(g(x)) = f(2x^2 - 1) = 5(2x^2 - 1) + 8 = 10x^2 + 3$

(b) $g(f(x)) = g(5x + 8) = 2(5x + 8)^2 - 1 = 50x^2 + 160x + 127$

27. (a) $f(g(x)) = f(x^3 - 5) = \sqrt[3]{x^3 - 5 + 5} = x$

(b) $g(f(x)) = g(\sqrt[3]{x + 5}) = [\sqrt[3]{x + 5}]^3 - 5 = x$

29. f is one-to-one.

$$y = \frac{x - 4}{5}$$

$$x = \frac{y - 4}{5}$$

$$5x + 4 = y \implies f^{-1}(x) = 5x + 4$$

31. f is one-to-one.

$$y = 2x^2 - 3, \quad x \geq 0$$

$$x = 2y^2 - 3, \quad y \geq 0$$

$$y^2 = \frac{(x + 3)}{2}$$

$$y = \sqrt{\frac{x + 3}{2}} \Rightarrow f^{-1}(x) = \sqrt{\frac{x + 3}{2}}$$

$$= \frac{\sqrt{2x + 6}}{2}, \quad x \geq -3$$

33.

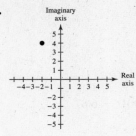

35.

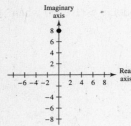

Review Exercises for Chapter 3

1.

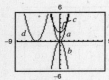

   (a) $y = 2x^2$ is a vertical stretch.

   (b) $y = -2x^2$ is a vertical stretch and reflection in the x-axis.

   (c) $y = x^2 + 2$ is a vertical shift two units upward.

   (d) $y = (x + 5)^2$ is a horizontal shift five units to the left.

3. $f(x) = \left(x + \frac{3}{2}\right)^2 + 1$

   Vertex: $\left(-\frac{3}{2}, 1\right)$

   y-intercept: $\left(0, \frac{13}{4}\right)$

   No x-intercepts

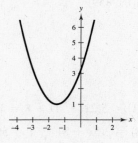

5. $f(x) = \frac{1}{3}(x^2 + 5x - 4)$

$$= \frac{1}{3}\left(x^2 + 5x + \frac{25}{4} - \frac{25}{4} - 4\right)$$

$$= \frac{1}{3}\left[\left(x + \frac{5}{2}\right)^2 - \frac{41}{4}\right]$$

$$= \frac{1}{3}\left(x + \frac{5}{2}\right)^2 - \frac{41}{12}$$

Vertex: $\left(-\frac{5}{2}, -\frac{41}{12}\right)$

y-intercept: $\left(0, -\frac{4}{3}\right)$

x-intercepts: $0 = \frac{1}{3}(x^2 + 5x - 4)$

$$0 = x^2 + 5x - 4$$

$$x = \frac{-5 \pm \sqrt{41}}{2} \quad \text{Use the Quadratic Formula.}$$

$$\left(\frac{-5 \pm \sqrt{41}}{2}, 0\right)$$

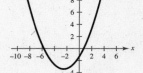

7. $f(x) = 3 - x^2 - 4x$

$\qquad = 3 - (x^2 + 4x + 4)$

$\qquad = 7 - (x + 2)^2$

Vertex: $(-2, 7)$

Intercepts: $(0, 3), \left(-2 \pm \sqrt{7}, 0\right)$

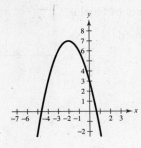

9. Vertex: $(1, -4) \Rightarrow f(x) = a(x - 1)^2 - 4$

$\qquad$ Point: $(2, -3) \Rightarrow -3 = a(2 - 1)^2 - 4$

$\qquad\qquad\qquad\qquad\qquad 1 = a$

$\qquad$ Thus, $f(x) = (x - 1)^2 - 4$.

11. Vertex: $(-2, -2) \Rightarrow f(x) = a(x + 2)^2 - 2$

$\qquad$ Point: $(-1, 0) \Rightarrow 0 = a(-1 + 2)^2 - 2$

$\qquad\qquad\qquad\qquad\qquad a = 2$

$\qquad$ Thus, $f(x) = 2(x + 2)^2 - 2$.

13. (a) $A = xy = x\left(\dfrac{8 - x}{2}\right)$, since $x + 2y - 8 = 0 \Rightarrow y = \dfrac{8 - x}{2}$.

Since the figure is in the first quadrant and x and y must be positive, the domain of

$A = x\left(\dfrac{8 - x}{2}\right)$ is $0 < x < 8$.

(b)

x	y	Area
1	$4 - \frac{1}{2}(1)$	$(1)\left[4 - \frac{1}{2}(1)\right] = \frac{7}{2}$
2	$4 - \frac{1}{2}(2)$	$(2)\left[4 - \frac{1}{2}(2)\right] = 6$
3	$4 - \frac{1}{2}(3)$	$(3)\left[4 - \frac{1}{2}(3)\right] = \frac{15}{2}$
4	$4 - \frac{1}{2}(4)$	$(4)\left[4 - \frac{1}{2}(4)\right] = 8$
5	$4 - \frac{1}{2}(5)$	$(5)\left[4 - \frac{1}{2}(5)\right] = \frac{15}{2}$
6	$4 - \frac{1}{2}(6)$	$(6)\left[4 - \frac{1}{2}(6)\right] = 6$

The dimensions that will produce a maximum area seem to be $x = 4$ and $y = 2$.

(c)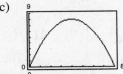

The maximum area of 8 occurs at the vertex when $x = 4$ and $y = \dfrac{8 - 4}{2} = 2$.

(d) $A = x\left(\dfrac{8 - x}{2}\right)$

$\qquad = \dfrac{1}{2}(8x - x^2)$

$\qquad = -\dfrac{1}{2}(x^2 - 8x)$

$\qquad = -\dfrac{1}{2}(x^2 - 8x + 16 - 16)$

$\qquad = -\dfrac{1}{2}[(x - 4)^2 - 16]$

$\qquad = -\dfrac{1}{2}(x - 4)^2 + 8$

(e) The answers are the same.

The maximum area of 8 occurs when $x = 4$ and $y = \dfrac{8 - 4}{2} = 2$.

15. $6x + 4y = 1500$ Total amount of fencing

    $A = 3xy$ Area enclosed

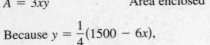

Because $y = \frac{1}{4}(1500 - 6x)$,

$$A = 3x\left(\frac{1}{4}\right)(1500 - 6x)$$

$$= -\frac{9}{2}x^2 + 1125x.$$

The vertex is at $x = \dfrac{-b}{2a} = \dfrac{-1125}{2(-9/2)} = 125$. Thus $x = 125$ feet, $y = \dfrac{1}{4}(1500 - 6(125)) = 187.5$

and the dimensions are 375 feet by 187.5 feet.

17. (a)   (b)   (c)   (d)

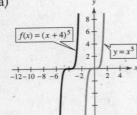 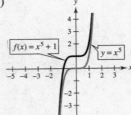 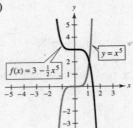 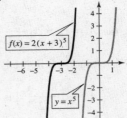

19. $f(x) = \frac{1}{2}x^3 - 2x + 1$; $g(x) = \frac{1}{2}x^3$

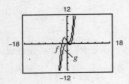

21. $f(x) = -x^2 + 6x + 9$

The degree is even and the leading coefficient is negative. The graph falls to the left and right.

23. $f(x) = \frac{3}{4}(x^4 + 3x^2 + 2)$

The degree is even and the leading coefficient is positive. The graph rises to the left and right.

25. (a) $x^4 - x^3 - 2x^2 = x^2(x^2 - x - 2)$

$$= x^2(x - 2)(x + 1) = 0$$

Zeros: $x = -1, 0, 2$

(b)

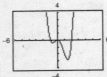

(c) Zeros: $x = -1, 0, 2$; the same

27. (a) $t^3 - 3t = t(t^2 - 3) = t\left(t + \sqrt{3}\right)\left(t - \sqrt{3}\right) = 0$

Zeros: $t = 0, \pm\sqrt{3}$

(b)

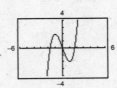

(c) Zeros: $t = 0, \pm 1.732$, the same

29. (a) $x(x + 3)^2 = 0$

Zeros: $x = 0, -3$

(b)

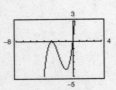

(c) Zeros: $x = -3, 0$, the same

31. $f(x) = (x + 2)(x - 1)(x - 5)$

$= x^3 - 4x^2 - 7x + 10$

33. $f(x) = (x - 3)\left(x - 2 + \sqrt{3}\right)\left(x - 2 - \sqrt{3}\right)$

$= x^3 - 7x^2 + 13x - 3$

35. (a) Degree is even and leading coefficient is $1 > 0$. Rises to the left and rises to the right.

(b) $x^4 - 2x^3 - 12x^2 + 18x + 27 = (x - 3)^2(x + 1)(x + 3)$

Zeros: $\pm 3, -1$

(c) and (d)

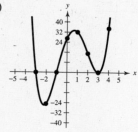

37. $f(x) = x^3 + 2x^2 - x - 1$

(a) $f(-3) < 0, f(-2) > 0 \implies$ zero in $[-3, -2]$

$f(-1) > 0, f(0) < 0 \implies$ zero in $[-1, 0]$

$f(0) < 0, f(1) > 0 \implies$ zero in $[0, 1]$

(b) Zeros: $-2.247, -0.555, 0.802$

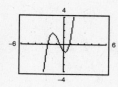

39. $f(x) = x^4 - 6x^2 - 4$

(a) $f(-3) > 0, f(-2) < 0 \implies$ zero in $[-3, -2]$

$f(2) < 0, f(3) > 0 \implies$ zero in $[2, 3]$

(b) Zeros: ± 2.570

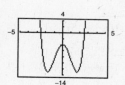

41. $y_1 = \dfrac{x^2}{x - 2}$

$y_2 = x + 2 + \dfrac{4}{x - 2}$

$= \dfrac{(x + 2)(x - 2)}{x - 2} + \dfrac{4}{x - 2}$

$= \dfrac{x^2 - 4}{x - 2} + \dfrac{4}{x - 2}$

$= \dfrac{x^2}{x - 2} = y_1$

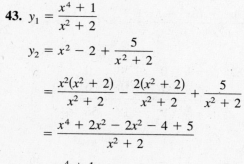

43. $y_1 = \dfrac{x^4 + 1}{x^2 + 2}$

$y_2 = x^2 - 2 + \dfrac{5}{x^2 + 2}$

$= \dfrac{x^2(x^2 + 2)}{x^2 + 2} - \dfrac{2(x^2 + 2)}{x^2 + 2} + \dfrac{5}{x^2 + 2}$

$= \dfrac{x^4 + 2x^2 - 2x^2 - 4 + 5}{x^2 + 2}$

$= \dfrac{x^4 + 1}{x^2 + 2} = y_1$

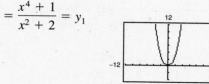

45.

$$
\begin{array}{r}
8x + 5 \\
3x - 2\overline{)24x^2 - x - 8} \\
\underline{24x^2 - 16x} \\
15x - 8 \\
\underline{15x - 10} \\
2
\end{array}
$$

Thus, $\dfrac{24x^2 - x - 8}{3x - 2} = 8x + 5 + \dfrac{2}{3x - 2}.$

47.

$$
\begin{array}{r}
x^2 - 2 \\
x^2 - 1\overline{)x^4 - 3x^2 + 2} \\
\underline{x^4 - x^2} \\
-2x^2 + 2 \\
\underline{-2x^2 + 2} \\
0
\end{array}
$$

Thus, $\dfrac{x^4 - 3x^2 + 2}{x^2 - 1} = x^2 - 2, \ (x \neq \pm 1).$

49.

$$
\begin{array}{r}
5x + 2 \\
x^2 - 3x + 1\overline{)5x^3 - 13x^2 - x + 2} \\
\underline{5x^3 - 15x^2 + 5x} \\
2x^2 - 6x + 2 \\
\underline{2x^2 - 6x + 2} \\
0
\end{array}
$$

Thus, $\dfrac{5x^3 - 13x^2 - x + 2}{x^2 - 3x + 1} = 5x + 2, \ x \neq \dfrac{1}{2}\left(3 \pm \sqrt{5}\right).$

51.

$$
\begin{array}{r}
3x^2 + 5x + 8 \\
2x^2 + 0x - 1\overline{)6x^4 + 10x^3 + 13x^2 - 5x + 2} \\
\underline{6x^4 + 0x^3 - 3x^2} \\
10x^3 + 16x^2 - 5x \\
\underline{10x^3 + 0x^2 - 5x} \\
16x^2 - 0 + 2 \\
\underline{16x^2 + 0 - 8} \\
10
\end{array}
$$

$$
\frac{6x^4 + 10x^3 + 13x^2 - 5x + 2}{2x^2 - 1} = 3x^2 + 5x + 8 + \frac{10}{2x^2 - 1}
$$

53.

$$
\begin{array}{r|rrrrr}
-2 & 0.25 & -4 & 0 & 0 & 0 \\
& & -\frac{1}{2} & 9 & -18 & 36 \\
\hline
& \frac{1}{4} & -\frac{9}{2} & 9 & -18 & 36
\end{array}
$$

Hence,

$$
\frac{0.25x^4 - 4x^3}{x + 2} = \frac{1}{4}x^3 - \frac{9}{2}x^2 + 9x - 18 + \frac{36}{x + 2}.
$$

55.

$$
\begin{array}{r|rrrrr}
\frac{2}{3} & 6 & -4 & -27 & 18 & 0 \\
& & 4 & 0 & -18 & 0 \\
\hline
& 6 & 0 & -27 & 0 & 0
\end{array}
$$

Thus,

$$
\frac{6x^4 - 4x^3 - 27x^2 + 18x}{x - (2/3)} = 6x^3 - 27x, \ x \neq \frac{2}{3}.
$$

57.

$$
\begin{array}{r|rrrr}
4 & 3 & -10 & 12 & -22 \\
& & 12 & 8 & 80 \\
\hline
& 3 & 2 & 20 & 58
\end{array}
$$

Thus,

$$
\frac{3x^3 - 10x^2 + 12x - 22}{x - 4} = 3x^2 + 2x + 20 + \frac{58}{x - 4}.
$$

59. (a)

$$
\begin{array}{r|rrrrr}
-3 & 1 & 10 & -24 & 20 & 44 \\
& & -3 & -21 & 135 & -465 \\
\hline
& 1 & 7 & -45 & 155 & -421 = f(-3)
\end{array}
$$

(b)

$$
\begin{array}{r|rrrrr}
-2 & 1 & 10 & -24 & 20 & 44 \\
& & -2 & -16 & 80 & -200 \\
\hline
& 1 & 8 & -40 & 100 & -156 = f(-2)
\end{array}
$$

61. $f(x) = x^3 + 4x^2 - 25x - 28$

(a)
$$
\begin{array}{r|rrrr}
4 & 1 & 4 & -25 & -28 \\
  &   & 4 & 32 & 28 \\
\hline
  & 1 & 8 & 7 & 0
\end{array}
$$

$(x - 4)$ is a factor.

(b) $x^2 + 8x + 7 = (x + 1)(x + 7)$

Remaining factors: $(x + 1), (x + 7)$

(c) $f(x) = (x - 4)(x + 1)(x + 7)$

(d) Zeros: $4, -1, -7$

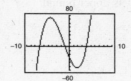

63. $f(x) = x^4 - 4x^3 - 7x^2 + 22x + 24$

(a)
$$
\begin{array}{r|rrrrr}
-2 & 1 & -4 & -7 & 22 & 24 \\
   &   & -2 & 12 & -10 & -24 \\
\hline
   & 1 & -6 & 5 & 12 & 0
\end{array}
$$

$(x + 2)$ is a factor.

$$
\begin{array}{r|rrrr}
3 & 1 & -6 & 5 & 12 \\
  &   & 3 & -9 & -12 \\
\hline
  & 1 & -3 & -4 & 0
\end{array}
$$

$(x - 3)$ is a factor.

(b) $x^2 - 3x - 4 = (x - 4)(x + 1)$

Remaining factors: $(x - 4), (x + 1)$

(c) $f(x) = (x + 2)(x - 3)(x - 4)(x + 1)$

(d) Zeros: $-2, 3, 4, -1$

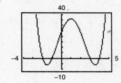

65. $f(x) = 4x^3 - 11x^2 + 10x - 3$

Possible rational zeros: $\pm 3, \pm \frac{3}{2}, \pm \frac{3}{4}, \pm 1, \pm \frac{1}{2}, \pm \frac{1}{4}$

Zeros: $1, 1, \frac{3}{4}$

67. $f(x) = 6x^3 - 5x^2 + 24x - 20$

$= (6x - 5)(x^2 + 4)$

Real zero: $\frac{5}{6}$

69. $f(x) = 6x^4 - 25x^3 + 14x^2 + 27x - 18$

Possible rational zeros: $\pm 1, \pm 2, \pm 3, \pm 6, \pm 9, \pm 18, \pm \frac{1}{2}, \pm \frac{3}{2}, \pm \frac{9}{2}, \pm \frac{1}{3}, \pm \frac{2}{3}, \pm \frac{1}{6}$

Use a graphing utility to see that $x = -1$ and $x = 3$ are probably zeros.

$$
\begin{array}{r|rrrrr}
-1 & 6 & -25 & 14 & 27 & -18 \\
   &   & -6 & 31 & -45 & 18 \\
\hline
   & 6 & -31 & 45 & -18 & 0
\end{array}
\qquad
\begin{array}{r|rrrr}
3 & 6 & -31 & 45 & -18 \\
  &   & 18 & -39 & 18 \\
\hline
  & 6 & -13 & 6 & 0
\end{array}
$$

$6x^4 - 25x^3 + 14x^2 + 27x - 18 = (x + 1)(x - 3)(6x^2 - 13x + 6)$

$= (x + 1)(x - 3)(3x - 2)(2x - 3)$

Thus, the zeros of f are $x = -1$, $x = 3$, $x = \frac{2}{3}$, and $x = \frac{3}{2}$.

71. $g(x) = 5x^3 - 6x + 9$ has two variations in sign $\Rightarrow$ 0 or 2 positive real zeros.

$g(-x) = -5x^3 + 6x + 9$ has one variation in sign $\Rightarrow$ 1 negative real zero.

73.

$$1 \,\big|\, \begin{array}{rrrr} 4 & -3 & 4 & -3 \\ & 4 & 1 & 5 \\ \hline 4 & 1 & 5 & 2 \end{array}$$

All entries positive; $x = 1$ is upper bound.

$$-\tfrac{1}{4} \,\big|\, \begin{array}{rrrr} 4 & -3 & 4 & -3 \\ & -1 & 1 & -\frac{5}{4} \\ \hline 4 & -4 & 5 & -\frac{17}{4} \end{array}$$

Alternating signs; $x = -\frac{1}{4}$ is lower bound.

75. $f(x) = 3x(x - 2)^2$

Zeros: $0, 2, 2$

77. $f(x) = 2x^4 - 5x^3 + 10x - 12$

$$2 \,\big|\, \begin{array}{rrrrr} 2 & -5 & 0 & 10 & -12 \\ & 4 & -2 & -4 & 12 \\ \hline 2 & -1 & -2 & 6 & 0 \end{array}$$

$x = 2$ is a zero.

$$-\tfrac{3}{2} \,\big|\, \begin{array}{rrrr} 2 & -1 & -2 & 6 \\ & -3 & 6 & -6 \\ \hline 2 & -4 & 4 & 0 \end{array}$$

$x = -\dfrac{3}{2}$ is a zero.

$$f(x) = (x - 2)\left(x + \frac{3}{2}\right)(2x^2 - 4x + 4)$$
$$= (x - 2)(2x + 3)(x^2 - 2x + 2)$$

By the Quadratic Formula, applied to $x^2 - 2x + 2$,

$$x = \frac{2 \pm \sqrt{4 - 4(2)}}{2} = 1 \pm i.$$

Zeros: $2, -\dfrac{3}{2}, 1 \pm i$

$$f(x) = (x - 2)(2x + 3)(x - 1 + i)(x - 1 - i)$$

79. $h(x) = x^3 - 7x^2 + 18x - 24$

$$4 \,\big|\, \begin{array}{rrrr} 1 & -7 & 18 & -24 \\ & 4 & -12 & 24 \\ \hline 1 & -3 & 6 & 0 \end{array}$$

$x = 4$ is a zero. Applying the Quadratic Formula on $x^2 - 3x + 6$,

$$x = \frac{3 \pm \sqrt{9 - 4(6)}}{2} = \frac{3}{2} \pm \frac{\sqrt{15}}{2}i.$$

Zeros: $4, \dfrac{3}{2} + \dfrac{\sqrt{15}}{2}i, \dfrac{3}{2} - \dfrac{\sqrt{15}}{2}i$

$$h(x) = (x - 4)\left(x - \frac{3 + \sqrt{15}i}{2}\right)\left(x - \frac{3 - \sqrt{15}i}{2}\right)$$

81. $f(x) = x^5 + x^4 + 5x^3 + 5x^2$
$$= x^2(x^3 + x^2 + 5x + 5)$$
$$= x^2[x^2(x + 1) + 5(x + 1)]$$
$$= x^2(x + 1)(x^2 + 5)$$
$$= x^2(x + 1)(x + \sqrt{5}i)(x - \sqrt{5}i)$$

Zeros: $0, 0, -1, \pm\sqrt{5}i$

83. $f(x) = x^3 - 4x^2 + 6x - 4$

    (a) $x^3 - 4x^2 + 6x - 4 = (x - 2)(x^2 - 2x + 2)$

       By the Quadratic Formula, for $x^2 - 2x + 2$,

$$x = \frac{2 \pm \sqrt{(-2)^2 - 4(2)}}{2} = 1 \pm i.$$

       Zeros: $2, 1 + i, 1 - i$

    (b) $f(x) = (x - 2)(x - 1 - i)(x - 1 + i)$

    (c) x-intercept: $(2, 0)$

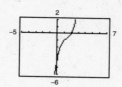

85. (a) $f(x) = -3x^3 - 19x^2 - 4x + 12$

       $\begin{array}{r|rrrr} -1 & -3 & -19 & -4 & 12 \\ & & 3 & 16 & -12 \\ \hline & -3 & -16 & 12 & 0 \end{array}$

    (b) $f(x) = -(x + 1)(3x^2 + 16x - 12)$

              $= -(x + 1)(3x - 2)(x + 6)$

    (c) x-intercepts: $(-1, 0), (-6, 0), \left(\dfrac{2}{3}, 0\right)$

87. $f(x) = x^4 + 34x^2 + 225$

    (a) $x^4 + 34x^2 + 225 = (x^2 + 9)(x^2 + 25)$

       Zeros: $\pm 3i, \pm 5i$

    (b) $(x + 3i)(x - 3i)(x + 5i)(x - 5i)$

    (c) No x-intercepts

    (d)

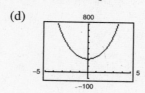

89. Since $5i$ is a zero, so is $-5i$.

$f(x) = (x - 4)(x + 2)(x - 5i)(x + 5i)$

        $= (x^2 - 2x - 8)(x^2 + 25)$

        $= x^4 - 2x^3 + 17x^2 - 50x - 200$

91. $f(x) = (x - 1)(x + 4)(x + 3 - 5i)(x + 3 + 5i)$

        $= (x^2 + 3x - 4)((x + 3)^2 + 25)$

        $= (x^2 + 3x - 4)(x^2 + 6x + 34)$

        $= x^4 + 9x^3 + 48x^2 + 78x - 136$

93. $f(x) = x^4 - 2x^3 + 8x^2 - 18x - 9$

    (a) $f(x) = (x^2 + 9)(x^2 - 2x - 1)$

       For the quadratic $x^2 - 2x - 1$, $x = \dfrac{2 \pm \sqrt{(-2)^2 - 4(-1)}}{2} = 1 \pm \sqrt{2}$.

    (b) $f(x) = (x^2 + 9)(x - 1 + \sqrt{2})(x - 1 - \sqrt{2})$

    (c) $f(x) = (x + 3i)(x - 3i)(x - 1 + \sqrt{2})(x - 1 - \sqrt{2})$

95. Zeros: $-2i, 2i$

    $(x + 2i)(x - 2i) = x^2 + 4$ is a factor.

    $f(x) = (x^2 + 4)(x + 3)$

    Zeros: $\pm 2i, -3$

97. (a) Domain: all $x \neq -3$

    (b) Horizontal asymptote: $y = -1$

       Vertical asymptote: $x = -3$

99. $f(x) = \dfrac{2}{x^2 - 3x - 18} = \dfrac{2}{(x - 6)(x + 3)}$

    (a) Domain: all $x \neq 6, -3$

    (b) Horizontal asymptote: $y = 0$

       Vertical asymptotes: $x = 6, x = -3$

101. $f(x) = \dfrac{7 + x}{7 - x}$

    (a) Domain: all $x \neq 7$

    (b) Horizontal asymptote: $y = -1$

       Vertical asymptote: $x = 7$

103. $f(x) = \dfrac{4x^2}{2x^2 - 3}$

   (a) Domain: all $x \neq \pm\sqrt{\dfrac{3}{2}} = \pm\dfrac{\sqrt{6}}{2}$

   (b) Horizontal asymptote: $y = 2$

      Vertical asymptotes: $x = \pm\sqrt{\dfrac{3}{2}} = \pm\dfrac{\sqrt{6}}{2}$

105. $f(x) = \dfrac{2x - 10}{x^2 - 2x - 15} = \dfrac{2(x - 5)}{(x - 5)(x + 3)} = \dfrac{2}{x + 3},$

   $x \neq 5$

   (a) Domain: all $x \neq 5, -3$

   (b) Vertical asymptote: $x = -3$

      (There is a hole at $x = 5$.)

      Horizontal asymptote: $y = 0$

107. $f(x) = \dfrac{x - 2}{|x| + 2}$

   (a) Domain: all real numbers

   (b) No vertical asymptotes

      Horizontal asymptotes: $y = 1, y = -1$

109. $C = \dfrac{528p}{100 - p}, \quad 0 \le p < 100$

   (a) When $p = 25, C = \dfrac{528(25)}{100 - 25} = 176$ million.

      When $p = 50, C = \dfrac{528(50)}{100 - 50} = 528$ million.

      When $p = 75, C = \dfrac{528(75)}{100 - 75} = 1584$ million.

   (b)

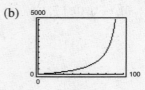

   (c) No. As $p \to 100, C$ tends to infinity.

111. $f(x) = \dfrac{x^2 - 5x + 4}{x^2 - 1}$

   $= \dfrac{(x - 4)(x - 1)}{(x - 1)(x + 1)}$

   $= \dfrac{x - 4}{x + 1}, \quad x \neq 1$

   Vertical asymptotes: $x = -1$

   Horizontal asymptote: $y = 1$

   No slant asymptotes

   Hole at $x = 1$: $\left(1, -\dfrac{3}{2}\right)$

113. $f(x) = \dfrac{2x^2 - 7x + 3}{2x^2 - 3x - 9}$

   $= \dfrac{(x - 3)(2x - 1)}{(x - 3)(2x + 3)} = \dfrac{2x - 1}{2x + 3}, x \neq 3$

   Vertical asymptote: $x = -\dfrac{3}{2}$

   Horizontal asymptote: $y = 1$

   No slant asymptotes

   Hole at $x = 3$: $\left(3, \dfrac{5}{9}\right)$

115. $f(x) = \dfrac{3x^2 - x^2 - 12x + 4}{x^2 + 3x + 2}$

$\qquad = \dfrac{(x-2)(x+2)(3x-1)}{(x+1)(x+2)}$

$\qquad = \dfrac{(x-2)(3x-1)}{x+1}, \quad x \neq -2$

$\qquad = 3x - 10 + \dfrac{12}{x+1}, \quad x \neq -2$

Vertical asymptote: $x = -1$

No horizontal asymptotes

Slant asymptote: $y = 3x - 10$

Hole at $x = -2$: $(-2, -28)$

117. $f(x) = \dfrac{2x-1}{x-5}$

Intercepts: $\left(0, \dfrac{1}{5}\right), \left(\dfrac{1}{2}, 0\right)$

Vertical asymptote: $x = 5$

Horizontal asymptote: $y = 2$

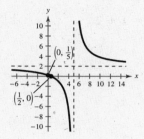

119. $f(x) = \dfrac{2x}{x^2+4}$

Intercept: $(0, 0)$

Origin symmetry

Horizontal asymptote: $y = 0$

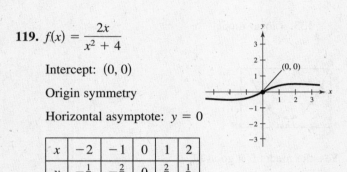

x	-2	-1	0	1	2
y	$-\frac{1}{2}$	$-\frac{2}{5}$	0	$\frac{2}{5}$	$\frac{1}{2}$

121. $f(x) = \dfrac{x^2}{x^2+1}$

Intercept: $(0, 0)$

y-axis symmetry

Horizontal asymptote:
$y = 1$

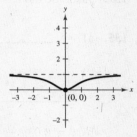

x	± 3	± 2	± 1	0
y	$\frac{9}{10}$	$\frac{4}{5}$	$\frac{1}{2}$	0

123. $f(x) = \dfrac{2}{(x+1)^2}$

Intercept: $(0, 2)$

Horizontal asymptote: $y = 0$

Vertical asymptote: $x = -1$

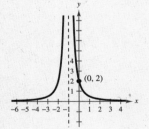

125. $f(x) = \dfrac{2x^3}{x^2+1} = 2x - \dfrac{2x}{x^2+1}$

Intercept: $(0, 0)$

Origin symmetry

Slant asymptote: $y = 2x$

x	-2	-1	0	1	2
y	$-\frac{16}{5}$	-1	0	1	$\frac{16}{5}$

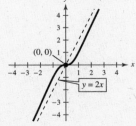

127. $f(x) = \dfrac{x^2 - x + 1}{x - 3}$

$\qquad = x + 2 + \dfrac{7}{x - 3}$

Intercept: $\left(0, -\dfrac{1}{3}\right)$

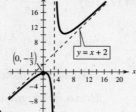

Vertical asymptote: $x = 3$

Slant asymptote: $y = x + 2$

129. $N = \dfrac{20(4 + 3t)}{1 + 0.05t}, \quad t \geq 0$

(a)
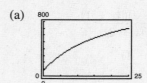

(b) $N(5) = 304{,}000$ fish

$\qquad N(10) \approx 453{,}333$ fish

$\qquad N(25) \approx 702{,}222$ fish

(c) The limit is

$\qquad \dfrac{60}{0.05} = 1{,}200{,}000$ fish, the horizontal asymptote.

131. Quadratic model

133. Linear model

135. (a)
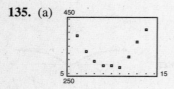

(b) $y = 8.03t^2 - 157.1t + 1041; \ 0.98348$

(c)
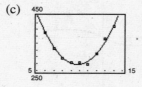

Yes, the model is a good fit.

(d) From the model, $y \geq 500$ when $t \approx 15.1$, or 2005.

(e) Answers will vary.

137. False. The degree of the numerator is two more than the degree of the denominator.

139. It means that the divisor is a factor of the dividend.

Chapter 3   Practice Test

1. Sketch the graph of $f(x) = x^2 - 6x + 5$ by hand and identify the vertex and the intercepts.

2. Find the number of units x that produce a minimum cost C if $C = 0.01x^2 - 90x + 15,000$.

3. Find the quadratic function that has a maximum at $(1, 7)$ and passes through the point $(2, 5)$.

4. Find two quadratic functions that have x-intercepts $(2, 0)$ and $\left(\frac{4}{3}, 0\right)$.

5. Use the Leading Coefficient Test to determine the right-hand and left-hand behavior of the graph of the polynomial function $f(x) = -3x^5 + 2x^3 - 17$.

6. Find all the real zeros of $f(x) = x^5 - 5x^3 + 4x$. Verify your answer with a graphing utility.

7. Find the polynomial function with 0, 3, and -2 as zeros.

8. Sketch $f(x) = x^3 - 12x$ by hand.

9. Divide $3x^4 - 7x^2 + 2x - 10$ by $x - 3$ using long division.

10. Divide $x^3 - 11$ by $x^2 + 2x - 1$.

11. Use synthetic division to divide $3x^5 + 13x^4 + 12x - 1$ by $x + 5$.

12. Use synthetic division to find $f(-6)$ when $f(x) = 7x^3 + 40x^2 - 12x + 15$.

13. Find the real zeros of $f(x) = x^3 - 19x - 30$.

14. Find the real zeros of $f(x) = x^4 + x^3 - 8x^2 - 9x - 9$.

15. List all possible rational zeros of the function $f(x) = 6x^3 - 5x^2 + 4x - 15$.

16. Find the rational zeros of the polynomial $f(x) = x^3 - \frac{20}{3}x^2 + 9x - \frac{10}{3}$.

17. Write $f(x) = x^4 + x^3 + 3x^2 + 5x - 10$ as a product of linear factors.

18. Find a polynomial with real coefficients that has 2, $3 + i$, and $3 - 2i$ as zeros.

19. Use synthetic division to show that $3i$ is a zero of $f(x) = x^3 + 4x^2 + 9x + 36$.

20. Find a mathematical model for the statement, "z varies directly as the square of x and inversely as the square root of y".

21. Sketch the graph of $f(x) = \dfrac{x - 1}{2x}$ and label all intercepts and asymptotes.

22. Sketch the graph of $f(x) = \dfrac{3x^2 - 4}{x}$ and label all intercepts and asymptotes.

23. Find all the asymptotes of $f(x) = \dfrac{8x^2 - 9}{x^2 + 1}$.

24. Find all the asymptotes of $f(x) = \dfrac{4x^2 - 2x + 7}{x - 1}$.

25. Sketch the graph of $f(x) = \dfrac{x - 5}{(x - 5)^2}$.

C H A P T E R   4
Exponential and Logarithmic Functions

CHAPTER 4
Exponential and Logarithmic Functions

Section 4.1   Exponential Functions and Their Graphs

- You should know that a function of the form $y = a^x$, where $a > 0$, $a \neq 1$, is called an exponential function with base a.
- You should be able to graph exponential functions.
- You should be familiar with the number e and the natural exponential function $f(x) = e^x$.
- You should know formulas for compound interest.
  - (a) For n compoundings per year: $A = P\left(1 + \dfrac{r}{n}\right)^{nt}$.
  - (b) For continuous compoundings: $A = Pe^{rt}$.

Vocabulary Check

1. algebraic

2. transcendental

3. natural exponential, natural

4. $A = P\left(1 + \dfrac{r}{n}\right)^{nt}$

5. $A = Pe^{rt}$

---

1. $(3.4)^{6.8} \approx 4112.033$

3. $5^{-\pi} \approx 0.006$

5. $g(x) = 5^x$

x	-2	-1	0	1	2
y	$\frac{1}{25}$	$\frac{1}{5}$	1	5	25

Asymptote: $y = 0$

Intercept: $(0, 1)$

Increasing

7. $f(x) = \left(\frac{1}{5}\right)^x = 5^{-x}$

x	-2	-1	0	1	2
y	25	5	1	$\frac{1}{5}$	$\frac{1}{25}$

Asymptote: $y = 0$

Intercept: $(0, 1)$

Decreasing

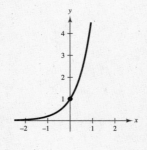

9. $h(x) = 5^{x-2}$

x	-1	0	1	2	3
y	$\frac{1}{125}$	$\frac{1}{25}$	$\frac{1}{5}$	1	5

Asymptote: $y = 0$

Intercept: $\left(0, \frac{1}{25}\right)$

Increasing

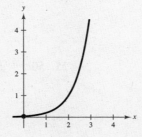

11. $g(x) = 5^{-x} - 3$

x	-1	0	1	2
y	2	-2	$-2\frac{4}{5}$	$-2\frac{24}{25}$

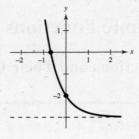

Asymptote: $y = -3$

Intercepts:

$(0, -2), (-0.683, 0)$

Decreasing

13. $f(x) = 2^{x-2}$ rises to the right.

Asymptote: $y = 0$

Intercept: $\left(0, \frac{1}{4}\right)$

Matches graph (d).

15. $f(x) = 2^x - 4$ rises to the right.

Asymptote: $y = -4$

Intercept: $(0, -3)$

Matches graph (c).

17. $f(x) = 3^x$

$g(x) = 3^{x-5} = f(x - 5)$

Horizontal shift five units to the right

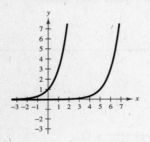

19. $f(x) = \left(\frac{3}{5}\right)^x$

$g(x) = -\left(\frac{3}{5}\right)^{x+4} = -f(x + 4)$

Horizontal shift four units to the left, followed by reflection in x-axis

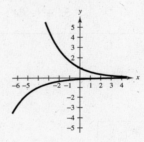

21. $f(x) = 4^x$

$g(x) = 4^{x-2} - 3 = f(x - 2) - 3$

Horizontal shift two units to the right followed by vertical shift three units downward

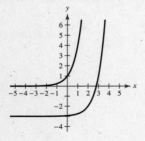

23. $e^{9.2} \approx 9897.129$

25. $50e^{4(0.02)} \approx 54.164$

27. $f(x) = \left(\frac{5}{2}\right)^x$

x	-2	-1	0	1	2
$f(x)$	0.16	0.4	1	2.5	6.25

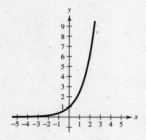

Asymptote: $y = 0$

29. $f(x) = 6^x$

x	-2	-1	0	1	2
$f(x)$	0.03	0.17	1	6	36

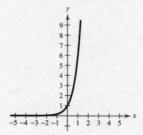

Asymptote: $y = 0$

31. $f(x) = 3^{x+2}$

x	-3	-2	-1	0	1
$f(x)$	0.33	1	3	9	27

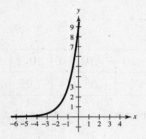

Asymptote: $y = 0$

33. $y = 2^{-x^2}$

x	-2	-1	0	1	2
y	0.06	0.5	1	0.5	0.06

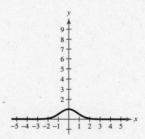

Asymptote: $y = 0$

35. $y = 3^{x-2} + 1$

x	-1	0	1	2	3	4
y	1.04	1.11	1.33	2	4	10

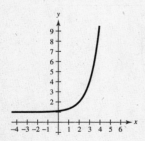

Asymptote: $y = 1$

37. $f(x) = e^{-x}$

x	-2	-1	0	1	2
$f(x)$	7.39	2.72	1	0.37	0.14

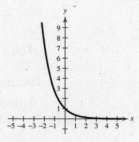

Asymptote: $y = 0$

39. $f(x) = 3e^{x+4}$

x	-6	-5	-4	-3	-2
$f(x)$	0.41	1.10	3	8.15	22.17

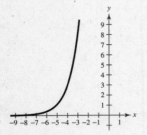

Asymptote: $y = 0$

41. $f(x) = 2 + e^{x-5}$

x	3	4	5	6	7
$f(x)$	2.14	2.37	3	4.72	9.39

Asymptote: $y = 2$

43. $s(t) = 2e^{0.12t}$

t	-2	-1	0	1	2
$s(t)$	1.57	1.77	2	2.26	2.54

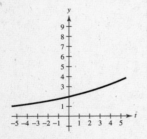

Asymptote: $y = 0$

45. $f(x) = \dfrac{8}{1 + e^{-0.5x}}$

(a)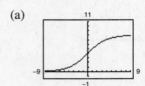

(b)

x	-30	-20	-10	0	10	20	30
$f(x)$	≈ 0	≈ 0	0.05	4	7.95	≈ 8	≈ 8

Horizontal asymptotes: $y = 0$, $y = 8$

47. $f(x) = \dfrac{-6}{2 - e^{0.2x}}$

(a)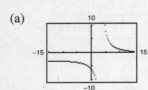

(b)

x	-20	-10	0	3	3.4	3.46
$f(x)$	-3.03	-3.22	-6	-34	-230	-2617

x	3.47	4	5	10	20
$f(x)$	3516	26.6	8.4	1.11	0.11

Horizontal asymptotes: $y = -3$, $y = 0$

Vertical asymptote: $x \approx 3.47$

49.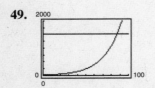

Intersection: (86.350, 1500)

51. $f(x) = x^2e^{-x}$

(a)

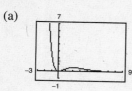

(b) Decreasing: $(-\infty, 0)$, $(2, \infty)$

Increasing: $(0, 2)$

(c) Relative maximum: $(2, 4e^{-2}) \approx (2, 0.541)$

Relative minimum: $(0, 0)$

53. $P = 2500$, $r = 2.5\% = 0.025$, $t = 10$

Compounded n times per year: $A = P\left(1 + \dfrac{r}{n}\right)^{nt} = 2500\left(1 + \dfrac{0.025}{n}\right)^{10n}$

Compounded continuously: $A = Pe^{rt} = 2500e^{(0.025)(10)}$

n	1	2	4	12	365	Continuous
A	3200.21	3205.09	3207.57	3209.23	3210.04	3210.06

55. $P = 2500$, $r = 4\% = 0.04$, $t = 20$

Compounded n times per year: $A = P\left(1 + \dfrac{r}{n}\right)^{nt} = 2500\left(1 + \dfrac{0.04}{n}\right)^{20n}$

Compounded continuously: $A = Pe^{rt} = 2500e^{(0.04)(20)}$

n	1	2	4	12	365	Continuous
A	5477.81	5520.10	5541.79	5556.46	5563.61	5563.85

57. $P = 12{,}000$, $r = 4\% = 0.04$

$A = Pe^{rt} = 12000e^{0.04t}$

t	1	10	20	30	40	50
A	12,489.73	17,901.90	26,706.49	39,841.40	59,436.39	88,668.67

59. $P = 12{,}000$, $r = 3.5\% = 0.035$

$A = Pe^{rt} = 12{,}000e^{0.035t}$

t	1	10	20	30	40	50
A	12,427.44	17,028.81	24,165.03	34,291.81	48,662.40	69,055.23

61. $A = 25\left[\dfrac{(1 + 0.12/12)^{48} - 1}{0.12/12}\right]$

$= 25\left[\dfrac{1.01^{48} - 1}{0.01}\right]$

$= \$1530.57$

63. $A = 200\left[\dfrac{(1 + 0.06/12)^{72} - 1}{0.06/12}\right]$

$= \$17{,}281.77$

65. $p = 5000\left(1 - \dfrac{4}{4 + e^{-0.002x}}\right)$

(a)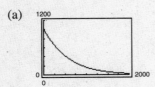

(b) If $x = 500$, $p \approx \$421.12$.

(c) For $x = 600$, $p \approx \$350.13$.

x	100	200	300	400	500	600	700
p	849.53	717.64	603.25	504.94	421.12	350.13	290.35

67. $Q = 25\left(\tfrac{1}{2}\right)^{t/1599}$

(a) When $t = 0$,

$Q = 25\left(\tfrac{1}{2}\right)^{0/1599} = 25(1) = 25$ grams.

(b) When $t = 1000$,

$Q = 25\left(\tfrac{1}{2}\right)^{1000/1599} \approx 16.21$ grams.

(c)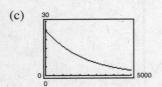

(d) Never. The graph has a horizontal asymptote $Q = 0$.

69. $P(t) = 100e^{0.2197t}$

(a)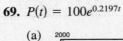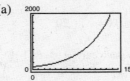

(b) $P(0) = 100$

$P(5) \approx 300$

$P(10) \approx 900$

(c) $P(0) = 100e^{0.2197(0)} = 100$

$P(5) = 100e^{0.2197(5)} = 299.966 \approx 300$

$P(10) = 100e^{0.2197(10)} = 899.798 \approx 900$

71. $C(t) = P(1.04)^t$

(a)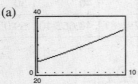

(b) $C(10) \approx 35.45$

(c) $C(10) = 23.95(1.04)^{10} \approx 35.45$

73. True. $f(x) = 1^x$ is not an exponential function.

75. The graph decreases for all x and has positive y-intercept. Matches (d).

77. $f(x) = \left(1 + \dfrac{0.5}{x}\right)^x$ and $g(x) = e^{0.5} \approx 1.6487$
(Horizontal line)

As $x \to \infty$, $f(x) \to g(x)$.

79. $e^{\pi} \approx 23.14$, $\pi^e \approx 22.46$

$e^{\pi} > \pi^e$

81. $5^{-3} = 0.008, 3^{-5} \approx 0.0041$

$5^{-3} > 3^{-5}$

83. f has an inverse because f is one-to-one.

$$y = 5x - 7$$
$$x = 5y - 7$$
$$x + 7 = 5y$$
$$f^{-1}(x) = \tfrac{1}{5}(x + 7)$$

85. f has an inverse because f is one-to-one.

$$y = \sqrt[3]{x + 8}$$
$$x = \sqrt[3]{y + 8}$$
$$x^3 = y + 8$$
$$x^3 - 8 = y$$
$$f^{-1}(x) = x^3 - 8$$

87. $f(x) = \dfrac{2x}{x - 7}$

Vertical asymptote: $x = 7$

Horizontal asymptote: $y = 2$

Intercept: $(0, 0)$

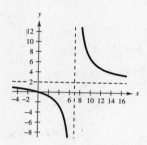

89. Answers will vary.

Section 4.2   Logarithmic Functions and Their Graphs

- You should know that a function of the form $y = \log_a x$, where $a > 0, a \neq 1$, and $x > 0$, is called a logarithm of x to base a.

- You should be able to convert from logarithmic form to exponential form and vice versa.

  $$y = \log_a x \iff a^y = x$$

- You should know the following properties of logarithms.

  (a) $\log_a 1 = 0$ since $a^0 = 1$.   (c) $\log_a a^x = x$ since $a^x = a^x$.

  (b) $\log_a a = 1$ since $a^1 = a$.   (d) If $\log_a x = \log_a y$, then $x = y$.

- You should know the definition of the natural logarithmic function.

  $$\log_e x = \ln x, x > 0$$

- You should know the properties of the natural logarithmic function.

  (a) $\ln 1 = 0$ since $e^0 = 1$.   (c) $\ln e^x = x$ since $e^x = e^x$.

  (b) $\ln e = 1$ since $e^1 = e$.   (d) If $\ln x = \ln y$, then $x = y$.

- You should be able to graph logarithmic functions.

Vocabulary Check

1. logarithmic function   **2.** 10   **3.** natural logarithmic

4. $a^{\log_a x} = x$ **5.** $x = y$

1. $\log_4 64 = 3 \implies 4^3 = 64$

3. $\log_7 \frac{1}{49} = -2 \implies 7^{-2} = \frac{1}{49}$

5. $\log_{32} 4 = \frac{2}{5} \implies 32^{2/5} = 4$

7. $\ln 1 = 0 \implies e^0 = 1$

9. $\ln e = 1 \implies e^1 = e$

11. $\ln \sqrt{e} = \frac{1}{2} \implies e^{1/2} = \sqrt{e}$

13. $5^3 = 125 \implies \log_5 125 = 3$

15. $81^{1/4} = 3 \implies \log_{81} 3 = \frac{1}{4}$

17. $6^{-2} = \frac{1}{36} \implies \log_6 \frac{1}{36} = -2$

19. $e^3 = 20.0855\ldots \implies \ln 20.0855\ldots = 3$

21. $e^{1.3} = 3.6692\ldots \implies \ln 3.6692\ldots = 1.3$

23. $\sqrt[3]{e} = 1.3956\ldots \implies \ln(1.3956\ldots) = \frac{1}{3}$

25. $\log_2 16 = \log_2 2^4 = 4$

27. $g\left(\frac{1}{1000}\right) = \log_{10}\left(\frac{1}{1000}\right)$
$$= \log_{10}(10^{-3})$$
$$= -3$$

29. $\log_{10} 345 \approx 2.538$

31. $6 \log_{10} 14.8 \approx 7.022$

33. $\log_7 x = \log_7 9$
$$x = 9$$

35. $\log_6 6^2 = x$
$$2 \log_6 6 = x$$
$$2 = x$$

37. $\log_8 x = \log_8 10^{-1}$
$$x = 10^{-1} = \frac{1}{10}$$

39. $\log_4 4^{3x} = (3x) \log_4 4 = 3x$

41. $3 \log_2\left(\frac{1}{2}\right) = 3 \log_2(2^{-1})$
$$= 3(-1) = -3$$

43. $f(x) = 3^x$ and $g(x) = \log_3 x$ are inverses of each other.

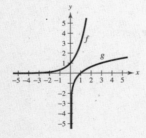

45. $f(x) = e^{2x}$ and $g(x) = \frac{1}{2} \ln x$ are inverses of each other.

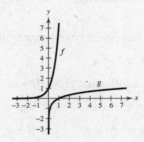

47. $y = \log_2(x + 2)$

Domain: $x + 2 > 0 \implies x > -2$
Vertical asymptote: $x = -2$
$\log_2(x + 2) = 0$
$\qquad x + 2 = 1$
$\qquad\qquad x = -1$
x-intercept: $(-1, 0)$

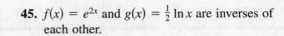

49. $y = 1 + \log_2 x$

Domain: $x > 0$
Vertical asymptote: $x = 0$
$1 + \log_2 x = 0$
$\qquad \log_2 x = -1$
$\qquad\qquad x = 2^{-1} = \frac{1}{2}$
x-intercept: $\left(\frac{1}{2}, 0\right)$

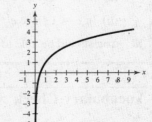

51. $y = 1 + \log_2(x - 2)$

Domain: $x - 2 > 0 \implies x > 2$

Vertical asymptote: $x = 2$

$1 + \log_2(x - 2) = 0$

$\log_2(x - 2) = -1$

$x - 2 = 2^{-1} = \frac{1}{2}$

$x = \frac{5}{2}$

x-intercept: $\left(\frac{5}{2}, 0\right)$

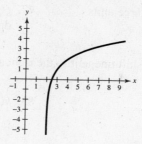

53. $f(x) = \log_3 x + 2$

Asymptote: $x = 0$

Point on graph: $(1, 2)$

Matches graph (b).

55. $f(x) = -\log_3(x + 2)$

Asymptote: $x = -2$

Point on graph: $(-1, 0)$

Matches graph (d).

57. $f(x) = \log_{10} x$

$g(x) = -\log_{10} x$ is a reflection in the x-axis of the graph of f.

59. $f(x) = \log_2 x$

$g(x) = 4 - \log_2 x$ is obtained from f by a reflection in the x-axis followed by a vertical shift four units upward.

61. Horizontal shift three units to the left and a vertical shift two units downward

63. $\ln \sqrt{42} \approx 1.869$

65. $-\ln\left(\frac{1}{2}\right) \approx 0.693$

67. $\ln e^2 = 2$

(Inverse Property)

69. $e^{\ln 1.8} = 1.8$

(Inverse Property)

71. $f(x) = \ln(x - 1)$

Domain: $x > 1$

Vertical asymptote: $x = 1$

x-intercept: $(2, 0)$

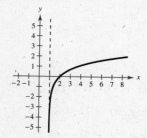

73. $g(x) = \ln(-x)$

Domain: $-x > 0 \implies x < 0$

The domain is $(-\infty, 0)$.

Vertical asymptote: $-x = 0 \implies x = 0$

x-intercept: $\quad 0 = \ln(-x)$

$e^0 = -x$

$-1 = x$

The x-intercept is $(-1, 0)$.

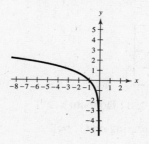

75. $g(x) = \ln(x + 3)$ is a horizontal shift three units to the left.

77. $g(x) = \ln x - 5$ is a vertical shift five units downward.

79. $g(x) = \ln(x - 1) + 2$ is a horizontal shift one unit to the right and a vertical shift two units upward.

81. $f(x) = \dfrac{x}{2} - \ln\dfrac{x}{4}$

(a)
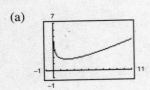

(b) Domain: $(0, \infty)$

(c) Increasing on $(2, \infty)$

Decreasing on $(0, 2)$

(d) Relative minimum: $(2, 1.693)$

83. $h(x) = 4x \ln x$

(a)
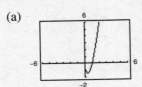

(b) Domain: $(0, \infty)$

(c) Increasing on $(0.368, \infty)$

Decreasing on $(0, 0.368)$

(d) Relative minimum: $(0.368, -1.472)$

85. $f(x) = \ln\left(\dfrac{x + 2}{x - 1}\right)$

(a)
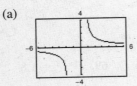

(b) $\dfrac{x + 2}{x - 1} > 0$; Critical numbers: $1, -2$

Test intervals: $(-\infty, -2), (-2, 1), (1, \infty)$

Testing these three intervals, we see that the domain is $(-\infty, -2) \cup (1, \infty)$.

(c) The graph is decreasing on $(-\infty, -2)$ and decreasing on $(1, \infty)$.

(d) There are no relative maximum or minimum values.

87. $f(x) = \ln\left(\dfrac{x^2}{10}\right)$

(a)
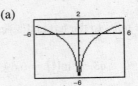

(b) $\dfrac{x^2}{10} > 0 \implies x \neq 0$; Domain: all $x \neq 0$

(c) The graph is increasing on $(0, \infty)$ and decreasing on $(-\infty, 0)$.

(d) There are no relative maximum or relative minimum values.

89. $f(x) = \sqrt{\ln x}$

(a)
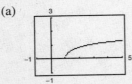

(b) $\ln x \geq 0 \implies x \geq 1$; Domain: $x \geq 1$

(c) The graph is increasing on $(1, \infty)$.

(d) There are no relative maximum or relative minimum values.

91. $f(t) = 80 - 17 \log_{10}(t + 1)$, $\quad 0 \leq t \leq 12$

(a) $f(0) = 80 - 17 \log_{10}(0 + 1) = 80$

(b) $f(4) = 80 - 17 \log_{10}(4 + 1) \approx 68.1$

(c) $f(10) = 80 - 17 \log_{10}(10 + 1) \approx 62.3$

(d)
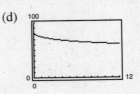

93. $t = \dfrac{\ln K}{0.055}$

(a)

K	1	2	4	6	8	10	12
t	0	12.6	25.2	32.6	37.8	41.9	45.2

As the amount increases, the time increases, but at a lesser rate.

(b)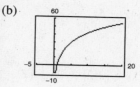

95. $\beta = 10 \log_{10}\left(\dfrac{I}{10^{-12}}\right)$

  (a) $I = 1$: $\beta = 10 \log_{10}\left(\dfrac{1}{10^{-12}}\right) = 10 \cdot \log_{10}(10^{12}) = 10(12) = 120$ decibels

  (b) $I = 10^{-2}$: $\beta = 10 \log_{10}\left(\dfrac{10^{-2}}{10^{-12}}\right) = 10 \log_{10}(10^{10}) = 10(10) = 100$ decibels

  (c) No, this is a logarithmic scale.

97. $y = 80.4 - 11 \ln x$

$y(300) = 80.4 - 11 \ln 300 \approx 17.66 \text{ ft}^3/\text{min}$

99. False. You would reflect $y = 6^x$ in the line $y = x$.

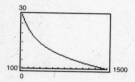

101. $5 = \log_b 32$

$b^5 = 32 = 2^5$

$b = 2$

103. $2 = \log_b\left(\tfrac{1}{16}\right)$

$b^2 = \tfrac{1}{16} = \left(\tfrac{1}{4}\right)^2$

$b = \tfrac{1}{4}$

105. The vertical asymptote is to the right of the y-axis, and the graph increases. Matches (b).

107. $f(x) = \log_a x$ is the inverse of $g(x) = a^x$, where $a > 0$, $a \neq 1$.

109. (a) False, y is not an exponential function of x. (y can never be 0.)

  (b) True, y could be $\log_2 x$.

  (c) True, x could be 2^y.

  (d) False, y is not linear. (The points are not collinear.)

111. $f(x) = \dfrac{\ln x}{x}$

(a)

x	1	5	10	10^2	10^4	10^6
$f(x)$	0	0.322	0.230	0.046	0.00092	0.0000138

(b) As x increases without bound, $f(x)$ approaches 0.

(c)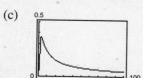

113. $x^2 + 2x - 3 = (x + 3)(x - 1)$

115. $12x^2 + 5x - 3 = (4x + 3)(3x - 1)$

117. $16x^2 - 25 = (4x + 5)(4x - 5)$

119. $2x^3 + x^2 - 45x = x(2x^2 + x - 45)$
$$= x(2x - 9)(x + 5)$$

121. $(f + g)(2) = f(2) + g(2) = [3(2) + 2] + [2^3 - 1] = 8 + 7 = 15$

123. $(fg)(6) = f(6)g(6) = [3(6) + 2][6^3 - 1] = [20][215] = 4300$

125. $5x - 7 = x + 4$

The graphs of $y = 5x - 7$ and $y_2 = x + 4$ intersect when $x = 2.75$ or $\frac{11}{4}$.

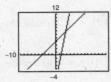

127. $\sqrt{3x - 2} = 9$

The graphs of $y_1 = \sqrt{3x - 2}$ and $y_2 = 9$ intersect when $x \approx 27.667$ or $\frac{83}{3}$.

Section 4.3   Properties of Logarithms

■   You should know the following properties of logarithms.

   (a) $\log_a x = \dfrac{\log_b x}{\log_b a}$

   (b) $\log_a(uv) = \log_a u + \log_a v$ $\ln(uv) = \ln u + \ln v$

   (c) $\log_a(u/v) = \log_a u - \log_a v$ $\ln(u/v) = \ln u - \ln v$

   (d) $\log_a u^n = n \log_a u$ $\ln u^n = n \ln u$

■   You should be able to rewrite logarithmic expressions using these properties.

Vocabulary Check

1. change-of-base   **2.** $\dfrac{\ln x}{\ln a}$ **3.** $\log_a u^n$ **4.** $\ln u + \ln v$

---

1. (a) $\log_5 x = \dfrac{\log_{10} x}{\log_{10} 5}$

    (b) $\log_5 x = \dfrac{\ln x}{\ln 5}$

3. (a) $\log_{1/5} x = \dfrac{\log_{10} x}{\log_{10} 1/5} = \dfrac{\log_{10} x}{-\log_{10} 5}$

    (b) $\log_{1/5} x = \dfrac{\ln x}{\ln 1/5} = \dfrac{\ln x}{-\ln 5}$

5. (a) $\log_a\left(\dfrac{3}{10}\right) = \dfrac{\log_{10}(3/10)}{\log_{10} a}$

    (b) $\log_a\left(\dfrac{3}{10}\right) = \dfrac{\ln(3/10)}{\ln a}$

7. (a) $\log_{2.6} x = \dfrac{\log_{10} x}{\log_{10} 2.6}$

    (b) $\log_{2.6} x = \dfrac{\ln x}{\ln 2.6}$

9. $\log_3 7 = \dfrac{\ln 7}{\ln 3} \approx 1.771$

11. $\log_{1/2} 4 = \dfrac{\ln 4}{\ln(1/2)} = -2$

13. $\log_9(0.8) = \dfrac{\ln(0.8)}{\ln 9} \approx -0.102$

15. $\log_{15} 1460 = \dfrac{\ln 1460}{\ln 15} \approx 2.691$

17. $\ln 20 = \ln(4 \cdot 5)$
$= \ln 4 + \ln 5$

19. $\ln \frac{5}{64} = \ln 5 - \ln 64$
$= \ln 5 - \ln 4^3$
$= \ln 5 - 3 \ln 4$

21. $\log_b 25 = \log_b 5^2$
$= 2 \log_b 5$
$\approx 2(0.8271) \approx 1.6542$

23. $\log_b \sqrt{3} = \frac{1}{2} \log_b 3$
$\approx \frac{1}{2}(0.5646)$
≈ 0.2823

25. $f(x) = \log_3(x + 2) = \dfrac{\ln(x + 2)}{\ln 3}$

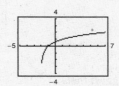

27. $f(x) = \log_{1/2}(x - 2) = \dfrac{\ln(x - 2)}{\ln(1/2)} = \dfrac{\ln(x - 2)}{-\ln 2}$

29. $f(x) = \log_{1/4}(x^2) = \dfrac{\ln x^2}{\ln(1/4)} = \dfrac{\ln x^2}{-\ln 4}$

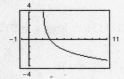

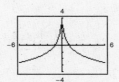

31. $\log_4 8 = \log_4 2^3 = 3 \log_4 2$
$= 3 \log_4 4^{1/2} = 3\left(\frac{1}{2}\right) \log_4 4$
$= \frac{3}{2}$

33. $\ln(5e^6) = \ln 5 + \ln e^6 = \ln 5 + 6 = 6 + \ln 5$

35. $\log_5 \frac{1}{250} = \log_5 1 - \log_5 250 = 0 - \log_5(125 \cdot 2)$
$= -\log_5(5^3 \cdot 2) = -[\log_5 5^3 + \log_5 2]$
$= -[3 \log_5 5 + \log_5 2] = -3 - \log_5 2$

37. $\log_{10} 5x = \log_{10} 5 + \log_{10} x$

39. $\log_{10} \dfrac{5}{x} = \log_{10} 5 - \log_{10} x$

41. $\log_8 x^4 = 4 \log_8 x$

43. $\ln \sqrt{z} = \ln z^{1/2} = \frac{1}{2} \ln z$

45. $\ln xyz = \ln x + \ln y + \ln z$

47. $\log_3\left(a^2 b c^3\right) = \log_3 a^2 + \log_3 b + \log_3 c^3$
$= 2 \log_3 a + \log_3 b + 3 \log_3 c$

49. $\ln\left(a^2 \sqrt{a - 1}\right) = \ln a^2 + \ln(a - 1)^{1/2}$
$= 2 \ln a + \frac{1}{2} \ln(a - 1), \ a > 1$

51. $\ln \sqrt[3]{\dfrac{x}{y}} = \frac{1}{3} \ln \dfrac{x}{y} = \frac{1}{3}[\ln x - \ln y] = \frac{1}{3} \ln x - \frac{1}{3} \ln y$

53. $\ln\left(\dfrac{x^2 - 1}{x^3}\right) = \ln(x^2 - 1) - \ln x^3$

$\qquad\qquad = \ln[(x - 1)(x + 1)] - 3\ln x$

$\qquad\qquad = \ln(x - 1) + \ln(x + 1) - 3\ln x, \; x > 1$

55. $\ln\left(\dfrac{x^4\sqrt{y}}{z^5}\right) = \ln x^4\sqrt{y} - \ln z^5$

$\qquad\qquad = \ln x^4 + \ln\sqrt{y} - \ln z^5$

$\qquad\qquad = 4\ln x + \dfrac{1}{2}\ln y - 5\ln z$

57. $y_1 = \ln[x^3(x + 4)]$

$\quad y_2 = 3\ln x + \ln(x + 4)$

(a)

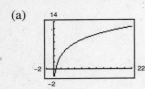

(b)

x	0.5	1	1.5	2	3	10
y_1	-0.5754	1.6094	2.9211	3.8712	5.2417	9.5468
y_2	-0.5754	1.6094	2.9211	3.8712	5.2417	9.5468

(c) The graphs and table suggest that
$y_1 = y_2$ for $x > 0$. In fact,

$\quad y_1 = \ln[x^3(x + 4)] = \ln x^3 + \ln(x + 4)$

$\qquad = 3\ln x + \ln(x + 4) = y_2.$

59. $\ln x + \ln 4 = \ln 4x$

61. $\log_4 z - \log_4 y = \log_4 \dfrac{z}{y}$

63. $2\log_2(x + 3) = \log_2(x + 3)^2$

65. $\dfrac{1}{2}\ln(x^2 + 4) = \ln(x^2 + 4)^{1/2}$

$\qquad\qquad = \ln\sqrt{x^2 + 4}$

67. $\ln x - 3\ln(x + 1) = \ln x - \ln(x + 1)^3$

$\qquad\qquad = \ln\dfrac{x}{(x + 1)^3}$

69. $\ln(x - 2) - \ln(x + 2) = \ln\left(\dfrac{x - 2}{x + 2}\right)$

71. $\ln x - 2[\ln(x + 2) + \ln(x - 2)] = \ln x - 2\ln[(x + 2)(x - 2)]$

$\qquad\qquad\qquad\qquad\qquad\qquad = \ln x - 2\ln(x^2 - 4)$

$\qquad\qquad\qquad\qquad\qquad\qquad = \ln x - \ln(x^2 - 4)^2$

$\qquad\qquad\qquad\qquad\qquad\qquad = \ln\dfrac{x}{(x^2 - 4)^2}$

73. $\dfrac{1}{3}[2\ln(x + 3) + \ln x - \ln(x^2 - 1)] = \dfrac{1}{3}[\ln(x + 3)^2 + \ln x - \ln(x^2 - 1)]$

$\qquad\qquad\qquad\qquad\qquad\qquad\qquad = \dfrac{1}{3}[\ln[x(x + 3)^2] - \ln(x^2 - 1)]$

$\qquad\qquad\qquad\qquad\qquad\qquad\qquad = \dfrac{1}{3}\ln\dfrac{x(x + 3)^2}{x^2 - 1}$

$\qquad\qquad\qquad\qquad\qquad\qquad\qquad = \ln\sqrt[3]{\dfrac{x(x + 3)^2}{x^2 - 1}}$

75. $\frac{1}{3}[\ln y + 2\ln(y + 4)] - \ln(y - 1) = \frac{1}{3}[\ln y + \ln(y + 4)^2] - \ln(y - 1)$

$$= \frac{1}{3}\ln[y(y + 4)^2] - \ln(y - 1)$$

$$= \ln\sqrt[3]{y(y + 4)^2} - \ln(y - 1)$$

$$= \ln\frac{\sqrt[3]{y(y + 4)^2}}{y - 1}$$

77. $y_1 = 2[\ln 8 - \ln(x^2 + 1)]$

$y_2 = \ln\left[\frac{64}{(x^2 + 1)^2}\right]$

(a)

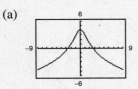

(c) The graphs and table suggest that $y_1 = y_2$. In fact,

$$y_1 = 2[\ln 8 - \ln(x^2 + 1)]$$

$$= 2\ln\frac{8}{x^2 + 1} = \ln\frac{64}{(x^2 + 1)^2} = y_2.$$

(b)

x	-8	-4	-2	0	2	4	8
y_1	-4.1899	-1.5075	0.9400	4.1589	0.9400	-1.5075	-4.1899
y_2	-4.1899	-1.5075	0.9400	4.1589	0.9400	-1.5075	-4.1899

79. $y_1 = \ln x^2$

$y_2 = 2\ln x$

(a)

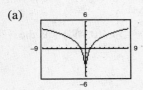

(The domain of y_2 is $x > 0$.)

(b)

x	-8	-4	1	2	4
y_1	4.1589	2.7726	0	1.3863	2.7726
y_2	undefined	undefined	0	1.3863	2.7726

(c) The graphs and table suggest that $y_1 = y_2$ for $x > 0$. The functions are not equivalent because the domains are different.

81. $\log_3 9 = 2\log_3 3 = 2$

83. $\log_4 16^{3.4} = 3.4\log_4(4^2) = 6.8\log_4 4 = 6.8$

85. $\log_2(-4)$ is undefined. -4 is not in the domain of $f(x) = \log_2 x$.

87. $\log_5 75 - \log_5 3 = \log_5 \frac{75}{3} = \log_5 25 = \log_5 5^2 = 2$

89. $\ln e^3 - \ln e^7 = 3 - 7 = -4$

91. $2\ln e^4 = 2(4)\ln e = 8$

93. $\ln\left(\frac{1}{\sqrt{e}}\right) = \ln(1) - \ln e^{1/2} = 0 - \frac{1}{2}\ln e = -\frac{1}{2}$

95. (a) $\beta = 10 \cdot \log_{10}\left(\dfrac{I}{10^{-12}}\right) = 10[\log_{10} I - \log_{10} 10^{-12}]$

$= 10[\log_{10} I - (-12) \log_{10} 10]$

$= 10[\log_{10} I + 12] = 120 + 10 \cdot \log_{10} I$

(b)

I	10^{-4}	10^{-6}	10^{-8}	10^{-10}	10^{-12}	10^{-14}
β	80	60	40	20	0	-20

(c) $\beta(10^{-4}) = 120 + 10 \cdot \log_{10} 10^{-4} = 120 - 40 = 80$

$\beta(10^{-6}) = 120 + 10 \cdot \log_{10} 10^{-6} = 120 - 60 = 60$

$\beta(10^{-8}) = 120 + 10 \cdot \log_{10} 10^{-8} = 120 - 80 = 40$

$\beta(10^{-10}) = 120 + 10 \cdot \log_{10} 10^{-10} = 120 - 100 = 20$

$\beta(10^{-12}) = 120 + 10 \cdot \log_{10} 10^{-12} = 120 - 120 = 0$

$\beta(10^{-14}) = 120 + 10 \cdot \log_{10} 10^{-14} = 120 - 140 = -20$

97. (a)

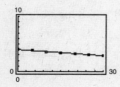

(b) $T - 21 = 54.4(0.964)^t$

$T = 21 + 54.4(0.964)^t$

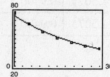

The data $(t, T - 21)$ fits the model
$T - 21 = 54.4(0.964)^t$.

The model $T = 21 + 54.4(0.964)^t$ fits the original data.

(c) $\ln(T - 21) = -0.0372t + 3.9971$, linear model

$T - 21 = e^{-0.0372t + 3.9971}$

$T = 21 + 54.4e^{-0.0372t}$

$= 21 + 54.4(0.964)^t$

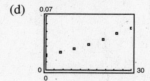

(d)

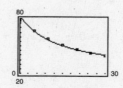

$\dfrac{1}{T - 21} = 0.00121t + 0.01615$, linear model

$T - 21 = \dfrac{1}{0.00121t + 0.01615}$

$T = 21 + \dfrac{1}{0.00121t + 0.01615}$

99. True

101. False. For example, let $x = 1$ and $a = 2$.

Then $f\left(\dfrac{x}{a}\right) = \ln\left(\dfrac{1}{2}\right)$. But $\dfrac{f(x)}{f(a)} = \dfrac{\ln 1}{\ln 2} = 0$.

103. False. $\sqrt{\ln x} \neq \dfrac{1}{2}\ln x$

In fact, $\ln x^{1/2} = \dfrac{1}{2}\ln x$.

105. True. In fact, if $\ln x < 0$, then $0 < x < 1$.

107. Let $y = \log_a x$ and $z = \log_{a/b} x$, then $a^y = x = \left(\dfrac{a}{b}\right)^z$ and

$$\left(\dfrac{1}{b}\right)^z = a^{y-z}$$

$$\dfrac{1}{b} = a^{(y-z)/z}$$

$$\log_a\left(\dfrac{1}{b}\right) = \dfrac{y-z}{z} = \dfrac{y}{z} - 1 \implies 1 + \log_a\left(\dfrac{1}{b}\right) = \dfrac{\log_a x}{\log_{a/b} x}.$$

109. $f(x) = \log_2 x = \dfrac{\ln x}{\ln 2}$

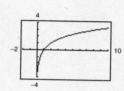

111. $f(x) = \log_3 \sqrt{x} = \dfrac{1}{2}\dfrac{\ln x}{\ln 3}$

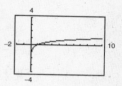

113. $f(x) = \log_5\left(\dfrac{x}{3}\right) = \dfrac{\ln(x/3)}{\ln 5}$

115. $\ln 1 = 0$, $\ln 2 \approx 0.6931$, $\ln 3 \approx 1.0986$, $\ln 5 \approx 1.6094$

$\ln 2 \approx 0.6931$

$\ln 3 \approx 1.0986$

$\ln 4 = \ln 2 + \ln 2 \approx 0.6931 + 0.6931 = 1.3862$

$\ln 5 \approx 1.6094$

$\ln 6 = \ln 2 + \ln 3 \approx 0.6931 + 1.0986 = 1.7917$

$\ln 8 = \ln 2^3 = 3\ln 2 \approx 3(0.6931) = 2.0793$

$\ln 9 = \ln 3^2 = 2\ln 3 \approx 2(1.0986) = 2.1972$

$\ln 10 = \ln 5 + \ln 2 \approx 1.6094 + 0.6931 = 2.3025$

$\ln 12 = \ln 2^2 + \ln 3 = 2\ln 2 + \ln 3 \approx 2(0.6931) + 1.0986 = 2.4848$

$\ln 15 = \ln 5 + \ln 3 \approx 1.6094 + 1.0986 = 2.7080$

$\ln 16 = \ln 2^4 = 4\ln 2 \approx 4(0.6931) = 2.7724$

$\ln 18 = \ln 3^2 + \ln 2 = 2\ln 3 + \ln 2 \approx 2(1.0986) + 0.6931 = 2.8903$

$\ln 20 = \ln 5 + \ln 2^2 = \ln 5 + 2\ln 2 \approx 1.6094 + 2(0.6931) = 2.9956$

117. $\left(\dfrac{2x^2}{3y}\right)^{-3} = \left(\dfrac{3y}{2x^2}\right)^3$

$= \dfrac{(3y)^3}{(2x^2)^3}$

$= \dfrac{27y^3}{8x^6}$

119. $xy(x^{-1} + y^{-1})^{-1} = \dfrac{xy}{x^{-1} + y^{-1}}$

$= \dfrac{xy}{\dfrac{1}{x} + \dfrac{1}{y}}$

$= \dfrac{xy}{\dfrac{y + x}{xy}}$

$= \dfrac{(xy)^2}{x + y}, \; x \neq 0, y \neq 0$

121. $2x^3 + 20x^2 + 50x = 0$

$2x(x^2 + 10x + 25) = 0$

$2x(x + 5)^2 = 0$

$x = 0, -5, -5$

123. $9x^4 - 37x^2 + 4 = 0$

$(x^2 - 4)(9x^2 - 1) = 0$

$(x - 2)(x + 2)(3x - 1)(3x + 1) = 0$

$x = \pm 2, \pm \frac{1}{3}$

125. $9x^4 - 226x^2 + 25 = 0$

$(x^2 - 25)(9x^2 - 1) = 0$

$(x - 5)(x + 5)(3x + 1)(3x - 1) = 0$

$x = \pm 5, \pm \frac{1}{3}$

Section 4.4 Solving Exponential and Logarithmic Equations

- To solve an exponential equation, isolate the exponential expression, then take the logarithm of both sides. Then solve for the variable.
    1. $\log_a a^x = x$
    2. $\ln e^x = x$
- To solve a logarithmic equation, rewrite it in exponential form. Then solve for the variable.
    1. $a^{\log_a x} = x$
    2. $e^{\ln x} = x$
- If $a > 0$ and $a \neq 1$ we have the following:
    1. $\log_a x = \log_a y \implies x = y$
    2. $a^x = a^y \implies x = y$
- Use your graphing utility to approximate solutions.

Vocabulary Check

1. solve

2. (a) $x = y$ (b) $x = y$ (c) x (d) x

3. extraneous

1. $4^{2x-7} = 64$

    (a) $x = 5$

        $4^{2(5)-7} = 4^3 = 64$

        Yes, $x = 5$ is a solution.

    (b) $x = 2$

        $4^{2(2)-7} = 4^{-3} = \frac{1}{64} \neq 64$

        No, $x = 2$ is not a solution.

3. $3e^{x+2} = 75$

    (a) $x = -2 + e^{25}$

        $3e^{(-2+e^{25})+2} = 3e^{e^{25}} \neq 75$

        No, $x = -2 + e^{25}$ is not a solution.

    (b) $x = -2 + \ln 25$

        $3e^{(-2+\ln 25)+2} = 3e^{\ln 25} = 3(25) = 75$

        Yes, $x = -2 + \ln 25$ is a solution.

    (c) $x \approx 1.2189$

        $3e^{1.2189+2} = 3e^{3.2189} \approx 75$

        Yes, $x \approx 1.2189$ is a solution.

5. $\log_4(3x) = 3$

        $4^3 = 3x$

        $x = \frac{64}{3} \approx 21.333$

    (a) $x = 21.3560$ is an approximate solution.

    (b) No, $x = -4$ is not a solution.

    (c) Yes, $x = \frac{64}{3}$ is a solution.

7. $\ln(x - 1) = 3.8$

    (a) $x = 1 + e^{3.8}$

        $\ln(1 + e^{3.8} - 1) = \ln e^{3.8} = 3.8$

        Yes, $x = 1 + e^{3.8}$ is a solution.

    (b) $x \approx 45.7012$

        $\ln(45.7012 - 1) = \ln(44.7012) \approx 3.8$

        Yes, $x \approx 45.7012$ is a solution.

    (c) $x = 1 + \ln 3.8$

        $\ln(1 + \ln 3.8 - 1) = \ln(\ln 3.8) \approx 0.289$

        No, $x = 1 + \ln 3.8$ is not a solution.

9.

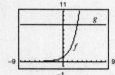

Point of intersection: $(3, 8)$

Algebraically: $2^x = 8$

        $2^x = 2^3$

        $x = 3 \implies y = 8 \implies (3, 8)$

11.

Point of intersection: $(4, 10)$

Algebraically: $5^{x-2} - 15 = 10$

        $5^{x-2} = 25 = 5^2$

        $x - 2 = 2$

        $x = 4$

$(4, 10)$

13.

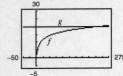

Point of intersection: $(243, 20)$

Algebraically: $4 \log_3 x = 20$

$$\log_3 x = 5$$

$$x = 3^5 = 243$$

$(243, 20)$

15.

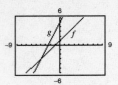

Point of intersection: $(-4, -3)$

Algebraically: $\ln e^{x+1} = 2x + 5$

$$x + 1 = 2x + 5$$

$$-4 = x$$

$(-4, -3)$

17. $4^x = 16$

$4^x = 4^2$

$x = 2$

19. $5^x = \dfrac{1}{625}$

$5^x = \dfrac{1}{5^4} = 5^{-4}$

$x = -4$

21. $\left(\dfrac{1}{8}\right)^x = 64$

$8^{-x} = 8^2$

$-x = 2$

$x = -2$

23. $\left(\dfrac{2}{3}\right)^x = \dfrac{81}{16}$

$\left(\dfrac{3}{2}\right)^{-x} = \left(\dfrac{3}{2}\right)^4$

$-x = 4$

$x = -4$

25. $6(10^x) = 216$

$10^x = 36$

$\log_{10} 10^x = \log_{10} 36$

$x = \log_{10} 36 \approx 1.5563$

27. $2^{x+3} = 256$

$2^x \cdot 2^3 = 256$

$2^x = 32$

$x = 5$

Alternate solution:

$2^{x+3} = 2^8$

$x + 3 = 8$

$x = 5$

29. $\ln x - \ln 5 = 0$

$\ln x = \ln 5$

$x = 5$

31. $\ln x = -7$

$x = e^{-7}$

33. $\log_x 625 = 4$

$x^4 = 625$

$x^4 = 5^4$

$x = 5$

35. $\log_{10} x = -1$

$x = 10^{-1}$

$x = \dfrac{1}{10}$

37. $\ln(2x - 1) = 5$

$2x - 1 = e^5$

$x = \dfrac{1 + e^5}{2} \approx 74.707$

39. $\ln e^{x^2} = x^2 \ln e^x = x^2$

41. $e^{\ln(5x+2)} = 5x + 2$

43. $-1 + \ln e^{2x} = -1 + 2x = 2x - 1$

45. $8^{3x} = 360$

$\ln 8^{3x} = \ln 360$

$3x \ln 8 = \ln 360$

$3x = \dfrac{\ln 360}{\ln 8}$

$x = \dfrac{1}{3} \dfrac{\ln 360}{\ln 8}$

$x \approx 0.944$

47. $5^{-t/2} = 0.20 = \dfrac{1}{5}$

$-\dfrac{t}{2} \ln 5 = \ln\left(\dfrac{1}{5}\right)$

$-\dfrac{t}{2} \ln 5 = -\ln 5$

$\dfrac{t}{2} = 1$

$t = 2$

49. $5(2^{3-x}) - 13 = 100$

$5(2^{3-x}) = 113$

$2^{3-x} = \dfrac{113}{5}$

$\ln 2^{3-x} = \ln\left(\dfrac{113}{5}\right)$

$3 - x = \dfrac{\ln(113/5)}{\ln 2}$

$x = 3 - \dfrac{\ln(113/5)}{\ln 2}$

$x \approx -1.498$

51. $\left(1 + \dfrac{0.10}{12}\right)^{12t} = 2$

$\left(\dfrac{12.1}{12}\right)^{12t} = 2$

$(12t) \ln\left(\dfrac{12.1}{12}\right) = \ln 2$

$t = \dfrac{1}{12} \dfrac{\ln 2}{\ln(12.1/12)}$

$t \approx 6.960$

53. $5000\left[\dfrac{(1 + 0.005)^x}{0.005}\right] = 250{,}000$

$5000(1.005)^x = 1250$

$1.005^x = 0.25$

$x \ln(1.005) = \ln 0.25$

$x = \dfrac{\ln 0.25}{\ln(1.005)}$

$x \approx -277.951$

55. $2e^{5x} = 18$

$e^{5x} = 9$

$5x = \ln 9$

$x = \dfrac{1}{5} \ln 9$

$x \approx 0.439$

57. $500e^{-x} = 300$

$e^{-x} = \dfrac{3}{5}$

$-x = \ln \dfrac{3}{5}$

$x = -\ln \dfrac{3}{5} = \ln \dfrac{5}{3} \approx 0.511$

59. $7 - 2e^x = 5$

$-2e^x = -2$

$e^x = 1$

$x = \ln 1 = 0$

61. $e^{2x} - 4e^x - 5 = 0$

$(e^x - 5)(e^x + 1) = 0$

$e^x = 5 \text{ or } e^x = -1$

$x = \ln 5 \approx 1.609$

$(e^x = -1 \text{ is impossible.})$

63. $250e^{0.02x} = 10{,}000$

$e^{0.02x} = 40$

$0.02x = \ln 40$

$x = \dfrac{\ln 40}{0.02}$

$x \approx 184.444$

65. $e^x = e^{x^2 - 2}$

$x = x^2 - 2$

$x^2 - x - 2 = 0$

$(x - 2)(x + 1) = 0$

$x = 2, -1$

67. $e^{x^2 - 3x} = e^{x-2}$

$x^2 - 3x = x - 2$

$x^2 - 4x + 2 = 0$

$x = \dfrac{4 \pm \sqrt{16 - 8}}{2}$

$x = 2 \pm \sqrt{2}$

$x \approx 3.414, 0.586$

69. $\dfrac{400}{1 + e^{-x}} = 350$

$1 + e^{-x} = \dfrac{400}{350} = \dfrac{8}{7}$

$e^{-x} = \dfrac{1}{7}$

$-x = \ln\left(\dfrac{1}{7}\right) = -\ln 7$

$x = \ln 7 \approx 1.946$

71. $\dfrac{40}{1 - 5e^{-0.01x}} = 200$

$1 - 5e^{-0.01x} = \dfrac{40}{200} = \dfrac{1}{5}$

$5e^{-0.01x} = \dfrac{4}{5}$

$e^{-0.01x} = \dfrac{4}{25}$

$-0.01x = \ln\left(\dfrac{4}{25}\right)$

$x = \dfrac{\ln(4/25)}{-0.01}$

$x \approx 183.258$

73. $e^{3x} = 12$

x	0.6	0.7	0.8	0.9	1.0
$f(x)$	6.05	8.17	11.02	14.88	20.09

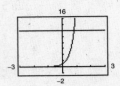

$x \approx 0.828$

75. $20(100 - e^{x/2}) = 500$

x	5	6	7	8	9
$f(x)$	1756	1598	1338	908	200

$x \approx 8.635$

77. $\left(1 + \dfrac{0.065}{365}\right)^{365t} = 4 \implies t = 21.330$

79. $\dfrac{3000}{2 + e^{2x}} = 2$

The zero of $y = \dfrac{3000}{2 + e^{2x}} - 2$

is $x \approx 3.656$.

81. $g(x) = 6e^{1-x} - 25$

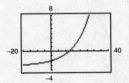

Zero at $x = -0.427$

83. $g(t) = e^{0.09t} - 3$

Zero at $t = 12.207$

85. $\ln x = -3$

$x = e^{-3} \approx 0.050$

87. $\ln 4x = 2.1$

$4x = e^{2.1}$

$x = \dfrac{1}{4}e^{2.1} \approx 2.042$

89. $-2 + 2\ln 3x = 17$

$2\ln 3x = 19$

$\ln 3x = \dfrac{19}{2}$

$3x = e^{19/2}$

$x = \dfrac{1}{3}e^{19/2}$

$x \approx 4453.242$

91. $\log_5(3x + 2) = \log_5(6 - x)$

$$3x + 2 = 6 - x$$
$$4x = 4$$
$$x = 1$$

93. $\log_{10}(z - 3) = 2$

$$z - 3 = 10^2$$
$$z = 10^2 + 3$$
$$= 103$$

95. $7 \log_4(0.6x) = 12$

$$\log_4(0.6x) = \frac{12}{7}$$
$$4^{12/7} = 0.6x = \frac{3}{5}x$$
$$x = \frac{5}{3} 4^{12/7}$$
$$\approx 17.945$$

97. $\ln \sqrt{x + 2} = 1$

$$\sqrt{x + 2} = e^1$$
$$x + 2 = e^2$$
$$x = e^2 - 2 \approx 5.389$$

99. $\ln(x + 1)^2 = 2$

$$e^{\ln(x+1)^2} = e^2$$
$$(x + 1)^2 = e^2$$
$$x + 1 = e \text{ or } x + 1 = -e$$
$$x = e - 1 \approx 1.718$$

or

$$x = -e - 1 \approx -3.718$$

101. $\log_4 x - \log_4(x - 1) = \frac{1}{2}$

$$\log_4\left(\frac{x}{x - 1}\right) = \frac{1}{2}$$
$$4^{\log_4(x/x-1)} = 4^{1/2}$$
$$\frac{x}{x - 1} = 2$$
$$x = 2(x - 1)$$
$$x = 2x - 2$$
$$2 = x$$

103. $\ln(x + 5) = \ln(x - 1) - \ln(x + 1)$

$$\ln(x + 5) = \ln\left(\frac{x - 1}{x + 1}\right)$$
$$x + 5 = \frac{x - 1}{x + 1}$$
$$(x + 5)(x + 1) = x - 1$$
$$x^2 + 6x + 5 = x - 1$$
$$x^2 + 5x + 6 = 0$$
$$(x + 2)(x + 3) = 0$$
$$x = -2 \text{ or } x = -3$$

Both of these solutions are extraneous, so the equation has no solution.

105. $\log_{10} 8x - \log_{10}\left(1 + \sqrt{x}\right) = 2$

$$\log_{10} \frac{8x}{1 + \sqrt{x}} = 2$$
$$\frac{8x}{1 + \sqrt{x}} = 10^2$$
$$8x = 100 + 100\sqrt{x}$$
$$8x - 100\sqrt{x} - 100 = 0$$
$$2x - 25\sqrt{x} - 25 = 0$$
$$\sqrt{x} = \frac{25 \pm \sqrt{25^2 - 4(2)(-25)}}{4}$$
$$= \frac{25 \pm 5\sqrt{33}}{4}$$

Choosing the positive value, we have
$\sqrt{x} \approx 13.431$ and $x \approx 180.384$.

107. $\ln 2x = 2.4$

x	2	3	4	5	6
$f(x)$	1.39	1.79	2.08	2.30	2.48

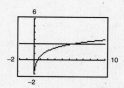

$$x \approx 5.512$$

109. $6 \log_3(0.5x) = 11$

x	12	13	14	15	16
$f(x)$	9.79	10.22	10.63	11.00	11.36

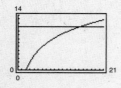

$x \approx 14.988$

111. $\log_{10} x = x^3 - 3$

Graphing $y = \log_{10} x - x^3 + 3$, you obtain two zeros, $x \approx 1.469$ and $x \approx 0.001$.

113. $\ln x + \ln(x - 2) = 1$

Graphing $y = \ln x + \ln(x - 2) - 1$, you obtain one zero, $x \approx 2.928$.

115. $\ln(x - 3) + \ln(x + 3) = 1$

Graphing $y = \ln(x - 3) + \ln(x + 3) - 1$, you obtain $x \approx 3.423$.

117. $y_1 = 7$

$y_2 = 2^{x-1} - 5$

Intersection: $(4.585, 7)$

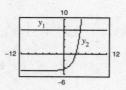

119. $y_1 = 80$

$y_2 = 4e^{-0.2x}$

Intersection: $(-14.979, 80)$

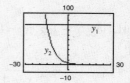

121. $y_1 = 3.25$

$y_2 = \frac{1}{2} \ln(x + 2)$

Intersection: $(663.142, 3.25)$

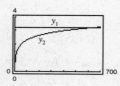

123. $2x^2e^{2x} + 2xe^{2x} = 0$

$(2x^2 + 2x)e^{2x} = 0$

$2x^2 + 2x = 0$ (since $e^{2x} \neq 0$)

$2x(x + 1) = 0$

$x = 0, -1$

125. $-xe^{-x} + e^{-x} = 0$

$(-x + 1)e^{-x} = 0$

$-x + 1 = 0$ (since $e^{-x} \neq 0$)

$x = 1$

127. $2x \ln x + x = 0$

$x(2 \ln x + 1) = 0$

$2 \ln x + 1 = 0$ (since $x > 0$)

$\ln x = -\frac{1}{2}$

$x = e^{-1/2} \approx 0.607$

129. $\dfrac{1 + \ln x}{2} = 0$

$1 + \ln x = 0$

$\ln x = -1$

$x = e^{-1} = \dfrac{1}{e} \approx 0.368$

131. (a) $2000 = 1000e^{0.075t}$

$2 = e^{0.075t}$

$\ln 2 = 0.075t$

$t = \dfrac{\ln 2}{0.075} \approx 9.24$ years

(b) $3000 = 1000e^{0.075t}$

$3 = e^{0.075t}$

$\ln 3 = 0.075t$

$t = \dfrac{\ln 3}{0.075} \approx 14.65$ years

133. (a) $2000 = 1000e^{0.025t}$

$$2 = e^{0.025t}$$

$$\ln 2 = 0.025t$$

$$t = \frac{\ln 2}{0.025} \approx 27.73 \text{ years}$$

(b) $3000 = 1000e^{0.025t}$

$$3 = e^{0.025t}$$

$$\ln 3 = 0.025t$$

$$t = \frac{\ln 3}{0.025} \approx 43.94 \text{ years}$$

135. $p = 500 - 0.5(e^{0.004x})$

(a) $p = 350$

$$350 = 500 - 0.5(e^{0.004x})$$

$$300 = e^{0.004x}$$

$$0.004x = \ln 300$$

$$x \approx 1426 \text{ units}$$

(b) $p = 300$

$$300 = 500 - 0.5(e^{0.004x})$$

$$400 = e^{0.004x}$$

$$0.004x = \ln 400$$

$$x \approx 1498 \text{ units}$$

137. $7247 - 596.5 \ln t = 5800$

$$-596.5 \ln t = -1447$$

$$\ln t \approx 2.4258$$

$$t \approx 11.3, \text{ or } 2001$$

139. (a)

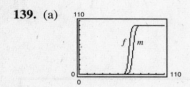

(b) From the graph we see horizontal asymptotes at $y = 0$ and $y = 100$. These represent the lower and upper percent bounds.

(c) Males: $50 = \dfrac{100}{1 + e^{-0.6114(x-69.71)}}$

$$1 + e^{-0.6114(x-69.71)} = 2$$

$$e^{-0.6114(x-69.71)} = 1$$

$$-0.6114(x - 69.71) = \ln 1$$

$$-0.6114(x - 69.71) = 0$$

$$x = 69.71 \text{ inches}$$

Females: $50 = \dfrac{100}{1 + e^{-0.66607(x-64.51)}}$

$$1 + e^{-0.66607(x-64.51)} = 2$$

$$e^{-0.66607(x-64.51)} = 1$$

$$-0.66607(x - 64.51) = \ln 1$$

$$-0.66607(x - 64.51) = 0$$

$$x = 64.51 \text{ inches}$$

141. $T = 20[1 + 7(2^{-h})]$

(a)

(b) We see a horizontal asymptote at $y = 20$. This represents the room temperature.

(c) $100 = 20[1 + 7(2^{-h})]$

$$5 = 1 + 7(2^{-h})$$

$$4 = 7(2^{-h})$$

$$\frac{4}{7} = 2^{-h}$$

$$\ln\left(\frac{4}{7}\right) = \ln 2^{-h}$$

$$\ln\left(\frac{4}{7}\right) = -h \ln 2$$

$$\frac{\ln(4/7)}{-\ln 2} = h$$

$$h \approx 0.81 \text{ hour}$$

143. False. The equation $e^x = 0$ has no solutions.   **145.** Answers will vary.

147. Yes. The doubling time is given by

$$2P = Pe^{rt}$$

$$2 = e^{rt}$$

$$\ln 2 = rt$$

$$t = \frac{\ln 2}{r}.$$

The time to quadruple is given by

$$4P = Pe^{rt}$$

$$4 = e^{rt}$$

$$\ln 4 = rt$$

$$t = \frac{\ln 4}{r} = \frac{\ln 2^2}{r} = \frac{2\ln 2}{r} = 2\left[\frac{\ln 2}{r}\right]$$

which is twice as long.

149. $f(x) = 3x^3 - 4$

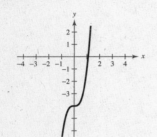

151. $f(x) = |x| + 9$

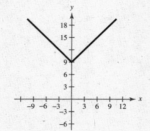

153. $f(x) = \begin{cases} 2x, & x < 0 \\ -x^2 + 4, & x \geq 0 \end{cases}$

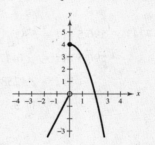

Section 4.5   Exponential and Logarithmic Models

■ You should be able to solve compound interest problems.

1. $A = P\left(1 + \dfrac{r}{n}\right)^{nt}$

2. $A = Pe^{rt}$

■ You should be able to solve growth and decay problems.

(a) Exponential growth if $b > 0$ and $y = ae^{bx}$.

(b) Exponential decay if $b > 0$ and $y = ae^{-bx}$.

■ You should be able to use the Gaussian model

$y = ae^{-(x-b)^2/c}$.

■ You should be able to use the logistics growth model

$y = \dfrac{a}{1 + be^{-(x-c)/d}}$.

■ You should be able to use the logarithmic models

$y = \ln(ax + b)$ and $y = \log_{10}(ax + b)$.

Vocabulary Check

1. (a) iv   (b) i   (c) vi   (d) iii   (e) vii   (f) ii   (g) v

2. Normally   **3.** Sigmoidal   **4.** Bell-shaped, mean

1. $y = 2e^{x/4}$

This is an exponential growth model.

Matches graph (c).

3. $y = 6 + \log_{10}(x + 2)$

This is a logarithmic model, and contains $(-1, 6)$.

Matches graph (b).

5. $y = \ln(x + 1)$

This is a logarithmic model.

Matches graph (d).

7. Since $A = 10,000e^{0.035t}$, the time to double is given by

$$20,000 = 10,000e^{0.035t}$$

$$2 = e^{0.035t}$$

$$\ln 2 = 0.035t$$

$$t = \frac{\ln 2}{0.035} \approx 19.8 \text{ years.}$$

Amount after 10 years:

$$A = 10,000e^{0.035(10)} \approx \$14,190.68$$

9. Since $A = 7500e^{rt}$ and $A = 15,000$ when $t = 21$, we have the following.

$$15,000 = 7500e^{21r}$$

$$2 = e^{21r}$$

$$\ln 2 = 21r$$

$$r = \frac{\ln 2}{21} \approx 0.033 = 3.3\%$$

Amount after 10 years:

$$A = 7500e^{0.033(10)} \approx \$10,432.26$$

11. Since $A = 5000e^{rt}$ and $A = 5665.74$ when $t = 10$, we have the following.

$$5665.74 = 5000e^{10r}$$

$$\frac{5665.74}{5000} = e^{10r}$$

$$\ln\left(\frac{5665.74}{5000}\right) = 10r$$

$$r = \frac{1}{10}\ln\left(\frac{5665.74}{5000}\right)$$

$$\approx 0.0125 = 1.25\%$$

The time to double is given by

$$10,000 = 5000e^{0.0125t}$$

$$2 = e^{0.0125t}$$

$$\ln 2 = 0.0125t$$

$$t = \frac{\ln 2}{0.0125} \approx 55.5 \text{ years.}$$

13. Since $A = Pe^{0.045t}$ and $A = 100,000$ when $t = 10$, we have the following.

$$100,000 = Pe^{0.045(10)}$$

$$\frac{100,000}{e^{0.45}} = P \approx 63,762.82$$

The time to double is given by

$$127,525.64 = 63,762.82e^{0.045t}$$

$$2 = e^{0.045t}$$

$$\ln 2 = 0.045t$$

$$t = \frac{\ln 2}{0.045} \approx 15.4 \text{ years.}$$

15. $3P = Pe^{rt}$

$3 = e^{rt}$

$\ln 3 = rt$

$\dfrac{\ln 3}{r} = t$

r	2%	4%	6%	8%	10%	12%
$t = \dfrac{\ln 3}{r}$	54.93	27.47	18.31	13.73	10.99	9.16

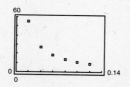

17.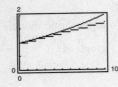

Continuous compounding results in faster growth.

$A = 1 + 0.075[\![t]\!]$

and $A = e^{0.07t}$

19. $\frac{1}{2}C = Ce^{k(1599)}$

$\frac{1}{2} = e^{1599k}$

$k = \frac{\ln(1/2)}{1599}$

$y = Ce^{kt}$

$= 10e^{[\ln(1/2)/1599]1000}$

≈ 6.48 g

21. $\frac{1}{2}C = Ce^{k(5715)}$

$\frac{1}{2} = e^{5715k}$

$k = \frac{\ln(1/2)}{5715}$

$y = Ce^{kt}$

$= 3e^{[\ln(1/2)/5715]1000}$

≈ 2.66 g

23. $y = ae^{bx}$

$1 = ae^{b(0)} \implies 1 = a$

$10 = e^{b(3)}$

$\ln 10 = 3b$

$\frac{\ln 10}{3} = b \implies b \approx 0.7675$

Thus, $y = e^{0.7675x}$.

25. $(0, 4) \implies a = 4$

$(5, 1) \implies 1 = 4e^{b(5)} \implies b = \frac{1}{5}\ln\left(\frac{1}{4}\right)$

$= -\frac{1}{5}\ln 4 \approx -0.2773$

$y = 4e^{-0.2773x}$

27. (a) Australia: $(0, 19.2), (10, 20.9)$

$a = 19.2$ and $20.9 = 19.2e^{b(10)} \implies b = 0.008484$

$y = 19.2e^{0.008484t}$

For 2030, $y \approx 24.8$ million.

Canada: $(0, 31.3), (10, 34.3)$

$a = 31.3$ and $34.3 = 31.3e^{b(10)} \implies b = 0.009153$

$y = 31.3e^{0.009153t}$

For 2030, $y \approx 41.2$ million.

Philippines: $(0, 79.7), (10, 95.9)$

$a = 79.7$ and $95.9 = 79.7e^{b(10)} \implies b = 0.0185$

$y = 79.7e^{0.0185t}$

For 2030, $y \approx 138.8$ million.

South Africa: $(0, 44.1), (10, 43.3)$

$a = 44.1$ and $43.3 = 44.1e^{b(10)} \implies b = -0.00183$

$y = 44.1e^{-0.00183t}$

For 2030, $y \approx 41.7$ million.

Turkey: $(0, 65.7), (10, 73.3)$

$a = 65.7$ and $73.3 = 65.7e^{b(10)} \implies b = 0.01095$

$y = 65.7e^{0.01095t}$

For 2030, $y \approx 91.2$ million.

(b) The constant b gives the growth rates.

(c) The constant b is negative for South Africa.

29. (a) $180 = 134.0e^{k(10)}$

$$10k = \ln \frac{180}{134.0}$$

$$k \approx 0.0295$$

(b) For 2010, $t = 20$ and

$$P = 134.0e^{0.0295(20)} \approx 241,734 \text{ people.}$$

31. $y = Ce^{kt}$

$$\frac{1}{2}C = Ce^{(1599)k}$$

$$\ln\left(\frac{1}{2}\right) = 1599k$$

$$k = \frac{\ln(1/2)}{1599}$$

When $t = 100$, we have

$$y = Ce^{[(\ln(1/2)\cdot 100)/1599]} \approx 0.958C, \text{ or } 95.8\%.$$

33. (a) $V = mt + b$, $V(0) = 30,788 \Rightarrow b = 30,788$

$$V(2) = 24,000 \Rightarrow 24,000 = 2m + 30,788$$

$$\Rightarrow m = -3394$$

$$V(t) = -3394t + 30,788$$

(b) $V = ae^{kt}$, $V(0) = 30,788 \Rightarrow b = 30,788$

$$V(2) = 24,000 \Rightarrow 24,000 = 30,788e^{2k}$$

$$\Rightarrow k = \frac{1}{2}\ln\left(\frac{24,000}{30,788}\right) \approx -0.1245$$

$$V = 30,788e^{-0.1245t}$$

(c)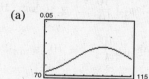

(d) The exponential model depreciates faster in the first year.

(e) Answers will vary.

35. $S(t) = 100(1 - e^{kt})$

(a) $15 = 100(1 - e^{k(1)})$

$$-85 = -100e^k$$

$$k = \ln 0.85$$

$$k \approx -0.1625$$

$$S(t) = 100(1 - e^{-0.1625t})$$

(b)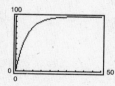

(c) $S(5) = 100(1 - e^{-0.1625(5)})$

$$\approx 55.625 = 55,625 \text{ units}$$

37. $y = 0.0266e^{-(x-100)^2/450}$, $70 \le x \le 115$

(a)

(b) Maximum point is $x = 100$, the average IQ score.

39. $p(t) = \dfrac{1000}{1 + 9e^{-0.1656t}}$

(a) $p(5) = \dfrac{1000}{1 + 9e^{-0.1656(5)}} \approx 203$ animals

(b) $500 = \dfrac{1000}{1 + 9e^{-0.1656t}}$

$$1 + 9e^{-0.1656t} = 2$$

$$9e^{-0.1656t} = 1$$

$$e^{-0.1656t} = \frac{1}{9}$$

$$t = \frac{-\ln(1/9)}{0.1656} \approx 13 \text{ months}$$

(c)

The horizontal asymptotes are $p = 0$ and $p = 1000$. The population will approach 1000 as time increases.

41. $R = \log_{10}\left(\dfrac{I}{I_0}\right) = \log_{10}(I) \Rightarrow I = 10^R$

(a) $I = 10^{6.1} \approx 1,258,925$

(b) $I = 10^{7.6} \approx 39,810,717$

(c) $I = 10^{9.0} \approx 1,000,000,000$

43. $\beta(I) = 10 \log_{10}(I/I_0)$, where $I_0 = 10^{-12}$ watt per square meter.

(a) $\beta(10^{-10}) = 10 \cdot \log_{10}\left(\dfrac{10^{-10}}{10^{-12}}\right) = 10 \log_{10} 10^2 = 20$ decibels

(b) $\beta(10^{-5}) = 10 \cdot \log_{10}\left(\dfrac{10^{-5}}{10^{-12}}\right) = 10 \log_{10} 10^7 = 70$ decibels

(c) $\beta(10^0) = 10 \cdot \log_{10}\left(\dfrac{10^0}{10^{-12}}\right) = 10 \log_{10} 10^{12} = 120$ decibels

45. $\beta = 10 \log_{10}\left(\dfrac{I}{I_0}\right)$

$$10^{\beta/10} = \frac{I}{I_0}$$

$$I = I_0 10^{\beta/10}$$

$$\% \text{ decrease} = \frac{I_0 10^{8.8} - I_0 10^{7.2}}{I_0 10^{8.8}} \times 100$$

$$= 97.5\%$$

47. $\text{pH} = -\log_{10}[\text{H}^+] = -\log_{10}[2.3 \times 10^{-5}] \approx 4.64$

49.

$$\text{pH} = -\log_{10}[\text{H}^+]$$

$$-\text{pH} = \log_{10}[\text{H}^+]$$

$$10^{-\text{pH}} = [\text{H}^+]$$

$$\frac{\text{Hydrogen ion concentration of grape}}{\text{Hydrogen ion concentration of milk of magnesia}} = \frac{10^{-3.5}}{10^{-10.5}} = 10^7$$

51. (a) $P = 120{,}000, r = 0.075, M = 839.06$

$$u = M - \left(M - \frac{Pr}{12}\right)\left(1 + \frac{r}{12}\right)^{12t}$$

$$= 839.06 - (839.06 - 750)(1 + 0.00625)^{12t}$$

$$v = (839.06 - 750)(1.00625)^{12t}$$

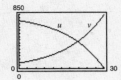

(b) In the early years, the majority of the monthly payment goes toward interest. The interest and principle are equal when $t \approx 20.729 \approx 21$ years.

(c) $P = 120{,}000, r = 0.075, M = 966.71$

$$u = 966.71 - (966.71 - 750)(1.00625)^{12t}$$

$$v = (966.71 - 750)(1.00625)^{12t}$$

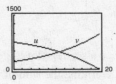

$u = v$ when $t \approx 10.73$ years.

53. $t = -10 \ln\left(\dfrac{T - 70}{98.6 - 70}\right)$

At 9:00 A.M. we have
$t = -10 \ln\left[(85.7 - 70)/(98.6 - 70)\right] \approx 6$ hours.

Thus, we can conclude that the person died 6 hours before 9 A.M., or 3:00 A.M.

55. False. The domain could be all real numbers.

57. True. For the Gaussian model, $y > 0$.

59. $4x - 3y - 9 = 0 \implies y = \frac{1}{3}(4x - 9)$

Slope: $\frac{4}{3}$

Matches (a).

Intercepts: $(0, -3), \left(\frac{9}{4}, 0\right)$

61. $y = 25 - 2.25x$

Slope: -2.25

Matches (d).

Intercepts: $(0, 25), \left(\frac{100}{9}, 0\right)$

63. $f(x) = 2x^3 - 3x^2 + x - 1$

The graph falls to the left and rises to the right.

65. $g(x) = -1.6x^5 + 4x^2 - 2$

The graph rises to the left and falls to the right.

67.

$$
\begin{array}{r|rrrr}
4 & 2 & -8 & 3 & -9 \\
  &   & 8 & 0 & 12 \\
\hline
  & 2 & 0 & 3 & 3
\end{array}
$$

$$\frac{2x^3 - 8x^2 + 3x - 9}{x - 4} = 2x^2 + 3 + \frac{3}{x - 4}$$

69. Answers will vary.

Section 4.6 Nonlinear Models

■ You should be able to use a graphing utility to find nonlinear models, including:

(a) Quadratic models

(b) Exponential models

(c) Power models

(d) Logarithmic models

(e) Logistic models

■ You should be able to use a scatter plot to determine which model is best.

■ You should be able to determine the sum of squared differences for a model.

Vocabulary Check

1. $y = ax + b$ **2.** quadratic   **3.** $y = ax^b$

4. sum, squared differences   **5.** $y = ab^x,\ ae^{cx}$

---

1. Logarithmic model   **3.** Quadratic model   **5.** Exponential model   **7.** Quadratic model

9.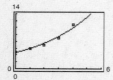

Logarithmic model

11.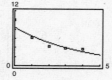

Exponential model

13.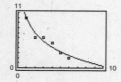

Linear model

15. $y = 4.752(1.2607)^x$

Coefficient of determination:
0.96773

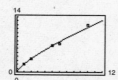

17. $y = 8.463(0.7775)^x$

Coefficient of determination:
0.86639

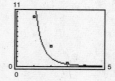

19. $y = 2.083 + 1.257 \ln x$

Coefficient of determination:
0.98672

21. $y = 9.826 - 4.097 \ln x$

Coefficient of determination:
0.93704

23. $y = 1.985x^{0.760}$

Coefficient of determination:
0.99686

25. $y = 16.103x^{-3.174}$

Coefficient of determination:
0.88161

27. (a) Quadratic model: $R = 0.031t^2 + 1.13t + 97.1$

Exponential model: $R = 94.435(1.0174)^t$

Power model: $R = 77.837t^{0.1918}$

(b)

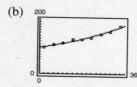 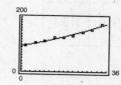 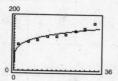

(c) The exponential model fits best. Answers will vary.

(d) For 2008, $t = 38$ and $R \approx 181.9$ million.

For 2012, $t = 42$ and $R \approx 194.9$ million.

Answers will vary.

29. (a) Linear model: $P = 3.11t + 250.9$

Coefficient of determination: 0.99942

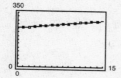

(b) Power model: $P = 246.52t^{0.0587}$

Coefficient of determination: 0.90955

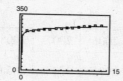

(c) Exponential model: $P = 251.57(1.0114)^t$

Coefficient of determination: 0.99811

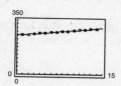

(d) Quadratic model:

$P = -0.020t^2 + 3.41t + 250.1$

Coefficient of determination: 0.99994

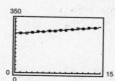

(e) The quadratic model is best because its coefficient of determination is closest to 1.

—CONTINUED—

29. **—CONTINUED—**

(f) Linear model:

Year	2005	2006	2007	2008	2009	2010
Population (in millions)	297.6	300.7	303.8	306.9	310.0	313.1

Power model:

Year	2005	2006	2007	2008	2009	2010
Population (in millions)	289.0	290.1	291.1	292.1	293.0	293.9

Exponential model:

Year	2005	2006	2007	2008	2009	2010
Population (in millions)	298.2	301.6	305.0	308.5	312.0	315.6

Quadratic model:

Year	2005	2006	2007	2008	2009	2010
Population (in millions)	296.8	299.5	302.3	305.0	307.7	310.3

(g) and (h) Answers will vary.

31. (a) $T = -1.239t + 73.02$
No, the data does not appear linear.

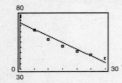

(b) $T = 0.034t^2 - 2.26t + 77.3$
Yes, the data appears quadratic. But, for $t = 60$, the graph is increasing, which is incorrect.

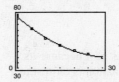

(c) Subtracting 21 from the T-values, the exponential model is $y = 54.438(0.9635)^t$. Adding back 21, $T = 54.438(0.9635)^t + 21$.

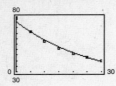

(d) Answers will vary.

33. (a) $P = \dfrac{162.4}{1 + 0.34e^{0.5609x}}$

(b)

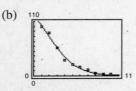

The model is a good fit.

35. (a) Linear model: $y = 15.71t + 51.0$

Logarithmic model: $y = 134.67 \ln t - 97.5$

Quadratic model: $y = -1.292t^2 + 38.96t - 45.0$

Exponential model: $y = 85.97(1.091)^t$

Power model: $y = 37.27t^{0.7506}$

(b)

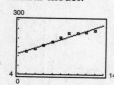

Linear model:

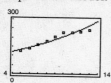

Logarithmic model:

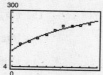

Quadratic model:

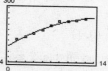

Exponential model:

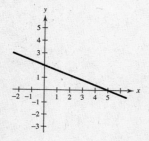

Power model:

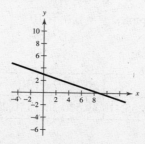

(c) Linear: 803.9

Logarithmic: 411.7

Quadratic: 289.8 (Best)

Exponential: 1611.4

Power: 667.1

(d) Linear: 0.9485

Logarithmic: 0.9736

Quadratic: 0.9814 (Best)

Exponential: 0.9274

Power: 0.9720

(e) Quadratic model is best.

37. True

39. $2x + 5y = 10$

$5y = -2x + 10$

$y = -\frac{2}{5}x + 2$

Slope: $-\frac{2}{5}$

y-intercept: $(0, 2)$

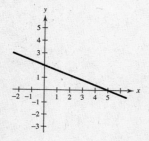

41. $1.2x + 3.5y = 10.5$

$35y = -12x + 105$

$y = -\frac{12}{35}x + \frac{105}{35}$

$= -\frac{12}{35}x + 3$

Slope: $-\frac{12}{35}$

y-intercept: $(0, 3)$

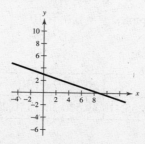

Review Exercises for Chapter 4

1. $(1.45)^{2\pi} \approx 10.3254$

3. $60^{2(-1.1)} = 60^{-2.2}$
$\approx 0.0001225 \approx 0.0$

5. $e^8 \approx 2980.958$

7. $e^{-(-2.1)} \approx e^{2.1} \approx 8.1662$

9. $f(x) = 4^x$

Intercept: $(0, 1)$

Horizontal asymptote: x-axis

Increasing on: $(-\infty, \infty)$

Matches graph (c).

11. $f(x) = -4^x$

Intercept: $(0, -1)$

Horizontal asymptote: x-axis

Decreasing on: $(-\infty, \infty)$

Matches graph (b).

13. $f(x) = 6^x$

Intercept: $(0, 1)$

Horizontal asymptote: x-axis

Increasing on: $(-\infty, \infty)$

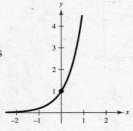

15. $g(x) = 1 + 6^{-x}$

Intercept: $(0, 2)$

Horizontal asymptote: $y = 1$

Decreasing on: $(-\infty, \infty)$

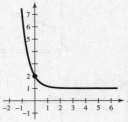

17.

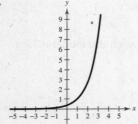

$h(x) = e^{x-1}$

Horizontal asymptote: $y = 0$

19.

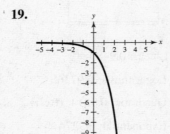

$h(x) = -e^x$

Horizontal asymptote: $y = 0$

21.

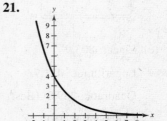

$f(x) = 4e^{-0.5x}$

Horizontal asymptote: $y = 0$

23. $f(x) = \dfrac{10}{1 + 2e^{-0.05x}}$

(a)

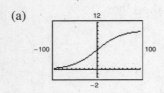

(b) Horizontal asymptotes: $y = 0$, $y = 10$

25. $A = Pe^{rt} = 10,000e^{0.08t}$

t	1	10	20	30	40	50
A	10,832.87	22,255.41	49,530.32	110,231.76	245,325.30	545,981.50

27. $V(t) = 26,000\left(\frac{3}{4}\right)^t$

(a)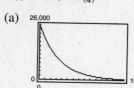

(b) For $t = 2$, $V(2) = \$14,625.$

(c) The car depreciates most rapidly at the beginning, which is realistic.

29. $\log_5 125 = 3$
$\qquad 5^3 = 125$

31. $\log_{64} 2 = \frac{1}{6}$
$\qquad 64^{1/6} = 2$

33. $\ln e^4 = 4$
$\qquad e^4 = e^4$

35. $\qquad 4^3 = 64$
$\qquad \log_4 64 = 3$

37. $\qquad 25^{3/2} = 125$
$\qquad \log_{25} 125 = \frac{3}{2}$

39. $\left(\frac{1}{2}\right)^{-3} = 8$
$\qquad \log_{1/2} 8 = -3$

41. $\qquad e^7 = 1096.6331\ldots$
$\qquad \ln 1096.6331\ldots = 7$

43. $\log_6 216 = \log_6 6^3$
$\qquad\qquad = 3 \log_6 6$
$\qquad\qquad = 3$

45. $\log_4\left(\frac{1}{4}\right) = \log_4(4^{-1})$
$\qquad\qquad = -\log_4 4$
$\qquad\qquad = -1$

47. $g(x) = -\log_2 x + 5 = 5 - \dfrac{\ln x}{\ln 2}$

Domain: $x > 0$

Vertical asymptote: $x = 0$

x-intercept: $(32, 0)$

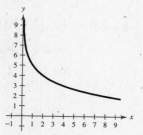

49. $f(x) = \log_2(x - 1) + 6 = 6 + \dfrac{\ln(x - 1)}{\ln (2)}$

Domain: $x > 1$

Vertical asymptote: $x = 1$

x-intercept: $(1.016, 0)$

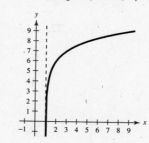

51. $\ln(21.5) \approx 3.068$

53. $\ln\sqrt{6} \approx 0.896$

55. $\log_5 3 = \log_5 x$
$\qquad 3 = x$

57. $\log_9 x = \log_9 3^{-2}$
$\qquad x = 3^{-2} = \frac{1}{9}$

59. $f(x) = \ln x + 3$

Domain: $(0, \infty)$

Vertical asymptote: $x = 0$

x-intercept: $(0.05, 0)$

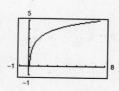

61. $h(x) = \frac{1}{2} \ln x$

Domain: $x > 0$

Vertical asymptote: $x = 0$

x-intercept: $(1, 0)$

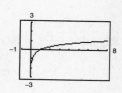

63. $t = 50 \log_{10} \dfrac{18,000}{18,000 - h}$

(a) $0 \le h < 18,000$

(b)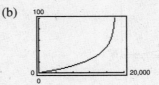

(c) The plane climbs at a faster rate as it approaches its absolute ceiling.

(d) If $h = 4000$, $t = 50 \log_{10} \dfrac{18,000}{18,000 - 4000} \approx 5.46$ minutes.

Vertical asymptote: $h = 18,000$

65. $\log_4 9 = \dfrac{\log_{10} 9}{\log_{10} 4} \approx 1.585$

$\log_4 9 = \dfrac{\ln 9}{\ln 4} \approx 1.585$

67. $\log_{12} 200 = \dfrac{\log_{10} 200}{\log_{10} 12} \approx 2.132$

$\log_{12} 200 = \dfrac{\ln 200}{\ln 12} \approx 2.132$

69. $f(x) = \log_2(x - 1) = \dfrac{\ln(x - 1)}{\ln 2}$

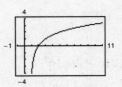

71. $f(x) = -\log_{1/2}(x + 2) = -\dfrac{\ln(x + 2)}{\ln(1/2)} = \dfrac{\ln(x + 2)}{\ln 2}$

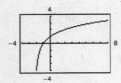

73. $\log_b 9 = \log_b 3^2$

$= 2 \log_b 3$

$= 2(0.5646)$

$= 1.1292$

75. $\log_b \sqrt{5} = \log_b 5^{1/2}$

$= \tfrac{1}{2} \log_b 5$

$= \tfrac{1}{2}(0.8271)$

$= 0.41355$

77. $\ln(5e^{-2}) = \ln 5 + \ln e^{-2}$

$= \ln 5 - 2 \ln e$

$= \ln 5 - 2$

79. $\log_{10} 200 = \log_{10}(2 \cdot 100)$

$= \log_{10} 2 + \log_{10} 10^2$

$= \log_{10} 2 + 2$

81. $\log_5 5x^2 = \log_5 5 + \log_5 x^2 = 1 + 2 \log_5 x$

83. $\log_{10} \dfrac{5\sqrt{y}}{x^2} = \log_{10} 5\sqrt{y} - \log_{10} x^2$

$= \log_{10} 5 + \log_{10}\sqrt{y} - \log_{10} x^2$

$= \log_{10} 5 + \dfrac{1}{2} \log_{10} y - 2 \log_{10} x$

85. $\ln\left(\dfrac{x + 3}{xy}\right) = \ln(x + 3) - \ln(xy)$

$= \ln(x + 3) - \ln x - \ln y$

87. $\log_2 5 + \log_2 x = \log_2 5x$

89. $\dfrac{1}{2} \ln(2x - 1) - 2 \ln(x + 1) = \ln\sqrt{2x - 1} - \ln(x + 1)^2$

$= \ln \dfrac{\sqrt{2x - 1}}{(x + 1)^2}$

91. $\ln 3 + \dfrac{1}{3}\ln(4 - x^2) - \ln x = \ln\left[\dfrac{3(4 - x^2)^{1/3}}{x}\right] = \ln\left[\dfrac{3\sqrt[3]{4 - x^2}}{x}\right]$

93. $s = 25 - \dfrac{13\ln(h/12)}{\ln 3}$

(a)

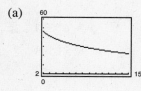

(b)

h	4	6	8	10	12	14
s	38	33.2	29.8	27.2	25	23.2

(c) As the depth increases, the number of miles of roads cleared decreases.

95. $8^x = 512 = 8^3 \implies x = 3$

97. $6^x = \dfrac{1}{216} = \dfrac{1}{6^3} = 6^{-3} \implies x = -3$

99. $2^{x+1} = \dfrac{1}{16}$

$2^{x+1} = 2^{-4}$

$x + 1 = -4$

$x = -5$

101. $\log_7 x = 4 \implies x = 7^4 = 2401$

103. $\log_2(x - 1) = 3$

$2^3 = x - 1$

$x = 9$

105. $\ln x = 4$

$x = e^4 \approx 54.598$

107. $\ln(x - 1) = 2$

$e^2 = x - 1$

$x = 1 + e^2$

109. $3e^{-5x} = 132$

$e^{-5x} = 44$

$-5x = \ln 44$

$x = -\dfrac{\ln 44}{5} \approx -0.757$

111. $2^x + 13 = 35$

$2^x = 22$

$x \ln 2 = \ln 22$

$x = \dfrac{\ln 22}{\ln 2} \approx 4.459$

113. $-4(5^x) = -68$

$5^x = 17$

$x \ln 5 = \ln 17$

$x = \dfrac{\ln 17}{\ln 5} \approx 1.760$

115. $2e^{x-3} - 1 = 4$

$2e^{x-3} = 5$

$e^{x-3} = \tfrac{5}{2}$

$x - 3 = \ln\left(\tfrac{5}{2}\right)$

$x = 3 + \ln\left(\tfrac{5}{2}\right) \approx 3.916$

117. $e^{2x} - 7e^x + 10 = 0$

$(e^x - 5)(e^x - 2) = 0$

$e^x = 5 \implies x = \ln 5 \approx 1.609$

$e^x = 2 \implies x = \ln 2 \approx 0.693$

119. $\ln 3x = 8.2$

$3x = e^{8.2}$

$x = \dfrac{e^{8.2}}{3} \approx 1213.650$

121. $\ln x - \ln 3 = 2$

$\ln \dfrac{x}{3} = 2$

$\dfrac{x}{3} = e^2$

$x = 3e^2 \approx 22.167$

123. $\ln \sqrt{x + 1} = 2$

$\dfrac{1}{2} \ln(x + 1) = 2$

$\ln(x + 1) = 4$

$x + 1 = e^4$

$x = e^4 - 1$

≈ 53.598

125. $\log_4(x - 1) = \log_4(x - 2) - \log_4(x + 2)$

$\log_4(x - 1) = \log_4\!\left(\dfrac{x - 2}{x + 2}\right)$

$x - 1 = \dfrac{x - 2}{x + 2}$

$(x - 1)(x + 2) = x - 2$

$x^2 + x - 2 = x - 2$

$x^2 = 0$

$x = 0$ (extraneous)

No solution

127. $\log_{10}(1 - x) = -1$

$10^{-1} = 1 - x$

$x = 1 - 10^{-1} = 0.9$

129. $xe^x + e^x = 0$

$(x + 1)e^x = 0$

$x + 1 = 0$ (since $e^x \neq 0$)

$x = -1$

131. $x \ln x + x = 0$

$x(\ln x + 1) = 0$

$\ln x + 1 = 0$ (since $x > 0$)

$\ln x = -1$

$x = e^{-1} = \dfrac{1}{e} \approx 0.368$

133. $3(7550) = 7550e^{0.0725t}$

$3 = e^{0.0725t}$

$\ln 3 = 0.0725t$

$t = \dfrac{\ln 3}{0.0725} \approx 15.2$ years

135. $y = 3e^{-2x/3}$

Decreasing exponential

Matches graph (e).

137. $y = \ln(x + 3)$

Logarithmic function shifted to left

Matches graph (f).

139. $y = 2e^{-(x+4)^2/3}$

Gaussian model

Matches graph (a).

141. $y = ae^{bx}$

$2 = ae^{b(0)} \implies a = 2$

$3 = 2e^{b(4)}$

$1.5 = e^{4b}$

$\ln 1.5 = 4b \implies b \approx 0.1014$

Thus, $y = 2e^{0.1014x}$.

143. $y = ae^{bx}$

$\dfrac{1}{2} = ae^{b(0)} \implies a = \dfrac{1}{2}$

$5 = \dfrac{1}{2}e^{b(5)}$

$10 = e^{5b}$

$\ln 10 = 5b \implies b \approx 0.4605$

Thus, $y = \dfrac{1}{2}e^{0.4605x}$.

145. $P = 361e^{kt}$

$t = 0$ corresponds to 2000.

$(-20, 215)$:

$215 = 361e^{k(-20)}$

$\dfrac{215}{361} = e^{-20k}$

$-20k = \ln\left(\dfrac{215}{361}\right)$

$k = -\dfrac{1}{20}\ln\left(\dfrac{215}{361}\right) = \dfrac{1}{20}\ln\left(\dfrac{361}{215}\right) \approx 0.02591$

$P = 361e^{0.02591t}$

For 2020, $P(20) = 361e^{0.02591(20)} \approx 606.1$ or

606,100 population in 2020.

147. (a) $20,000 = 10,000e^{r(12)}$

$2 = e^{12r}$

$\ln 2 = 12r$

$r = \dfrac{\ln 2}{12} \approx 0.0578$ or 5.78%

(b) $10,000e^{0.0578(1)} \approx \$10,595.03$

149. (a) $50 = \dfrac{158}{1 + 5.4e^{-0.12t}}$

$1 + 5.4e^{-0.12t} = \dfrac{158}{50}$

$5.4e^{-0.12t} = \dfrac{108}{50}$

$e^{-0.12t} = \dfrac{108}{50(5.4)}$

$-0.12t = \ln\dfrac{108}{270}$

$t = \dfrac{\ln(108/270)}{-0.12} \approx 7.6$ weeks

(b) Similarly:

$75 = \dfrac{158}{1 + 5.4e^{-0.12t}}$

$1 + 5.4e^{-0.12t} = \dfrac{158}{75}$

$e^{-0.12t} = 0.20494$

$-0.12t = \ln(0.20494)$

$t \approx 13.2$ weeks

151. Logistic model

153. Logarithmic model

155. (a) Linear model: $y = 297.8t + 739$; 0.97653

Quadratic model: $y = 11.79t^2 + 38.5t + 2118$; 0.98112

Exponential model: $y = 1751.5(1.077)^t$; 0.98225

Logarithmic model: $y = 3169.8 \ln t - 3532$; 0.95779

Power model: $y = 598.1t^{0.7950}$; 0.97118

—CONTINUED—

155. **—CONTINUED—**

(b)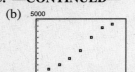

Linear model:

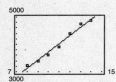

Quadratic model:

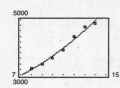

Exponential model:

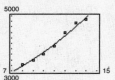

Logarithmic model:

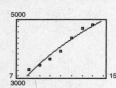

Power model:

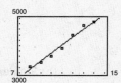

(c) The exponential model is best because its coefficient of determination is closest to 1. Answers will vary.

(d) For 2010, $t = 20$ and $y \approx \$7722$ million.

(e) $y = 5250$ when $t \approx 14.8$, or 2004.

157. (a) $P = \dfrac{9999.887}{1 + 19.0e^{-0.2x}}$

(c) The model is a good fit.

(d) The limiting size is $\dfrac{9999.887}{1 + 0} \approx 10,000$ fish.

(b)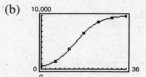

159. True; by the Inverse Properties, $\log_b b^{2x} = 2x$.

161. False; $\ln x + \ln y = \ln(xy) \neq \ln(x + y)$

163. False. The domain of $f(x) = \ln(x)$ is $x > 0$.

165. Since $1 < \sqrt{2} < 2$,
$$2^1 < 2^{\sqrt{2}} < 2^2 \implies 2 < 2^{\sqrt{2}} < 4.$$

Chapter 4   Practice Test

1. Solve for x: $x^{3/5} = 8$

2. Solve for x: $3^{x-1} = \frac{1}{81}$

3. Graph $f(x) = 2^{-x}$ by hand.

4. Graph $g(x) = e^x + 1$ by hand.

5. If $5000 is invested at 9% interest, find the amount after three years if the interest is compounded
   (a) monthly.   (b) quarterly.   (c) continuously.

6. Write the equation in logarithmic form: $7^{-2} = \frac{1}{49}$

7. Solve for x: $x - 4 = \log_2 \frac{1}{64}$

8. Given $\log_b 2 = 0.3562$ and $\log_b 5 = 0.8271$, evaluate $\log_b \sqrt[4]{8/25}$.

9. Write $5 \ln x - \frac{1}{2} \ln y + 6 \ln z$ as a single logarithm.

10. Using your calculator and the change of base formula, evaluate $\log_9 28$.

11. Use your calculator to solve for N: $\log_{10} N = 0.6646$

12. Graph $y = \log_4 x$ by hand.

13. Determine the domain of $f(x) = \log_3(x^2 - 9)$.

14. Graph $y = \ln(x - 2)$ by hand.

15. True or false: $\dfrac{\ln x}{\ln y} = \ln(x - y)$

16. Solve for x: $5^x = 41$

17. Solve for x: $x - x^2 = \log_5 \frac{1}{25}$

18. Solve for x: $\log_2 x + \log_2(x - 3) = 2$

19. Solve for x: $\dfrac{e^x + e^{-x}}{3} = 4$

20. Six thousand dollars is deposited into a fund at an annual percentage rate of 13%.
    Find the time required for the investment to double if the interest is compounded continuously.

21. Use a graphing utility to find the points of intersection of the graphs of $y = \ln(3x)$ and $y = e^x - 4$.

22. Use a graphing utility to find the power model $y = ax^b$ for the data $(1, 1)$, $(2, 5)$, $(3, 8)$, and $(4, 17)$.

C H A P T E R   5
Linear Systems and Matrices

C H A P T E R   5
Linear Systems and Matrices

Section 5.1   Solving Systems of Equations

■ You should be able to solve systems of equations by the method of substitution.

    1. Solve one of the equations for one of the variables.

    2. Substitute this expression into the other equation and solve.

    3. Back-substitute into the first equation to find the value of the other variable.

    4. Check your answer in each of the original equations.

■ You should be able to find solutions graphically. (See Example 5 in textbook.)

Vocabulary Check

1. system, equations   **2.** solution   **3.** method, substitution

4. point, intersection   **5.** break-even point

---

1. (a) $4(0) - (-3) \overset{?}{=} 1$

        $6(0) + (-3) \overset{?}{=} -6$

           $3 \neq 1$

          $-3 \neq -6$

    No, $(0, -3)$ is not a solution.

   (b) $4(-1) - (-5) \overset{?}{=} 1$

        $6(-1) + (-5) \overset{?}{=} -6$

           $1 = 1$

         $-11 \neq -6$

    No, $(-1, -5)$ is not a solution.

   (c) $4\left(-\frac{3}{2}\right) - (3) \overset{?}{=} 1$

        $6\left(-\frac{3}{2}\right) + (3) \overset{?}{=} -6$

          $-9 \neq 1$

          $-6 = -6$

    No, $\left(-\frac{3}{2}, 3\right)$ is not a solution.

   (d) $4\left(-\frac{1}{2}\right) - (-3) \overset{?}{=} 1$

        $6\left(-\frac{1}{2}\right) + (-3) \overset{?}{=} -6$

          $1 = 1$

          $-6 = -6$

    Yes, $\left(-\frac{1}{2}, -3\right)$ is a solution.

3. (a) $0 \overset{?}{=} -2e^{-2}$

       $3(-2) - 0 \overset{?}{=} 2$

           $0 \neq -2e^{-2}$

         $-6 \neq 2$

    No, $(-2, 0)$ is not a solution.

   (b) $-2 \overset{?}{=} -2e^{0}$

       $3(0) - (-2) \overset{?}{=} 2$

          $-2 = -2$

          $2 = 2$

    Yes, $(0, -2)$ is a solution.

—CONTINUED—

3. —CONTINUED—

(c) $\qquad -3 \overset{?}{=} -2e^0$

$\qquad 3(0) - (-3) \overset{?}{=} 2$

$\qquad\qquad -3 \neq -2$

$\qquad\qquad\quad 3 \neq 2$

No, $(0, -3)$ is not a solution.

(d) $\qquad -5 \overset{?}{=} -2e^{-1}$

$\qquad 3(-1) - (-5) \overset{?}{=} 2$

$\qquad\qquad -5 \neq -2e^{-1}$

$\qquad\qquad\quad 2 = 2$

No, $(-1, -5)$ is not a solution.

5. $\begin{cases} 2x + y = 6 & \text{Equation 1} \\ -x + y = 0 & \text{Equation 2} \end{cases}$

Solve for y in Equation 1: $y = 6 - 2x$

Substitute for y in Equation 2: $-x + (6 - 2x) = 0$

Solve for x: $-3x + 6 = 0 \implies x = 2$

Back-substitute $x = 2$: $y = 6 - 2(2) = 2$

Answer: $(2, 2)$

7. $\begin{cases} x - y = -4 & \text{Equation 1} \\ x^2 - y = -2 & \text{Equation 2} \end{cases}$

Solve for y in Equation 1: $y = x + 4$

Substitute for y in Equation 2: $x^2 - (x + 4) = -2$

Solve for x: $x^2 - x - 2 = 0$

$\qquad\qquad\qquad \implies (x + 1)(x - 2) = 0$

$\qquad\qquad\qquad \implies x = -1, 2$

Back-substitute $x = -1$: $y = -1 + 4 = 3$

Back-substitute $x = 2$: $y = 2 + 4 = 6$

Answer: $(-1, 3), (2, 6)$

9. $\begin{cases} 3x + y = 2 & \text{Equation 1} \\ x^3 - 2 + y = 0 & \text{Equation 2} \end{cases}$

Solve for y in Equation 1: $y = 2 - 3x$

Substitute for y in Equation 2: $x^3 - 2 + (2 - 3x) = 0$

Solve for x: $x^3 - 3x = 0 \implies x(x^2 - 3) = 0 \implies x = 0, \pm\sqrt{3}$

Back-substitute: $x = 0$: $y = 2$

$\qquad\qquad\quad x = \sqrt{3}$: $y = 2 - 3\sqrt{3}$

$\qquad\qquad\quad x = -\sqrt{3}$: $y = 2 + 3\sqrt{3}$

Answer: $(0, 2), \left(\sqrt{3}, 2 - 3\sqrt{3}\right), \left(-\sqrt{3}, 2 + 3\sqrt{3}\right)$

11. $\begin{cases} -\frac{7}{2}x - y = -18 & \text{Equation 1} \\ 8x^2 - 2y^3 = 0 & \text{Equation 2} \end{cases}$

Solve for x in Equation 1: $-\frac{7}{2}x = y - 18 \implies x = -\frac{2}{7}y + \frac{36}{7}$

Substitute for x in Equation 2: $8\left(-\frac{2}{7}y + \frac{36}{7}\right)^2 - 2y^3 = 0$

Solve for x: $\quad -2y^3 + 8\left(\frac{4}{49}y^2 - \frac{144}{49}y + \frac{36^2}{49}\right) = 0$

$\qquad\qquad 49y^3 - 16y^2 + 576y - 5184 = 0$

$\qquad\qquad (y - 4)(49y^2 + 180y + 1296) = 0$

Hence, $y = 4$ and $x = -\frac{2}{7}(4) + \frac{36}{7} = 4$.

Answer: $(4, 4)$

13. $\begin{cases} x - y = 0 & \text{Equation 1} \\ 5x - 3y = 10 & \text{Equation 2} \end{cases}$

Solve for y in Equation 1: $y = x$

Substitute for y in Equation 2: $5x - 3x = 10$

Solve for x: $2x = 10 \implies x = 5$

Back-substitute in Equation 1: $y = x = 5$

Answer: $(5, 5)$

15. $\begin{cases} 2x - y + 2 = 0 & \text{Equation 1} \\ 4x + y - 5 = 0 & \text{Equation 2} \end{cases}$

Solve for y in Equation 1: $y = 2x + 2$

Substitute for y in Equation 2:
$4x + (2x + 2) - 5 = 0$

Solve for x:
$4x + (2x + 2) - 5 = 0 \implies 6x - 3 = 0 \implies x = \frac{1}{2}$

Back-substitute $x = \frac{1}{2}$: $y = 2x + 2 = 2\left(\frac{1}{2}\right) + 2 = 3$

Answer: $\left(\frac{1}{2}, 3\right)$

17. $\begin{cases} 1.5x + 0.8y = 2.3 \implies 15x + 8y = 23 \\ 0.3x - 0.2y = 0.1 \implies 3x - 2y = 1 \end{cases}$

Solve for y in Equation 2: $-2y = 1 - 3x$

$$y = \frac{3x - 1}{2}$$

Substitute for y in Equation 1:

$$15x + 8\left(\frac{3x - 1}{2}\right) = 23$$

$$15x + 12x - 4 = 23$$

$$27x = 27$$

$$x = 1$$

Then, $y = \dfrac{3x - 1}{2} = \dfrac{3(1) - 1}{2} = 1.$

Answer: $(1, 1)$

19. $\begin{cases} \frac{1}{5}x + \frac{1}{2}y = 8 & \text{Equation 1} \\ x + y = 20 & \text{Equation 2} \end{cases}$

Solve for x in Equation 2: $x = 20 - y$

Substitute for x in Equation 1: $\frac{1}{5}(20 - y) + \frac{1}{2}y = 8$

Solve for y: $4 + \frac{3}{10}y = 8 \implies y = \frac{40}{3}$

Back-substitute $y = \frac{40}{3}$: $x = 20 - y$

$$= 20 - \frac{40}{3} = \frac{20}{3}$$

Answer: $\left(\frac{20}{3}, \frac{40}{3}\right)$

21. $\begin{cases} -\frac{5}{3}x + y = 5 & \text{Equation 1} \\ -5x + 3y = 6 & \text{Equation 2} \end{cases}$

Solve for y in Equation 1: $y = 5 + \frac{5}{3}x$

Substitute for y in Equation 2: $-5x + 3\left(5 + \frac{5}{3}x\right) = 6$

Solve for x: $-5x + 15 + 5x = 6$

$\qquad\qquad\qquad\qquad 15 \neq 6 \quad$ Inconsistent

No solution

23. $\begin{cases} x^2 - 2x + y = 8 & \text{Equation 1} \\ x - y = -2 & \text{Equation 2} \end{cases}$

Solve for y in Equation 2: $y = x + 2$

Substitute for y in Equation 1:

$$x^2 - 2x + (x + 2) = 8$$

$$x^2 - x - 6 = 0$$

$$(x - 3)(x + 2) = 0$$

$$x = 3, -2$$

$x = 3 \implies y = 5$

$x = -2 \implies y = 0$

Answer: $(3, 5), (-2, 0)$

25. $\begin{cases} 2x^2 - y = 1 & \text{Equation 1} \\ x - y = 2 & \text{Equation 2} \end{cases}$

Solve for y in Equation 2: $y = x - 2$

Substitute for y in Equation 1: $2x^2 - (x - 2) = 1$

$$2x^2 - x + 1 = 0$$

No real solution

27. $\begin{cases} x^3 - y = 0 & \text{Equation 1} \\ x - y = 0 & \text{Equation 2} \end{cases}$

Solve for y in Equation 2: $y = x$

Substitute for y in Equation 1: $x^3 - x = 0$

Solve for x:
$x(x - 1)(x + 1) = 0 \implies x = 0, 1, -1$

Back-substitute: $x = 0 \implies y = 0$

$$x = 1 \implies y = 1$$

$$x = -1 \implies y = -1$$

Answer: $(0, 0), (1, 1), (-1, -1)$

29. $\begin{cases} -x + 2y = 2 \\ 3x + y = 15 \end{cases}$

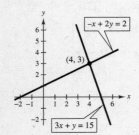

Point of intersection: $(4, 3)$

31. $\begin{cases} x - 3y = -2 \\ 5x + 3y = 17 \end{cases}$

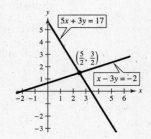

Point of intersection: $\left(\frac{5}{2}, \frac{3}{2}\right)$

33. $\begin{cases} x^2 + y = 1 \\ x + y = 2 \end{cases}$

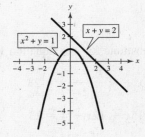

No solution

35. $\begin{cases} -x + y = 3 \implies y_1 = x + 3 \\ x^2 - 6x - 27 + y^2 = 0 \implies y_2 = \sqrt{6x - x^2 + 27} \end{cases}$

$$y_3 = -\sqrt{6x - x^2 + 27}$$

Points of intersection: $(-3, 0), (3, 6)$

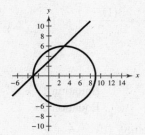

37. $\begin{cases} 7x + 8y = 24 \implies y_1 = -\frac{7}{8}x + 3 \\ x - 8y = 8 \implies y_2 = \frac{1}{8}x - 1 \end{cases}$

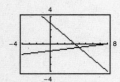

Point of intersection: $\left(4, -\frac{1}{2}\right)$

39. $x - y^2 = -1 \implies y^2 = x + 1 \implies y_1 = \sqrt{x + 1}$

$$y_2 = -\sqrt{x + 1}$$

$x - y = 5 \implies y_3 = x - 5$

Points of intersection: $(8, 3), (3, -2)$

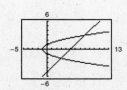

41. $x^2 + y^2 = 8 \implies y_1 = \sqrt{8 - x^2}, y_2 = -\sqrt{8 - x^2}$

$\quad\quad y = x^2 \implies y_3 = x^2$

Points of intersection:

$(1.540, 2.372), (-1.540, 2.372)$

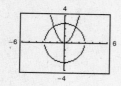

43. $\begin{cases} y = e^x \\ x - y + 1 = 0 \implies y = x + 1 \end{cases}$

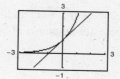

Point of intersection: $(0, 1)$

45. $\begin{cases} x + 2y = 8 \quad\quad \implies y_1 = 4 - x/2 \\ \quad\quad y = 2 + \ln x \implies y_2 = 2 + \ln x \end{cases}$

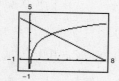

Point of intersection: $\approx (2.318, 2.841)$

47. $\begin{cases} y = \sqrt{x} + 4 \\ y = 2x + 1 \end{cases}$

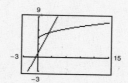

Point of intersection: $\left(\frac{9}{4}, \frac{11}{2}\right)$

49. $\begin{cases} x^2 + y^2 = 169 \implies y_1 = \sqrt{169 - x^2} \text{ and} \\ \quad\quad\quad\quad\quad y_2 = -\sqrt{169 - x^2} \\ x^2 - 8y = 104 \implies y_3 = \frac{1}{8}x^2 - 13 \end{cases}$

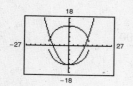

Points of intersection: $(0, -13), (\pm 12, 5)$

51. $\begin{cases} y = 2x \quad\quad\quad\quad \text{Equation 1} \\ y = x^2 + 1 \quad\quad\quad \text{Equation 2} \end{cases}$

Substitute for y in Equation 2: $2x = x^2 + 1$

Solve for x:

$x^2 - 2x + 1 = (x - 1)^2 = 0 \implies x = 1$

Back-substitute $x = 1$ in Equation 1: $y = 2x = 2$

Answer: $(1, 2)$

53. $\begin{cases} 3x - 7y + 6 = 0 \quad\quad \text{Equation 1} \\ \quad\quad x^2 - y^2 = 4 \quad\quad\quad \text{Equation 2} \end{cases}$

Solve for y in Equation 1: $y = \dfrac{3x + 6}{7}$

Substitute for y in Equation 2: $x^2 - \left(\dfrac{3x + 6}{7}\right)^2 = 4$

Solve for x: $\quad x^2 - \left(\dfrac{9x^2 + 36x + 36}{49}\right) = 4$

$\quad\quad\quad\quad 49x^2 - (9x^2 + 36x + 36) = 196$

$\quad\quad\quad\quad\quad 40x^2 - 36x - 232 = 0$

$10x^2 - 9x - 58 = 0 \implies x = \dfrac{9 \pm \sqrt{81 + 40(58)}}{20} \implies x = \dfrac{29}{10}, -2$

Back-substitute $x = \dfrac{29}{10}$: $y = \dfrac{3x + 6}{7} = \dfrac{3(29/10) + 6}{7} = \dfrac{21}{10}$

Back-substitute $x = -2$: $y = \dfrac{3x + 6}{7} = 0$

Answer: $\left(\frac{29}{10}, \frac{21}{10}\right), (-2, 0)$

55. $x^2 + y^2 = 1$

$x + y = 4$

Graphing $y_1 = \sqrt{1 - x^2}$, $y_2 = -\sqrt{1 - x^2}$ and $y_3 = 4 - x$, you see that there are no points of intersection.

No solution

57. $\begin{cases} y = 2x + 1 \\ y = \sqrt{x} + 2 \end{cases}$

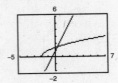

Point of intersection: $\left(\frac{1}{4}, \frac{3}{2}\right)$

59. $\begin{cases} y - e^{-x} = 1 \implies y = e^{-x} + 1 \\ y - \ln x = 3 \implies y = \ln x + 3 \end{cases}$

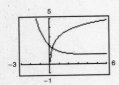

Point of intersection: Approximately $(0.287, 1.751)$

61. $\begin{cases} y = x^3 - 2x^2 + 1 & \text{Equation 1} \\ y = 1 - x^2 & \text{Equation 2} \end{cases}$

Substitute for y in Equation 2:

$x^3 - 2x^2 + 1 = 1 - x^2$

Solve for x: $x^3 - x^2 = 0$

$x^2(x - 1) = 0 \implies x = 0, 1$

Back-substitute: $x = 0 \implies y = 1$

$x = 1 \implies y = 0$

Answer: $(0, 1), (1, 0)$

63. $\begin{cases} xy - 1 = 0 & \text{Equation 1} \\ 2x - 4y + 7 = 0 & \text{Equation 2} \end{cases}$

Solve for y in Equation 1: $y = \dfrac{1}{x}$

Substitute for y in Equation 2: $2x - 4\left(\dfrac{1}{x}\right) + 7 = 0$

Solve for x: $2x^2 - 4 + 7x = 0$

$\implies (2x - 1)(x + 4) = 0 \implies x = \dfrac{1}{2}, -4$

Back-substitute $x = \dfrac{1}{2}$: $y = \dfrac{1}{1/2} = 2$

Back-substitute $x = -4$: $y = \dfrac{1}{-4} = -\dfrac{1}{4}$

Two points of intersection: $\left(\dfrac{1}{2}, 2\right), \left(-4, -\dfrac{1}{4}\right)$

65. $C = 8650x + 250,000$, $R = 9950x$

$R = C$

$9950x = 8650x + 250,000$

$1300x = 250,000$

$x \approx 192$ units

$R \approx \$1,910,400$ (Answers will vary.)

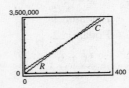

67. $C = 5.5\sqrt{x} + 10,000$, $R = 3.29x$

$R = C$

$3.29x = 5.5\sqrt{x} + 10,000$

$3.29x - 10,000 = 5.5\sqrt{x}$

$10.8241x^2 - 65,800x + 100,000,000 = 30.25x$

$10.8241x^2 - 65,830.25x + 100,000,000 = 0$

$x \approx 3133$ units

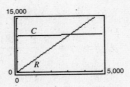

In order for the revenue to break even with the cost, 3133 units must be sold, $R \approx \$10,308$.

69. $N = 360 - 24x$ Animated film

$N = 24 + 18x$ Horror film

(a)

Week x	1	2	3	4	5	6	7	8	9	10	11	12
Animated	336	312	288	264	240	216	192	168	144	120	96	72
Horror	42	60	78	96	114	132	150	168	186	204	222	240

(b) For $x = 8$, $N = 168$

(c) $360 - 24x = 24 + 18x$

(d) The answers are the same.

(e) During week 8 the same number (168) were rented.

$336 = 42x$

$x = 8$

$N = 24 + 18(8) = 168$

71. (a) $C = 35.45x + 16,000$

$R = 55.95x$

(b)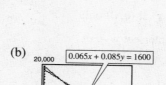

$C = R$ for $x \approx 780$ units

Algebraically,

$35.45x + 16,000 = 55.95x$

$16,000 = 20.5x$

$x = \dfrac{16,000}{20.5} \approx 780$ units

73. $0.06x = 0.03x + 350$

$0.03x = 350$

$x \approx \$11,666.67$

To make the straight commission offer better, you would have to sell more than \$11,666.67 per week.

75. (a) $\begin{cases} x + y = 20,000 \\ 0.065x + 0.085y = 1600 \end{cases}$

(b)

(c) The curves intersect at $x = 5000$. Thus, \$5000 should be invested at 6.5%.

As x increases, y decreases and the amount of interest decreases.

77. $M = 47.4t + 5104$

$P = 76.5t + 4875$

(a)

Year	1990	1994	1998	2002	2006	2010
Missouri	5104	5294	5483	5673	5862	6052
Tennessee	4875	5181	5487	5793	6099	6405

(b) From 1998 to 2010

—CONTINUED—

77. —CONTINUED—

(c)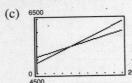

Point of intersection: $(7.87, 5477.0)$

(d) $47.4t + 5104 = 76.5t + 4875$

$229 = 29.1t$

$t = 229/29.1 \approx 7.87$

(e) The population of Tennessee surpassed that of Missouri during 1997.

79. $2l + 2w = 30 \Rightarrow l + w = 15$

$l = w + 3 \Rightarrow (w + 3) + w = 15$

$2w = 12$

$w = 6$

$l = w + 3 = 9$

Dimensions: 6 meters $\times$ 9 meters

81. $2l + 2w = 40 \Rightarrow l + w = 20 \Rightarrow w = 20 - l$

$lw = 96 \Rightarrow l(20 - l) = 96$

$20l - l^2 = 96$

$0 = l^2 - 20l + 96$

$0 = (l - 8)(l - 12)$

$l = 8$ or $l = 12$

$l = 12, w = 8$

If the length is supposed to be greater than the width, we have $l = 12$ miles and $w = 8$ miles.

83. False. You could solve for x first.

85. The system has no solution if you arrive at a false statement, ie. $4 = 8$, or you have a quadratic equation with a negative discriminant, which would yield imaginary roots.

87. (a) The line $y = 2x$ intersects the parabola $y = x^2$ at two points, $(0, 0)$ and $(2, 4)$.

(b) The line $y = 0$ intersects $y = x^2$ at $(0, 0)$ only.

(c) The line $y = x - 2$ does not intersect $y = x^2$.

(Other answers possible.)

89. Answers will vary. For example,

$y = x - 3$

$2y = x - 4$

91. $(-2, 7), (5, 5)$

$m = \dfrac{5 - 7}{5 - (-2)} = -\dfrac{2}{7}$

$y - 7 = -\dfrac{2}{7}(x - (-2))$

$7y - 49 = -2x - 4$

$2x + 7y - 45 = 0$

93. $(6, 3), (10, 3)$

$m = \dfrac{3 - 3}{10 - 6} = 0$

The line is horizontal.

$y = 3 \Rightarrow y - 3 = 0$

95. $\left(\dfrac{3}{5}, 0\right), (4, 6)$

$m = \dfrac{6 - 0}{4 - \frac{3}{5}} = \dfrac{6}{\frac{17}{5}} = \dfrac{30}{17}$

$y - 6 = \dfrac{30}{17}(x - 4)$

$17y - 102 = 30x - 120$

$0 = 30x - 17y - 18$

97. Domain: all $x \neq 6$

Vertical asymptotes: $x = 6$

Horizontal asymptote: $y = 0$

99. Domain: all $x \neq \pm 4$

Vertical asymptotes: $x = \pm 4$

Horizontal asymptote: $y = 1$

101. Domain: all real numbers x

Horizontal asymptote: $y = 0$

Section 5.2   Systems of Linear Equations in Two Variables

■ You should be able to solve a linear system by the method of elimination.

1. Obtain coefficients for either x or y that differ only in sign. This is done by multiplying all the terms of one or both equations by appropriate constants.

2. Add the equations to eliminate one of the variables and then solve for the remaining variable.

3. Use back-substitution into either original equation and solve for the other variable.

4. Check your answer.

■ You should know that for a system of two linear equations, one of the following is true.

(a) There are infinitely many solutions; the lines are identical. The system is consistent.

(b) There is no solution; the lines are parallel. The system is inconsistent.

(c) There is one solution; the lines intersect at one point. The system is consistent.

Vocabulary Check

1. method, elimination   **2.** equivalent   **3.** consistent, inconsistent

---

1. $\begin{cases} 2x + y = 5 & \text{Equation 1} \\ x - y = 1 & \text{Equation 2} \end{cases}$

Add to eliminate y: $3x = 6 \implies x = 2$

Substitute $x = 2$ in Equation 2: $2 - y = 1 \implies y = 1$

Answer: $(2, 1)$

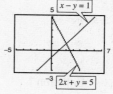

3. $\begin{cases} x + y = 0 & \text{Equation 1} \\ 3x + 2y = 1 & \text{Equation 2} \end{cases}$

Multiply Equation 1 by -2: $-2x - 2y = 0$

Add this to Equation 2 to eliminate y: $x = 1$

Substitute $x = 1$ in Equation 1: $1 + y = 0 \implies y = -1$

Answer: $(1, -1)$

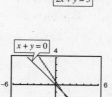

5. $\begin{cases} x - y = 2 & \text{Equation 1} \\ -2x + 2y = 5 & \text{Equation 2} \end{cases}$

Multiply Equation 1 by 2: $2x - 2y = 4$

Add this to Equation 2: $0 = 9$

There are no solutions.

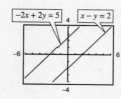

7. $\begin{cases} x + 2y = 4 & \text{Equation 1} \\ x - 2y = 1 & \text{Equation 2} \end{cases}$

Add to eliminate y:

$$2x = 5$$

$$x = \frac{5}{2}$$

Substitute $x = \frac{5}{2}$ in Equation 1:

$$\frac{5}{2} + 2y = 4 \implies y = \frac{3}{4}$$

Answer: $\left(\frac{5}{2}, \frac{3}{4}\right)$

9. $\begin{cases} 2x + 3y = 18 & \text{Equation 1} \\ 5x - y = 11 & \text{Equation 2} \end{cases}$

Multiply Equation 2 by 3: $15x - 3y = 33$

Add this to Equation 1 to eliminate y:

$$17x = 51 \implies x = 3$$

Substitute $x = 3$ in Equation 1:

$$6 + 3y = 18 \implies y = 4$$

Answer: $(3, 4)$

11. $\begin{cases} 3r + 2s = 10 & \text{Equation 1} \\ 2r + 5s = 3 & \text{Equation 2} \end{cases}$

Multiply Equation 1 by 2 and Equation 2 by -3:

$$6r + 4s = 20$$

$$-6r - 15s = -9$$

Add to eliminate r:

$$-11s = 11 \implies s = -1$$

Substitute $s = -1$ in Equation 1:

$$3r + 2(-1) = 10 \implies r = 4$$

Answer: $(4, -1)$

13. $\begin{cases} 5u + 6v = 24 & \text{Equation 1} \\ 3u + 5v = 18 & \text{Equation 2} \end{cases}$

Multiply Equation 1 by 3 and Equation 2 by (-5):

$$15u + 18v = 72$$

$$-15u - 25v = -90$$

Add to eliminate u: $-7v = -18 \implies v = \frac{18}{7}$

Substitute $v = \frac{18}{7}$ in Equation 2:

$$3u + 5\left(\frac{18}{7}\right) = 18 \implies u = \frac{12}{7}$$

Answer: $\left(\frac{12}{7}, \frac{18}{7}\right)$

15. $\begin{cases} 1.8x + 1.2y = 4 & \text{Equation 1} \\ 9x + 6y = 3 & \text{Equation 2} \end{cases}$

Multiply Equation 1 by (-5): $-9x - 6y = -20$

Add this to Equation 2: $0 = -17$

Inconsistent; no solution

17. $2x - 5y = 0$

$x - y = 3 \implies y = x - 3$

$2x - 5(x - 3) = 0$

$$-3x = -15$$

$$x = 5, y = 2$$

Matches (b).

One solution; consistent

19. $\left. \begin{array}{l} 2x - 5y = 0 \\ 2x - 3y = -4 \end{array} \right\} \begin{array}{l} -2y = 4 \\ y = -2, x = -5 \end{array}$

One solution; consistent

Matches (c).

21. $\begin{cases} 4x + 3y = 3 & \text{Equation 1} \\ 3x + 11y = 13 & \text{Equation 2} \end{cases}$

Multiply Equation 1 by 3 and Equation 2 by -4:

$$\begin{cases} 12x + 9y = 9 \\ -12x - 44y = -52 \end{cases}$$

Add to eliminate x: $-35y = -43 \implies y = \frac{43}{35}$

Substitute $y = \frac{43}{35}$ into Equation 1:

$$4x + 3\left(\frac{43}{35}\right) = 3 \implies x = -\frac{6}{35}$$

Answer: $\left(-\frac{6}{35}, \frac{43}{35}\right)$

23. $\begin{cases} \frac{2}{5}x - \frac{3}{2}y = 4 & \text{Equation 1} \\ \frac{1}{5}x - \frac{3}{4}y = -2 & \text{Equation 2} \end{cases}$

Multiply Equation 2 by -2 and add to Equation 1:

$$0 = 8$$

Inconsistent; no solution

25. $\begin{cases} \frac{3}{4}x - y = \frac{1}{8} & \text{Equation 1} \\ \frac{9}{4}x - 3y = \frac{3}{8} & \text{Equation 2} \end{cases}$

Multiply Equation 1 by -3: $-\frac{9}{4}x - 3y = -\frac{3}{8}$

Add this to Equation 2: $0 = 0$

There are an infinite number of solutions.

The solutions consist of all (x, y) satisfying $\frac{3}{4}x + y = \frac{1}{8}$, or $6x + 8y = 1$.

27. $\begin{cases} \dfrac{x+3}{4} + \dfrac{y-1}{3} = 1 & \text{Equation 1} \\ 2x - y = 12 & \text{Equation 2} \end{cases}$

Multiply Equation 1 by 12 and Equation 2 by 4:

$$\begin{cases} 3x + 4y = 7 \\ 8x - 4y = 48 \end{cases}$$

Add to eliminate y: $11x = 55 \implies x = 5$

Substitute $x = 5$ into Equation 2:

$$2(5) - y = 12 \implies y = -2$$

Answer: $(5, -2)$

29. $\begin{cases} \dfrac{x-1}{2} + \dfrac{y+2}{3} = 4 & \text{Equation 1} \\ x - 2y = 5 & \text{Equation 2} \end{cases}$

Multiply Equation 1 by 6:

$$3(x - 1) + 2(y + 2) = 24 \implies 3x + 2y = 23$$

Add this to Equation 2 to eliminate y:

$$4x = 28 \implies x = 7$$

Substitute $x = 7$ in Equation 2:

$$7 - 2y = 5 \implies y = 1$$

Answer: $(7, 1)$

31. $\begin{cases} 2.5x - 3y = 1.5 & \text{Equation 1} \\ 10x - 12y = 6 & \text{Equation 2 multiplied by 5} \end{cases}$

Multiply Equation 1 by (-4):

$$-10x + 12y = -6$$

Add this to Equation 2 to eliminate x: $0 = 0$

The solution set consists of all points lying on the line $10x - 12y = 6$.

All points on the line $5x - 6y = 3$

Let $x = a$, then $y = \frac{5}{6}a - \frac{1}{2}$.

Answer: $\left(a, \frac{5}{6}a - \frac{1}{2}\right)$, where a is any real number.

33. $\begin{cases} 0.2x - 0.5y = -27.8 & \text{Equation 1} \\ 0.3x + 0.4y = 68.7 & \text{Equation 2} \end{cases}$

Multiply Equation 1 by 40 and Equation 2 by 50:

$$\begin{cases} 8x - 20y = -1112 \\ 15x + 20y = 3435 \end{cases}$$

Adding the equation eliminates y:

$$23x = 2323 \implies x = 101$$

Substitute $x = 101$ into Equation 1:

$$8(101) - 20y = -1112 \implies y = 96$$

Answer: $(101, 96)$

35. $\begin{cases} 0.05x - 0.03y = 0.21 & \text{Equation 1} \\ 0.07x + 0.02y = 0.16 & \text{Equation 2} \end{cases}$

Multiply Equation 1 by 200 and Equation 2 by 300:

$$\begin{cases} 10x - 6y = 42 \\ 21x + 6y = 48 \end{cases}$$

Add to eliminate y: $31x = 90$

$$x = \frac{90}{31}$$

Substitute $x = \frac{90}{31}$ in Equation 2:

$$0.07\left(\frac{90}{31}\right) + 0.02y = 0.16$$

$$y = -\frac{67}{31}$$

Answer: $\left(\frac{90}{31}, -\frac{67}{31}\right)$

37. Let $X = \dfrac{1}{x}$ and $Y = \dfrac{1}{y}$.

$$\begin{cases} X + 3Y = 2 & \text{Equation 1} \\ 4X - Y = -5 & \text{Equation 2} \end{cases}$$

Multiply Equation 1 by 4:

$$\begin{cases} 4X + 12Y = 8 \\ 4X - Y = -5 \end{cases}$$

Subtract to eliminates X:

$$13Y = 13 \implies Y = 1.$$

Hence,

$$X = 2 - 3Y = 2 - 3(1) = -1$$

$$x = \frac{1}{X} = -1, \quad y = \frac{1}{Y} = 1$$

Answer: $(-1, 1)$

39. Let $X = \dfrac{1}{x}$ and $Y = \dfrac{1}{y}$.

$$\begin{cases} X + 2Y = 5 & \text{Equation 1} \\ 3X - 4Y = -5 & \text{Equation 2} \end{cases}$$

Multiply Equation 1 by 2:

$$\begin{cases} 2X + 4Y = 10 \\ 3X - 4Y = -5 \end{cases}$$

Adding the equations eliminates Y:

$$5X = 5 \implies X = 1.$$

Hence,

$$2Y = 5 - X = 4 \implies Y = 2.$$

$$x = \frac{1}{X} = 1, \quad y = \frac{1}{Y} = \frac{1}{2}$$

Answer: $\left(1, \frac{1}{2}\right)$

41. $\begin{cases} 2x - 5y = 0 \implies y = \frac{2}{5}x \\ x - y = 3 \implies y = x - 3 \end{cases}$

The system is consistent.
There is one solution, $(5, 2)$.

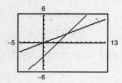

43. $\begin{cases} \frac{3}{5}x - y = 3 \implies y = \frac{3}{5}x - 3 \\ -3x + 5y = 9 \implies y = \frac{1}{5}(3x + 9) = \frac{3}{5}x + \frac{9}{5} \end{cases}$

The lines are parallel.
The system is inconsistent.

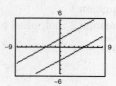

45. $\begin{cases} 8x - 14y = 5 \implies y = \dfrac{8x - 5}{14} = \dfrac{4}{7}x - \dfrac{5}{14} \\ 2x - 3.5y = 1.25 \implies y = \dfrac{2x - 1.25}{3.5} = \dfrac{4}{7}x - \dfrac{5}{14} \end{cases}$

The system is consistent. The solution set consists of all points on the line $y = \frac{4}{7}x - \frac{5}{14}$, or $8x - 14y = 5$.

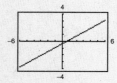

47. $\begin{cases} 6y = 42 \implies y = 7 \\ 6x - y = 16 \implies y = 6x - 16 \end{cases}$

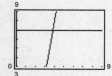

Answer: $\left(\frac{23}{6}, 7\right) \approx (3.833, 7)$

49. $\begin{cases} \frac{3}{2}x - \frac{1}{5}y = 8 \implies y = 5\left(\frac{3}{2}x - 8\right) \\ -2x + 3y = 3 \implies y = \frac{1}{3}(3 + 2x) \end{cases}$

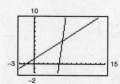

Answer: $(6, 5)$

51. $\frac{1}{3}x + y = -\frac{1}{3} \implies y = -\frac{1}{3} - \frac{1}{3}x$

$5x - 3y = 7 \implies y = \frac{1}{3}(5x - 7)$

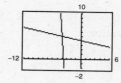

Answer: $(1, -0.667)$

53. $\begin{cases} 0.5x + 2.2y = 9 \implies y = \dfrac{1}{2.2(9 - 0.5x)} \\ 6x + 0.4y = -22 \implies y = \dfrac{1}{0.4(-22 - 6x)} \end{cases}$

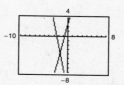

Answer: $(-4, 5)$

55. $\begin{cases} 3x - 5y = 7 & \text{Equation 1} \\ 2x + y = 9 & \text{Equation 2} \end{cases}$

Multiply Equation 2 by 5:

$10x + 5y = 45$

Add this to Equation 1:

$13x = 52 \implies x = 4$

Back-substitute $x = 4$ into Equation 2:

$2(4) + y = 9 \implies y = 1$

Answer: $(4, 1)$

57. $\begin{cases} y = 4x + 3 \\ y = -5x - 12 \end{cases}$

The lines intersect at

$\left(-\frac{5}{3}, -\frac{11}{3}\right) \approx (-1.667, -3.667)$.

59. $\begin{cases} x - 5y = 21 \\ 6x + 5y = 21 \end{cases}$

Adding the equations, $7x = 42 \implies x = 6$.

Back-substituting, $x - 5y = 6 - 5y = 21 \implies$

$-5y = 15 \implies y = -3$.

Answer: $(6, -3)$

61. $\begin{cases} -2x + 8y = 19 & \text{Equation 1} \\ y = x - 3 & \text{Equation 2} \end{cases}$

Substitute into Equation 1,

$-2x + 8(x - 3) = 19 \implies 6x = 43$

$\implies x = \frac{43}{6}$.

Back-substituting, $y = x - 3 = \frac{43}{6} - 3 = \frac{25}{6}$.

Answer: $\left(\frac{43}{6}, \frac{25}{6}\right)$

63. There are infinitely many systems that have the solution $(0, 8)$. One possible system is:

$\begin{cases} x + y = 8 \\ -x + y = 8 \end{cases}$

65. There are infinitely many systems that have the solution $\left(3, \frac{5}{2}\right)$. One possible system:

$2(3) + 2\left(\frac{5}{2}\right) = 11 \implies 2x + 2y = 11$

$3 - 4\left(\frac{5}{2}\right) = -7 \implies x - 4y = -7$

67. Demand $=$ Supply

$50 - 0.5x = 0.125x$

$50 = 0.625x$

$x = 80$ units

$p = \$10$

Answer: $(80, 10)$

69. Demand $=$ Supply

$140 - 0.00002x = 80 + 0.00001x$

$60 = 0.00003x$

$x = 2{,}000{,}000$ units

$p = \$100.00$

Answer: $(2{,}000{,}000, 100)$

71. Let $x = $ the ground speed and $y = $ the wind speed.

$$\begin{cases} 3.6(x - y) = 1800 & \text{Equation 1} \\ 3(x + y) = 1800 & \text{Equation 2} \end{cases}$$

$$\begin{aligned} x - y &= 500 \\ x + y &= 600 \\ \hline 2x &= 1100 \\ x &= 550 \end{aligned}$$

Substituting $x = 550$ in Equation 2:

$$\begin{aligned} 550 + y &= 600 \\ y &= 50 \end{aligned}$$

Answer: $x = 550$ mph, $y = 50$ mph

73. (a) $\begin{cases} A + C = 1175 & \text{Equation 1} \\ 5A + 3.5C = 5087.5 & \text{Equation 2} \end{cases}$

(b) Multiply Equation 1 by 5:

$$5A + 5C = 5875$$

$$5A + 3.5C = 5087.5$$

Subtracting eliminates A:

$$1.5C = 787.5$$

$$C = 525$$

Hence, $A = 1175 - 525 = 650$.

650 adult tickets and 525 child tickets

(c)

Let $C = y_1 = 1175 - x$

$$C = y_2 = \frac{1}{3.5}(5087.5 - 5x)$$

Point of intersection: $(A, C) = (650, 525)$

75. Let $M = $ number of oranges and let $R = $ number of grapefruit.

$$\begin{cases} M + R = 16 & \text{Equation 1} \\ 0.95M + 1.05R = 15.90 & \text{Equation 2} \end{cases}$$

Solving for R in Equation 1: $R = 16 - M$.

Substituting into Equation 2:

$$0.95M + 1.05(16 - M) = 15.9$$

$$0.95M + 16.8 - 1.05M = 15.9$$

$$0.9 = 0.1M$$

$$M = 9$$

Hence, $R = 16 - 9 = 7$.

9 oranges and 7 grapefruit

77. Let $m = $ number of movies and let $v = $ number of videos.

$$\begin{cases} m + v = 310 & \text{Equation 1} \\ 3m + 2.5v = 867.5 & \text{Equation 2} \end{cases}$$

Solve for m in Equation 1: $m = 310 - v$.

Substitute for m Equation 2:

$$3(310 - v) + 2.5v = 867.5$$

$$62.5 = 0.5v$$

$$v = 125$$

$$m = 310 - 125 = 185$$

185 movies and 125 videos

79. $\begin{cases} 5b + 10a = 20.2 \implies -10b - 20a = -40.4 \\ 10b + 30a = 50.1 \implies 10b + 30a = 50.1 \end{cases}$

$$\begin{aligned} 10a &= 9.7 \\ a &= 0.97 \\ b &= 2.10 \end{aligned}$$

Least squares regression line: $y = 0.97x + 2.10$

81. $\begin{cases} 5b + 10a = 2.7 & \text{Equation 1} \\ 10b + 30a = -19.6 & \text{Equation 2} \end{cases}$

Multiply Equation 1 by -2:

$-10b - 20a = -5.4$

$10b + 30a = -19.6$

Adding,

$10a = -25.0$

$a = -2.5$

$b = (2.7 + 10(2.5))/5 = 5.54$

$y = -2.5x + 5.54$

83. (a) $\begin{cases} 4b + 7a = 174 \implies 28b + 49a = 1218 \\ 7b + 13.5a = 322 \implies -28b - 54a = -1288 \end{cases}$

Adding, $-5a = -70 \implies a = 14, \ b = 19$.

Thus, $y = 14x + 19$

(b) Using a graphing utility, you obtain $y = 14x + 19$.

(c)

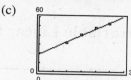

(d) If $x = 1.6$, (160 pounds/acre),

$y = 14(1.6) + 19 = 41.4$ bushels per acre.

85. True. A consistent linear system has either one solution or an infinite number of solutions.

87. $\begin{cases} 100y - x = 200 & \text{Equation 1} \\ 99y - x = -198 & \text{Equation 2} \end{cases}$

Subtract Equation 2 from Equation 1 to eliminate x: $y = 398$

Substitute $y = 398$ into Equation 1: $100(398) - x = 200 \implies x = 39,600$

Answer: (39,600, 398)

The lines are not parallel. The scale on the axes must be changed to see the point of intersection.

89. No, it is not possible for a consistent system of linear equations to have exactly two solutions. Either the lines will intersect once or they will coincide and then the system would have infinite solutions.

91. $\begin{cases} 4x + 8y = -3 & \text{Equation 1} \\ 2x + ky = 16 & \text{Equation 2} \end{cases}$

Multiply Equation 2 by -2: $-4x - 2ky = -32$

Add this to Equation 1: $-8y - 2ky = -35$

The system is inconsistent if $-8y - 2ky = 0$.

This occurs when $k = -4$. Note that for $k = -4$, the two original equations represent parallel lines.

93. Subtracting the two equations:

$vxe^x - v(x + 1)e^x = -e^x \ln x$

$vxe^x - vxe^x - ve^x = -e^x \ln x$

$ve^x = e^x \ln x$

$v = \ln x$

Finally, $ue^x + vxe^x = ue^x + \ln x \cdot x \cdot e^x = 0$

$\implies ue^x = -x \ln x \cdot e^x$

$\implies u = -x \ln x$.

95. $-11 - 6x \geq 33$

$-6x \geq 44$

$x \leq -\frac{44}{6} = -\frac{22}{3}$

97. $|x - 8| < 10$

$-10 < x - 8 < 10$

$-2 < x < 18$

99. $2x^2 + 3x - 35 < 0$

$(2x - 7)(x + 5) < 0$

Critical numbers: $\frac{7}{2}, -5$.

Testing the three intervals,

$-5 < x < \frac{7}{2}$.

101. $\ln x + \ln 6 = \ln 6x$

103. $\log_9 12 - \log_9 x = \log_9 \dfrac{12}{x}$

105. $2 \ln x - \ln(x + 2) = \ln x^2 - \ln(x + 2)$

$$= \ln\left[\dfrac{x^2}{x + 2}\right]$$

107. Answers will vary.

Section 5.3   Multivariable Linear Systems

■   You should know the operations that lead to equivalent systems of linear equations:

(a) Interchange any two equations.

(b) Multiply all terms of an equation by a nonzero constant.

(c) Replace an equation by the sum of itself and a constant multiple of any other equation in the system.

■   You should be able to use the method of elimination.

Vocabulary Check

1. row-echelon

2. ordered triple

3. Gaussian

4. independent, dependent

5. nonsquare

6. three-dimensional

7. partial fraction decomposition

---

1. (a) $3(3) - 5 + (-3) \overset{?}{=} 1$ Yes

$2(3) - 3(-3) \overset{?}{=} -14$ No

$5(5) + 2(-3) \overset{?}{=} 8$ No

No, $(3, 5, -3)$ is not a solution.

(b) $3(-1) - (0) + 4 \overset{?}{=} 1$ Yes

$2(-1) - 3(4) \overset{?}{=} -14$ Yes

$5(0) + 2(4) \overset{?}{=} 8$ Yes

Yes, $(-1, 0, 4)$ is a solution.

(c) $3(-4) - 1 + 2 \overset{?}{=} 1$ No

$2(-4) - 3(2) \overset{?}{=} -14$ Yes

$5(1) + 2(2) \overset{?}{=} 8$ No

No, $(-4, 1, 2)$ is not a solution.

(d) $3(1) - 0 + 4 \overset{?}{=} 1$ No

$2(1) - 3(4) \overset{?}{=} -14$ No

$5(0) + 2(4) \overset{?}{=} 8$ Yes

No, $(1, 0, 4)$ is not a solution.

3. (a) $4(0) + 1 - 1 \overset{?}{=} 0$ Yes

$-8(0) - 6(1) + 1 \overset{?}{=} -\frac{7}{4}$ No

$3(0) - 1 \overset{?}{=} -\frac{9}{4}$ No

No, $(0, 1, 1)$ is not a solution.

(b) $4\left(-\frac{3}{2}\right) + \frac{5}{4} - \left(-\frac{5}{4}\right) \overset{?}{=} 0$ No

$-8\left(-\frac{3}{2}\right) - 6\left(\frac{5}{4}\right) + \left(-\frac{5}{4}\right) \overset{?}{=} -\frac{7}{4}$ No

$3\left(-\frac{3}{2}\right) - \left(\frac{5}{4}\right) \overset{?}{=} -\frac{9}{4}$ No

No, $\left(-\frac{3}{2}, \frac{5}{4}, -\frac{5}{4}\right)$ is not a solution.

(c) $4\left(-\frac{1}{2}\right) + \frac{3}{4} - \left(-\frac{5}{4}\right) \overset{?}{=} 0$ Yes

$-8\left(-\frac{1}{2}\right) - 6\left(\frac{3}{4}\right) - \frac{5}{4} \overset{?}{=} -\frac{7}{4}$ Yes

$3\left(-\frac{1}{2}\right) - \frac{3}{4} \overset{?}{=} -\frac{9}{4}$ Yes

Yes, $\left(-\frac{1}{2}, \frac{3}{4}, -\frac{5}{4}\right)$ is a solution.

(d) $4\left(-\frac{1}{2}\right) + 2 - 0 \overset{?}{=} 0$ Yes

$-8\left(-\frac{1}{2}\right) - 6(2) + 0 \overset{?}{=} -\frac{7}{2}$ No

$3\left(-\frac{1}{2}\right) - 2 \overset{?}{=} -\frac{9}{4}$ No

No, $\left(-\frac{1}{2}, 2, 0\right)$ is not a solution.

5. $\begin{cases} 2x - y + 5z = 16 & \text{Equation 1} \\ y + 2z = 2 & \text{Equation 2} \\ z = 2 & \text{Equation 3} \end{cases}$

Back-substitute $z = 2$ into Equation 2:

$$y + 2(2) = 2$$
$$y = -2$$

Back-substitute $z = 2$ and $y = -2$ into Equation 1:

$$2x - (-2) + 5(2) = 16$$
$$2x = 4$$
$$x = 2$$

Answer: $(2, -2, 2)$

9. $\begin{cases} 4x - 2y + z = 8 & \text{Equation 1} \\ - y + z = 4 & \text{Equation 2} \\ z = 2 & \text{Equation 3} \end{cases}$

Back-substitute $z = 2$ into Equation 2:

$$-y + 2 = 4$$
$$y = -2$$

Back-substitute $y = -2$ and $z = 2$ into Equation 1:

$$4x - 2(-2) + 2 = 8$$
$$4x = 2$$
$$x = \tfrac{1}{2}$$

Answer: $\left(\tfrac{1}{2}, -2, 2\right)$

13. $\begin{cases} x - 2y + z = 1 & \text{Equation 1} \\ 2x - y + 3z = 0 & \text{Equation 2} \\ 3x - y - 4z = 1 & \text{Equation 3} \end{cases}$

Add -2 times Equation 1 to Equation 2.

$$3y + z = -2 \quad \text{New Equation 2}$$

This is the first step in putting the system in row-echelon form.

7. $\begin{cases} 2x + y - 3z = 10 & \text{Equation 1} \\ y + z = 12 & \text{Equation 2} \\ z = 2 & \text{Equation 3} \end{cases}$

Back-substitute $z = 2$ into Equation 2:

$$y + 2 = 12$$
$$y = 10$$

Back-substitute $y = 10$ and $z = 2$ into Equation 1:

$$2x + 10 - 3(2) = 10$$
$$2x = 6$$
$$x = 3$$

Answer: $(3, 10, 2)$

11. $\begin{cases} x - 2y + 3z = 5 & \text{Equation 1} \\ -x + 3y - 5z = 4 & \text{Equation 2} \\ 2x - 3z = 0 & \text{Equation 3} \end{cases}$

Add Equation 1 to Equation 2.

$$y - 2z = 9 \quad \text{New Equation 2}$$

This is the first step in putting the system in row-echelon form.

$$\begin{cases} x - 2y + 3z = 5 \\ y - 2z = 9 \\ 2x - 3z = 0 \end{cases}$$

15. $\begin{cases} x + y + z = 6 & \text{Equation 1} \\ 2x - y + z = 3 & \text{Equation 2} \\ 3x - z = 0 & \text{Equation 3} \end{cases}$

$$\begin{cases} x + y + z = 6 \\ - 3y - z = -9 & (-2)\ \text{Eq. 1} + \text{Eq. 2} \\ - 3y - 4z = -18 & (-3)\ \text{Eq. 1} + \text{Eq. 3} \end{cases}$$

$$\begin{cases} x + y + z = 6 \\ - 3y - z = -9 \\ - 3z = -9 & (-1)\ \text{Eq. 2} + \text{Eq. 3} \end{cases}$$

$$-3z = -9 \implies z = 3$$
$$-3y - 3 = -9 \implies y = 2$$
$$x + 2 + 3 = 6 \implies x = 1$$

Answer: $(1, 2, 3)$

17. $\begin{cases} 2x \qquad + 2z = 2 \\ 5x + 3y \qquad = 4 \\ \qquad 3y - 4z = 4 \end{cases}$ Equation 1
Equation 2
Equation 3

$\begin{cases} x \qquad + z = 1 \\ 5x + 3y \qquad = 4 \\ \qquad 3y - 4z = 4 \end{cases}$ $\left(\frac{1}{2}\right)$ Eq. 1

$\begin{cases} x \qquad + z = 1 \\ \quad 3y - 5z = -1 \\ \quad 3y - 4z = 4 \end{cases}$ (-5) Eq. 1 + Eq. 2

$\begin{cases} x \qquad + z = 1 \\ \quad 3y - 5z = -1 \\ \qquad z = 5 \end{cases}$ (-1) Eq. 2 + Eq. 3

$3y - 5(5) = -1 \implies y = 8$

$x + 5 = 1 \implies x = -4$

Answer: $(-4, 8, 5)$

19. $\begin{cases} 4x + y - 3z = 11 \\ 2x - 3y + 2z = 9 \\ x + y + z = -3 \end{cases}$ Equation 1
Equation 2
Equation 3

$\begin{cases} x + y + z = -3 \\ 2x - 3y + 2z = 9 \\ 4x + y - 3z = 11 \end{cases}$ Interchange Equations 1 and 3.

$\begin{cases} x + y + z = -3 \\ \quad -5y = 15 \\ \quad -3y - 7z = 23 \end{cases}$ (-2) Eq. 1 + Eq. 2
(-4) Eq. 1 + Eq. 3

$y = -3 \implies -3(-3) - 7z = 23$

$\implies -7z = 14$

$\implies z = -2$

$x + (-3) + (-2) = -3 \implies x = 2$

Answer: $(2, -3, -2)$

21. $\begin{cases} x + y - 2z = 3 \\ 3x - 2y + 4z = 1 \\ 2x - 3y + 6z = 8 \end{cases}$ Interchange
the equations.

$\begin{cases} x + y - 2z = 3 \\ \quad -5y + 10z = -8 \\ \quad -5y + 10z = 2 \end{cases}$ -3 Eq. 1 + Eq. 2
-2 Eq. 1 + Eq. 3

$\begin{cases} x + y - 2z = 3 \\ \quad -5y + 10z = -8 \\ \qquad 0 = 10 \end{cases}$ $-$Eq. 2 + Eq. 3

Inconsistent; no solution.

23. $\begin{cases} 3x + 3y + 5z = 1 \\ 3x + 5y + 9z = 0 \\ 5x + 9y + 17z = 0 \end{cases}$

$\begin{cases} 6x + 6y + 10z = 2 \\ 3x + 5y + 9z = 0 \\ 5x + 9y + 17z = 0 \end{cases}$ 2 Eq. 1

$\begin{cases} x - 3y - 7z = 2 \\ 3x + 5y + 9z = 0 \\ 5x + 9y + 17z = 0 \end{cases}$ $-$Eq. 3 + Eq. 1

$\begin{cases} x - 3y - 7z = 2 \\ \quad 14y + 30z = -6 \\ \quad 24y + 52z = -10 \end{cases}$ -3 Eq. 1 + Eq. 2
-5 Eq. 1 + Eq. 3

$\begin{cases} x - 3y - 7z = 2 \\ \quad 84y + 180z = -36 \\ \quad 84y + 182z = -35 \end{cases}$ 6 Eq. 2
3.5 Eq. 3

$\begin{cases} x - 3y - 7z = 2 \\ \quad 84y + 180z = -36 \\ \qquad 2z = 1 \end{cases}$ $-$Eq. 2 + Eq. 3

$2z = 1 \implies z = \frac{1}{2}$

$84y + 180\left(\frac{1}{2}\right) = -36 \implies y = -\frac{3}{2}$

$x - 3\left(-\frac{3}{2}\right) - 7\left(\frac{1}{2}\right) = 2 \implies x = 1$

Answer: $\left(1, -\frac{3}{2}, \frac{1}{2}\right)$

25. $\begin{cases} x + 2y - 7z = -4 \\ 2x + y + z = 13 \\ 3x + 9y - 36z = -33 \end{cases}$ Equation 1
Equation 2
Equation 3

$\begin{cases} x + 2y - 7z = -4 \\ -3y + 15z = 21 \\ 3y - 15z = -21 \end{cases}$ -2 Eq. 1 + Eq. 2
-3 Eq. 1 + Eq. 3

$\begin{cases} x + 2y - 7z = -4 \\ -3y + 15z = 21 \\ 0 = 0 \end{cases}$ Eq. 2 + Eq. 3

$\begin{cases} x + 2y - 7z = -4 \\ y - 5z = -7 \end{cases}$ $-\frac{1}{3}$ Eq. 2

$\begin{cases} x + 3z = 10 \\ y - 5z = -7 \end{cases}$ -2 Eq. 2 + Eq. 1

Let $z = a$, then:

$y = 5a - 7$

$x = -3a + 10$

Answer: $(-3a + 10, 5a - 7, a)$

27. $\begin{cases} x - y + 2z = 6 \\ 2x + y + z = 3 \\ x + y + z = 2 \end{cases}$ Equation 1
Equation 2
Equation 3

$\begin{cases} x - y + 2z = 6 \\ 3y - 3z = -9 \\ 2y - z = -4 \end{cases}$ (-2) Eq. 1 + Eq. 2
(-1) Eq. 1 + Eq. 3

$\begin{cases} x - y + 2z = 6 \\ y - z = -3 \\ 2y - z = -4 \end{cases}$ $\left(\frac{1}{3}\right)$ Eq. 2

$\begin{cases} x - y + 2z = 6 \\ y - z = -3 \\ z = 2 \end{cases}$ (-2) Eq. 2 + Eq. 3

$y - z = -3 \implies y = 2 - 3 = -1$

$x - y + 2z = 6 \implies x = 6 + (-1) - 2(2) = 1$

Answer: $(1, -1, 2)$

29. $\begin{cases} 3x - 3y + 6z = 6 \\ x + 2y - z = 5 \\ 5x - 8y + 13z = 7 \end{cases}$ Equation 1
Equation 2
Equation 3

$\begin{cases} x - y + 2z = 2 \\ x + 2y - z = 5 \\ 5x - 8y + 13z = 7 \end{cases}$ $\left(\frac{1}{3}\right)$ Equation 1

$\begin{cases} x - y + 2z = 2 \\ 3y - 3z = 3 \\ -3y + 3z = -3 \end{cases}$ (-1) Eq. 1 + Eq. 2
(-5) Eq. 1 + Eq. 3

$\begin{cases} x - y + 2z = 2 \\ y - z = 1 \\ 0 = 0 \end{cases}$

$\begin{cases} x + z = 3 \\ y - z = 1 \end{cases}$

Let $z = a$, then:

$y = a + 1$

$x = -a + 3$

Answer: $(-a + 3, a + 1, a)$

31. $\begin{cases} x - 2y + 3z = 4 \\ 3x - y + 2z = 0 \\ x + 3y - 4z = -2 \end{cases}$ Equation 1
Equation 2
Equation 3

$\begin{cases} x - 2y + 3z = 4 \\ 5y - 7z = -12 \\ 5y - 7z = -6 \end{cases}$ -3 Eq. 1 + Eq. 2
-1 Eq. 1 + Eq. 3

$\begin{cases} x - 2y + 3z = 4 \\ 5y - 7z = -12 \\ 0 = 6 \end{cases}$ $-$Eq. 2 + Eq. 3

No solution; inconsistent

33. $\begin{cases} x \quad\;\; + 4z = \quad 1 \\ x + y + 10z = \quad 10 \\ 2x - y + \;\; 2z = -5 \end{cases}$

$\begin{cases} x \quad\;\; + 4z = \quad 1 \\ \quad y + 6z = \quad 9 \\ \quad - y - 6z = -7 \end{cases}$ $\quad$ $-$ Eq. 1 + Eq. 2

$\qquad\qquad\qquad\qquad\qquad\;\;$ -2 Eq. + Eq. 3

$\begin{cases} x \;\; + 4z = 1 \\ \quad y + 6z = 9 \\ \qquad\quad 0 = 2 \end{cases}$ $\quad$ Eq. 2 + Eq. 3

No solution; inconsistent

35. $\begin{cases} x + 2y + \;\; z = \quad 1 \\ x - 2y + 3z = -3 \\ 2x + \;\; y + \;\; z = -1 \end{cases}$ $\quad$ Equation 1

$\qquad\qquad\qquad\qquad\qquad\;$ Equation 2

$\qquad\qquad\qquad\qquad\qquad\;$ Equation 3

$\begin{cases} x + 2y + \;\; z = \quad 1 \\ \quad -4y + 2z = -4 \\ \quad -3y - \;\; z = -3 \end{cases}$ $\quad$ (-1) Eq. 1 + Eq. 2

$\qquad\qquad\qquad\qquad\qquad$ (-2) Eq. 1 + Eq. 3

$\begin{cases} x + 2y + \;\; z = 1 \\ \quad y - \frac{1}{2}z = 1 \\ \quad 3y + \;\; z = 3 \end{cases}$ $\quad$ $\left(-\frac{1}{4}\right)$ Eq. 2

$\qquad\qquad\qquad\qquad\;$ (-1) Eq. 3

$\begin{cases} x + 2y + \;\; z = 1 \\ \quad y - \frac{1}{2}z = 1 \\ \qquad\quad \frac{5}{2}z = 0 \end{cases}$ $\quad$ (-3) Eq. 2 + Eq. 3

$z = 0$

$y = 1 - 0 = 1$

$x + 2y + z = 1 \implies x = 1 - 2 = -1$

Answer: $(-1, 1, 0)$

37. $\begin{cases} x - 2y + 5z = 2 \\ 4x \quad\;\; - \;\; z = 0 \end{cases}$

$\begin{cases} x - 2y + \;\; 5z = \quad 2 \\ \quad 8y - 21z = -8 \end{cases}$ $\quad$ -4 Eq. 1 + Eq. 2

$\begin{cases} x - 2y + \;\; 5z = \quad 2 \\ \quad y - \frac{21}{8}z = -1 \end{cases}$ $\quad$ $\frac{1}{8}$ Eq. 2

$\begin{cases} x \qquad - \frac{1}{4}z = \quad 0 \\ \quad y - \frac{21}{8}z = -1 \end{cases}$ $\quad$ 2 Eq. 2 + Eq. 1

Let $z = a$. Then $y = \frac{21}{8}a - 1$ and $x = \frac{1}{4}a$.

Answer: $\left(\frac{1}{4}a, \frac{21}{8}a - 1, a\right)$

39. $\begin{cases} 2x - 3y + z = -2 \\ -4x + 9y \qquad = \quad 7 \end{cases}$

$\begin{cases} 2x - 3y + \;\; z = -2 \\ \quad 3y + 2z = \quad 3 \end{cases}$ $\quad$ 2 Eq. 1 + Eq. 2

$\begin{cases} 2x \qquad + 3z = 1 \\ \quad 3y + 2z = 3 \end{cases}$ $\quad$ Eq. 2 + Eq. 1

Let $z = a$, then:

$y = -\frac{2}{3}a + 1$

$x = -\frac{3}{2}a + \frac{1}{2}$

Answer: $\left(-\frac{3}{2}a + \frac{1}{2}, -\frac{2}{3}a + 1, a\right)$

41. $\begin{cases} x - \;\; 3y + \;\; 2z = 18 \\ 5x - 13y + 12z = 80 \end{cases}$ $\quad$ Equation 1

$\qquad\qquad\qquad\qquad\;$ Equation 2

$\begin{cases} x - 3y + 2z = \quad 18 \\ \quad 2y + 2z = -10 \end{cases}$ $\quad$ -5 Eq. 1 + Eq. 2

$\begin{cases} x - 3y + 2z = \quad 18 \\ \quad y + \;\; z = -5 \end{cases}$ $\quad$ $\frac{1}{2}$ Eq. 2

$\begin{cases} x \qquad + 5z = \quad 3 \\ \quad y + \;\; z = -5 \end{cases}$ $\quad$ 3 Eq. 2 + Eq. 1

Let $z = a$, then $y = -a - 5$, and $x = -5a + 3$.

Answer: $(-5a + 3, -a - 5, a)$

43. $\begin{cases} x - y + 2z - w = 0 & \text{Equation 1} \\ 2x + y + z - w = 0 & \text{Equation 2} \\ x + y \quad - w = -1 & \text{Equation 3} \\ x + y - 2z + w = 1 & \text{Equation 4} \end{cases}$

$\begin{cases} x - y + 2z - w = 0 & \\ 3y - 3z + w = 0 & (-2)\text{ Eq. 1 + Eq. 2} \\ 2y - 2z \quad = -1 & (-1)\text{ Eq. 1 + Eq. 3} \\ 2y - 4z + 2w = 1 & (-1)\text{ Eq. 1 + Eq. 4} \end{cases}$

$\begin{cases} x - y + 2z - w = 0 & \\ 2y - 2z \quad = -1 & \\ 3y - 3z + w = 0 & \text{Interchange} \\ 2y - 4z + 2w = 1 & \text{Eq. 2 and 3} \end{cases}$

$\begin{cases} x - y + 2z - w = 0 & \\ 2y - 2z \quad = -1 & \\ \quad\quad w = \frac{3}{2} & \left(-\frac{3}{2}\right)\text{ Eq. 2 + Eq. 3} \\ -2z + 2w = 2 & (-1)\text{ Eq. 2 + Eq. 4} \end{cases}$

$w = \frac{3}{2}$

$-2z + 2w = 2 \implies -2z = 2 - 2\left(\frac{3}{2}\right) = -1 \implies z = \frac{1}{2}$

$2y - 2z = -1 \implies 2y = -1 + 2\left(\frac{1}{2}\right) = 0 \implies y = 0$

$x - y + 2z - w = 0 \implies x = -2\left(\frac{1}{2}\right) + \frac{3}{2} = \frac{1}{2}$

Answer: $\left(\frac{1}{2}, 0, \frac{1}{2}, \frac{3}{2}\right)$

45. Let $X = \dfrac{1}{x}, Y = \dfrac{1}{y}, Z = \dfrac{1}{z}$.

$\begin{cases} X + 2Y - 3Z = 3 & \text{Equation 1} \\ X - 2Y + Z = 1 & \text{Equation 2} \\ 2X + 2Y - 3Z = 4 & \text{Equation 3} \end{cases}$

$\begin{cases} X + 2Y - 3Z = 3 & \\ -4Y + 4Z = -2 & (-1)\text{ Eq. 1 + Eq. 2} \\ -2Y + 3Z = -2 & (-2)\text{ Eq. 1 + Eq. 3} \end{cases}$

$\begin{cases} X + 2Y - 3Z = 3 & \\ 2Y - 2Z = 1 & \left(-\frac{1}{2}\right)\text{ Eq. 2} \\ -2Y + 3Z = -2 & \end{cases}$

$\begin{cases} X + 2Y - 3Z = 3 & \\ 2Y - 2Z = 1 & \\ Z = -1 & \text{Eq. 2 + Eq. 3} \end{cases}$

$2Y - 2Z = 1 \implies 2Y = 1 + 2Z = -1 \implies Y = -\dfrac{1}{2}$

$X + 2Y - 3Z = 3 \implies X = 3 - 2\left(-\dfrac{1}{2}\right) + 3(-1) = 1$

$x = \dfrac{1}{X} = 1, y = \dfrac{1}{Y} = -2, z = \dfrac{1}{Z} = -1$

Answer: $(1, -2, -1)$

47. Let $X = \dfrac{1}{x}, Y = \dfrac{1}{y}, Z = \dfrac{1}{z}$.

$$\begin{cases} 2X - Y + 2Z = 4 \\ X + 2Y - 2Z = -2 \\ 3X + 3Y + 4Z = 2 \end{cases}$$
  Equation 1
  Equation 2
  Equation 3

$$\begin{cases} X + 2Y - 2Z = -2 \\ 2X - Y + 2Z = 4 \\ 3X + 3Y + 4Z = 2 \end{cases}$$
  Interchange
  Eq. 1 and 2

$$\begin{cases} X + 2Y - 2Z = -2 \\ -5Y + 6Z = 8 \\ -3Y + 10Z = 8 \end{cases}$$
  (-2) Eq. 1 + Eq. 2
  (-3) Eq. 1 + Eq. 3

$$\begin{cases} X + 2Y - 2Z = -2 \\ -5Y + 6Z = 8 \\ \frac{32}{5}Z = \frac{16}{5} \end{cases}$$
  $\left(-\frac{3}{5}\right)$ Eq. 2 + Eq. 3

$$Z = \frac{1}{2}$$

$$-5Y + 6Z = 8 \implies -5Y = 8 - 6\left(\frac{1}{2}\right) = 5 \implies Y = -1$$

$$X + 2Y - 2Z = -2 \implies X = -2 - 2(-1) + 2\left(\frac{1}{2}\right) = 1$$

$$x = \frac{1}{X} = 1, y = \frac{1}{Y} = -1, z = \frac{1}{Z} = 2$$

Answer: $(1, -1, 2)$

49. There are an infinite number of linear systems that have $(4, -1, 2)$ as their solution. One such system is as follows:

$$\begin{aligned} 3(4) + (-1) - (2) = 9 &\implies 3x + y - z = 9 \\ (4) + 2(-1) - (2) = 0 &\implies x + 2y - z = 0 \\ -(4) + (-1) + 3(2) = 1 &\implies -x + y + 3z = 1 \end{aligned}$$

51. There are an infinite numbers of linear systems that have $\left(3, -\frac{1}{2}, \frac{7}{4}\right)$ as their solution. One such system is:

$$\begin{aligned} 1(3) + 2\left(-\frac{1}{2}\right) + 4\left(\frac{7}{4}\right) = 9 &\implies x + 2y + 4z = 9 \\ 4\left(-\frac{1}{2}\right) + 8\left(\frac{7}{4}\right) = 12 &\implies 4y + 8z = 12 \\ 4\left(\frac{7}{4}\right) = 7 &\implies 4z = 7 \end{aligned}$$

53. $2x + 3y + 4z = 12$

$(6, 0, 0), (0, 4, 0), (0, 0, 3), (4, 0, 1)$

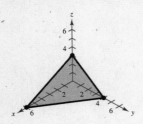

55. $2x + y + z = 4$

$(2, 0, 0), (0, 4, 0), (0, 0, 4), (0, 2, 2)$

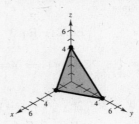

57. $\dfrac{7}{x^2 - 14x} = \dfrac{7}{x(x - 14)} = \dfrac{A}{x} + \dfrac{B}{x - 14}$

59. $\dfrac{12}{x^3 - 10x^2} = \dfrac{12}{x^2(x - 10)} = \dfrac{A}{x} + \dfrac{B}{x^2} + \dfrac{C}{x - 10}$

61. $\dfrac{4x^2 + 3}{(x - 5)^3} = \dfrac{A}{(x - 5)} + \dfrac{B}{(x - 5)^2} + \dfrac{C}{(x - 5)^3}$

63. $\dfrac{1}{x^2 - 1} = \dfrac{A}{x + 1} + \dfrac{B}{x - 1}$

$1 = A(x - 1) + B(x + 1) = (A + B)x + (B - A)$

$\begin{cases} A + B = 0 \\ -A + B = 1 \end{cases}$

$2B = 1 \Rightarrow B = \frac{1}{2} \Rightarrow A = -\frac{1}{2}$

$\dfrac{1}{x^2 - 1} = \dfrac{-1/2}{x + 1} + \dfrac{1/2}{x - 1} = \dfrac{1}{2}\left[\dfrac{1}{x - 1} - \dfrac{1}{x + 1}\right]$

65. $\dfrac{1}{x^2 + x} = \dfrac{1}{x(x + 1)} = \dfrac{A}{x} + \dfrac{B}{x + 1}$

$1 = A(x + 1) + Bx = (A + B)x + A$

$\begin{cases} A + B = 0 \\ \phantom{A + {}}A = 1 \Rightarrow B = -1 \end{cases}$

$\dfrac{1}{x^2 + x} = \dfrac{1}{x} + \dfrac{-1}{x + 1} = \dfrac{1}{x} - \dfrac{1}{x + 1}$

67. $\dfrac{1}{2x^2 + x} = \dfrac{1}{x(2x + 1)} = \dfrac{A}{2x + 1} + \dfrac{B}{x}$

$1 = Ax + B(2x + 1) = (A + 2B)x + B$

$\begin{cases} A + 2B = 0 \\ B = 1 \Rightarrow A = -2 \end{cases}$

$\dfrac{1}{2x^2 + x} = \dfrac{-2}{2x + 1} + \dfrac{1}{x} = \dfrac{1}{x} - \dfrac{2}{2x + 1}$

69. $\dfrac{5 - x}{2x^2 + x - 1} = \dfrac{5 - x}{(2x - 1)(x + 1)}$

$\qquad = \dfrac{A}{2x - 1} + \dfrac{B}{x + 1}$

$5 - x = A(x + 1) + B(2x - 1)$

$\qquad = (A + 2B)x + (A - B)$

$\begin{cases} A + 2B = -1 \Rightarrow A = -1 - 2B \\ A - B = 5 \end{cases}$

$(-1 - 2B) - B = 5 \Rightarrow B = -2 \quad \text{and} \quad A = 3$

$\dfrac{5 - x}{2x^2 + x - 1} = \dfrac{3}{2x - 1} + \dfrac{-2}{x + 1}$

71. $\dfrac{x^2 + 12x + 12}{x^3 - 4x} = \dfrac{x^2 + 12x + 12}{x(x - 2)(x + 2)} = \dfrac{A}{x} + \dfrac{B}{x + 2} + \dfrac{C}{x - 2}$

$x^2 + 12x + 12 = A(x + 2)(x - 2) + Bx(x - 2) + Cx(x + 2)$

$\qquad\qquad\qquad = (A + B + C)x^2 + (-2B + 2C)x + (-4A)$

$\begin{cases} A + + \,C = 1 \\ \, -2B + 2C = 12 \\ -4A = 12 \Rightarrow A = -3 \end{cases}$

$\begin{cases} B + C = 4 \\ -B + C = 6 \end{cases}$

$2C = 10 \Rightarrow C = 5 \Rightarrow B = -1$

$\dfrac{x^2 + 12x + 12}{x^3 - 4x} = \dfrac{-3}{x} + \dfrac{-1}{x + 2} + \dfrac{5}{x - 2}$

73. $\dfrac{4x^2 + 2x - 1}{x^2(x + 1)} = \dfrac{A}{x} + \dfrac{B}{x^2} + \dfrac{C}{x + 1}$

$4x^2 + 2x - 1 = Ax(x + 1) + B(x + 1) + Cx^2 = (A + C)x^2 + (A + B)x + B$

$\begin{cases} A \, + C = 4 \\ A + B = 2 \\ \phantom{A + {}}B = -1 \end{cases}$

$B = -1 \Rightarrow A = 3 \Rightarrow C = 1$

$\dfrac{4x^2 + 2x - 1}{x^2(x + 1)} = \dfrac{3}{x} + \dfrac{-1}{x^2} + \dfrac{1}{x + 1}$

75. $\dfrac{27 - 7x}{x(x - 3)^2} = \dfrac{A}{x} + \dfrac{B}{x - 3} + \dfrac{C}{(x - 3)^2}$

$27 - 7x = A(x - 3)^2 + Bx(x - 3) + Cx$

$\qquad = (A + B)x^2 + (-6A - 3B + C)x + 9A$

$\begin{cases} A + B \qquad\qquad = \quad 0 \\ -6A - 3B + C = -7 \\ 9A \qquad\qquad = \quad 27 \end{cases}$

$A = 3 \Rightarrow B = -3 \Rightarrow C = -7 + 18 - 9 = 2$

$\dfrac{27 - 7x}{x(x - 3)^2} = \dfrac{3}{x} + \dfrac{-3}{x - 3} + \dfrac{2}{(x - 3)^2}$

77. $\dfrac{2x^3 - x^2 + x + 5}{x^2 + 3x + 2} = 2x - 7 + \dfrac{18x + 19}{(x + 1)(x + 2)}$

$\dfrac{18x + 19}{(x + 1)(x + 2)} = \dfrac{A}{x + 1} + \dfrac{B}{x + 2}$

$18x + 19 = A(x + 2) + B(x + 1)$

$\qquad = (A + B)x + (2A + B)$

$\begin{cases} A + B = 18 \\ 2A + B = 19 \end{cases}$

$A = 1 \Rightarrow B = 17$

$\dfrac{2x^3 - x^2 + x + 5}{x^2 + 3x + 2} = 2x - 7 + \dfrac{1}{x + 1} + \dfrac{17}{x + 2}$

79. $\dfrac{x^4}{(x - 1)^3} = x + 3 + \dfrac{6x^2 - 8x + 3}{(x - 1)^3}$

$\dfrac{6x^2 - 8x + 3}{(x - 1)^3} = \dfrac{A}{x - 1} + \dfrac{B}{(x - 1)^2} + \dfrac{C}{(x - 1)^3}$

$6x^2 - 8x + 3 = A(x - 1)^2 + B(x - 1) + C = Ax^2 + (-2A + B)x + (A - B + C)$

$\begin{cases} A \qquad\qquad = \quad 6 \\ -2A + B \qquad = -8 \\ A - B + C = \quad 3 \end{cases}$

$A = 6 \Rightarrow B = -8 + 2(6) = 4 \Rightarrow C = 3 - 6 + 4 = 1$

$\dfrac{x^4}{(x - 1)^3} = \dfrac{6}{x - 1} + \dfrac{4}{(x - 1)^2} + \dfrac{1}{(x - 1)^3} + x + 3$

81. $\dfrac{x - 12}{x(x - 4)} = \dfrac{A}{x} + \dfrac{B}{x - 4}$

$x - 12 = A(x - 4) + Bx$

$\begin{cases} A + B = \quad 1 \\ -4A \quad = -12 \end{cases} \Rightarrow A = 3, B = -2$

$\dfrac{x - 12}{x(x - 4)} = \dfrac{3}{x} - \dfrac{2}{x - 4}$

$y = \dfrac{x - 12}{x(x - 4)} \qquad\qquad y = \dfrac{3}{x}, y = -\dfrac{2}{x - 4}$

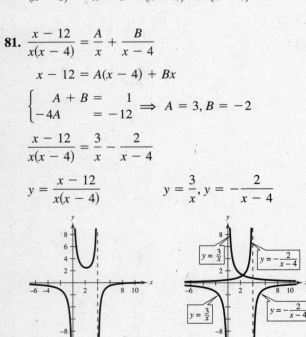

Vertical asymptotes: Vertical asymptotes:

$x = 0$ and $x = 4$ $x = 0$ and $x = 4$

The combination of the vertical asymptotes of the terms of the decompositions are the same as the vertical asymptotes of the rational function.

83. $s = \frac{1}{2}at^2 + v_0t + s_0$

$(1, 128), (2, 80), (3, 0)$

$$\begin{cases} 128 = \frac{1}{2}a + v_0 + s_0 \implies a + 2v_0 + 2s_0 = 256 \\ 80 = 2a + 2v_0 + s_0 \implies 2a + 2v_0 + s_0 = 80 \\ 0 = \frac{9}{2}a + 3v_0 + s_0 \implies 9a + 6v_0 + 2s_0 = 0 \end{cases}$$

Solving the system, $a = -32$, $v_0 = 0$, $s_0 = 144$.

Thus, $s = \frac{1}{2}(-32)t^2 + (0)t + 144$

$\qquad = -16t^2 + 144$.

85. $s = \frac{1}{2}at^2 + v_0t + a_0$

$(1, 452), (2, 372), (3, 260)$

$$\begin{cases} 452 = \frac{1}{2}a + v_0 + s_0 \implies a + 2v_0 + 2s_0 = 904 \\ 372 = 2a + 2v_0 + s_0 \implies 2a + 2v_0 + s_0 = 372 \\ 260 = \frac{9}{2}a + 3v_0 + s_0 \implies 9a + 6v_0 + 2s_0 = 520 \end{cases}$$

Solving the system, $a = -32$, $v_0 = -32$, $s_0 = 500$

Thus, $s = \frac{1}{2}(-32)t^2 - 32t + 500$

$\qquad = -16t^2 - 32t + 500$.

87. $y = ax^2 + bx + c$ passing through $(0, 0), (2, -2), (4, 0)$

$$\begin{cases} (0, 0): & 0 = 4a + 2b + c \implies c = -4a - 2b \\ (2, -2): -2 = 4a + 2b + c \implies -1 = 2a + b \\ (4, 0): & 0 = 16a + 4b + c \implies 0 = 4a + b \end{cases}$$

Answer: $a = \frac{1}{2}$, $b = -2$, $c = 0$

The equation of the parabola is $y = \frac{1}{2}x^2 - 2x$.

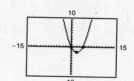

89. $y = ax^2 + bx + c$ passing through $(2, 0), (3, -1), (4, 0)$

$$\begin{cases} (2, 0): & 0 = 4a + 2b + c \implies c = -4a - 2b \\ (3, -1): -1 = 9a + 3b + c \implies -1 = 5a + b \\ (4, 0): & 0 = 16a + 4b + c \implies 0 = 12a + 2b \end{cases}$$

Answer: $a = 1$, $b = -6$, $c = 8$

The equation of the parabola is $y = x^2 - 6x + 8$.

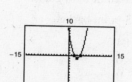

91. $x^2 + y^2 + Dx + Ey + F = 0$ passing through $(0, 0), (2, 2), (4, 0)$

$(0, 0)$: $\qquad\qquad\qquad F = 0$

$(2, 2)$: $8 + 2D + 2E + F = 0 \implies D + E = -4$

$(4, 0)$: $16 + 4D \qquad + F = 0 \implies D = -4$ and $E = 0$

The equation of the circle is $x^2 + y^2 - 4x = 0$.

To graph, let $y_1 = \sqrt{4x - x^2}$ and $y_2 = -\sqrt{4x - x^2}$.

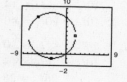

93. $x^2 + y^2 + Dx + Ey + F = 0$ passes through $(-3, -1), (2, 4), (-6, 8)$.

$(-3, -1)$: $10 - 3D - E + F = 0 \implies 10 = 3D + E - F$

$(\ 2,\ \ 4)$: $20 + 2D + 4E + F = 0 \implies 20 = -2D - 4E - F$

$(-6,\ \ 8)$: $100 - 6D + 8E + F = 0 \implies 100 = 6D - 8E - F$

Answer: $D = 6$, $E = -8$, $F = 0$

The equation of the circle is $x^2 + y^2 + 6x - 8y = 0$. To graph, complete the squares first, then solve for y.

$(x^2 + 6x + 9) + (y^2 - 8y + 16) = 0 + 9 + 16$

$\qquad\qquad (x + 3)^2 + (y - 4)^2 = 25$

$\qquad\qquad\qquad\qquad (y - 4)^2 = 25 - (x + 3)^2$

$\qquad\qquad\qquad\qquad\quad y - 4 = \pm\sqrt{25 - (x + 3)^2}$

$\qquad\qquad\qquad\qquad\qquad\ y = 4 \pm \sqrt{25 - (x + 3)^2}$

Let $y_1 = 4 + \sqrt{25 - (x + 3)^2}$ and $y_2 = 4 - \sqrt{25 - (x + 3)^2}$.

95. Let x = amount at 8%.

Let y = amount at 9%.

Let z = amount at 10%.

$$\begin{cases} x + y + z = 775{,}000 \\ 0.08x + 0.09y + 0.10z = 67{,}000 \\ x = 4z \end{cases}$$

$$\begin{cases} x + y + z = 775{,}000 \\ 8x + 9y + 10z = 6{,}700{,}000 \\ x - 4z = 0 \end{cases}$$

Solving the system, $x = \$366{,}666.67$ at 8%,

$y = \$316{,}666.67$ at 9%, $z = \$91{,}666.67$ at 10%.

97. Let C = amount in certificates of deposit.

Let M = amount in municipal bonds.

Let B = amount in blue chip stocks.

Let G = amount in growth or speculative stocks.

$$\begin{cases} C + M + B + G = 500{,}000 \\ 0.08C + 0.09M + 0.12B + 0.15G = 0.10(500{,}000) \\ M = \frac{1}{4}(500{,}000) \end{cases}$$

Solving the system

$C = 156{,}250 + 0.75s$

$M = 125{,}000$

$B = 218{,}750 - 1.75s$

$G = s$

99. Let x = number of 1-point free throws.

Let y = number of 2-point field goals.

Let z = number of 3-point basket.

$$\begin{cases} x + 2y + 3z = 84 \\ -x + y = 6 \\ y - 4z = 0 \end{cases}$$

Solving the system, $x = 18$, $y = 24$, $z = 6$.

18 free throws, 24 2-point field goals, 6 3-point baskets

101. Let x = number of touchdowns.

Let y = number of extra-point kicks.

Let z = number of field goals.

$$\begin{cases} x + y + z = 9 \\ 6x + y + 3z = 31 \\ x - 4z = 0 \\ x - y = 0 \end{cases}$$

Solving the system, $x = 4$, $y = 4$, $z = 1$.

4 touchdowns, 4 extra-points and 1 field goal

103. $\begin{cases} I_1 - I_2 + I_3 = 0 \\ 3I_1 + 2I_2 = 7 \\ 2I_2 + 4I_3 = 8 \end{cases}$ Equation 1
Equation 2
Equation 3

$\begin{cases} I_1 - I_2 + I_3 = 0 \\ 5I_2 - 3I_3 = 7 \\ 2I_2 + 4I_3 = 8 \end{cases}$ -3 Eq. 1 + Eq. 2

$\begin{cases} I_1 - I_2 + I_3 = 0 \\ 10I_2 - 6I_3 = 14 \\ 10I_2 + 20I_3 = 40 \end{cases}$ 2 Eq. 2
5 Eq. 3

$\begin{cases} I_1 - I_2 + I_3 = 0 \\ 10I_2 - 6I_3 = 14 \\ 26I_3 = 26 \end{cases}$ $-$Eq. 2 + Eq. 3

$26I_3 = 26 \implies I_3 = 1$

$10I_2 - 6(1) = 14 \implies I_2 = 2$

$I_1 - 2 + 1 = 0 \implies I_1 = 1$

Answer: $I_1 = 1$ ampere, $I_2 = 2$ amperes,
$I_3 = 1$ ampere

105. Let x = number of par-3 holes.

Let y = number of par-4 holes.

Let z = number of par-5 holes.

$$\begin{cases} 3x + 4y + 5z = 72 \\ y - 2z = 2 \\ x - z = 0 \end{cases}$$

Solving the system, $x = 4$, $y = 10$, $z = 4$.

4 par-3 holes, 10 par-4 holes, and 4 par-5 holes

107. Least squares regression parabola through
$(-4, 5)$, $(-2, 6)$, $(2, 6)$, $(4, 2)$

$$\begin{cases} 4c & + 40a = 19 \\ 40b & = -12 \\ 40c & + 544a = 160 \end{cases}$$

Solving the system, $a = -\frac{5}{24}$, $b = -\frac{3}{10}$, and $c = \frac{41}{6}$.

Thus, $y = -\frac{5}{24}x^2 - \frac{3}{10}x + \frac{41}{6}$.

109. Least squares regression parabola through
$(0, 0)$, $(2, 2)$, $(3, 6)$, $(4, 12)$

$$\begin{cases} 4c + 9b + 29a = 20 \\ 9c + 29b + 99a = 70 \\ 29c + 99b + 353a = 254 \end{cases}$$

Solving the system, $a = 1$, $b = -1$, and $c = 0$.
Thus, $y = x^2 - x$.

111. (a) $$\begin{cases} a(30)^2 + b(30) + c = 55 \\ a(40)^2 + b(40) + c = 105 \\ a(50)^2 + b(50) + c = 188 \end{cases}$$

Solving the system, $a = 0.165$, $b = -6.55$ and $c = 103$.

$$y = 0.165x^2 - 6.55x + 103$$

(c) For $x = 70$, $y = 453$ feet.

(b)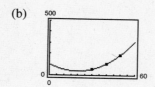

113. (a) $$\frac{2000(4 - 3x)}{(11 - 7x)(7 - 4x)} = \frac{A}{11 - 7x} + \frac{B}{7 - 4x}, \quad 0 \le x \le 1$$

$$2000(4 - 3x) = A(7 - 4x) + B(11 - 7x)$$

$$\begin{cases} -6000 = -4A - 7B \\ 8000 = 7A + 11B \end{cases} \Rightarrow \begin{array}{l} A = -2000 \\ B = 2000 \end{array}$$

$$\frac{2000(4 - 3x)}{(11 - 7x)(7 - 4x)} = \frac{-2000}{11 - 7x} + \frac{2000}{7 - 4x}$$

$$= \frac{2000}{7 - 4x} - \frac{2000}{11 - 7x}$$

(b) $y_1 = \dfrac{2000}{7 - 4x}$

$y_2 = \dfrac{2000}{11 - 7x}$

115. False. The coefficient of y in the second equation is not 1.

117. False. The correct form is

$$\frac{A}{x + 10} + \frac{B}{x - 10} + \frac{C}{(x - 10)^2}.$$

119. $\dfrac{1}{a^2 - x^2} = \dfrac{1}{(a + x)(a - x)} = \dfrac{A}{a + x} + \dfrac{B}{a - x}$

$$1 = A(a - x) + B(a + x)$$

$$= (-A + B)x + (Aa + Ba)$$

$$\begin{cases} -A + B = 0 \\ Aa + Ba = 1 \end{cases} \Rightarrow A = \frac{1}{2a}, B = \frac{1}{2a}$$

$$\frac{1}{a^2 - x^2} = \frac{1/2a}{a + x} + \frac{1/2a}{a - x} = \frac{1}{2a}\left[\frac{1}{a + x} + \frac{1}{a - x}\right]$$

121. $\dfrac{1}{y(a - y)} = \dfrac{A}{y} + \dfrac{B}{a - y}$

$$1 = A(a - y) + By = (-A + B)y + aA$$

$$A = \frac{1}{a}, B = \frac{1}{a}$$

$$\frac{1}{y(a - y)} = \frac{1}{a}\left(\frac{1}{y} + \frac{1}{a - y}\right)$$

123. No, they are not equivalent. In the second system, the constant in the second equation should be -11 and the coefficient of z in the third equation should be 2.

125. $$\begin{cases} y + \lambda = \\ x + \lambda = \\ x + y - 10 = \end{cases} \Rightarrow \begin{array}{l} x = y = -\lambda \\ \Rightarrow 2x - 10 = 0 \\ x = 5 \\ y = 5 \\ \lambda = -5 \end{array}$$

127.
$$\begin{cases} 2x - 2x\lambda = 0 \implies x = x\lambda \\ -2y + \lambda = 0 \implies 2y = \lambda \\ y - x^2 = 0 \implies y = x^2 \end{cases}$$

From the first equation, $x = 0$ or $\lambda = 1$.

If $x = 0$, then $y = 0^2 = 0$ and $\lambda = 0$.

If $x \neq 0$, then $\lambda = 1 \implies y = \frac{1}{2}$ and $x = \pm\sqrt{\frac{1}{2}}$.

Thus, the solutions are:

(1) $x = y = \lambda = 0$

(2) $x = \dfrac{\sqrt{2}}{2}$, $y = \dfrac{1}{2}$, $\lambda = 1$

(3) $x = -\dfrac{\sqrt{2}}{2}$, $y = \dfrac{1}{2}$, $\lambda = 1$

129. $y = -3x + 7$

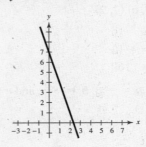

131. $y = -2x^2$

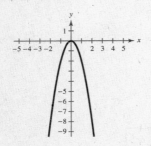

133. $y = -x^2(x - 3)$

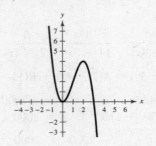

135. (a) $f(x) = x^3 + x^2 - 12x$

$\qquad = x(x^2 + x - 12) = x(x + 4)(x - 3)$

$\qquad \implies x = 0, -4, 3$

(b)

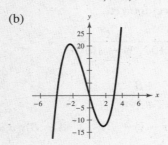

137. (a) $f(x) = 2x^3 + 5x^2 - 21x - 36$

$\qquad = (2x + 3)(x + 4)(x - 3)$

$\qquad \implies x = -\frac{3}{2}, -4, 3$

(b)

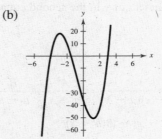

139.

x	-2	-1	0	1	2
y	16	8	4	2	1

$y = \left(\frac{1}{2}\right)^{x-2}$

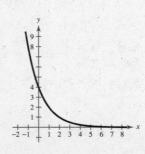

141.

x	-2	-1	0	1	2
y	$-\frac{1}{2}$	0	1	3	7

$y = 2^{x+1} - 1$

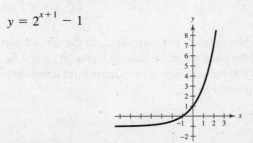

143. Answers will vary.

Section 5.4   Matrices and Systems of Equations

> ■ You should be able to use elementary row operations to produce a row-echelon form (or reduced row-echelon form) of a matrix.
>
> 1. Interchange two rows.
>
> 2. Multiply a row by a nonzero constant.
>
> 3. Add a multiple of one row to another row.
>
> ■ You should be able to use either Gaussian elimination with back-substitution or Gauss-Jordan elimination to solve a system of linear equations.

Vocabulary Check

1. matrix   **2.** square   **3.** row matrix, column matrix

4. augmented matrix   **5.** coefficient matrix   **6.** row-equivalent

7. reduced row-echelon form   **8.** Gauss-Jordan elimination

---

1. Since the matrix has one row and two columns, its order is 1×2.

3. Since the matrix has three rows and one column, its order is 3×1.

5. Since the matrix has two rows and two columns, its order is 2×2.

7.
$$
\begin{aligned}
6x - 7y &= 11 \\
-2x + 5y &= -1
\end{aligned}
$$
$$
\begin{bmatrix}
6 & -7 & \vdots & 11 \\
-2 & 5 & \vdots & -1
\end{bmatrix}
$$

9.
$$
\begin{cases}
x + 10y - 2z = 2 \\
5x - 3y + 4z = 0 \\
2x + y = 6
\end{cases}
$$
$$
\begin{bmatrix}
1 & 10 & -2 & \vdots & 2 \\
5 & -3 & 4 & \vdots & 0 \\
2 & 1 & 0 & \vdots & 6
\end{bmatrix}
$$

11.
$$
\begin{bmatrix}
3 & 4 & \vdots & 9 \\
1 & -1 & \vdots & -3
\end{bmatrix}
$$
$$
\begin{aligned}
3x + 4y &= 9 \\
x - y &= -3
\end{aligned}
$$

13.
$$
\begin{bmatrix}
9 & 12 & 3 & \vdots & 0 \\
-2 & 18 & 5 & \vdots & 10 \\
1 & 7 & -8 & \vdots & -4
\end{bmatrix}
$$
$$
\begin{cases}
9x + 12y + 3z = 0 \\
-2x + 18y + 5z = 10 \\
x + 7y - 8z = -4
\end{cases}
$$

15.
$$
\begin{bmatrix}
1 & 4 & 3 \\
2 & 10 & 5
\end{bmatrix}
$$
$$
-2R_1 + R_2 \to
\begin{bmatrix}
1 & 4 & 3 \\
0 & \boxed{2} & -1
\end{bmatrix}
$$

17.
$$
\begin{bmatrix}
1 & 1 & 4 & -1 \\
3 & 8 & 10 & 3 \\
-2 & 1 & 12 & 6
\end{bmatrix}
$$
$$
\begin{aligned}
-3R_1 + R_2 \to \\
2R_1 + R_3 \to
\end{aligned}
\begin{bmatrix}
1 & 1 & 4 & -1 \\
0 & 5 & \boxed{-2} & \boxed{6} \\
0 & 3 & \boxed{20} & \boxed{4}
\end{bmatrix}
$$
$$
\tfrac{1}{5}R_2 \to
\begin{bmatrix}
1 & 1 & 4 & -1 \\
0 & 1 & -\tfrac{2}{5} & \tfrac{6}{5} \\
0 & 3 & \boxed{20} & \boxed{4}
\end{bmatrix}
$$

19. Add -3 times Row 2 to Row 1.

21. Interchange Rows 1 and 2.

23. $\begin{bmatrix} 1 & 0 & 0 & 0 \\ 0 & 1 & 1 & 5 \\ 0 & 0 & 0 & 0 \end{bmatrix}$

This matrix is in reduced row-echelon form.

25. $\begin{bmatrix} 3 & 0 & 3 & 7 \\ 0 & -2 & 0 & 4 \\ 0 & 0 & 1 & 5 \end{bmatrix}$

The first nonzero entries in rows one and two are not one. The matrix is not in row-echelon form.

27. $\begin{bmatrix} 1 & 0 & 0 & 1 \\ 0 & 1 & 0 & -1 \\ 0 & 0 & 0 & 2 \end{bmatrix}$

The first nonzero entry in row three is two, not one. The matrix is not in row-echelon form.

29. $\begin{bmatrix} 1 & 2 & 3 \\ 2 & -1 & -4 \\ 3 & 1 & -1 \end{bmatrix}$

(a) $\begin{bmatrix} 1 & 2 & 3 \\ 0 & -5 & -10 \\ 3 & 1 & -1 \end{bmatrix}$
(b) $\begin{bmatrix} 1 & 2 & 3 \\ 0 & -5 & -10 \\ 0 & -5 & -10 \end{bmatrix}$
(c) $\begin{bmatrix} 1 & 2 & 3 \\ 0 & -5 & -10 \\ 0 & 0 & 0 \end{bmatrix}$

(d) $\begin{bmatrix} 1 & 2 & 3 \\ 0 & 1 & 2 \\ 0 & 0 & 0 \end{bmatrix}$
(e) $\begin{bmatrix} 1 & 0 & -1 \\ 0 & 1 & 2 \\ 0 & 0 & 0 \end{bmatrix}$ This matrix is in reduced row-echelon form.

31. (See Exercise 29.) (Answer is a series of screens.)

(a)
```
*row+(-2,[A],1,2
)→[B]
    [[1 2  3   ]
     [0 -5 -10]
     [3 1  -1 ]]
```

(b)
```
*row+(-3,[B],1,3
)→[C]
    [[1 2  3   ]
     [0 -5 -10]
     [0 -5 -10]]
```

(c)
```
*row+(-1,[C],2,3
)→[D]
    [[1 2  3  ]
     [0 -5 -10]
     [0 0  0  ]]
```

(d)
```
*row(-1/5,[D],2)
→[E]
    [[1 2 3]
     [0 1 2]
     [0 0 0]]
```

(e)
```
*row+(-2,[E],2,1
)
    [[1 0 -1]
     [0 1 2 ]
     [0 0 0 ]]
```

33.

$\begin{bmatrix} 1 & 2 & 3 & 0 \\ -1 & 4 & 0 & -5 \\ 2 & 6 & 3 & 10 \end{bmatrix}$
$\begin{matrix} R_1 + R_2 \rightarrow \\ -2R_1 + R_3 \rightarrow \end{matrix}$
$\begin{bmatrix} 1 & 2 & 3 & 0 \\ 0 & 6 & 3 & -5 \\ 0 & 2 & -3 & 10 \end{bmatrix}$

$\frac{1}{6}R_2 \rightarrow$
$\begin{bmatrix} 1 & 2 & 3 & 0 \\ 0 & 1 & \frac{1}{2} & -\frac{5}{6} \\ 0 & 2 & -3 & 10 \end{bmatrix}$
$-2R_2 + R_3 \rightarrow$
$\begin{bmatrix} 1 & 2 & 3 & 0 \\ 0 & 1 & \frac{1}{2} & -\frac{5}{6} \\ 0 & 0 & -4 & \frac{35}{3} \end{bmatrix}$

$-\frac{1}{4}R_3 \rightarrow$
$\begin{bmatrix} 1 & 2 & 3 & 0 \\ 0 & 1 & \frac{1}{2} & -\frac{5}{6} \\ 0 & 0 & 1 & -\frac{35}{12} \end{bmatrix}$

(Answers may vary.)

35.
$$\begin{bmatrix} 1 & -1 & -1 & 1 \\ 5 & -4 & 1 & 8 \\ -6 & 8 & 18 & 0 \end{bmatrix}$$

$$\begin{matrix} -5R_1 + R_2 \rightarrow \\ 6R_1 + R_3 \rightarrow \end{matrix} \begin{bmatrix} 1 & -1 & -1 & 1 \\ 0 & 1 & 6 & 3 \\ 0 & 2 & 12 & 6 \end{bmatrix}$$

$$-2R_2 + R_3 \rightarrow \begin{bmatrix} 1 & -1 & -1 & 1 \\ 0 & 1 & 6 & 3 \\ 0 & 0 & 0 & 0 \end{bmatrix}$$

37.
$$\begin{bmatrix} 3 & 3 & 3 \\ -1 & 0 & -4 \\ 2 & 4 & -2 \end{bmatrix}$$

$$\tfrac{1}{3}R_1 \rightarrow \begin{bmatrix} 1 & 1 & 1 \\ -1 & 0 & -4 \\ 2 & 4 & -2 \end{bmatrix}$$

$$\begin{matrix} R_1 + R_2 \rightarrow \\ -2R_1 + R_3 \rightarrow \end{matrix} \begin{bmatrix} 1 & 1 & 1 \\ 0 & 1 & -3 \\ 0 & 2 & -4 \end{bmatrix}$$

$$\begin{matrix} -R_2 + R_1 \rightarrow \\ \\ -2R_2 + R_3 \rightarrow \end{matrix} \begin{bmatrix} 1 & 0 & 4 \\ 0 & 1 & -3 \\ 0 & 0 & 2 \end{bmatrix}$$

$$\tfrac{1}{2}R_3 \rightarrow \begin{bmatrix} 1 & 0 & 4 \\ 0 & 1 & -3 \\ 0 & 0 & 1 \end{bmatrix}$$

$$\begin{matrix} -4R_3 + R_1 \rightarrow \\ 3R_3 + R_2 \rightarrow \end{matrix} \begin{bmatrix} 1 & 0 & 0 \\ 0 & 1 & 0 \\ 0 & 0 & 1 \end{bmatrix}$$

39.
$$\begin{bmatrix} -4 & 1 & 0 & 6 \\ 1 & -2 & 3 & -4 \end{bmatrix}$$

$$\begin{matrix} R_1 \rightarrow \\ R_2 \rightarrow \end{matrix} \begin{bmatrix} 1 & -2 & 3 & -4 \\ -4 & 1 & 0 & 6 \end{bmatrix}$$

$$4R_1 + R_2 \rightarrow \begin{bmatrix} 1 & -2 & 3 & -4 \\ 0 & -7 & 12 & -10 \end{bmatrix}$$

$$-\tfrac{1}{7}R_2 \rightarrow \begin{bmatrix} 1 & -2 & 3 & -4 \\ 0 & 1 & -\tfrac{12}{7} & \tfrac{10}{7} \end{bmatrix}$$

$$2R_2 + R_1 \rightarrow \begin{bmatrix} 1 & 0 & -\tfrac{3}{7} & -\tfrac{8}{7} \\ 0 & 1 & -\tfrac{12}{7} & \tfrac{10}{7} \end{bmatrix}$$

41. $x - 2y = 4$

$$y = -3$$

$$x = 2y + 4 = 2(-3) + 4 = -2$$

Answer: $(x, y) = (-2, -3)$

43. $\begin{cases} x - y + 2z = 4 \\ y - z = 2 \\ z = -2 \end{cases}$

$$y - (-2) = 2$$

$$y = 0$$

$$x - 0 + 2(-2) = 4$$

$$x = 8$$

Answer: $(8, 0, -2)$

45. $\begin{bmatrix} 1 & 0 & \vdots & 7 \\ 0 & 1 & \vdots & -5 \end{bmatrix}$

$$x = 7$$

$$y = -5$$

Answer: $(7, -5)$

47. $\begin{bmatrix} 1 & 0 & 0 & \vdots & -4 \\ 0 & 1 & 0 & \vdots & -8 \\ 0 & 0 & 1 & \vdots & 2 \end{bmatrix}$

$$x = -4$$

$$y = -8$$

$$z = 2$$

Answer: $(-4, -8, 2)$

49. $\begin{cases} x + 2y = 7 \\ 2x + y = 8 \end{cases}$

$$\begin{bmatrix} 1 & 2 & \vdots & 7 \\ 2 & 1 & \vdots & 8 \end{bmatrix}$$

$$-2R_1 + R_2 \rightarrow \begin{bmatrix} 1 & 2 & \vdots & 7 \\ 0 & -3 & \vdots & -6 \end{bmatrix}$$

$$-\tfrac{1}{3}R_2 \rightarrow \begin{bmatrix} 1 & 2 & \vdots & 7 \\ 0 & 1 & \vdots & 2 \end{bmatrix}$$

$$y = 2$$

$$x + 2(2) = 7 \implies x = 3$$

Answer: $(3, 2)$

51. $\begin{cases} -x + y = -22 \\ 3x + 4y = 4 \\ 4x - 8y = 32 \end{cases}$

$$\begin{bmatrix} -1 & 1 & \vdots & -22 \\ 3 & 4 & \vdots & 4 \\ 4 & -8 & \vdots & 32 \end{bmatrix}$$

$$\begin{matrix} 3R_1 + R_2 \\ 4R_1 + R_3 \end{matrix} \begin{bmatrix} -1 & 1 & \vdots & -22 \\ 0 & 7 & \vdots & -62 \\ 0 & -4 & \vdots & -56 \end{bmatrix}$$

$$\begin{matrix} R_2 \\ R_3 \end{matrix} \begin{bmatrix} -1 & 1 & -22 \\ 0 & -4 & -56 \\ 0 & 7 & -62 \end{bmatrix}$$

$$\begin{matrix} -\tfrac{1}{4}R_2 \\ -7R_2 + R_3 \end{matrix} \begin{bmatrix} -1 & 1 & -22 \\ 0 & 1 & 14 \\ 0 & 0 & -160 \end{bmatrix}$$

No solution, inconsistent

53. $\begin{bmatrix} 3 & 2 & -1 & 1 & \vdots & 0 \\ 1 & -1 & 4 & 2 & \vdots & 25 \\ -2 & 1 & 2 & -1 & \vdots & 2 \\ 1 & 1 & 1 & 1 & \vdots & 6 \end{bmatrix}$

$$\begin{bmatrix} 1 & -1 & 4 & 2 & \vdots & 25 \\ 0 & 5 & -13 & -5 & \vdots & -75 \\ 0 & -1 & 10 & 3 & \vdots & 52 \\ 0 & 2 & -3 & -1 & \vdots & -19 \end{bmatrix}$$

$$\begin{bmatrix} 1 & -1 & 4 & 2 & \vdots & 25 \\ 0 & 1 & -10 & -3 & \vdots & -52 \\ 0 & 0 & 37 & 10 & \vdots & 185 \\ 0 & 0 & 17 & 5 & \vdots & 85 \end{bmatrix}$$

$$\begin{bmatrix} 1 & -1 & 4 & 2 & \vdots & 25 \\ 0 & 1 & -10 & -3 & \vdots & -52 \\ 0 & 0 & 37 & 10 & \vdots & 185 \\ 0 & 0 & 0 & \tfrac{15}{37} & \vdots & 0 \end{bmatrix}$$

$w = 0, z = \frac{185}{37} = 5, y = -52 + 10(5) = -2$

$x = 25 + (-2) - 4(5) = 3$

Answer: $(3, -2, 5, 0)$

55. $\begin{cases} x - 3z = -2 \\ 3x + y - 2z = 5 \\ 2x + 2y + z = 4 \end{cases}$

$$\begin{bmatrix} 1 & 0 & -3 & \vdots & -2 \\ 3 & 1 & -2 & \vdots & 5 \\ 2 & 2 & 1 & \vdots & 4 \end{bmatrix}$$

$$\begin{matrix} -3R_1 + R_2 \rightarrow \\ -2R_1 + R_3 \rightarrow \end{matrix} \begin{bmatrix} 1 & 0 & -3 & \vdots & -2 \\ 0 & 1 & 7 & \vdots & 11 \\ 0 & 2 & 7 & \vdots & 8 \end{bmatrix}$$

$$-2R_2 + R_3 \rightarrow \begin{bmatrix} 1 & 0 & -3 & \vdots & -2 \\ 0 & 1 & 7 & \vdots & 11 \\ 0 & 0 & -7 & \vdots & -14 \end{bmatrix}$$

$$-\tfrac{1}{7}R_3 \rightarrow \begin{bmatrix} 1 & 0 & -3 & \vdots & -2 \\ 0 & 1 & 7 & \vdots & 11 \\ 0 & 0 & 1 & \vdots & 2 \end{bmatrix}$$

$$\begin{matrix} 3R_3 + R_1 \rightarrow \\ -7R_3 + R_2 \rightarrow \end{matrix} \begin{bmatrix} 1 & 0 & 0 & \vdots & 4 \\ 0 & 1 & 0 & \vdots & -3 \\ 0 & 0 & 1 & \vdots & 2 \end{bmatrix}$$

Answer: $(4, -3, 2)$

57. $\begin{cases} x + y - 5z = 3 \\ x \quad\;\; - 2z = 1 \\ 2x - y - \; z = 0 \end{cases}$

$$\begin{bmatrix} 1 & 1 & -5 & \vdots & 3 \\ 1 & 0 & -2 & \vdots & 1 \\ 2 & -1 & -1 & \vdots & 0 \end{bmatrix}$$

$\begin{matrix} \\ -R_1 + R_2 \to \\ -2R_1 + R_3 \to \end{matrix} \begin{bmatrix} 1 & 1 & -5 & \vdots & 3 \\ 0 & -1 & 3 & \vdots & -2 \\ 0 & -3 & 9 & \vdots & -6 \end{bmatrix}$

$\begin{matrix} \\ \\ -3R_2 + R_3 \to \end{matrix} \begin{bmatrix} 1 & 1 & -5 & \vdots & 3 \\ 0 & -1 & 3 & \vdots & -2 \\ 0 & 0 & 0 & \vdots & 0 \end{bmatrix}$

$\begin{matrix} R_2 + R_1 \to \\ -R_2 \to \\ \\ \end{matrix} \begin{bmatrix} 1 & 0 & -2 & \vdots & 1 \\ 0 & 1 & -3 & \vdots & 2 \\ 0 & 0 & 0 & \vdots & 0 \end{bmatrix}$

Let $z = a$, any real number.

$y - 3a = 2 \implies y = 3a + 2$

$x - 2a = 1 \implies x = 2a + 1$

Answer: $(2a + 1, 3a + 2, a)$

59. $\begin{cases} -x + \; y - z = -14 \\ 2x - \; y + z = \quad 21 \\ 3x + 2y + z = \quad 19 \end{cases}$

$$\begin{bmatrix} -1 & 1 & -1 & \vdots & -14 \\ 2 & -1 & 1 & \vdots & 21 \\ 3 & 2 & 1 & \vdots & 19 \end{bmatrix}$$

$\begin{matrix} \\ 2R_1 + R_2 \\ 3R_1 + R_3 \end{matrix} \begin{bmatrix} -1 & 1 & -1 & \vdots & -14 \\ 0 & 1 & -1 & \vdots & -7 \\ 0 & 5 & -2 & \vdots & -23 \end{bmatrix}$

$\begin{matrix} -R_1 \\ \\ -5R_2 + R_3 \end{matrix} \begin{bmatrix} 1 & -1 & 1 & \vdots & 14 \\ 0 & 1 & -1 & \vdots & -7 \\ 0 & 0 & 3 & \vdots & 12 \end{bmatrix}$

$\begin{matrix} \\ \\ \frac{1}{3}R_3 \end{matrix} \begin{bmatrix} 1 & -1 & 1 & \vdots & 14 \\ 0 & 1 & -1 & \vdots & -7 \\ 0 & 0 & 1 & \vdots & 4 \end{bmatrix}$

$\begin{matrix} -R_3 + R_1 \\ R_3 + R_2 \\ \\ \end{matrix} \begin{bmatrix} 1 & -1 & 0 & \vdots & 10 \\ 0 & 1 & 0 & \vdots & -3 \\ 0 & 0 & 1 & \vdots & 4 \end{bmatrix}$

$\begin{matrix} R_2 + R_1 \\ \\ \\ \end{matrix} \begin{bmatrix} 1 & 0 & 0 & \vdots & 7 \\ 0 & 1 & 0 & \vdots & -3 \\ 0 & 0 & 1 & \vdots & 4 \end{bmatrix}$

Answer: $(7, -3, 4)$

61. $\begin{cases} 3x + 3y + 12z = \;\; 6 \\ x + \; y + \; 4z = \;\; 2 \\ 2x + 5y + 20z = 10 \\ -x + 2y + \; 8z = \;\; 4 \end{cases}$

$$\begin{bmatrix} 3 & 3 & 12 & \vdots & 6 \\ 1 & 1 & 4 & \vdots & 2 \\ 2 & 5 & 20 & \vdots & 10 \\ -1 & 2 & 8 & \vdots & 4 \end{bmatrix} \implies \begin{bmatrix} 1 & 0 & 0 & \vdots & 0 \\ 0 & 1 & 4 & \vdots & 2 \\ 0 & 0 & 0 & \vdots & 0 \\ 0 & 0 & 0 & \vdots & 0 \end{bmatrix}$$

Let $z = a$, any real number.

$y = -4a + 2$

$x = 0$

Answer: $(0, -4a + 2, a)$

63. $\begin{cases} 2x + 10y + 2z = \;\; 6 \\ x + \; 5y + 2z = \;\; 6 \\ x + \; 5y + \; z = \;\; 3 \\ -3x - 15y - 3z = -9 \end{cases}$

$$\begin{bmatrix} 2 & 10 & 2 & \vdots & 6 \\ 1 & 5 & 2 & \vdots & 6 \\ 1 & 5 & 1 & \vdots & 3 \\ -3 & -15 & -3 & \vdots & -9 \end{bmatrix} \implies \begin{bmatrix} 1 & 5 & 0 & \vdots & 0 \\ 0 & 0 & 1 & \vdots & 3 \\ 0 & 0 & 0 & \vdots & 0 \\ 0 & 0 & 0 & \vdots & 0 \end{bmatrix}$$

$z = 3, y = a, x = -5a$

Answer: $(-5a, a, 3)$

65. Yes, the systems yield the same solutions.

(a) $z = -3; y = 5(-3) + 16 = 1;$
    $x = 2(1) - (-3) - 6 = -1$

   Answer: $(-1, 1, -3)$

(b) $z = -3, y = -3(-3) - 8 = 1,$
    $x = -1 + 2(-3) + 6 = -1$

   Answer: $(-1, 1, -3)$

67. No, solutions are different.

(a) $z = 8, y = 7(8) - 54 = 2,$
    $x = 4(2) - 5(8) + 27 = -5$

   Answer: $(-5, 2, 8)$

(b) $z = 8, y = -5(8) + 42 = 2,$
    $x = 6(2) - 8 + 15 = 19$

   Answer: $(19, 2, 8)$

69. $f(x) = ax^2 + bx + c$

$$\begin{cases} f(1) = a + b + c = 8 \\ f(2) = 4a + 2b + c = 13 \\ f(3) = 9a + 3b + c = 20 \end{cases}$$

$$\begin{bmatrix} 1 & 1 & 1 & \vdots & 8 \\ 4 & 2 & 1 & \vdots & 13 \\ 9 & 3 & 1 & \vdots & 20 \end{bmatrix}$$

$$\begin{matrix} \\ -4R_1 + R_2 \longrightarrow \\ -9R_1 + R_3 \longrightarrow \end{matrix} \begin{bmatrix} 1 & 1 & 1 & \vdots & 8 \\ 0 & -2 & -3 & \vdots & -19 \\ 0 & -6 & -8 & \vdots & -52 \end{bmatrix}$$

$$\begin{matrix} \\ -\frac{1}{2}R_2 \longrightarrow \\ -3R_2 + R_3 \longrightarrow \end{matrix} \begin{bmatrix} 1 & 1 & 1 & \vdots & 8 \\ 0 & 1 & \frac{3}{2} & \vdots & \frac{19}{2} \\ 0 & 0 & 1 & \vdots & 5 \end{bmatrix}$$

$$c = 5$$

$$b + \tfrac{3}{2}(5) = \tfrac{19}{2} \implies b = 2$$

$$a + 2 + 5 = 8 \implies a = 1$$

Answer: $y = x^2 + 2x + 5$

71. $f(x) = ax^2 + bx + c$

$$\begin{cases} f(1) = a + b + c = 2 \\ f(-2) = 4a - 2b + c = 11 \\ f(3) = 9a + 3b + c = 16 \end{cases}$$

$$\begin{bmatrix} 1 & 1 & 1 & \vdots & 2 \\ 4 & -2 & 1 & \vdots & 11 \\ 9 & 3 & 1 & \vdots & 16 \end{bmatrix}$$

$$\begin{matrix} \\ -4R_1 + R_2 \longrightarrow \\ -9R_1 + R_3 \longrightarrow \end{matrix} \begin{bmatrix} 1 & 1 & 1 & \vdots & 2 \\ 0 & -6 & -3 & \vdots & 3 \\ 0 & -6 & -8 & \vdots & -2 \end{bmatrix}$$

$$\begin{matrix} \\ \\ (-1)R_2 + R_3 \longrightarrow \end{matrix} \begin{bmatrix} 1 & 1 & 1 & \vdots & 2 \\ 0 & -6 & -3 & \vdots & 3 \\ 0 & 0 & -5 & \vdots & -5 \end{bmatrix}$$

$$-5c = -5 \implies c = 1$$

$$-6b - 3c = 3 \implies -6b = 3 + 3 = 6 \implies b = -1$$

$$a + b + c = 2 \implies a = 2 + 1 - 1 = 2$$

Answer: $y = 2x^2 - x + 1$

73. $f(x) = ax^2 + bx + c$

$$\begin{aligned} f(-2) &= 4a - 2b + c = -15 \\ f(-1) &= a - b + c = 7 \\ f(1) &= a + b + c = -3 \end{aligned}$$

Solving the system, $a = -9, b = -5, c = 11$.

$$f(x) = -9x^2 - 5x + 11$$

75. $f(x) = ax^3 + bx^2 + cx + d$

$$\begin{aligned} f(-2) &= -8a + 4b - 2c + d = -7 \\ f(-1) &= -a + b - c + d = 2 \\ f(1) &= a + b + c + d = -4 \\ f(2) &= 8a + 4b + 2c + d = -7 \end{aligned}$$

Solving the system,
$a = 1, b = -2, c = -4, d = 1.$

$$f(x) = x^3 - 2x^2 - 4x + 1$$

77. $x =$ amount at 7%, $y =$ amount at 8%, $z =$ amount at 10%

$$\begin{cases} x + y + z = 1{,}500{,}000 \\ 0.07x + 0.08y + 0.1z = 130{,}500 \\ 4x - z = 0 \end{cases}$$

$$\begin{bmatrix} 1 & 1 & 1 & \vdots & 1{,}500{,}000 \\ 0.07 & 0.08 & 0.1 & \vdots & 130{,}500 \\ 4 & 0 & -1 & \vdots & 0 \end{bmatrix}$$

$$\begin{matrix} \\ -0.07R_1 + R_2 \\ -4R_1 + R_3 \end{matrix} \begin{bmatrix} 1 & 1 & 1 & \vdots & 1{,}500{,}000 \\ 0 & 0.01 & 0.03 & \vdots & 25{,}500 \\ 0 & -4 & -5 & \vdots & -6{,}000{,}000 \end{bmatrix}$$

$$\begin{matrix} \\ 100R_2 \\ 4R_2 + R_3 \end{matrix} \begin{bmatrix} 1 & 1 & 1 & \vdots & 1{,}500{,}000 \\ 0 & 1 & 3 & \vdots & 2{,}550{,}000 \\ 0 & 0 & 7 & \vdots & 4{,}200{,}000 \end{bmatrix}$$

$$7z = 4{,}200{,}000 \implies z = 600{,}00$$

$$y + 3(600{,}000) = 2{,}550{,}000 \implies y = 750{,}000$$

$$x + 750{,}000 + 600{,}000 = 1{,}500{,}000 \implies x = 150{,}000$$

Answers: $150{,}000 at 7%, $750{,}000 at 8%, $600{,}000 at 10%

79. $\begin{cases} I_1 - I_2 + I_3 = 0 \\ 2I_1 + 2I_2 \quad\quad = 7 \\ \quad\quad 2I_2 + 4I_3 = 8 \end{cases}$

$$\begin{bmatrix} 1 & -1 & 1 & \vdots & 0 \\ 2 & 2 & 0 & \vdots & 7 \\ 0 & 2 & 4 & \vdots & 8 \end{bmatrix}$$

$-2R_1 + R_2 \rightarrow \begin{bmatrix} 1 & -1 & 1 & \vdots & 0 \\ 0 & 4 & -2 & \vdots & 7 \\ 0 & 2 & 4 & \vdots & 8 \end{bmatrix}$

$\begin{matrix} R_3 \rightarrow \\ R_2 \rightarrow \end{matrix} \begin{bmatrix} 1 & -1 & 1 & \vdots & 0 \\ 0 & 2 & 4 & \vdots & 8 \\ 0 & 4 & -2 & \vdots & 7 \end{bmatrix}$

$\frac{1}{2}R_2 \rightarrow \begin{bmatrix} 1 & -1 & 1 & \vdots & 0 \\ 0 & 1 & 2 & \vdots & 4 \\ 0 & 4 & -2 & \vdots & 7 \end{bmatrix}$

$-4R_2 + R_3 \rightarrow \begin{bmatrix} 1 & -1 & 1 & \vdots & 0 \\ 0 & 1 & 2 & \vdots & 4 \\ 0 & 0 & -10 & \vdots & -9 \end{bmatrix}$

$-\frac{1}{10}R_3 \rightarrow \begin{bmatrix} 1 & -1 & 1 & \vdots & 0 \\ 0 & 1 & 2 & \vdots & 4 \\ 0 & 0 & 1 & \vdots & \frac{9}{10} \end{bmatrix}$

$I_3 = \frac{9}{10}$ amperes; $I_2 + 2\left(\frac{9}{10}\right) = 4 \implies I_2 = \frac{11}{5}$ amperes; $I_1 - \frac{11}{5} + \frac{9}{10} = 0 \implies I_1 = \frac{13}{10}$ amperes

81. (a) $(2, 55.37)$
$(3, 59.52)$
$(4, 63.59)$
$\begin{cases} 4a + 2b + c = 55.37 \\ 9a + 3b + c = 59.52 \\ 16a + 4b + c = 63.59 \end{cases}$

(b)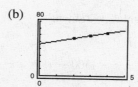

Solving the system using matrices,

$$\begin{bmatrix} 4 & 2 & 1 & \vdots & 55.37 \\ 9 & 3 & 1 & \vdots & 59.52 \\ 16 & 4 & 1 & \vdots & 63.59 \end{bmatrix} \implies \begin{bmatrix} 1 & 0 & 0 & \vdots & -0.04 \\ 0 & 1 & 0 & \vdots & 4.35 \\ 0 & 0 & 1 & \vdots & 46.83 \end{bmatrix}$$

you obtain $a = -0.04$, $b = 4.35$, $c = 46.83$.

$y = -0.04t^2 + 4.35t + 46.83$

(c) For 2005, $t = 5$ and $y \approx 67.58$ dollars.

For 2010, $t = 10$ and $y \approx 86.33$ dollars.

For 2015, $t = 15$ and $y \approx 103.08$ dollars.

83. Let x = number of pounds of glossy.

Let y = number of pounds of semi-glossy.

Let z = number of pounds of matte.

$\begin{cases} x + y + z = 100 \\ 5.5x + 4.25y + 3.75z = 480 \\ y + z = 50 \end{cases}$

Solving the system, $x = 50$, $y = 35$ and $z = 15$.

50 pounds of glossy, 35 pounds of semi-glossy and 15 pounds of matte

85. (a)

$$\begin{cases} x_1 + x_3 = 600 \\ x_1 = x_2 + x_4 \Longrightarrow x_1 - x_2 - x_4 = 0 \\ x_2 + x_5 = 500 \\ x_3 + x_6 = 600 \\ x_4 + x_7 = x_6 \Longrightarrow x_4 - x_6 + x_7 = 0 \\ x_5 + x_7 = 500 \end{cases}$$

$$\begin{bmatrix} 1 & 0 & 1 & 0 & 0 & 0 & 0 & \vdots & 600 \\ 1 & -1 & 0 & -1 & 0 & 0 & 0 & \vdots & 0 \\ 0 & 1 & 0 & 0 & 1 & 0 & 0 & \vdots & 500 \\ 0 & 0 & 1 & 0 & 0 & 1 & 0 & \vdots & 600 \\ 0 & 0 & 0 & 1 & 0 & -1 & 1 & \vdots & 0 \\ 0 & 0 & 0 & 0 & 1 & 0 & 1 & \vdots & 500 \end{bmatrix}$$

$$\begin{matrix} \\ -R_1 + R_2 \to \\ R_2 + R_3 \to \\ R_3 + R_4 \to \\ R_4 + R_5 \to \\ -R_5 + R_6 \to \end{matrix} \begin{bmatrix} 1 & 0 & 1 & 0 & 0 & 0 & 0 & \vdots & 600 \\ 0 & -1 & -1 & -1 & 0 & 0 & 0 & \vdots & -600 \\ 0 & 0 & -1 & -1 & 1 & 0 & 0 & \vdots & -100 \\ 0 & 0 & 0 & -1 & 1 & 1 & 0 & \vdots & 500 \\ 0 & 0 & 0 & 0 & 1 & 0 & 1 & \vdots & 500 \\ 0 & 0 & 0 & 0 & 0 & 0 & 0 & \vdots & 0 \end{bmatrix}$$

$$\begin{matrix} \\ -R_3 + R_2 \to \\ -R_4 + R_3 \to \\ -R_4 \to \\ \\ \end{matrix} \begin{bmatrix} 1 & 0 & 1 & 0 & 0 & 0 & 0 & \vdots & 600 \\ 0 & -1 & 0 & 0 & -1 & 0 & 0 & \vdots & -500 \\ 0 & 0 & -1 & 0 & 0 & -1 & 0 & \vdots & -600 \\ 0 & 0 & 0 & 1 & -1 & -1 & 0 & \vdots & -500 \\ 0 & 0 & 0 & 0 & 1 & 0 & 1 & \vdots & 500 \\ 0 & 0 & 0 & 0 & 0 & 0 & 0 & \vdots & 0 \end{bmatrix}$$

Let $x_7 = t$ and $x_6 = s$, then:

$x_5 = 500 - t$

$x_4 = -500 + s + (500 - t) = s - t$

$x_3 = 600 - s$

$x_2 = 500 - (500 - t) = t$

$x_1 = 600 - (600 - s) = s$

(b) If $x_6 = x_7 = 0$, then $s = t = 0$, and

$x_1 = 0$

$x_2 = 0$

$x_3 = 600$

$x_4 = 0$

$x_5 = 500$

$x_6 = x_7 = 0.$

(c) If $x_5 = 1000$ and $x_6 = 0$, then $s = 0$ and $t = -500$.

Thus:

$x_1 = 0$

$x_2 = -500$

$x_3 = 600$

$x_4 = 500$

$x_5 = 1000$

$x_6 = 0$

$x_7 = -500$

87. False. It is a 2×4 matrix.

89. $\begin{cases} x + 3z = -2 & \text{Equation 1} \\ y + 4z = 1 & \text{Equation 2} \end{cases}$

(Equation 1) + (Equation 2) → new Equation 1

(Equation 1) + 2(Equation 2) → new Equation 2

2(Equation 1) + (Equation 2) → new Equation 3

$\begin{cases} x + y + 7z = -1 \\ x + 2y + 11z = 0 \\ 2x + y + 10z = -3 \end{cases}$

91. The row operation $-2R_1 + R_2$ was not performed on the last column. Nor was $-R_2 + R_1$.

93. $f(x) = \dfrac{7}{-x - 1}$

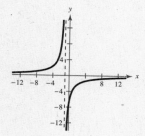

Asymptotes: $x = -1, y = 0$

95. $f(x) = \dfrac{x^2 - 2x - 3}{x - 4} = x + 2 + \dfrac{5}{x - 4}$

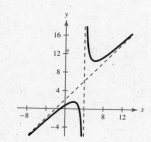

Asymptotes: $x = 4, y = x + 2$

Section 5.5   Operations with Matrices

- $A = B$ if and only if they have the same order and $a_{ij} = b_{ij}$.
- You should be able to perform the operations of matrix addition, scalar multiplication, and matrix multiplication.
- Some properties of matrix addition and scalar multiplication are:
  - (a) $A + B = B + A$
  - (b) $A + (B + C) = (A + B) + C$
  - (c) $(cd)A = c(dA)$
  - (d) $1A = A$
  - (e) $c(A + B) = cA + cB$
  - (f) $(c + d)A = cA + dA$
- Some properties of matrix multiplication are:
  - (a) $A(BC) = (AB)C$
  - (b) $A(B + C) = AB + AC$
  - (c) $(A + B)C = AC + BC$
  - (d) $c(AB) = (cA)B = A(cB)$
- You should remember that $AB \neq BA$ in general.

Vocabulary Check

1. equal

2. scalars

3. zero, *0*

4. identity

5. (a) iii   (b) i   (c) iv   (d) v   (e) ii

6. (a) ii   (b) iv   (c) i   (d) iii

1. $x = -4$, $y = 22$

3. $2x + 7 = 5 \implies x = -1$

$3y = 12 \implies y = 4$

$3z - 14 = 4 \implies z = 6$

5. (a) $A + B = \begin{bmatrix} 1 & -1 \\ 2 & -1 \end{bmatrix} + \begin{bmatrix} 2 & -1 \\ -1 & 8 \end{bmatrix} = \begin{bmatrix} 1+2 & -1-1 \\ 2-1 & -1+8 \end{bmatrix} = \begin{bmatrix} 3 & -2 \\ 1 & 7 \end{bmatrix}$

(b) $A - B = \begin{bmatrix} 1 & -1 \\ 2 & -1 \end{bmatrix} - \begin{bmatrix} 2 & -1 \\ -1 & 8 \end{bmatrix} = \begin{bmatrix} 1-2 & -1+1 \\ 2+1 & -1-8 \end{bmatrix} = \begin{bmatrix} -1 & 0 \\ 3 & -9 \end{bmatrix}$

(c) $3A = 3\begin{bmatrix} 1 & -1 \\ 2 & -1 \end{bmatrix} = \begin{bmatrix} 3(1) & 3(-1) \\ 3(2) & 3(-1) \end{bmatrix} = \begin{bmatrix} 3 & -3 \\ 6 & -3 \end{bmatrix}$

(d) $3A - 2B = \begin{bmatrix} 3 & -3 \\ 6 & -3 \end{bmatrix} - 2\begin{bmatrix} 2 & -1 \\ -1 & 8 \end{bmatrix} = \begin{bmatrix} 3 & -3 \\ 6 & -3 \end{bmatrix} + \begin{bmatrix} -4 & 2 \\ 2 & -16 \end{bmatrix} = \begin{bmatrix} -1 & -1 \\ 8 & -19 \end{bmatrix}$

7. $A = \begin{bmatrix} 8 & -1 \\ 2 & 3 \\ -4 & 5 \end{bmatrix}$, $B = \begin{bmatrix} 1 & 6 \\ -1 & -5 \\ 1 & 10 \end{bmatrix}$

(a) $A + B = \begin{bmatrix} 9 & 5 \\ 1 & -2 \\ -3 & 15 \end{bmatrix}$

(b) $A - B = \begin{bmatrix} 7 & -7 \\ 3 & 8 \\ -5 & -5 \end{bmatrix}$

(c) $3A = \begin{bmatrix} 24 & -3 \\ 6 & 9 \\ -12 & 15 \end{bmatrix}$

(d) $3A - 2B = \begin{bmatrix} 24 & -3 \\ 6 & 9 \\ -12 & 15 \end{bmatrix} - \begin{bmatrix} 2 & 12 \\ -2 & -10 \\ 2 & 20 \end{bmatrix} = \begin{bmatrix} 22 & -15 \\ 8 & 19 \\ -14 & -5 \end{bmatrix}$

9. $A = \begin{bmatrix} 4 & 5 & -1 & 3 & 4 \\ 1 & 2 & -2 & -1 & 0 \end{bmatrix}$, $B = \begin{bmatrix} 1 & 0 & -1 & 1 & 0 \\ -6 & 8 & 2 & -3 & -7 \end{bmatrix}$

(a) $A + B = \begin{bmatrix} 5 & 5 & -2 & 4 & 4 \\ -5 & 10 & 0 & -4 & -7 \end{bmatrix}$

(b) $A - B = \begin{bmatrix} 3 & 5 & 0 & 2 & 4 \\ 7 & -6 & -4 & 2 & 7 \end{bmatrix}$

(c) $3A = \begin{bmatrix} 12 & 15 & -3 & 9 & 12 \\ 3 & 6 & -6 & -3 & 0 \end{bmatrix}$

(d) $3A - 2B = \begin{bmatrix} 12 & 15 & -3 & 9 & 12 \\ 3 & 6 & -6 & -3 & 0 \end{bmatrix} - \begin{bmatrix} 2 & 0 & -2 & 2 & 0 \\ -12 & 16 & 4 & -6 & -14 \end{bmatrix}$

$= \begin{bmatrix} 10 & 15 & -1 & 7 & 12 \\ 15 & -10 & -10 & 3 & 14 \end{bmatrix}$

11. $A = \begin{bmatrix} 6 & 0 & 3 \\ -1 & -4 & 0 \end{bmatrix}$, $B = \begin{bmatrix} 8 & -1 \\ 4 & -3 \end{bmatrix}$

(a) $A + B$ is not possible.

(b) $A - B$ is not possible.

(c) $3A = \begin{bmatrix} 18 & 0 & 9 \\ -3 & -12 & 0 \end{bmatrix}$

(d) $3A - 2B$ is not possible.

13. $\begin{bmatrix} -5 & 0 \\ 3 & -6 \end{bmatrix} + \begin{bmatrix} 7 & 1 \\ -2 & -1 \end{bmatrix} + \begin{bmatrix} -10 & -8 \\ 14 & 6 \end{bmatrix} = \begin{bmatrix} -5 & 0 \\ 3 & -6 \end{bmatrix} + \begin{bmatrix} -3 & -7 \\ 12 & 5 \end{bmatrix} = \begin{bmatrix} -8 & -7 \\ 15 & -1 \end{bmatrix}$

15. $4\left(\begin{bmatrix} -4 & 0 & 1 \\ 0 & 2 & 3 \end{bmatrix} - \begin{bmatrix} 2 & 1 & -2 \\ 3 & -6 & 0 \end{bmatrix}\right) = 4\begin{bmatrix} -6 & -1 & 3 \\ -3 & 8 & 3 \end{bmatrix} = \begin{bmatrix} -24 & -4 & 12 \\ -12 & 32 & 12 \end{bmatrix}$

17. $\begin{bmatrix} 2 & 5 \\ -1 & -4 \end{bmatrix} + \begin{bmatrix} -3 & 0 \\ 2 & 2 \end{bmatrix} = \begin{bmatrix} -1 & 5 \\ 1 & -2 \end{bmatrix}$

19. $-\frac{1}{2}\begin{bmatrix} 3.211 & 6.829 \\ -1.004 & 4.914 \\ 0.055 & -3.889 \end{bmatrix} - 8\begin{bmatrix} 1.630 & -3.090 \\ 5.256 & 8.335 \\ -9.768 & 4.251 \end{bmatrix} = \begin{bmatrix} -14.645 & 21.305 \\ -41.546 & -69.137 \\ 78.117 & -32.064 \end{bmatrix}$

21. $X = 3\begin{bmatrix} -2 & -1 \\ 1 & 0 \\ 3 & -4 \end{bmatrix} - 2\begin{bmatrix} 0 & 3 \\ 2 & 0 \\ -4 & -1 \end{bmatrix} = \begin{bmatrix} -6 & -3 \\ 3 & 0 \\ 9 & -12 \end{bmatrix} - \begin{bmatrix} 0 & 6 \\ 4 & 0 \\ -8 & -2 \end{bmatrix} = \begin{bmatrix} -6 & -9 \\ -1 & 0 \\ 17 & -10 \end{bmatrix}$

23. $X = -\frac{3}{2}A + \frac{1}{2}B = -\frac{3}{2}\begin{bmatrix} -2 & -1 \\ 1 & 0 \\ 3 & -4 \end{bmatrix} + \frac{1}{2}\begin{bmatrix} 0 & 3 \\ 2 & 0 \\ -4 & -1 \end{bmatrix} = \begin{bmatrix} 3 & 3 \\ -\frac{1}{2} & 0 \\ -\frac{13}{2} & \frac{11}{2} \end{bmatrix}$

25. A is 3×2 and B is $3 \times 3 \implies AB$ is not defined.

27. $AB = \begin{bmatrix} -1 & 6 \\ -4 & 5 \\ 0 & 3 \end{bmatrix}\begin{bmatrix} 2 & 3 \\ 0 & 9 \end{bmatrix} = \begin{bmatrix} -2 & 51 \\ -8 & 33 \\ 0 & 27 \end{bmatrix}$

29. A is 3×3, B is $3 \times 3 \implies AB$ is 3×3.

$AB = \begin{bmatrix} 5 & 0 & 0 \\ 0 & -8 & 0 \\ 0 & 0 & 7 \end{bmatrix}\begin{bmatrix} \frac{1}{5} & 0 & 0 \\ 0 & -\frac{1}{8} & 0 \\ 0 & 0 & \frac{1}{2} \end{bmatrix} = \begin{bmatrix} 1 & 0 & 0 \\ 0 & 1 & 0 \\ 0 & 0 & \frac{7}{2} \end{bmatrix}$

31. $AB = \begin{bmatrix} 5 \\ 6 \end{bmatrix}\begin{bmatrix} -3 & -1 & -5 & -9 \end{bmatrix} = \begin{bmatrix} -15 & -5 & -25 & -45 \\ -18 & -6 & -30 & -54 \end{bmatrix}$

33. (a) $AB = \begin{bmatrix} 1 & 2 \\ 5 & 2 \end{bmatrix}\begin{bmatrix} 2 & -1 \\ -1 & 8 \end{bmatrix} = \begin{bmatrix} 2-2 & -1+16 \\ 10-2 & -5+16 \end{bmatrix} = \begin{bmatrix} 0 & 15 \\ 8 & 11 \end{bmatrix}$

(b) $BA = \begin{bmatrix} 2 & -1 \\ -1 & 8 \end{bmatrix}\begin{bmatrix} 1 & 2 \\ 5 & 2 \end{bmatrix} = \begin{bmatrix} 2-5 & 4-2 \\ -1+40 & -2+16 \end{bmatrix} = \begin{bmatrix} -3 & 2 \\ 39 & 14 \end{bmatrix}$

(c) $A^2 = \begin{bmatrix} 1 & 2 \\ 5 & 2 \end{bmatrix}\begin{bmatrix} 1 & 2 \\ 5 & 2 \end{bmatrix} = \begin{bmatrix} 1+10 & 2+4 \\ 5+10 & 10+4 \end{bmatrix} = \begin{bmatrix} 11 & 6 \\ 15 & 14 \end{bmatrix}$

35. (a) $AB = \begin{bmatrix} 3 & -1 \\ 1 & 3 \end{bmatrix}\begin{bmatrix} 1 & -3 \\ 3 & 1 \end{bmatrix} = \begin{bmatrix} 3-3 & -9-1 \\ 1+9 & -3+3 \end{bmatrix} = \begin{bmatrix} 0 & -10 \\ 10 & 0 \end{bmatrix}$

(b) $BA = \begin{bmatrix} 1 & -3 \\ 3 & 1 \end{bmatrix}\begin{bmatrix} 3 & -1 \\ 1 & 3 \end{bmatrix} = \begin{bmatrix} 3-3 & -1-9 \\ 9+1 & -3+3 \end{bmatrix} = \begin{bmatrix} 0 & -10 \\ 10 & 0 \end{bmatrix}$

(c) $A^2 = \begin{bmatrix} 3 & -1 \\ 1 & 3 \end{bmatrix}\begin{bmatrix} 3 & -1 \\ 1 & 3 \end{bmatrix} = \begin{bmatrix} 9-1 & -3-3 \\ 3+3 & -1+9 \end{bmatrix} = \begin{bmatrix} 8 & -6 \\ 6 & 8 \end{bmatrix}$

37. (a) $AB = \begin{bmatrix} 7 \\ 8 \\ -1 \end{bmatrix} \begin{bmatrix} 1 & 1 & 2 \end{bmatrix} = \begin{bmatrix} 7 & 7 & 14 \\ 8 & 8 & 16 \\ -1 & -1 & -2 \end{bmatrix}$ (b) $BA = \begin{bmatrix} 1 & 1 & 2 \end{bmatrix} \begin{bmatrix} 7 \\ 8 \\ -1 \end{bmatrix} = [7 + 8 - 2] = [13]$

(c) A^2 is not defined.

39. $AB = \begin{bmatrix} 70 & -17 & 73 \\ 32 & 11 & 6 \\ 16 & -38 & 70 \end{bmatrix}$

41. $\begin{bmatrix} -3 & 8 & -6 & 8 \\ -12 & 15 & 9 & 6 \\ 5 & -1 & 1 & 5 \end{bmatrix} \begin{bmatrix} 3 & 1 & 6 \\ 24 & 15 & 14 \\ 16 & 10 & 21 \\ 8 & -4 & 10 \end{bmatrix} = \begin{bmatrix} 151 & 25 & 48 \\ 516 & 279 & 387 \\ 47 & -20 & 87 \end{bmatrix}$

43. A is 2×4 and B is $2 \times 4 \implies AB$ is not defined.

45. $\left(\begin{bmatrix} 3 & 1 \\ 0 & -2 \end{bmatrix} \begin{bmatrix} 1 & 0 \\ -2 & 2 \end{bmatrix} \right) \begin{bmatrix} 1 & 0 \\ 2 & 4 \end{bmatrix} = \begin{bmatrix} 1 & 2 \\ 4 & -4 \end{bmatrix} \begin{bmatrix} 1 & 0 \\ 2 & 4 \end{bmatrix} = \begin{bmatrix} 5 & 8 \\ -4 & -16 \end{bmatrix}$

47. $\begin{bmatrix} 0 & 2 & -2 \\ 4 & 1 & 2 \end{bmatrix} \left(\begin{bmatrix} 4 & 0 \\ 0 & -1 \\ -1 & 2 \end{bmatrix} + \begin{bmatrix} -2 & 3 \\ -3 & 5 \\ 0 & -3 \end{bmatrix} \right) = \begin{bmatrix} 0 & 2 & -2 \\ 4 & 1 & 2 \end{bmatrix} \begin{bmatrix} 2 & 3 \\ -3 & 4 \\ -1 & -1 \end{bmatrix} = \begin{bmatrix} -4 & 10 \\ 3 & 14 \end{bmatrix}$

49. $\begin{bmatrix} 1 & 2 & \vdots & 4 \\ 3 & 2 & \vdots & 0 \end{bmatrix}$

(a) $\begin{bmatrix} 1 & 2 \\ 3 & 2 \end{bmatrix} \begin{bmatrix} 2 \\ 1 \end{bmatrix} = \begin{bmatrix} 4 \\ 8 \end{bmatrix} \implies \begin{bmatrix} 2 \\ 1 \end{bmatrix}$ is not a solution.

(b) $\begin{bmatrix} 1 & 2 \\ 3 & 2 \end{bmatrix} \begin{bmatrix} -2 \\ 3 \end{bmatrix} = \begin{bmatrix} 4 \\ 0 \end{bmatrix} \implies \begin{bmatrix} -2 \\ 3 \end{bmatrix}$ is a solution.

(c) $\begin{bmatrix} 1 & 2 \\ 3 & 2 \end{bmatrix} \begin{bmatrix} -4 \\ 4 \end{bmatrix} = \begin{bmatrix} 4 \\ -4 \end{bmatrix} \implies \begin{bmatrix} -4 \\ 4 \end{bmatrix}$ is not a solution.

(d) $\begin{bmatrix} 1 & 2 \\ 3 & 2 \end{bmatrix} \begin{bmatrix} 2 \\ -3 \end{bmatrix} = \begin{bmatrix} -4 \\ 0 \end{bmatrix} \implies \begin{bmatrix} 2 \\ -3 \end{bmatrix}$ is not a solution.

51. $\begin{bmatrix} -2 & -3 & \vdots & -6 \\ 4 & 2 & \vdots & 20 \end{bmatrix}$

(a) $\begin{bmatrix} -2 & -3 \\ 4 & 2 \end{bmatrix} \begin{bmatrix} 3 \\ 0 \end{bmatrix} = \begin{bmatrix} -6 \\ 12 \end{bmatrix} \implies \begin{bmatrix} 3 \\ 0 \end{bmatrix}$ is not a solution.

(b) $\begin{bmatrix} -2 & -3 \\ 4 & 2 \end{bmatrix} \begin{bmatrix} 6 \\ -2 \end{bmatrix} = \begin{bmatrix} -6 \\ 20 \end{bmatrix} \implies \begin{bmatrix} 6 \\ -2 \end{bmatrix}$ is a solution.

(c) $\begin{bmatrix} -2 & -3 \\ 4 & 2 \end{bmatrix} \begin{bmatrix} -6 \\ 6 \end{bmatrix} = \begin{bmatrix} -6 \\ -12 \end{bmatrix} \implies \begin{bmatrix} -6 \\ 6 \end{bmatrix}$ is not a solution.

(d) $\begin{bmatrix} -2 & -3 \\ 4 & 2 \end{bmatrix} \begin{bmatrix} 4 \\ 2 \end{bmatrix} = \begin{bmatrix} -14 \\ 20 \end{bmatrix} \implies \begin{bmatrix} 4 \\ 2 \end{bmatrix}$ is not a solution.

53. (a) $A = \begin{bmatrix} -1 & 1 \\ -2 & 1 \end{bmatrix}$, $X = \begin{bmatrix} x_1 \\ x_2 \end{bmatrix}$, $B = \begin{bmatrix} 4 \\ 0 \end{bmatrix}$

(b) By Gauss-Jordan elimination on

$$\begin{bmatrix} -1 & 1 & \vdots & 4 \\ -2 & 1 & \vdots & 0 \end{bmatrix}$$

$$\begin{matrix} -R_1 \to \\ 2R_1 + R_2 \to \end{matrix} \begin{bmatrix} 1 & -1 & \vdots & -4 \\ 0 & -1 & \vdots & -8 \end{bmatrix}$$

$$\begin{matrix} -R_2 + R_1 \to \\ -R_2 \to \end{matrix} \begin{bmatrix} 1 & 0 & \vdots & 4 \\ 0 & 1 & \vdots & 8 \end{bmatrix},$$

we have $x_1 = 4$ and $x_2 = 8$. Thus, $X = \begin{bmatrix} 4 \\ 8 \end{bmatrix}$.

55. (a) $A = \begin{bmatrix} -2 & -3 \\ 6 & 1 \end{bmatrix}$, $X = \begin{bmatrix} x_1 \\ x_2 \end{bmatrix}$, $B = \begin{bmatrix} -4 \\ -36 \end{bmatrix}$

(b)
$$\begin{bmatrix} -2 & -3 & \vdots & -4 \\ 6 & 1 & \vdots & -36 \end{bmatrix}$$

$$3R_1 + R_2 \begin{bmatrix} -2 & -3 & \vdots & -4 \\ 0 & -8 & \vdots & -48 \end{bmatrix}$$

$$\left(-\tfrac{1}{8}\right)R_2 \begin{bmatrix} -2 & -3 & \vdots & -4 \\ 0 & 1 & \vdots & 6 \end{bmatrix}$$

$$3R_2 + R_1 \begin{bmatrix} -2 & 0 & \vdots & 14 \\ 0 & 1 & \vdots & 6 \end{bmatrix}$$

$$-\tfrac{1}{2}R_1 \begin{bmatrix} 1 & 0 & \vdots & -7 \\ 0 & 1 & \vdots & 6 \end{bmatrix}$$

$$x_1 = -7, x_2 = 6$$

Answer: $X = \begin{bmatrix} -7 \\ 6 \end{bmatrix}$

57. (a) $A = \begin{bmatrix} 1 & -2 & 3 \\ -1 & 3 & -1 \\ 2 & -5 & 5 \end{bmatrix}$, $X = \begin{bmatrix} x_1 \\ x_2 \\ x_3 \end{bmatrix}$, $B = \begin{bmatrix} 9 \\ -6 \\ 17 \end{bmatrix}$

(b)
$$\begin{bmatrix} 1 & -2 & 3 & \vdots & 9 \\ -1 & 3 & -1 & \vdots & -6 \\ 2 & -5 & 5 & \vdots & 17 \end{bmatrix}$$

$$\begin{matrix} R_1 + R_2 \to \\ -2R_1 + R_3 \to \end{matrix} \begin{bmatrix} 1 & -2 & 3 & \vdots & 9 \\ 0 & 1 & 2 & \vdots & 3 \\ 0 & -1 & -1 & \vdots & -1 \end{bmatrix}$$

$$\begin{matrix} 2R_2 + R_1 \to \\ \\ R_2 + R_3 \to \end{matrix} \begin{bmatrix} 1 & 0 & 7 & \vdots & 15 \\ 0 & 1 & 2 & \vdots & 3 \\ 0 & 0 & 1 & \vdots & 2 \end{bmatrix}$$

$$\begin{matrix} -7R_3 + R_1 \to \\ -2R_3 + R_2 \to \\ \\ \end{matrix} \begin{bmatrix} 1 & 0 & 0 & \vdots & 1 \\ 0 & 1 & 0 & \vdots & -1 \\ 0 & 0 & 1 & \vdots & 2 \end{bmatrix}$$

$$x_1 = 1, x_2 = -1, x_3 = 2$$

Answer: $X = \begin{bmatrix} 1 \\ -1 \\ 2 \end{bmatrix}$

59. (a) $A = \begin{bmatrix} 1 & -5 & 2 \\ -3 & 1 & -1 \\ 0 & -2 & 5 \end{bmatrix}$, $X = \begin{bmatrix} x_1 \\ x_2 \\ x_3 \end{bmatrix}$, $B = \begin{bmatrix} -20 \\ 8 \\ -16 \end{bmatrix}$

(b)
$$\begin{bmatrix} 1 & -5 & 2 & \vdots & -20 \\ -3 & 1 & -1 & \vdots & 8 \\ 0 & -2 & 5 & \vdots & -16 \end{bmatrix}$$

$$3R_1 + R_2 \begin{bmatrix} 1 & -5 & 2 & \vdots & -20 \\ 0 & -14 & 5 & \vdots & -52 \\ 0 & -2 & 5 & \vdots & -16 \end{bmatrix}$$

$$\begin{matrix} R_2 \\ R_3 \end{matrix} \begin{bmatrix} 1 & -5 & 2 & \vdots & -20 \\ 0 & -2 & 5 & \vdots & -16 \\ 0 & -14 & 5 & \vdots & -52 \end{bmatrix}$$

$$-7R_2 + R_3 \begin{bmatrix} 1 & -5 & 2 & \vdots & -20 \\ 0 & -2 & 5 & \vdots & -16 \\ 0 & 0 & -30 & \vdots & 60 \end{bmatrix}$$

$$-\tfrac{1}{30}R_3 \begin{bmatrix} 1 & -5 & 2 & \vdots & -20 \\ 0 & -2 & 5 & \vdots & -16 \\ 0 & 0 & 1 & \vdots & -2 \end{bmatrix}$$

$$\begin{matrix} -2R_3 + R_1 \\ -5R_3 + R_2 \\ \\ \end{matrix} \begin{bmatrix} 1 & -5 & 0 & \vdots & -16 \\ 0 & -2 & 0 & \vdots & -6 \\ 0 & 0 & 1 & \vdots & -2 \end{bmatrix}$$

$$\left(-\tfrac{1}{2}\right)R_2 \begin{bmatrix} 1 & -5 & 0 & \vdots & -16 \\ 0 & 1 & 0 & \vdots & 3 \\ 0 & 0 & 1 & \vdots & -2 \end{bmatrix}$$

$$5R_2 + R_1 \begin{bmatrix} 1 & 0 & 0 & \vdots & -1 \\ 0 & 1 & 0 & \vdots & 3 \\ 0 & 0 & 1 & \vdots & -2 \end{bmatrix}$$

$$x_1 = -1, x_2 = 3, x_3 = -2$$

Answer: $X = \begin{bmatrix} -1 \\ 3 \\ -2 \end{bmatrix}$

61. (a) $A(B + C) = \begin{bmatrix} 7 & -2 & 5 \\ -6 & 13 & -8 \\ 16 & 11 & -3 \end{bmatrix}$

(b) $AB + AC = \begin{bmatrix} 7 & -2 & 5 \\ -6 & 13 & -8 \\ 16 & 11 & -3 \end{bmatrix}$

The answers are the same.

63. (a) $(A + B)^2 = \begin{bmatrix} 26 & 11 & 0 \\ 11 & 20 & -3 \\ 11 & 14 & 0 \end{bmatrix}$

(b) $A^2 + AB + BA + B^2 = \begin{bmatrix} 26 & 11 & 0 \\ 11 & 20 & -3 \\ 11 & 14 & 0 \end{bmatrix}$

The answers are the same.

65. (a) $A(BC) = \begin{bmatrix} 25 & -34 & 28 \\ -53 & 34 & -7 \\ -76 & 30 & 21 \end{bmatrix}$

(b) $(AB)C = \begin{bmatrix} 25 & -34 & 28 \\ -53 & 34 & -7 \\ -76 & 30 & 21 \end{bmatrix}$

The answers are the same.

67. (a) $A + cB = \begin{bmatrix} -1 & 10 & -4 \\ -5 & -1 & 0 \end{bmatrix}$

(b) $A + cB = \begin{bmatrix} 1 & 2 & -2 \\ -1 & 1 & 0 \end{bmatrix} + 2\begin{bmatrix} -1 & 4 & -1 \\ -2 & -1 & 0 \end{bmatrix} = \begin{bmatrix} -1 & 10 & -4 \\ -5 & -1 & 0 \end{bmatrix}$

69. (*a*), (*b*) $c(AB)$, Not possible

Number of columns of A (3) does not equal the number of rows of B (2).

71. (*a*), (*b*) $CA - BC$, Not possible

CA is 3×3, BC is 2×2.

73. (a) $cd\,A = \begin{bmatrix} -6 & -12 & 12 \\ 6 & -6 & 0 \end{bmatrix}$

(b) $cd\,A = 2(-3)\begin{bmatrix} 1 & 2 & -2 \\ -1 & 1 & 0 \end{bmatrix} = \begin{bmatrix} -6 & -12 & 12 \\ 6 & -6 & 0 \end{bmatrix}$

75. $A = \begin{bmatrix} 2 & 0 \\ 4 & 5 \end{bmatrix}$

$f(A) = A^2 - 5A + 2I = \begin{bmatrix} 2 & 0 \\ 4 & 5 \end{bmatrix}\begin{bmatrix} 2 & 0 \\ 4 & 5 \end{bmatrix} - 5\begin{bmatrix} 2 & 0 \\ 4 & 5 \end{bmatrix} + 2\begin{bmatrix} 1 & 0 \\ 0 & 1 \end{bmatrix} = \begin{bmatrix} -4 & 0 \\ 8 & 2 \end{bmatrix}$

77. $1.20\begin{bmatrix} 70 & 50 & 25 \\ 35 & 100 & 70 \end{bmatrix} = \begin{bmatrix} 84 & 60 & 30 \\ 42 & 120 & 84 \end{bmatrix}$

79. $BA = \begin{bmatrix} 3.50 & 6.00 \end{bmatrix}\begin{bmatrix} 125 & 100 & 75 \\ 100 & 175 & 125 \end{bmatrix} = \begin{bmatrix} 1037.50 & 1400 & 1012.50 \end{bmatrix}$

The entries in the last matrix BA represent the profit for both crops at each of the three outlets.

81. $ST = \begin{bmatrix} 3 & 2 & 2 & 3 & 0 \\ 0 & 2 & 3 & 4 & 3 \\ 4 & 2 & 1 & 3 & 2 \end{bmatrix} \begin{bmatrix} 840 & 1100 \\ 1200 & 1350 \\ 1450 & 1650 \\ 2650 & 3000 \\ 3050 & 3200 \end{bmatrix} = \begin{bmatrix} \$15,770 & \$18,300 \\ \$26,500 & \$29,250 \\ \$21,260 & \$24,150 \end{bmatrix}$

The entries represent the wholesale and retail prices of the inventory at each outlet.

83. $P^2 = \begin{bmatrix} 0.6 & 0.1 & 0.1 \\ 0.2 & 0.7 & 0.1 \\ 0.2 & 0.2 & 0.8 \end{bmatrix} \begin{bmatrix} 0.6 & 0.1 & 0.1 \\ 0.2 & 0.7 & 0.1 \\ 0.2 & 0.2 & 0.8 \end{bmatrix} = \begin{bmatrix} 0.40 & 0.15 & 0.15 \\ 0.28 & 0.53 & 0.17 \\ 0.32 & 0.32 & 0.68 \end{bmatrix}.$

This product represents the changes in party affiliation after *two* elections.

85. True

For 87–93, A is of order 2×3, B is of order 2×3, C is of order 3×2 and D is of order 2×2.

87. $A + 2C$ is not possible. A and C are not of the same order.

89. AB is not possible. The number of columns of A does not equal the number of rows of B.

91. $BC - D$ is possible. The resulting order is 2×2.

93. $D(A - 3B)$ is possible. The resulting order is 2×3.

95. $(A + B)^2 = \begin{bmatrix} 1 & 0 \\ 2 & 1 \end{bmatrix}$

$A^2 + 2AB + B^2 = \begin{bmatrix} 0 & 0 \\ 3 & 2 \end{bmatrix}$

97. $(A + B)(A - B) = \begin{bmatrix} 3 & -2 \\ 4 & 3 \end{bmatrix}$

$A^2 - B^2 = \begin{bmatrix} 2 & -2 \\ 5 & 4 \end{bmatrix}$

99. $AC = \begin{bmatrix} 0 & 1 \\ 0 & 1 \end{bmatrix} \begin{bmatrix} 2 & 3 \\ 2 & 3 \end{bmatrix} = \begin{bmatrix} 2 & 3 \\ 2 & 3 \end{bmatrix}$

$BC = \begin{bmatrix} 1 & 0 \\ 1 & 0 \end{bmatrix} \begin{bmatrix} 2 & 3 \\ 2 & 3 \end{bmatrix} = \begin{bmatrix} 2 & 3 \\ 2 & 3 \end{bmatrix}$

$AC = BC$, but $A \neq B$.

101. (a) $A^2 = \begin{bmatrix} i & 0 \\ 0 & i \end{bmatrix} \begin{bmatrix} i & 0 \\ 0 & i \end{bmatrix} = \begin{bmatrix} -1 & 0 \\ 0 & -1 \end{bmatrix}$ and $i^2 = -1$

$A^3 = A^2A = \begin{bmatrix} -1 & 0 \\ 0 & -1 \end{bmatrix} \begin{bmatrix} i & 0 \\ 0 & i \end{bmatrix} = \begin{bmatrix} -i & 0 \\ 0 & -i \end{bmatrix}$ and $i^3 = -i$

$A^4 = A^3A = \begin{bmatrix} -i & 0 \\ 0 & -i \end{bmatrix} \begin{bmatrix} i & 0 \\ 0 & i \end{bmatrix} = \begin{bmatrix} 1 & 0 \\ 0 & 1 \end{bmatrix}$ and $i^4 = 1$

(b) $B^2 = \begin{bmatrix} 0 & -i \\ i & 0 \end{bmatrix} \begin{bmatrix} 0 & -i \\ i & 0 \end{bmatrix} = \begin{bmatrix} 1 & 0 \\ 0 & 1 \end{bmatrix}$, the identity matrix

103. (a) $A = \begin{bmatrix} 0 & 2 \\ 0 & 0 \end{bmatrix}$, $B = \begin{bmatrix} 0 & 2 & 3 \\ 0 & 0 & 4 \\ 0 & 0 & 0 \end{bmatrix}$

(b) A^2 and B^3 are both zero matrices.

(c) If A is 4×4, then A^4 will be the zero matrix.

(d) If A is $n \times n$, then A^n is the zero matrix.

105. $3 \ln 4 - \dfrac{1}{3}\ln(x^2 + 3) = \ln 4^3 - \ln(x^2 + 3)^{1/3} = \ln\left[\dfrac{64}{(x^2 + 3)^{1/3}}\right]$

107. $\dfrac{1}{2}[2 \ln(x + 5) + \ln x - \ln(x - 8)] = \ln(x + 5) + \ln x^{1/2} - \ln(x - 8)^{1/2}$

$$= \ln\left[\dfrac{(x + 5)\sqrt{x}}{\sqrt{x - 8}}\right]$$

Section 5.6   The Inverse of a Square Matrix

- ■ You should be able to find the inverse, if it exists, of a square matrix.
  - (a) Write the $n \times 2n$ matrix that consists of the given matrix A on the left and the $n \times n$ identity matrix I on the right to obtain $[A \; \vdots \; I]$. Note that we separate the matrices A and I by a dotted line. We call this process **adjoining** the matrices A and I.
  - (b) If possible, row reduce A to I using elementary row operations on the *entire* matrix $[A \; \vdots \; I]$. The result will be the matrix $[I \; \vdots \; A^{-1}]$. If this is not possible, then A is not invertible.
  - (c) Check your work by multiplying to see that $AA^{-1} = I = A^{-1}A$.
- ■ You should be able to use inverse matrices to solve systems of equation.
- ■ You should be able to find inverses using a graphing utility.

Vocabulary Check

1. square   **2.** inverse   **3.** nonsingular, singular

1. $AB = \begin{bmatrix} 2 & 1 \\ 5 & 3 \end{bmatrix}\begin{bmatrix} 3 & -1 \\ -5 & 2 \end{bmatrix} = \begin{bmatrix} 2(3) + 1(-5) & 2(-1) + 1(2) \\ 5(3) + 3(-5) & 5(-1) + 3(2) \end{bmatrix} = \begin{bmatrix} 1 & 0 \\ 0 & 1 \end{bmatrix}$

$BA = \begin{bmatrix} 3 & -1 \\ -5 & 2 \end{bmatrix}\begin{bmatrix} 2 & 1 \\ 5 & 3 \end{bmatrix} = \begin{bmatrix} 3(2) + (-1)(5) & 3(1) + (-1)(3) \\ -5(2) + 2(5) & -5(1) + 2(3) \end{bmatrix} = \begin{bmatrix} 1 & 0 \\ 0 & 1 \end{bmatrix}$

3. $AB = \begin{bmatrix} 1 & 2 \\ 3 & 4 \end{bmatrix}\begin{bmatrix} -2 & 1 \\ \frac{3}{2} & -\frac{1}{2} \end{bmatrix} = \begin{bmatrix} -2 + 3 & 1 - 1 \\ -6 + 6 & 3 - 2 \end{bmatrix} = \begin{bmatrix} 1 & 0 \\ 0 & 1 \end{bmatrix}$

$BA = \begin{bmatrix} -2 & 1 \\ \frac{3}{2} & -\frac{1}{2} \end{bmatrix}\begin{bmatrix} 1 & 2 \\ 3 & 4 \end{bmatrix} = \begin{bmatrix} -2 + 3 & -4 + 4 \\ \frac{3}{2} - \frac{3}{2} & 3 - 2 \end{bmatrix} = \begin{bmatrix} 1 & 0 \\ 0 & 1 \end{bmatrix}$

5. $AB = \begin{bmatrix} 2 & -17 & 11 \\ -1 & 11 & -7 \\ 0 & 3 & -2 \end{bmatrix}\begin{bmatrix} 1 & 1 & 2 \\ 2 & 4 & -3 \\ 3 & 6 & -5 \end{bmatrix}$

$= \begin{bmatrix} 2 - 34 + 33 & 2 - 68 + 66 & 4 + 51 - 55 \\ -1 + 22 - 21 & -1 + 44 - 42 & -2 - 33 + 35 \\ 6 - 6 & 12 - 12 & -9 + 10 \end{bmatrix} = \begin{bmatrix} 1 & 0 & 0 \\ 0 & 1 & 0 \\ 0 & 0 & 1 \end{bmatrix}$

$BA = \begin{bmatrix} 1 & 1 & 2 \\ 2 & 4 & -3 \\ 3 & 6 & -5 \end{bmatrix}\begin{bmatrix} 2 & -17 & 11 \\ -1 & 11 & -7 \\ 0 & 3 & -2 \end{bmatrix} = \begin{bmatrix} 2 - 1 & -17 + 11 + 6 & 11 - 7 - 4 \\ 4 - 4 & -34 + 44 - 9 & 22 - 28 + 6 \\ 6 - 6 & -51 + 66 - 15 & 33 - 42 + 10 \end{bmatrix} = \begin{bmatrix} 1 & 0 & 0 \\ 0 & 1 & 0 \\ 0 & 0 & 1 \end{bmatrix}$

7. $AB = \begin{bmatrix} -1 & -4 \\ 1 & 2 \end{bmatrix} \begin{bmatrix} 1 & 2 \\ -\frac{1}{2} & -\frac{1}{2} \end{bmatrix} = \begin{bmatrix} 1 & 0 \\ 0 & 1 \end{bmatrix}; BA = \begin{bmatrix} 1 & 0 \\ 0 & 1 \end{bmatrix}$

9. $AB = \begin{bmatrix} 1.6 & 2 \\ -3.5 & -4.5 \end{bmatrix} \begin{bmatrix} 22.5 & 10 \\ -17.5 & -8 \end{bmatrix} = \begin{bmatrix} 1 & 0 \\ 0 & 1 \end{bmatrix}; BA = \begin{bmatrix} 1 & 0 \\ 0 & 1 \end{bmatrix}$

11. $[A \,\vdots\, I] = \begin{bmatrix} 2 & 0 & \vdots & 1 & 0 \\ 0 & 3 & \vdots & 0 & 1 \end{bmatrix}$

$\begin{array}{c} \frac{1}{2}R_1 \to \\ \frac{1}{3}R_2 \to \end{array} \begin{bmatrix} 1 & 0 & \vdots & \frac{1}{2} & 0 \\ 0 & 1 & \vdots & 0 & \frac{1}{3} \end{bmatrix} = [I \,\vdots\, A^{-1}]$

$A^{-1} = \begin{bmatrix} \frac{1}{2} & 0 \\ 0 & \frac{1}{3} \end{bmatrix} = \frac{1}{6}\begin{bmatrix} 3 & 0 \\ 0 & 2 \end{bmatrix}$

13. $[A \,\vdots\, I] = \begin{bmatrix} 1 & -2 & \vdots & 1 & 0 \\ 2 & -3 & \vdots & 0 & 1 \end{bmatrix}$

$-2R_1 + R_2 \to \begin{bmatrix} 1 & -2 & \vdots & 1 & 0 \\ 0 & 1 & \vdots & -2 & 1 \end{bmatrix}$

$2R_2 + R_1 \to \begin{bmatrix} 1 & 0 & \vdots & -3 & 2 \\ 0 & 1 & \vdots & -2 & 1 \end{bmatrix}$

$A^{-1} = \begin{bmatrix} -3 & 2 \\ -2 & 1 \end{bmatrix}$

15. $A = \begin{bmatrix} 2 & 7 & 1 \\ -3 & -9 & 2 \end{bmatrix}$

A has no inverse because it is not square.

17. $[A \,\vdots\, I] = \begin{bmatrix} 1 & 1 & 1 & \vdots & 1 & 0 & 0 \\ 3 & 5 & 4 & \vdots & 0 & 1 & 0 \\ 3 & 6 & 5 & \vdots & 0 & 0 & 1 \end{bmatrix}$

$\begin{array}{c} -3R_1 + R_2 \to \\ -3R_1 + R_3 \to \end{array} \begin{bmatrix} 1 & 1 & 1 & \vdots & 1 & 0 & 0 \\ 0 & 2 & 1 & \vdots & -3 & 1 & 0 \\ 0 & 3 & 2 & \vdots & -3 & 0 & 1 \end{bmatrix}$

$\begin{array}{c} -R_2 + R_1 \to \\ \frac{1}{2}R_2 \to \\ -3R_2 + R_3 \to \end{array} \begin{bmatrix} 1 & 0 & \frac{1}{2} & \vdots & \frac{5}{2} & -\frac{1}{2} & 0 \\ 0 & 1 & \frac{1}{2} & \vdots & -\frac{3}{2} & \frac{1}{2} & 0 \\ 0 & 0 & \frac{1}{2} & \vdots & \frac{3}{2} & -\frac{3}{2} & 1 \end{bmatrix}$

$\begin{array}{c} -R_3 + R_1 \to \\ -R_3 + R_2 \to \\ 2R_3 \to \end{array} \begin{bmatrix} 1 & 0 & 0 & \vdots & 1 & 1 & -1 \\ 0 & 1 & 0 & \vdots & -3 & 2 & -1 \\ 0 & 0 & 1 & \vdots & 3 & -3 & 2 \end{bmatrix}$

$= [I \,\vdots\, A^{-1}]$

$A^{-1} = \begin{bmatrix} 1 & 1 & -1 \\ -3 & 2 & -1 \\ 3 & -3 & 2 \end{bmatrix}$

19. $[A \,\vdots\, I] = \begin{bmatrix} -5 & 0 & 0 & \vdots & 1 & 0 & 0 \\ 2 & 0 & 0 & \vdots & 0 & 1 & 0 \\ -1 & 5 & 7 & \vdots & 0 & 0 & 1 \end{bmatrix}$

$\begin{array}{c} (-\frac{1}{5})R_1 \\ (-2)R_1 + R_2 \end{array} \begin{bmatrix} 1 & 0 & 0 & \vdots & -\frac{1}{5} & 0 & 0 \\ 0 & 0 & 0 & \vdots & \frac{2}{5} & 1 & 0 \\ -1 & 5 & 7 & \vdots & 0 & 0 & 1 \end{bmatrix}$

Not invertible,   (row of zeros)

A^{-1} does not exist.

21. Not invertible

A^{-1} does not exist.

23. $A = \begin{bmatrix} -\frac{1}{2} & \frac{3}{4} & \frac{1}{4} \\ 1 & 0 & -\frac{3}{2} \\ 0 & -1 & \frac{1}{2} \end{bmatrix}$

$A^{-1} = \begin{bmatrix} -12 & -5 & -9 \\ -4 & -2 & -4 \\ -8 & -4 & -6 \end{bmatrix}$

25. $A = \begin{bmatrix} 0.1 & 0.2 & 0.3 \\ -0.3 & 0.2 & 0.2 \\ 0.5 & 0.4 & 0.4 \end{bmatrix}$

$A^{-1} = \frac{5}{11}\begin{bmatrix} 0 & -4 & 2 \\ -22 & 11 & 11 \\ 22 & -6 & -8 \end{bmatrix}$

27. $A = \begin{bmatrix} -1 & 0 & 1 & 0 \\ 0 & 2 & 0 & -1 \\ 2 & 0 & -1 & 0 \\ 0 & -1 & 0 & 1 \end{bmatrix}$

$A^{-1} = \begin{bmatrix} 1 & 0 & 1 & 0 \\ 0 & 1 & 0 & 1 \\ 2 & 0 & 1 & 0 \\ 0 & 1 & 0 & 2 \end{bmatrix}$

29. $\begin{bmatrix} 5 & 1 \\ -2 & -2 \end{bmatrix}^{-1} = \frac{1}{5(-2)-(-2)(1)}\begin{bmatrix} -2 & -1 \\ 2 & 5 \end{bmatrix} = \frac{1}{-8}\begin{bmatrix} -2 & -1 \\ 2 & 5 \end{bmatrix} = \begin{bmatrix} \frac{1}{4} & \frac{1}{8} \\ -\frac{1}{4} & -\frac{5}{8} \end{bmatrix}$

31. $\begin{bmatrix} \frac{7}{2} & -\frac{3}{4} \\ \frac{1}{5} & \frac{4}{5} \end{bmatrix}^{-1} = \frac{1}{\left(\frac{7}{2}\right)\left(\frac{4}{5}\right)-\left(-\frac{3}{4}\right)\left(\frac{1}{5}\right)}\begin{bmatrix} \frac{4}{5} & \frac{3}{4} \\ -\frac{1}{5} & \frac{7}{2} \end{bmatrix} = \frac{20}{59}\begin{bmatrix} \frac{4}{5} & \frac{3}{4} \\ -\frac{1}{5} & \frac{7}{2} \end{bmatrix} = \frac{1}{59}\begin{bmatrix} 16 & 15 \\ -4 & 70 \end{bmatrix}$

33. $\begin{bmatrix} 2 & 3 \\ -1 & 5 \end{bmatrix}^{-1} = \frac{1}{2(5)-(3)(-1)}\begin{bmatrix} 5 & -3 \\ 1 & 2 \end{bmatrix} = \frac{1}{13}\begin{bmatrix} 5 & -3 \\ 1 & 2 \end{bmatrix} = \begin{bmatrix} \frac{5}{13} & -\frac{3}{13} \\ \frac{1}{13} & \frac{2}{13} \end{bmatrix}$

35. $\begin{bmatrix} -1 & 0 \\ 3 & -2 \end{bmatrix}^{-1} = \frac{1}{(-1)(-2)-0(3)}\begin{bmatrix} -2 & 0 \\ -3 & -1 \end{bmatrix} = \frac{1}{2}\begin{bmatrix} -2 & 0 \\ -3 & -1 \end{bmatrix} = \begin{bmatrix} -1 & 0 \\ -\frac{3}{2} & -\frac{1}{2} \end{bmatrix}$

37. $\begin{bmatrix} 1 & 2 \\ -2 & 0 \end{bmatrix}^{-1} = \begin{bmatrix} 0 & -\frac{1}{2} \\ \frac{1}{2} & \frac{1}{4} \end{bmatrix} \Rightarrow k = 0$

39. $\begin{bmatrix} -1 & 2 \\ -3 & 1 \end{bmatrix}^{-1} = \begin{bmatrix} \frac{1}{5} & -\frac{2}{5} \\ \frac{3}{5} & -\frac{1}{5} \end{bmatrix} \Rightarrow k = \frac{3}{5}$

41. $\begin{bmatrix} x \\ y \end{bmatrix} = \begin{bmatrix} -3 & 2 \\ -2 & 1 \end{bmatrix}\begin{bmatrix} 5 \\ 10 \end{bmatrix} = \begin{bmatrix} 5 \\ 0 \end{bmatrix}$

Answer: $(5, 0)$

43. $\begin{bmatrix} x \\ y \end{bmatrix} = \begin{bmatrix} -3 & 2 \\ -2 & 1 \end{bmatrix}\begin{bmatrix} 4 \\ 2 \end{bmatrix}\begin{bmatrix} -8 \\ -6 \end{bmatrix}$

Answer: $(-8, -6)$

45. $\begin{bmatrix} x \\ y \\ z \end{bmatrix} = \begin{bmatrix} 1 & 1 & -1 \\ -3 & 2 & -1 \\ 3 & -3 & 2 \end{bmatrix}\begin{bmatrix} 0 \\ 5 \\ 2 \end{bmatrix} = \begin{bmatrix} 3 \\ 8 \\ -11 \end{bmatrix}$

Answer: $(3, 8, -11)$

47. $\begin{bmatrix} x_1 \\ x_2 \\ x_3 \\ x_4 \end{bmatrix} = \begin{bmatrix} -24 & 7 & 1 & -2 \\ -10 & 3 & 0 & -1 \\ -29 & 7 & 3 & -2 \\ 12 & -3 & -1 & 1 \end{bmatrix}\begin{bmatrix} 0 \\ 1 \\ -1 \\ 2 \end{bmatrix} = \begin{bmatrix} 2 \\ 1 \\ 0 \\ 0 \end{bmatrix}$

Answer: $(2, 1, 0, 0)$

49. $A = \begin{bmatrix} 3 & 4 \\ 5 & 3 \end{bmatrix}$

$A^{-1} = \dfrac{1}{9 - 20} \begin{bmatrix} 3 & -4 \\ -5 & 3 \end{bmatrix}$

$\begin{bmatrix} x \\ y \end{bmatrix} = -\dfrac{1}{11} \begin{bmatrix} 3 & -4 \\ -5 & 3 \end{bmatrix} \begin{bmatrix} -2 \\ 4 \end{bmatrix} = -\dfrac{1}{11} \begin{bmatrix} -22 \\ 22 \end{bmatrix} = \begin{bmatrix} 2 \\ -2 \end{bmatrix}$

Answer: $(2, -2)$

51. $A = \begin{bmatrix} -0.4 & 0.8 \\ 2 & -4 \end{bmatrix}$

$A^{-1} = \dfrac{1}{1.6 - 1.6} \begin{bmatrix} -4 & -0.8 \\ -2 & -0.4 \end{bmatrix}$

A^{-1} does not exist.

[The system actually has no solution.]

53. $A = \begin{bmatrix} -\frac{1}{4} & \frac{3}{8} \\ \frac{3}{2} & \frac{3}{4} \end{bmatrix}$

$A^{-1} = \begin{bmatrix} -1 & \frac{1}{2} \\ 2 & \frac{1}{3} \end{bmatrix}$

$\begin{bmatrix} x \\ y \end{bmatrix} = A^{-1}b = \begin{bmatrix} -1 & \frac{1}{2} \\ 2 & \frac{1}{3} \end{bmatrix} \begin{bmatrix} -2 \\ -12 \end{bmatrix} = \begin{bmatrix} -4 \\ -8 \end{bmatrix}$

Answer: $(-4, -8)$

55. $A = \begin{bmatrix} 4 & -1 & 1 \\ 2 & 2 & 3 \\ 5 & -2 & 6 \end{bmatrix}$

$A^{-1} = \dfrac{1}{55} \begin{bmatrix} 18 & 4 & -5 \\ 3 & 19 & -10 \\ -14 & 3 & 10 \end{bmatrix}$

$\begin{bmatrix} x \\ y \\ z \end{bmatrix} = \dfrac{1}{55} \begin{bmatrix} 18 & 4 & -5 \\ 3 & 19 & -10 \\ -14 & 3 & 10 \end{bmatrix} \begin{bmatrix} -5 \\ 10 \\ 1 \end{bmatrix}$

$= \dfrac{1}{55} \begin{bmatrix} -55 \\ 165 \\ 110 \end{bmatrix} = \begin{bmatrix} -1 \\ 3 \\ 2 \end{bmatrix}$

Answer: $(-1, 3, 2)$

57. $A = \begin{bmatrix} 5 & -3 & 2 \\ 2 & 2 & -3 \\ -1 & 7 & -8 \end{bmatrix}$

A^{-1} does not exist.

The system actually has an infinite number of solutions of the form

$x = 0.3125t + 0.8125$

$y = 1.1875t + 0.6875$

$z = t$

where t is any real number.

59. $\begin{bmatrix} 7 & -3 & 0 & 2 & \vdots & 41 \\ -2 & 1 & 0 & -1 & \vdots & -13 \\ 4 & 0 & 1 & -2 & \vdots & 12 \\ -1 & 1 & 0 & -1 & \vdots & -8 \end{bmatrix}$ row reduces to $\begin{bmatrix} 1 & 0 & 0 & 0 & \vdots & 5 \\ 0 & 1 & 0 & 0 & \vdots & 0 \\ 0 & 0 & 1 & 0 & \vdots & -2 \\ 0 & 0 & 0 & 1 & \vdots & 3 \end{bmatrix}$.

Answer: $(5, 0, -2, 3)$

61. (a) $(x, y) = (-3, 2)$ (b) $(x + h, y + k) = (-3 + 2, 2 + (-1)) = (-1, 1)$

(c) $B = AX = \begin{bmatrix} -1 \\ 1 \\ 1 \end{bmatrix}$ (d) $A^{-1} = \begin{bmatrix} 1 & 0 & -2 \\ 0 & 1 & 1 \\ 0 & 0 & 1 \end{bmatrix}$ (e) $A^{-1}B = \begin{bmatrix} -3 \\ 2 \\ 1 \end{bmatrix}$

The point $(-1, 1)$ has been translated back to $(-3, 2)$.

63. (a) $(x, y) = (2, -4)$ 　　　(b) $(x + h, y + k) = (2 - 3, -4 + 4) = (-1, 0)$

(c) $B = AX = \begin{bmatrix} -1 \\ 0 \\ 1 \end{bmatrix}$ 　　(d) $A^{-1} = \begin{bmatrix} 1 & 0 & 3 \\ 0 & 1 & -4 \\ 0 & 0 & 1 \end{bmatrix}$ 　　(e) $A^{-1}B = \begin{bmatrix} 2 \\ -4 \\ 1 \end{bmatrix}$

The point $(-1, 0)$ has been translated back to $(2, -4)$.

For Exercises 65 and 67 use $A = \begin{bmatrix} 1 & 1 & 1 \\ 0.065 & 0.07 & 0.09 \\ 0 & 2 & -1 \end{bmatrix}$. **Using the methods of this section, we have**

$$A^{-1} = \tfrac{1}{11} \begin{bmatrix} 50 & -600 & -4 \\ -13 & 200 & 5 \\ -26 & 400 & -1 \end{bmatrix}.$$

65. $X = A^{-1}B = \tfrac{1}{11} \begin{bmatrix} 50 & -600 & -4 \\ -13 & 200 & 5 \\ -26 & 400 & -1 \end{bmatrix} \begin{bmatrix} 25{,}000 \\ 1900 \\ 0 \end{bmatrix} = \begin{bmatrix} 10{,}000 \\ 5000 \\ 10{,}000 \end{bmatrix}$

Answer: $10,000 in AAA bonds, $5000 in A-bonds and $10,000 in B-bonds.

67. $X = A^{-1}B = \tfrac{1}{11} \begin{bmatrix} 50 & -600 & -4 \\ -13 & 200 & 5 \\ -26 & 400 & -1 \end{bmatrix} \begin{bmatrix} 65{,}000 \\ 5050 \\ 0 \end{bmatrix} = \begin{bmatrix} 20{,}000 \\ 15{,}000 \\ 30{,}000 \end{bmatrix}$

Answer: $20,000 in AAA bonds, $15,000 in A-bonds and $30,000 in B-bonds.

69. $A = \begin{bmatrix} 2 & 0 & 4 \\ 0 & 1 & 4 \\ 1 & 1 & -1 \end{bmatrix}$ 　$A^{-1} = \tfrac{1}{14} \begin{bmatrix} 5 & -4 & 4 \\ -4 & 6 & 8 \\ 1 & 2 & -2 \end{bmatrix}$

$\begin{bmatrix} I_1 \\ I_2 \\ I_3 \end{bmatrix} = \tfrac{1}{14} \begin{bmatrix} 5 & -4 & 4 \\ -4 & 6 & 8 \\ 1 & 2 & -2 \end{bmatrix} \begin{bmatrix} 14 \\ 28 \\ 0 \end{bmatrix} = \begin{bmatrix} -3 \\ 8 \\ 5 \end{bmatrix}$

Answer: $I_1 = -3$ amps, $I_2 = 8$ amps, $I_3 = 5$ amps

For Exercises 71 and 73, let $x =$ **number of muffins,** $y =$ **number of bones,** $z =$ **number of cookies.**

$$\underbrace{\begin{bmatrix} 2 & 1 & 2 \\ 3 & 1 & 1 \\ 2 & 1 & 1.5 \end{bmatrix}}_{A} \underbrace{\begin{bmatrix} x \\ y \\ z \end{bmatrix}}_{X} = \begin{bmatrix} \text{Beef} \\ \text{Chicken} \\ \text{Liver} \end{bmatrix}$$

$$A^{-1} = \begin{bmatrix} 1 & 1 & -2 \\ -5 & -2 & 8 \\ 2 & 0 & -2 \end{bmatrix}$$

71. $A^{-1} \begin{bmatrix} 700 \\ 500 \\ 600 \end{bmatrix} = \begin{bmatrix} 0 \\ 300 \\ 200 \end{bmatrix}$

300 units of bones

200 units of cookies

73. $A^{-1} \begin{bmatrix} 800 \\ 750 \\ 725 \end{bmatrix} = \begin{bmatrix} 100 \\ 300 \\ 150 \end{bmatrix}$

100 units of muffins

300 units of bones

150 units of cookies

75. (a)
$$f + \quad h + \quad s = 10$$
$$2f + 2.5h + 3s = 26$$
$$h - \quad s = 0$$

(b)
$$\begin{bmatrix} 1 & 1 & 1 \\ 2 & 2.5 & 3 \\ 0 & 1 & -1 \end{bmatrix} \begin{bmatrix} f \\ h \\ s \end{bmatrix} = \begin{bmatrix} 10 \\ 26 \\ 0 \end{bmatrix}$$
$$\quad A \quad\quad X \;=\; B$$

(c) $X = A^{-1}B = \begin{bmatrix} \frac{11}{3} & -\frac{4}{3} & -\frac{1}{3} \\ -\frac{4}{3} & \frac{2}{3} & \frac{2}{3} \\ -\frac{4}{3} & \frac{2}{3} & -\frac{1}{3} \end{bmatrix} \begin{bmatrix} 10 \\ 26 \\ 0 \end{bmatrix} = \begin{bmatrix} 2 \\ 4 \\ 4 \end{bmatrix}$

2 pounds French vanilla

4 pounds Hazelnut

4 pounds Swiss chocolate

77. (a)
$$\begin{array}{l} (2, 5343) : \\ (3, 5589) : \\ (4, 6309) : \end{array} \begin{cases} 4a + 2b + c = 5343 \\ 9a + 3b + c = 5589 \\ 16a + 4b + c = 6309 \end{cases}$$

(b) $A = \begin{bmatrix} 4 & 2 & 1 \\ 9 & 3 & 1 \\ 16 & 4 & 1 \end{bmatrix},\; A^{-1} = \begin{bmatrix} \frac{1}{2} & -1 & \frac{1}{2} \\ -\frac{7}{2} & 6 & -\frac{5}{2} \\ 6 & -8 & 3 \end{bmatrix}$

$$A^{-1} \begin{bmatrix} 5343 \\ 5589 \\ 6309 \end{bmatrix} = \begin{bmatrix} a \\ b \\ c \end{bmatrix} = \begin{bmatrix} 237 \\ -939 \\ 6273 \end{bmatrix}$$

$$y = 237t^2 - 939t + 6273$$

(c)
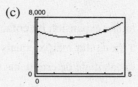

(d) For 2005, $t = 5$ and $y = 7503$ thousand.

For 2010, $t = 10$ and $y = 20{,}583$ thousand.

For 2015, $t = 15$ and $y = 45{,}513$ thousand.

(e) Answers will vary.

79. True. $AA^{-1} = A^{-1}A = I$

81. $AA^{-1} = \begin{bmatrix} a & b \\ c & d \end{bmatrix} \left(\dfrac{1}{ad - bc} \right) \begin{bmatrix} d & -b \\ -c & a \end{bmatrix} = \dfrac{1}{ad - bc} \begin{bmatrix} a & b \\ c & d \end{bmatrix} \begin{bmatrix} d & -b \\ -c & a \end{bmatrix}$

$$= \frac{1}{ad - bc} \begin{bmatrix} ad - bc & 0 \\ 0 & ad - bc \end{bmatrix} = \begin{bmatrix} 1 & 0 \\ 0 & 1 \end{bmatrix}$$

$$A^{-1}A = \frac{1}{ad - bc} \begin{bmatrix} d & -b \\ -c & a \end{bmatrix} \begin{bmatrix} a & b \\ c & d \end{bmatrix} = \frac{1}{ad - bc} \begin{bmatrix} ad - bc & 0 \\ 0 & ad - bc \end{bmatrix} = \begin{bmatrix} 1 & 0 \\ 0 & 1 \end{bmatrix}$$

83. $\dfrac{\left(\dfrac{9}{x} \right)}{\left(\dfrac{6}{x} + 2 \right)} = \dfrac{\left(\dfrac{9}{x} \right)}{\left(\dfrac{6 + 2x}{x} \right)} = \dfrac{9}{x} \cdot \dfrac{x}{6 + 2x} = \dfrac{9}{6 + 2x}, \quad x \neq 0$

85. $\dfrac{\dfrac{4}{x^2 - 9} + \dfrac{2}{x - 2}}{\dfrac{1}{x + 3} + \dfrac{1}{x - 3}} \cdot \dfrac{(x^2 - 9)(x - 2)}{(x^2 - 9)(x - 2)} = \dfrac{4(x - 2) + 2(x^2 - 9)}{(x - 3)(x - 2) + (x + 3)(x - 2)}$

$$= \frac{2x^2 + 4x - 26}{2x^2 - 4x}$$

$$= \frac{x^2 + 2x - 13}{x(x - 2)}, \quad x \neq \pm 3$$

87. $e^{2x} + 2e^x - 15 = (e^x + 5)(e^x - 3) = 0 \Rightarrow e^x = 3 \Rightarrow x = \ln 3 \approx 1.099$

89. $7 \ln 3x = 12$

$\ln 3x = \frac{12}{7}$

$3x = e^{12/7}$

$x = \frac{1}{3}e^{12/7} \approx 1.851$

91. Answers will vary.

Section 5.7 The Determinant of a Square Matrix

- ■ You should be able to determine the determinant of a matrix of order 2×2 by using the products of the diagonals.
- ■ You should be able to use expansion by cofactors to find the determinant of a matrix of order 3 or greater.
- ■ The determinant of a triangular matrix equals the product of the entries on the main diagonal.
- ■ You should be able to calculate determinants using a graphing utility.

Vocabulary Check

1. determinant

2. minor

3. cofactor

4. expanding by cofactors

5. triangular

6. diagonal

1. $|4| = 4$

3. $\begin{vmatrix} 8 & 4 \\ 2 & 3 \end{vmatrix} = 8(3) - 4(2) = 24 - 8 = 16$

5. $\begin{vmatrix} 6 & 2 \\ -5 & 3 \end{vmatrix} = 6(3) - (2)(-5) = 18 + 10 = 28$

7. $\begin{vmatrix} -7 & 6 \\ \frac{1}{2} & 3 \end{vmatrix} = -7(3) - 6\left(\frac{1}{2}\right) = -21 - 3 = -24$

9. $\begin{vmatrix} 2 & -1 & 0 \\ 4 & 2 & 1 \\ 4 & 2 & 1 \end{vmatrix} = 2\begin{vmatrix} 2 & 1 \\ 2 & 1 \end{vmatrix} - 4\begin{vmatrix} -1 & 0 \\ 2 & 1 \end{vmatrix} + 4\begin{vmatrix} -1 & 0 \\ 2 & 1 \end{vmatrix} = 2(0) - 4(-1) + 4(-1) = 0$

11. $\begin{vmatrix} -1 & 2 & -5 \\ 0 & 3 & 4 \\ 0 & 0 & 3 \end{vmatrix} = (-1)(3)(3) = -9$ (Upper Triangular)

13. $\begin{vmatrix} 0.3 & 0.2 & 0.2 \\ 0.2 & 0.2 & 0.2 \\ -0.4 & 0.4 & 0.3 \end{vmatrix} = -0.002$

15. $\begin{bmatrix} 3 & 4 \\ 2 & -5 \end{bmatrix}$

(a) $M_{11} = -5$

$M_{12} = 2$

$M_{21} = 4$

$M_{22} = 3$

(b) $C_{11} = M_{11} = -5$

$C_{12} = -M_{12} = -2$

$C_{21} = -M_{21} = -4$

$C_{22} = M_{22} = 3$

17. $\begin{bmatrix} -4 & 6 & 3 \\ 7 & -2 & 8 \\ 1 & 0 & -5 \end{bmatrix}$

(a) $M_{11} = \begin{vmatrix} -2 & 8 \\ 0 & -5 \end{vmatrix} = 10$ $M_{23} = \begin{vmatrix} -4 & 6 \\ 1 & 0 \end{vmatrix} = -6$ (b) $C_{11} = 10$

$M_{12} = \begin{vmatrix} 7 & 8 \\ 1 & -5 \end{vmatrix} = -43$ $M_{31} = \begin{vmatrix} 6 & 3 \\ -2 & 8 \end{vmatrix} = 54$ $C_{12} = 43$

$\qquad\qquad\qquad\qquad\qquad\qquad\qquad\qquad\qquad\qquad\qquad C_{13} = 2$

$M_{13} = \begin{vmatrix} 7 & -2 \\ 1 & 0 \end{vmatrix} = 2$ $M_{32} = \begin{vmatrix} -4 & 3 \\ 7 & 8 \end{vmatrix} = -53$ $C_{21} = 30$

$\qquad\qquad\qquad\qquad\qquad\qquad\qquad\qquad\qquad\qquad\qquad C_{22} = 17$

$M_{21} = \begin{vmatrix} 6 & 3 \\ 0 & -5 \end{vmatrix} = -30$ $M_{33} = \begin{vmatrix} -4 & 6 \\ 7 & -2 \end{vmatrix} = -34$ $C_{23} = 6$

$\qquad\qquad\qquad\qquad\qquad\qquad\qquad\qquad\qquad\qquad\qquad C_{31} = 54$

$M_{22} = \begin{vmatrix} -4 & 3 \\ 1 & -5 \end{vmatrix} = 17$ $\qquad\qquad\qquad\qquad\qquad C_{32} = 53$

$\qquad\qquad\qquad\qquad\qquad\qquad\qquad\qquad\qquad\qquad\qquad C_{33} = -34$

19. (a) $\begin{vmatrix} -3 & 2 & 1 \\ 4 & 5 & 6 \\ 2 & -3 & 1 \end{vmatrix} = -3\begin{vmatrix} 5 & 6 \\ -3 & 1 \end{vmatrix} - 2\begin{vmatrix} 4 & 6 \\ 2 & 1 \end{vmatrix} + \begin{vmatrix} 4 & 5 \\ 2 & -3 \end{vmatrix} = -3(23) - 2(-8) - 22 = -75$

(b) $\begin{vmatrix} -3 & 2 & 1 \\ 4 & 5 & 6 \\ 2 & -3 & 1 \end{vmatrix} = -2\begin{vmatrix} 4 & 6 \\ 2 & 1 \end{vmatrix} + 5\begin{vmatrix} -3 & 1 \\ 2 & 1 \end{vmatrix} + 3\begin{vmatrix} -3 & 1 \\ 4 & 6 \end{vmatrix} = -2(-8) + 5(-5) + 3(-22) = -75$

21. (a) $\begin{vmatrix} 6 & 0 & -3 & 5 \\ 4 & 13 & 6 & -8 \\ -1 & 0 & 7 & 4 \\ 8 & 6 & 0 & 2 \end{vmatrix} = -4\begin{vmatrix} 0 & -3 & 5 \\ 0 & 7 & 4 \\ 6 & 0 & 2 \end{vmatrix} + 13\begin{vmatrix} 6 & -3 & 5 \\ -1 & 7 & 4 \\ 8 & 0 & 2 \end{vmatrix} - 6\begin{vmatrix} 6 & 0 & 5 \\ -1 & 0 & 4 \\ 8 & 6 & 2 \end{vmatrix} - 8\begin{vmatrix} 6 & 0 & -3 \\ -1 & 0 & 7 \\ 8 & 6 & 0 \end{vmatrix}$

$= -4(-282) + 13(-298) - 6(-174) - 8(-234) = 170$

(b) $\begin{vmatrix} 6 & 0 & -3 & 5 \\ 4 & 13 & 6 & -8 \\ -1 & 0 & 7 & 4 \\ 8 & 6 & 0 & 2 \end{vmatrix} = 0\begin{vmatrix} 4 & 6 & -8 \\ -1 & 7 & 4 \\ 8 & 0 & 2 \end{vmatrix} + 13\begin{vmatrix} 6 & -3 & 5 \\ -1 & 7 & 4 \\ 8 & 0 & 2 \end{vmatrix} + 0\begin{vmatrix} 6 & -3 & 5 \\ 4 & 6 & -8 \\ 8 & 0 & 2 \end{vmatrix} + 6\begin{vmatrix} 6 & -3 & 5 \\ 4 & 6 & -8 \\ -1 & 7 & 4 \end{vmatrix}$

$= 0 + 13(-298) + 0 + 6(674) = 170$

23. Expand by Column 3.

$\begin{vmatrix} 1 & 4 & -2 \\ 3 & 2 & 0 \\ -1 & 4 & 3 \end{vmatrix} = -2\begin{vmatrix} 3 & 2 \\ -1 & 4 \end{vmatrix} + 3\begin{vmatrix} 1 & 4 \\ 3 & 2 \end{vmatrix} = -2(14) + 3(-10) = -58$

25. Expand by Column 3.

$\begin{vmatrix} 2 & 6 & 6 & 2 \\ 2 & 7 & 3 & 6 \\ 1 & 5 & 0 & 1 \\ 3 & 7 & 0 & 7 \end{vmatrix} = 6\begin{vmatrix} 2 & 7 & 6 \\ 1 & 5 & 1 \\ 3 & 7 & 7 \end{vmatrix} - 3\begin{vmatrix} 2 & 6 & 2 \\ 1 & 5 & 1 \\ 3 & 7 & 7 \end{vmatrix} = 6(-20) - 3(16) = -168$

27. Expand by Column 2.

$$\begin{vmatrix} 3 & 2 & 4 & -1 & 5 \\ -2 & 0 & 1 & 3 & 2 \\ 1 & 0 & 0 & 4 & 0 \\ 6 & 0 & 2 & -1 & 0 \\ 3 & 0 & 5 & 1 & 0 \end{vmatrix} = -2 \begin{vmatrix} -2 & 1 & 3 & 2 \\ 1 & 0 & 4 & 0 \\ 6 & 2 & -1 & 0 \\ 3 & 5 & 1 & 0 \end{vmatrix} = (-2)(-2) \begin{vmatrix} 1 & 0 & 4 \\ 6 & 2 & -1 \\ 3 & 5 & 1 \end{vmatrix} = 4(103) = 412$$

29. $\begin{vmatrix} 4 & 0 & 0 & 0 \\ 6 & -5 & 0 & 0 \\ 1 & 3 & 1 & 0 \\ 1 & -2 & 7 & 3 \end{vmatrix} = (4)(-5)(1)(3) = -60$ (Lower Triangular)

31. $\det(A) = (-6)(-1)(-7)(-2)(-2) = -168$

(Upper Triangular)

33. $\begin{vmatrix} 1 & -1 & 8 & 4 \\ 2 & 6 & 0 & -4 \\ 2 & 0 & 2 & 6 \\ 0 & 2 & 8 & 0 \end{vmatrix} = -336$

35. $\begin{vmatrix} 3 & -2 & 4 & 3 & 1 \\ -1 & 0 & 2 & 1 & 0 \\ 5 & -1 & 0 & 3 & 2 \\ 4 & 7 & -8 & 0 & 0 \\ 1 & 2 & 3 & 0 & 2 \end{vmatrix} = 410$

37. (a) $\begin{vmatrix} -1 & 0 \\ 0 & 3 \end{vmatrix} = -3$

(b) $\begin{vmatrix} 2 & 0 \\ 0 & -1 \end{vmatrix} = -2$

(c) $\begin{bmatrix} -1 & 0 \\ 0 & 3 \end{bmatrix}\begin{bmatrix} 2 & 0 \\ 0 & -1 \end{bmatrix} = \begin{bmatrix} -2 & 0 \\ 0 & -3 \end{bmatrix}$

(d) $\begin{vmatrix} -2 & 0 \\ 0 & -3 \end{vmatrix} = 6$ **[Note:** $|AB| = |A|\,|B|]$

39. (a) $\begin{vmatrix} -1 & 2 & 1 \\ 1 & 0 & 1 \\ 0 & 1 & 0 \end{vmatrix} = 2$

(b) $\begin{vmatrix} -1 & 0 & 0 \\ 0 & 2 & 0 \\ 0 & 0 & 3 \end{vmatrix} = -6$

(c) $\begin{bmatrix} -1 & 2 & 1 \\ 1 & 0 & 1 \\ 0 & 1 & 0 \end{bmatrix}\begin{bmatrix} -1 & 0 & 0 \\ 0 & 2 & 0 \\ 0 & 0 & 3 \end{bmatrix} = \begin{bmatrix} 1 & 4 & 3 \\ -1 & 0 & 3 \\ 0 & 2 & 0 \end{bmatrix}$

(d) $\begin{vmatrix} 1 & 4 & 3 \\ -1 & 0 & 3 \\ 0 & 2 & 0 \end{vmatrix} = -12$ **[Note:** $|AB| = |A|\,|B|]$

41. (a) $|A| = -25$

(b) $|B| = -220$

(c) $AB = \begin{bmatrix} -7 & -16 & -1 & -28 \\ -4 & -14 & -11 & 8 \\ 13 & 4 & 4 & -4 \\ -2 & 3 & 2 & 2 \end{bmatrix}$

(d) $|AB| = 5500$ [Note: $|AB| = |A|\,|B|]$

43. $\begin{vmatrix} w & x \\ y & z \end{vmatrix} = wz - xy$

$-\begin{vmatrix} y & z \\ w & x \end{vmatrix} = -(xy - wz) = wz - xy$

Thus, $\begin{vmatrix} w & x \\ y & z \end{vmatrix} = -\begin{vmatrix} y & z \\ w & x \end{vmatrix}$.

45. $\begin{vmatrix} w & x \\ y & z \end{vmatrix} = wz - xy$

$\begin{vmatrix} w & x + cw \\ y & z + cy \end{vmatrix} = w(z + cy) - y(x + cw) = wz - xy$

Thus, $\begin{vmatrix} w & x \\ y & z \end{vmatrix} = \begin{vmatrix} w & x + cw \\ y & z + cy \end{vmatrix}$.

47. $\begin{vmatrix} 1 & x & x^2 \\ 1 & y & y^2 \\ 1 & z & z^2 \end{vmatrix} = \begin{vmatrix} y & y^2 \\ z & z^2 \end{vmatrix} - \begin{vmatrix} x & x^2 \\ z & z^2 \end{vmatrix} + \begin{vmatrix} x & x^2 \\ y & y^2 \end{vmatrix}$

$\qquad\qquad = (yz^2 - y^2z) - (xz^2 - x^2z) + (xy^2 - x^2y)$

$\qquad\qquad = yz^2 - xz^2 - y^2z + x^2z + xy(y - x)$

$\qquad\qquad = z^2(y - x) - z(y^2 - x^2) + xy(y - x)$

$\qquad\qquad = z^2(y - x) - z(y - x)(y + x) + xy(y - x)$

$\qquad\qquad = (y - x)[z^2 - z(y + x) + xy]$

$\qquad\qquad = (y - x)[z^2 - zy - zx + xy]$

$\qquad\qquad = (y - x)[z^2 - zx - zy + xy]$

$\qquad\qquad = (y - x)[z(z - x) - y(z - x)]$

$\qquad\qquad = (y - x)(z - x)(z - y)$

49. $\begin{vmatrix} x & 2 \\ 1 & x \end{vmatrix} = 2$

$\qquad x^2 - 2 = 2$

$\qquad\quad x^2 = 4$

$\qquad\quad x = \pm 2$

51. $\begin{vmatrix} 2x & -3 \\ -2 & 2x \end{vmatrix} = 3$

$\qquad 4x^2 - 6 = 3$

$\qquad\quad 4x^2 = 9$

$\qquad\quad\; x^2 = \frac{9}{4}$

$\qquad\quad\; x = \pm \frac{3}{2}$

53. $\begin{vmatrix} x & 1 \\ 2 & x - 2 \end{vmatrix} = -1$

$\qquad x^2 - 2x - 2 = -1$

$\qquad x^2 - 2x - 1 = 0$

$\qquad\quad x = \dfrac{2 \pm \sqrt{4 + 4}}{2}$

$\qquad\quad x = 1 \pm \sqrt{2}$

55. $\begin{vmatrix} x + 3 & 2 \\ 1 & x + 2 \end{vmatrix} = 0$

$\qquad (x + 3)(x + 2) - 2 = 0$

$\qquad\quad x^2 + 5x + 4 = 0$

$\qquad\quad (x + 4)(x + 1) = 0$

$\qquad\qquad\qquad x = -4, -1$

57. $\begin{vmatrix} 2x & 1 \\ -1 & x - 1 \end{vmatrix} = x$

$\qquad 2x^2 - 2x + 1 = x$

$\qquad 2x^2 - 3x + 1 = 0$

$\qquad (x - 1)(2x - 1) = 0$

$\qquad\qquad\quad x = 1, \frac{1}{2}$

59. $\begin{vmatrix} 1 & 2 & x \\ -1 & 3 & 2 \\ 3 & -2 & 1 \end{vmatrix} = 0$

$1 \begin{vmatrix} 3 & 2 \\ -2 & 1 \end{vmatrix} - 2 \begin{vmatrix} -1 & 2 \\ 3 & 1 \end{vmatrix} + x \begin{vmatrix} -1 & 3 \\ 3 & -2 \end{vmatrix} = 0$

$7 - 2(-7) + x(-7) = 0$

$\qquad\qquad 21 = 7x$

$\qquad\qquad\; x = 3$

61. $\begin{vmatrix} 4u & -1 \\ -1 & 2v \end{vmatrix} = 8uv - 1$

63. $\begin{vmatrix} e^{2x} & e^{3x} \\ 2e^{2x} & 3e^{3x} \end{vmatrix} = 3e^{5x} - 2e^{5x} = e^{5x}$

65. $\begin{vmatrix} x & \ln x \\ 1 & \dfrac{1}{x} \end{vmatrix} = 1 - \ln x$

67. True. Expand along the row of zeros.

69. Let $A = \begin{bmatrix} 1 & 3 \\ -2 & 4 \end{bmatrix}$ and $B = \begin{bmatrix} -4 & 0 \\ 3 & 5 \end{bmatrix}$.

$|A| = \begin{vmatrix} 1 & 3 \\ -2 & 4 \end{vmatrix} = 10, \quad |B| = \begin{vmatrix} -4 & 0 \\ 3 & 5 \end{vmatrix} = -20$

$A + B = \begin{bmatrix} -3 & 3 \\ 1 & 9 \end{bmatrix}, \quad |A + B| = \begin{vmatrix} -3 & 3 \\ 1 & 9 \end{vmatrix} = -30$

Thus, $|A + B| \neq |A| + |B|$. Your answer may differ, depending on how you choose A and B.

71. (a) $|A| = 6$

(b) $A^{-1} = \begin{bmatrix} \frac{1}{3} & -\frac{1}{3} \\ \frac{1}{3} & \frac{1}{6} \end{bmatrix}$

(c) $\det(A^{-1}) = \dfrac{1}{6}$

(d) In general, $\det(A^{-1}) = \dfrac{1}{\det A}$.

73. (a) $|A| = 2$

(b) $A^{-1} = \begin{bmatrix} -4 & -5 & 1.5 \\ -1 & -1 & 0.5 \\ -1 & -1 & 0 \end{bmatrix}$

(c) $\det(A^{-1}) = \dfrac{1}{2}$

(d) In general, $\det(A^{-1}) = \dfrac{1}{\det A}$.

75. (a) Columns 2 and 3 are interchanged.

(b) Rows 1 and 3 are interchanged.

77. (a) 5 is factored out of the first row of A.

(b) 4 and 3 are factored out of columns 2 and 3.

79. $x^2 - 3x + 2 = (x - 2)(x - 1)$

81. $4y^2 - 12y + 9 = (2y - 3)^2$

83. $3x - 10y = 46$

$x + y = -2$

$y = -x - 2$

$3x - 10(-x - 2) = 46$

$13x = 26$

$x = 2$

$y = -2 - 2 = -4$

Answer: $(2, -4)$

Section 5.8   Applications of Matrices and Determinants

■ You should be able to find the area of a triangle with vertices (x_1, y_1), (x_2, y_2), and (x_3, y_3).

$$\text{Area} = \pm\frac{1}{2}\begin{vmatrix} x_1 & y_1 & 1 \\ x_2 & y_2 & 1 \\ x_3 & y_3 & 1 \end{vmatrix}$$

The $\pm$ symbol indicates that the appropriate sign should be chosen so that the area is positive.

■ You should be able to test to see if three points, (x_1, y_1), (x_2, y_2), and (x_3, y_3), are collinear.

$$\begin{vmatrix} x_1 & y_1 & 1 \\ x_2 & y_2 & 1 \\ x_3 & y_3 & 1 \end{vmatrix} = 0, \text{ if and only if they are collinear.}$$

■ You should be able to use Cramer's Rule to solve a system of linear equations.

■ Now you should be able to solve a system of linear equations by substitution, elimination, elementary row operations on an augmented matrix, using the inverse matrix, or Cramer's Rule.

■ You should be able to encode and decode messages by using an invertible $n \times n$ matrix.

Vocabulary Check

1. collinear   **2.** Cramer's Rule   **3.** cryptogram   **4.** uncoded, coded

---

1. Vertices: $(-2, 4), (2, 3), (-1, 5)$

$$\frac{1}{2}\begin{vmatrix} -2 & 4 & 1 \\ 2 & 3 & 1 \\ -1 & 5 & 1 \end{vmatrix} = \frac{1}{2}\left[-2\begin{vmatrix} 3 & 1 \\ 5 & 1 \end{vmatrix} - 4\begin{vmatrix} 2 & 1 \\ -1 & 1 \end{vmatrix} + \begin{vmatrix} 2 & 3 \\ -1 & 5 \end{vmatrix} \right]$$

$$= \frac{1}{2}\left[-2(-2) - 4(3) + 13 \right] = \frac{1}{2}(5) = \frac{5}{2}$$

Area $= \frac{5}{2}$ square units

3. Vertices: $\left(0, \frac{1}{2}\right), \left(\frac{5}{2}, 0\right), (4, 3)$

$$\frac{1}{2}\begin{vmatrix} 0 & \frac{1}{2} & 1 \\ \frac{5}{2} & 0 & 1 \\ 4 & 3 & 1 \end{vmatrix} = \frac{1}{2}\left[-\frac{1}{2}\left(-\frac{3}{2}\right) + \frac{15}{2} \right] = \frac{1}{2}\left[\frac{33}{4}\right] = \frac{33}{8}.$$

Area $= \frac{33}{8}$ square units

5. $\begin{vmatrix} -3 & 2 & 1 \\ 1 & 2 & 1 \\ -1 & -4 & 1 \end{vmatrix} = -3(6) - 2(2) + 1(-2) = -24$

Area rhombus $= |-24| = 24$ square units

7. $4 = \pm\frac{1}{2}\begin{vmatrix} -1 & 5 & 1 \\ -2 & 0 & 1 \\ x & 2 & 1 \end{vmatrix}$

$8 = \pm\left[(-1)(-2) - 5(-2 - x) + 1(-4) \right]$

$8 = \pm\left[5x + 8\right]$

$5x + 8 = 8 \quad \text{or} \quad 5x + 8 = -8$

$\quad x = 0 \quad \text{or} \qquad x = -\frac{16}{5}$

$x = 0, -\frac{16}{5}$

9. Points: $(3, -1), (0, -3), (12, 5)$

$$\begin{vmatrix} 3 & -1 & 1 \\ 0 & -3 & 1 \\ 12 & 5 & 1 \end{vmatrix} = 3(-8) + 12(2) = 0$$

The points are collinear.

11. Points: $\left(2, -\frac{1}{2}\right)$, $(-4, 4)$, $(6, -3)$

$$\begin{vmatrix} 2 & -\frac{1}{2} & 1 \\ -4 & 4 & 1 \\ 6 & -3 & 1 \end{vmatrix} = 2(7) + \frac{1}{2}(-10) + 1(-12)$$

$$= -3 \neq 0$$

The points are not collinear.

13. $\begin{vmatrix} 1 & -2 & 1 \\ x & 2 & 1 \\ 5 & 6 & 1 \end{vmatrix} = 0$

$$1(-4) + 2(x - 5) + 1(6x - 10) = 0$$
$$8x - 24 = 0$$
$$x = 3$$

15. $\begin{cases} -7x + 11y = -1 \\ 3x - 9y = 9 \end{cases}$

$$x = \dfrac{\begin{vmatrix} -1 & 11 \\ 9 & -9 \end{vmatrix}}{\begin{vmatrix} -7 & 11 \\ 3 & -9 \end{vmatrix}} = \dfrac{-90}{30} = -3$$

$$y = \dfrac{\begin{vmatrix} -7 & -1 \\ 3 & 9 \end{vmatrix}}{\begin{vmatrix} -7 & 11 \\ 3 & -9 \end{vmatrix}} = \dfrac{-60}{30} = -2$$

Answer: $(-3, -2)$

17. $\begin{cases} 3x + 2y = -2 \\ 6x + 4y = 4 \end{cases}$

$$\begin{vmatrix} 3 & 2 \\ 6 & 4 \end{vmatrix} = 12 - 17 = 0$$

Cramer's rule cannot be used.

(In fact, the system is inconsistent.)

19. $\begin{cases} -0.4x + 0.8y = 1.6 \\ 0.2x + 0.3y = 2.2 \end{cases}$

$$D = \begin{vmatrix} -0.4 & 0.8 \\ 0.2 & 0.3 \end{vmatrix} = -0.28$$

$$x = \dfrac{\begin{vmatrix} 1.6 & 0.8 \\ 2.2 & 0.3 \end{vmatrix}}{-0.28} = \dfrac{-1.28}{-0.28} = \dfrac{32}{7}$$

$$y = \dfrac{\begin{vmatrix} -0.4 & 1.6 \\ 0.2 & 2.2 \end{vmatrix}}{-0.28} = \dfrac{-1.20}{-0.28} = \dfrac{30}{7}$$

Answer: $\left(\dfrac{32}{7}, \dfrac{30}{7}\right)$

21. $\begin{cases} 4x - y + z = -5 \\ 2x + 2y + 3z = 10 \\ 5x - 2y + 6z = 1 \end{cases}$

$$D = \begin{vmatrix} 4 & -1 & 1 \\ 2 & 2 & 3 \\ 5 & -2 & 6 \end{vmatrix} = 55$$

$$x = \dfrac{\begin{vmatrix} -5 & -1 & 1 \\ 10 & 2 & 3 \\ 1 & -2 & 6 \end{vmatrix}}{55} = \dfrac{-55}{55} = -1, \quad y = \dfrac{\begin{vmatrix} 4 & -5 & 1 \\ 2 & 10 & 3 \\ 5 & 1 & 6 \end{vmatrix}}{55} = \dfrac{165}{55} = 3, \quad z = \dfrac{\begin{vmatrix} 4 & -1 & -5 \\ 2 & 2 & 10 \\ 5 & -2 & 1 \end{vmatrix}}{55} = \dfrac{110}{55} = 2$$

Answer: $(-1, 3, 2)$

23. (a)
$$\begin{cases} 3x + 3y + 5z = 1 \\ 3x + 5y + 9z = 2 \\ 5x + 9y + 17z = 4 \end{cases}$$

$$\begin{cases} 3x + 3y + 5z = 1 \\ 2y + 4z = 1 \\ 4y + \frac{26}{3}z = \frac{7}{3} \end{cases}$$

$$\begin{cases} 3x + 3y + 5z = 1 \\ 2y + 4z = 1 \\ \frac{2}{3}z = \frac{1}{3} \end{cases}$$

$z = \frac{1}{2}$

$2y + 4\left(\frac{1}{2}\right) = 1 \implies y = -\frac{1}{2}$

$3x + 3\left(-\frac{1}{2}\right) + 5\left(\frac{1}{2}\right) = 1 \implies x = 0$

Answer: $\left(0, -\frac{1}{2}, \frac{1}{2}\right)$

(b) $D = \begin{vmatrix} 3 & 3 & 5 \\ 3 & 5 & 9 \\ 5 & 9 & 17 \end{vmatrix} = 4$

$x = \dfrac{\begin{vmatrix} 1 & 3 & 5 \\ 2 & 5 & 9 \\ 4 & 9 & 17 \end{vmatrix}}{4} = 0$

$y = \dfrac{\begin{vmatrix} 3 & 1 & 5 \\ 3 & 2 & 9 \\ 5 & 4 & 17 \end{vmatrix}}{4} = -\dfrac{1}{2}$

$z = \dfrac{\begin{vmatrix} 3 & 3 & 1 \\ 3 & 5 & 2 \\ 5 & 9 & 4 \end{vmatrix}}{4} = \dfrac{1}{2}$

Answer: $\left(0, -\frac{1}{2}, \frac{1}{2}\right)$

25. (a)
$$\begin{cases} 2x - y + z = 5 \\ x - 2y - z = 1 \\ 3x + y + z = 4 \end{cases}$$

$$\begin{cases} x - 2y - z = 1 \\ 3y + 3z = 3 \\ 7y + 4z = 1 \end{cases}$$

$$\begin{cases} x - 2y - z = 1 \\ y + z = 1 \\ -3z = -6 \end{cases}$$

$z = 2$

$y + 2 = 1 \implies y = -1$

$x - 2(-1) - 2 = 1 \implies x = 1$

Answer: $(1, -1, 2)$

(b) $D = \begin{vmatrix} 2 & -1 & 1 \\ 1 & -2 & -1 \\ 3 & 1 & 1 \end{vmatrix} = 9$

$x = \dfrac{\begin{vmatrix} 5 & -1 & 1 \\ 1 & -2 & -1 \\ 4 & 1 & 1 \end{vmatrix}}{9} = 1$

$y = \dfrac{\begin{vmatrix} 2 & 5 & 1 \\ 1 & 1 & -1 \\ 3 & 4 & 1 \end{vmatrix}}{9} = -1$

$z = \dfrac{\begin{vmatrix} 2 & -1 & 5 \\ 1 & -2 & 1 \\ 3 & 1 & 4 \end{vmatrix}}{9} = 2$

Answer: $(1, -1, 2)$

27. (a) $D = \begin{vmatrix} 5 & 10 & 30 \\ 10 & 30 & 100 \\ 30 & 100 & 354 \end{vmatrix} = 700$

$$c = \frac{\begin{vmatrix} 5543 & 10 & 30 \\ 12{,}447 & 30 & 100 \\ 38{,}333 & 100 & 354 \end{vmatrix}}{700} = \frac{548{,}580}{700} \approx 784$$

$$b = \frac{\begin{vmatrix} 5 & 5543 & 30 \\ 10 & 12{,}447 & 100 \\ 30 & 38{,}333 & 354 \end{vmatrix}}{700} = \frac{169{,}070}{700} \approx 241.5$$

$$a = \frac{\begin{vmatrix} 5 & 10 & 5543 \\ 10 & 30 & 12{,}447 \\ 30 & 100 & 38{,}333 \end{vmatrix}}{700} = \frac{-18{,}450}{700} \approx -26.36$$

$$y = -26.36t^2 + 241.5t + 784$$

(b)

(c) No, because of the negative t^2 coefficient

29. The uncoded row matrices are the rows of the 6×3 matrix on the left.

$$\begin{matrix} C & A & L \\ L & & M \\ E & & T \\ O & M & O \\ R & R & O \\ W & & \end{matrix} \begin{bmatrix} 3 & 1 & 12 \\ 12 & 0 & 13 \\ 5 & 0 & 20 \\ 15 & 13 & 15 \\ 18 & 18 & 15 \\ 23 & 0 & 0 \end{bmatrix} \begin{bmatrix} 1 & -1 & 0 \\ 1 & 0 & -1 \\ -6 & 2 & 3 \end{bmatrix} = \begin{bmatrix} -68 & 21 & 35 \\ -66 & 14 & 39 \\ -115 & 35 & 60 \\ -62 & 15 & 32 \\ -54 & 12 & 27 \\ 23 & -23 & 0 \end{bmatrix}$$

Answer: $[-68, 21, 35], [-66, 14, 39], [-115, 35, 60]$

$\qquad\quad [-62, 15, 32], [-54, 12, 27], [23, -23, 0]$

31.
$$\begin{matrix} G & O & N \\ E & & F \\ I & S & H \\ I & N & G \end{matrix} \begin{bmatrix} 7 & 15 & 14 \\ 5 & 0 & 6 \\ 9 & 19 & 8 \\ 9 & 14 & 7 \end{bmatrix} \begin{bmatrix} 1 & 2 & 2 \\ 3 & 7 & 9 \\ -1 & -4 & -7 \end{bmatrix} = \begin{bmatrix} 38 & 63 & 51 \\ -1 & -14 & -32 \\ 58 & 119 & 133 \\ 44 & 88 & 95 \end{bmatrix}$$

Cryptogram: 38 63 51 −1 −14 −32 58 119 133 44 88 95

33. $A^{-1} = \begin{bmatrix} 1 & 2 \\ 3 & 5 \end{bmatrix}^{-1} = \begin{bmatrix} -5 & 2 \\ 3 & -1 \end{bmatrix}$

$$\begin{bmatrix} 11 & 21 \\ 64 & 112 \\ 25 & 50 \\ 29 & 53 \\ 23 & 46 \\ 40 & 75 \\ 55 & 92 \end{bmatrix} \begin{bmatrix} -5 & 2 \\ 3 & -1 \end{bmatrix} = \begin{bmatrix} 8 & 1 \\ 16 & 16 \\ 25 & 0 \\ 14 & 5 \\ 23 & 0 \\ 25 & 5 \\ 1 & 18 \end{bmatrix} \begin{matrix} H & A \\ P & P \\ Y & \\ N & E \\ W & \\ Y & E \\ A & R \end{matrix}$$

Message: HAPPY NEW YEAR

35. $A^{-1} = \begin{bmatrix} 1 & 2 & 1 \\ -1 & 0 & 2 \\ 1 & -1 & -2 \end{bmatrix}^{-1} = \begin{bmatrix} \frac{2}{3} & 1 & \frac{4}{3} \\ 0 & -1 & -1 \\ \frac{1}{3} & 1 & \frac{2}{3} \end{bmatrix}$

$\begin{bmatrix} 38 & 36 & -1 \\ 11 & 17 & 11 \\ 42 & 15 & -27 \\ -5 & 18 & 37 \\ 26 & 28 & 17 \\ 8 & 24 & 20 \\ 32 & 20 & -7 \\ 23 & -1 & -19 \end{bmatrix} \begin{bmatrix} \frac{2}{3} & 1 & \frac{4}{3} \\ 0 & -1 & -1 \\ \frac{1}{3} & 1 & \frac{2}{3} \end{bmatrix} = \begin{bmatrix} 25 & 1 & 14 \\ 11 & 5 & 5 \\ 19 & 0 & 23 \\ 9 & 14 & 0 \\ 23 & 15 & 18 \\ 12 & 4 & 0 \\ 19 & 5 & 18 \\ 9 & 5 & 19 \end{bmatrix} \quad \begin{matrix} Y & A & N \\ K & E & E \\ S & - & W \\ I & N & - \\ W & O & R \\ L & D & - \\ S & E & R \\ I & E & S \end{matrix}$

YANKEES WIN WORLD SERIES

37. True. Cramer's Rule requires that the determinant of the coefficient matrix be nonzero.

39. Answers will vary.

41.
$$y - 5 = \frac{5 - 3}{-1 - 7}(x + 1) = \frac{-1}{4}(x + 1)$$
$$4y - 20 = -x - 1$$
$$4y + x = 19$$
$$x + 4y - 19 = 0$$

43.
$$y + 3 = \frac{-3 + 1}{3 - 10}(x - 3) = \frac{2}{7}(x - 3)$$
$$7y + 21 = 2x - 6$$
$$7y - 2x = -27$$
$$2x - 7y - 27 = 0$$

45. $f(x) = \frac{2x^2}{x^2 + 4}$

Horizontal asymptote: $y = 2$

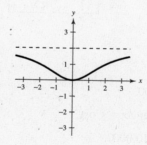

Review Exercises for Chapter 5

1. $\begin{cases} x + y = 2 \implies & y = 2 - x \\ x - y = 0 \implies & x - (2 - x) = 0 \end{cases}$

$$2x - 2 = 0$$
$$x = 1$$
$$y = 2 - 1 = 1$$

Answer: $(1, 1)$

3. $\begin{cases} x^2 - y^2 = 9 \\ x - y = 1 \implies & x = y + 1 \end{cases}$

$$(y + 1)^2 - y^2 = 9$$
$$2y + 1 = 9$$
$$y = 4$$
$$x = 5$$

Answer: $(5, 4)$

5. $\begin{cases} y = 2x^2 \\ y = x^4 - 2x^2 \end{cases} \Rightarrow \; 2x^2 = x^4 - 2x^2$

$\qquad\qquad\qquad\qquad 0 = x^4 - 4x^2$

$\qquad\qquad\qquad\qquad 0 = x^2(x^2 - 4)$

$\qquad\qquad\qquad\qquad 0 = x^2(x + 2)(x - 2)$

$\qquad\qquad\qquad\qquad x = 0, x = -2, x = 2$

$\qquad\qquad\qquad\qquad y = 0, y = 8, y = 8$

Answer: $(0, 0), (-2, 8), (2, 8)$

7. $\begin{cases} 5x + 6y = 7 \;\Rightarrow\; y_1 = \dfrac{1}{6}(7 - 5x) \\ -x - 4y = 0 \;\Rightarrow\; y_2 = -\dfrac{x}{4} \end{cases}$

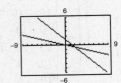

Answer: $\left(2, -\dfrac{1}{2}\right)$

9. $\begin{cases} y^2 - 4x = 0 \;\Rightarrow\; y^2 = 4x \Rightarrow y = \pm 2\sqrt{x} \\ x + y = 0 \;\Rightarrow\; y = -x \end{cases}$

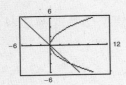

Points of intersection: $(0, 0), (4, -4)$

11. $\begin{cases} y = 3 - x^2 \\ y = 2x^2 + x + 1 \end{cases}$

Points of intersection:

$(0.67, 2.56), (-1, 2)$

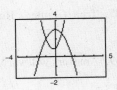

13. $\begin{cases} y = 2(6 - x) \\ y = 2^{x-2} \end{cases}$

Point of intersection: $(4, 4)$

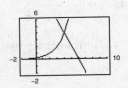

15. $\begin{cases} y = \ln(x + 2) + 1 \\ y = -x \end{cases}$

Point of intersection: $(-1, 1)$

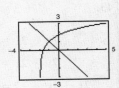

17. Revenue $= 4.95x$

Cost $= 2.85x + 10,000$

Break even when Revenue $=$ Cost

$\qquad 4.95x = 2.85x + 10,000$

$\qquad 2.10x = 10,000$

$\qquad\quad x \approx 4762$ units

19. $\begin{cases} 2l + 2 = 480 \\ \quad l = 1.50w \end{cases}$

$\qquad 2(1.50w) + 2w = 480$

$\qquad\qquad\qquad 5w = 480$

$\qquad\qquad\qquad w = 96$

$\qquad\qquad\qquad l = 144$

The dimensions are 96×144 meters.

21. $\begin{cases} 2x - y = 2 \;\Rightarrow\; 16x - 8y = 16 \\ 6x + 8y = 39 \;\Rightarrow\; 6x + 8y = 39 \end{cases}$

$\qquad\qquad\qquad 22x \qquad\quad = 55$

$\qquad\qquad\qquad\quad x = \dfrac{55}{22} = \dfrac{5}{2}$

$\qquad\qquad\qquad\quad y = 3$

Answer: $\left(\dfrac{5}{2}, 3\right)$

23. $\begin{cases} 1/5x + 3/10y = 7/50 \\ 2/5x + 1/2y = 1/5 \end{cases} \Rightarrow \begin{array}{l} 20x + 30y = 14 \\ 4x + 5y = 2 \end{array} \Rightarrow \begin{array}{r} 20x + 30y = 14 \\ -20x - 25y = -10 \end{array}$

$$\begin{array}{r} 5y = 4 \\ y = \tfrac{4}{5} \\ x = -\tfrac{1}{2} \end{array}$$

Answer: $\left(-\tfrac{1}{2}, \tfrac{4}{5}\right)$ or $(-0.5, 0.8)$

25. $\begin{cases} 3x - 2y = 0 \\ 3x + 2(y + 5) = 10 \end{cases} \Rightarrow \begin{array}{l} 3x - 2y = 0 \\ 3x + 2y = 0 \end{array}$

$$\begin{array}{r} 6x \qquad = 0 \\ x = 0 \\ y = 0 \end{array}$$

Answer: $(0, 0)$

27. $\begin{cases} 1.25x - 2y = 3.5 \\ 5x - 8y = 14 \end{cases} \Rightarrow \begin{array}{r} 5x - 8y = 14 \\ -5x + 8y = -14 \end{array}$

$$0 = 0$$

Infinite number of solutions

Let $y = a$, then $5x - 8a = 14 \Rightarrow x = \tfrac{14}{5} + \tfrac{8}{5}a$.

Answer: $\left(\tfrac{14}{5} + \tfrac{8}{5}a, a\right)$

29. $\begin{cases} 3x + 2y = 0 \\ x - y = 4 \end{cases} \Rightarrow \begin{array}{l} y = -\tfrac{3}{2}x \\ y = x - 4 \end{array}$

Consistent

Answer: $(1.6, -2.4)$

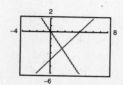

31. $\begin{cases} \tfrac{1}{4}x - \tfrac{1}{5}y = 2 \\ -5x + 4y = 8 \end{cases} \Rightarrow \begin{array}{l} y = \tfrac{5}{4}x - 10 \\ y = \tfrac{1}{4}(8 + 5x) = \tfrac{5}{4}x + 2 \end{array}$

Inconsistent; Lines are parallel.

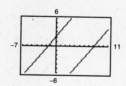

33. $\begin{cases} 2x - 2y = 8 \\ 4x - 1.5y = -5.5 \end{cases} \Rightarrow \begin{array}{l} y = x - 4 \\ y = \tfrac{8}{3}x + \tfrac{11}{3} \end{array}$

Consistent

Answer: $(-4.6, -8.6)$

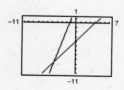

35. Demand = Supply

$$37 - 0.0002x = 22 + 0.00001x$$

$$15 = 0.00021x$$

$$x = \frac{500,000}{7}, p = \frac{159}{7}$$

Point of equilibrium: $\left(\dfrac{500,000}{7}, \dfrac{159}{7}\right)$

37. Let $x =$ speed of the slower plane.

Let $y =$ speed of the faster plane.

Then, distance of first plane + distance of second plane = 275 miles.

(rate of first plane)(time) + (rate of second plane)(time) = 275 miles

$$\begin{cases} x\left(\frac{40}{60}\right) + y\left(\frac{40}{60}\right) = 275 \\ \qquad\qquad y = x + 25 \end{cases}$$

$$\frac{2}{3}x + \frac{2}{3}(x + 25) = 275$$

$$4x + 50 = 825$$

$$4x = 775$$

$$x = 193.75 \text{ mph}$$

$$y = x + 25 = 218.75 \text{ mph}$$

39. $z = -2$

$-y + z = -5 \implies -y - 2 = -5 \implies y = 3$

$x - 4y + 3z = -14$

$$\implies x = 4(3) - 3(-2) - 14 = 4$$

Answer: $(4, 3, -2)$

41. $\begin{cases} x + 3y - z = 13 \\ 2x \qquad - 5z = 23 \\ 4x - y - 2z = 14 \end{cases}$

$\begin{cases} x + 3y - z = 13 \\ \quad - 6y - 3z = -3 \\ \quad - 13y + 2z = -38 \end{cases}$

$\begin{cases} x + 3y - z = 13 \\ \quad - 6y - 3z = -3 \\ \qquad\quad \frac{17}{2}z = -\frac{63}{2} \end{cases}$

$\frac{17}{2}z = -\frac{63}{2} \implies z = -\frac{63}{17}$

$-6y - 3\left(-\frac{63}{17}\right) = -3 \implies y = \frac{40}{17}$

$x + 3\left(\frac{40}{17}\right) - \left(-\frac{63}{17}\right) = 13 \implies x = \frac{38}{17}$

Answer: $\left(\frac{38}{17}, \frac{40}{17}, -\frac{63}{17}\right)$

43. $\begin{cases} x - 2y + z = -6 \\ 2x - 3y \qquad = -7 \\ -x + 3y - 3z = 11 \end{cases}$

$\begin{cases} x - 2y + z = -6 \\ \quad y - 2z = 5 \\ \quad y - 2z = 5 \end{cases}$ $\quad\begin{matrix} -2\text{ Eq.1 } + \text{ Eq. 2} \\ \text{Eq. 1 } + \text{ Eq. 3} \end{matrix}$

$\begin{cases} x - 2y + z = -6 \\ \quad y - 2z = 5 \\ \qquad\quad 0 = 0 \end{cases}$ $\quad -\text{Eq. 2 } + \text{ Eq. 3}$

Let $z = a$, then $y = 2a + 5$.

$x - 2(2a + 5) + a = -6$

$x - 3a - 10 = -6$

$x = 3a + 4$

Answer: $(3a + 4, 2a + 5, a)$ where a is any real number.

45. $\begin{cases} x - 2y + 3z = -5 \\ 2x + 4y + 5z = 1 \\ x + 2y + z = 0 \end{cases}$

$\begin{cases} x - 2y + 3z = -5 \\ \quad 8y - z = 11 \\ \quad 4y - 2z = 5 \end{cases}$

$\begin{cases} x - 2y + 3z = -5 \\ \quad 4y - 2z = 5 \\ \qquad\quad 3z = 1 \end{cases}$

$z = \frac{1}{3}$

$4y = 5 + 2\left(\frac{1}{3}\right) = \frac{17}{3} \implies y = \frac{17}{12}$

$x = 2\left(\frac{17}{12}\right) - 3\left(\frac{1}{3}\right) - 5 = -\frac{19}{6}$

Answer: $\left(-\frac{19}{6}, \frac{17}{12}, \frac{1}{3}\right)$

47. $\begin{cases} 5x - 12y + 7z = 16 & \text{Equation 1} \\ 3x - 7y + 4z = 9 & \text{Equation 2} \end{cases}$

3 times Eq. 1 and (-5) times Eq. 2:

$\begin{cases} 15x - 36y + 21z = 48 \\ -15x + 35y - 20z = -45 \end{cases}$

Adding, $-y + z = 3 \implies y = z - 3.$

$$5x - 12(z - 3) + 7z = 16$$
$$5x - 5z + 36 = 16$$
$$5x = 5z - 20$$
$$x = z - 4$$

Let $z = a$, then $x = a - 4$ and $y = a - 3$.

Answer: $(a - 4, a - 3, a)$ where a is any real number.

49.

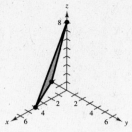

$(4, 0, 0), (0, -2, 0), (0, 0, 8), (1, 0, 6)$

51. $\dfrac{4 - x}{x^2 + 6x + 8} = \dfrac{A}{x + 2} + \dfrac{B}{x + 4}$

$4 - x = A(x + 4) + B(x + 2)$

$ = (A + B)x + (4A + 2B)$

$\begin{cases} A + B = -1 \\ 4A + 2B = 4 \end{cases} \implies A = 3, B = -4$

$\dfrac{4 - x}{x^2 + 6x + 8} = \dfrac{3}{x + 2} - \dfrac{4}{x + 4}$

53. $\dfrac{x^2 + 2x}{x^3 - x^2 + x - 1} = \dfrac{A}{x - 1} + \dfrac{Bx + C}{x^2 + 1}$

$x^2 + 2x = A(x^2 + 1) + (Bx + C)(x - 1)$

$ = (A + B)x^2 + (C - B)x + (A - C)$

$\begin{cases} A + B = 1 \\ -B + C = 2 \\ A - C = 0 \end{cases} \implies A = \tfrac{3}{2}, B = -\tfrac{1}{2}, C = \tfrac{3}{2}$

$\dfrac{x^2 + 2x}{x^3 - x^2 + x - 1} = \dfrac{3/2}{x - 1} + \dfrac{-(1/2)x + 3/2}{x^2 + 1}$

$\phantom{\dfrac{x^2 + 2x}{x^3 - x^2 + x - 1}} = \dfrac{1}{2}\left(\dfrac{3}{x - 1} - \dfrac{x - 3}{x^2 + 1}\right)$

55. $\dfrac{x^2 + 3x - 3}{x^3 + 2x^2 + x + 2} = \dfrac{x^2 + 3x - 3}{(x + 2)(x^2 + 1)}$

$\phantom{\dfrac{x^2 + 3x - 3}{x^3 + 2x^2 + x + 2}} = \dfrac{A}{x + 2} + \dfrac{Bx + C}{x^2 + 1}$

$x^2 + 3x - 3 = A(x^2 + 1) + (Bx + C)(x + 2)$

$\begin{cases} A + B = 1 \\ 2B + C = 3 \\ A + 2C = -3 \end{cases}$

$\begin{cases} A + B = 1 \\ 2B + C = 3 \\ -B + 2C = -4 \end{cases}$

$\begin{cases} A + B = 1 \\ B - 2C = 4 \\ 5C = -5 \end{cases}$

$C = -1, B = 2(-1) + 4 = 2, A = 1 - 2 = -1$

$\dfrac{x^2 + 3x - 3}{x^3 + 2x^2 + x + 2} = \dfrac{-1}{x + 2} + \dfrac{2x - 1}{x^2 + 1}$

57. $y = ax^2 + bx + c$

$\begin{array}{lrrrrrr} (-1, -4): & a & - & b & + & c & = -4 \\ (1, -2): & a & + & b & + & c & = -2 \\ (2, 5): & 4a & + & 2b & + & c & = 5 \end{array}$

Solving the system, $a = 2, b = 1, c = -5$.

$y = 2x^2 + x - 5$

59. Let x = gallons of spray X

Let y = gallons of spray Y

Let z = gallons of spray Z

$$\begin{cases} \text{Chemical A: } \frac{1}{5}x + \qquad \frac{1}{3}z = 6 \\ \text{Chemical B: } \frac{2}{5}x + \qquad \frac{1}{3}z = 8 \\ \text{Cehmical C: } \frac{2}{5}x + y + \frac{1}{3}z = 13 \end{cases}$$

Subtracting Eq. 2 − Eq. 1 gives $\frac{1}{5}x = 2 \implies x = 10$.

Then $z = 12$ and $y = 5$.

Answer: 10 gallons of spray X

5 gallons of spray Y

12 gallons of spray Z

61. Order 3×1

63. Order 1×1

65. $\begin{bmatrix} 3 & -10 & \vdots & 15 \\ 5 & 4 & \vdots & 22 \end{bmatrix}$

67. $\begin{bmatrix} 8 & -7 & 4 & \vdots & 12 \\ 3 & -5 & 2 & \vdots & 20 \\ 5 & 3 & -3 & \vdots & 26 \end{bmatrix}$

69. $\begin{bmatrix} 5 & 1 & 7 & \vdots & -9 \\ 4 & 2 & 0 & \vdots & 10 \\ 9 & 4 & 2 & \vdots & 3 \end{bmatrix}$

$$\begin{cases} 5x + y + 7z = -9 \\ 4x + 2y \qquad = 10 \\ 9x + 4y + 2z = 3 \end{cases}$$

71.
$$\begin{bmatrix} 0 & 1 & 1 \\ 1 & 2 & 3 \\ 2 & 2 & 2 \end{bmatrix}$$

$$\begin{matrix} R_1 + R_2 \to \\ -R_1 + R_2 \to \\ -2R_1 + R_3 \to \end{matrix} \begin{bmatrix} 1 & 3 & 4 \\ 0 & -1 & -1 \\ 0 & -4 & -6 \end{bmatrix}$$

$$\begin{matrix} 3R_2 + R_1 \to \\ -R_2 \to \\ -4R_2 + R_3 \to \end{matrix} \begin{bmatrix} 1 & 0 & 1 \\ 0 & 1 & 1 \\ 0 & 0 & -2 \end{bmatrix}$$

73. $\begin{bmatrix} 3 & -2 & 1 & 0 \\ 4 & -3 & 0 & 1 \end{bmatrix} \implies \begin{bmatrix} 1 & 0 & 3 & -2 \\ 0 & 1 & 4 & -3 \end{bmatrix}$

75. $\begin{bmatrix} 1.5 & 3.6 & 4.2 \\ 0.2 & 1.4 & 1.8 \\ 2.0 & 4.4 & 6.4 \end{bmatrix} \implies \begin{bmatrix} 1 & 0 & 0 \\ 0 & 1 & 0 \\ 0 & 0 & 1 \end{bmatrix}$

77.
$$\begin{bmatrix} 5 & 4 & \vdots & 2 \\ -1 & 1 & \vdots & -22 \end{bmatrix}$$

$$\begin{matrix} 4R_2 + R_1 \to \\ R_1 + R_2 \to \end{matrix} \begin{bmatrix} 1 & 8 & \vdots & -86 \\ 0 & 9 & \vdots & -108 \end{bmatrix}$$

$9y = -108$

$y = -12$

$x = -8(-12) - 86 = 10$

Answer: $(10, -12)$

79.
$$\begin{bmatrix} 2 & 1 & \vdots & 0.3 \\ 3 & -1 & \vdots & -1.3 \end{bmatrix}$$

$$-R_1 + R_2 \to \begin{bmatrix} 2 & 1 & \vdots & 0.3 \\ 1 & -2 & \vdots & -1.6 \end{bmatrix}$$

$$\begin{bmatrix} 1 & -2 & \vdots & -1.6 \\ 2 & 1 & \vdots & 0.3 \end{bmatrix}$$

$$-2R_1 + R_2 \to \begin{bmatrix} 1 & -2 & \vdots & -1.6 \\ 0 & 5 & \vdots & 3.5 \end{bmatrix}$$

$5y = 3.5 \implies y = 0.7$

$x = 2(0.7) - 1.6 = -0.2$

$x = -0.2, \ y = 0.7$

Answer: $(-0.2, 0.7)$

81.

$$\begin{bmatrix} 2 & 3 & 3 & \vdots & 3 \\ 6 & 6 & 12 & \vdots & 13 \\ 12 & 9 & -1 & \vdots & 2 \end{bmatrix}$$

$$\begin{matrix} \\ -3R_1 + R_2 \rightarrow \\ -6R_1 + R_3 \rightarrow \end{matrix} \begin{bmatrix} 2 & 3 & 3 & \vdots & 3 \\ 0 & -3 & 3 & \vdots & 4 \\ 0 & -9 & -19 & \vdots & -16 \end{bmatrix}$$

$$\begin{matrix} R_2 + R_1 \rightarrow \\ \\ -3R_2 + R_3 \rightarrow \end{matrix} \begin{bmatrix} 2 & 0 & 6 & \vdots & 7 \\ 0 & -3 & 3 & \vdots & 4 \\ 0 & 0 & -28 & \vdots & -28 \end{bmatrix}$$

$$\begin{matrix} \frac{1}{2}R_1 \rightarrow \\ -\frac{1}{3}R_2 \rightarrow \\ -\frac{1}{28}R_3 \rightarrow \end{matrix} \begin{bmatrix} 1 & 0 & 3 & \vdots & \frac{7}{2} \\ 0 & 1 & -1 & \vdots & -\frac{4}{3} \\ 0 & 0 & 1 & \vdots & 1 \end{bmatrix}$$

$z = 1$

$y - 1 = -\frac{4}{3} \implies y = -\frac{1}{3}$

$x + 3(1) = \frac{7}{2} \implies x = \frac{1}{2}$

Answer: $\left(\frac{1}{2}, -\frac{1}{3}, 1\right)$

83. $\begin{cases} x + 2y - z = 1 \\ \quad\; y + z = 0 \end{cases}$

$$\begin{bmatrix} 1 & 2 & -1 & \vdots & 1 \\ 0 & 1 & 1 & \vdots & 0 \end{bmatrix} \Rightarrow \begin{bmatrix} 1 & 0 & -3 & \vdots & 1 \\ 0 & 1 & 1 & \vdots & 0 \end{bmatrix}$$

$y = -z$

$x = 1 + 3z$

Answer: $(1 + 3a, -a, a)$, a is a real number.

85.

$$\begin{bmatrix} -1 & 1 & 2 & \vdots & 1 \\ 2 & 3 & 1 & \vdots & -2 \\ 5 & 4 & 2 & \vdots & 4 \end{bmatrix}$$

$$\begin{matrix} -R_1 \rightarrow \\ 2R_1 + R_2 \rightarrow \\ 5R_1 + R_3 \rightarrow \end{matrix} \begin{bmatrix} 1 & -1 & -2 & \vdots & -1 \\ 0 & 5 & 5 & \vdots & 0 \\ 0 & 9 & 12 & \vdots & 9 \end{bmatrix}$$

$$\begin{matrix} \\ \frac{1}{5}R_2 \rightarrow \\ \\ \end{matrix} \begin{bmatrix} 1 & -1 & -2 & \vdots & -1 \\ 0 & 1 & 1 & \vdots & 0 \\ 0 & 9 & 12 & \vdots & 9 \end{bmatrix}$$

$$\begin{matrix} R_2 + R_1 \rightarrow \\ \\ -9R_2 + R_3 \rightarrow \end{matrix} \begin{bmatrix} 1 & 0 & -1 & \vdots & -1 \\ 0 & 1 & 1 & \vdots & 0 \\ 0 & 0 & 3 & \vdots & 9 \end{bmatrix}$$

$$\begin{matrix} \\ \\ \frac{1}{3}R_3 \rightarrow \end{matrix} \begin{bmatrix} 1 & 0 & -1 & \vdots & -1 \\ 0 & 1 & 1 & \vdots & 0 \\ 0 & 0 & 1 & \vdots & 3 \end{bmatrix}$$

$$\begin{matrix} R_3 + R_1 \rightarrow \\ -R_3 + R_2 \rightarrow \\ \\ \end{matrix} \begin{bmatrix} 1 & 0 & 0 & \vdots & 2 \\ 0 & 1 & 0 & \vdots & -3 \\ 0 & 0 & 1 & \vdots & 3 \end{bmatrix}$$

$x = 2, y = -3, z = 3$

Answer: $(2, -3, 3)$

87. $\begin{cases} x + y + 2z = 4 \\ x - y + 4z = 1 \\ 2x - y + 2z = 1 \end{cases}$ $\begin{bmatrix} 1 & 1 & 2 & \vdots & 4 \\ 1 & -1 & 4 & \vdots & 1 \\ 2 & -1 & 2 & \vdots & 1 \end{bmatrix} \Rightarrow \begin{bmatrix} 1 & 0 & 0 & \vdots & 1 \\ 0 & 1 & 0 & \vdots & 2 \\ 0 & 0 & 1 & \vdots & \frac{1}{2} \end{bmatrix}$

Answer: $\left(1, 2, \frac{1}{2}\right)$

89. $\begin{bmatrix} 1 & 2 & -1 & \vdots & 7 \\ 0 & -1 & -1 & \vdots & 4 \\ 4 & 0 & -1 & \vdots & 16 \end{bmatrix}$ reduces to $\begin{bmatrix} 1 & 0 & 0 & \vdots & 3 \\ 0 & 1 & 0 & \vdots & 0 \\ 0 & 0 & 1 & \vdots & -4 \end{bmatrix}$.

Answer: $(3, 0, -4)$

91. $\begin{bmatrix} 3 & -1 & 5 & -2 & \vdots & -44 \\ 1 & 6 & 4 & -1 & \vdots & 1 \\ 5 & -1 & 1 & 3 & \vdots & -15 \\ 0 & 4 & -1 & -8 & \vdots & 58 \end{bmatrix}$ reduces to $\begin{bmatrix} 1 & 0 & 0 & 0 & \vdots & 2 \\ 0 & 1 & 0 & 0 & \vdots & 6 \\ 0 & 0 & 1 & 0 & \vdots & -10 \\ 0 & 0 & 0 & 1 & \vdots & -3 \end{bmatrix}$.

Answer: $(2, 6, -10, -3)$

93. $x = 12$

$y = -7$

95. $x + 3 = 5x - 1 \Rightarrow x = 1$

$-4y = -44 \quad \Rightarrow y = 11$

$y + 5 = 16 \quad \Rightarrow y = 11$

$6x = 6 \quad \Rightarrow x = 1$

Answer: $x = 1, y = 11$

97. (a) $A + B = \begin{bmatrix} 7 & 3 \\ -1 & 5 \end{bmatrix} + \begin{bmatrix} 10 & -20 \\ 14 & -3 \end{bmatrix} = \begin{bmatrix} 17 & -17 \\ 13 & 2 \end{bmatrix}$

(b) $A - B = \begin{bmatrix} -3 & 23 \\ -15 & 8 \end{bmatrix}$

(c) $4A = \begin{bmatrix} 28 & 12 \\ -4 & 20 \end{bmatrix}$

(d) $A + 3B = \begin{bmatrix} 7 & 3 \\ -1 & 5 \end{bmatrix} + \begin{bmatrix} 30 & -60 \\ 42 & -9 \end{bmatrix} = \begin{bmatrix} 37 & -57 \\ 41 & -4 \end{bmatrix}$

99. (a) $A + B = \begin{bmatrix} 6 & 0 & 7 \\ 5 & -1 & 2 \\ 3 & 2 & 3 \end{bmatrix} + \begin{bmatrix} 0 & 5 & 1 \\ -4 & 8 & 6 \\ 2 & -1 & 1 \end{bmatrix} = \begin{bmatrix} 6 & 5 & 8 \\ 1 & 7 & 8 \\ 5 & 1 & 4 \end{bmatrix}$

(b) $A - B = \begin{bmatrix} 6 & -5 & 6 \\ 9 & -9 & -4 \\ 1 & 3 & 2 \end{bmatrix}$

(c) $4A = \begin{bmatrix} 24 & 0 & 28 \\ 20 & -4 & 8 \\ 12 & 8 & 12 \end{bmatrix}$

(d) $A + 3B = \begin{bmatrix} 6 & 0 & 7 \\ 5 & -1 & 2 \\ 3 & 2 & 3 \end{bmatrix} + \begin{bmatrix} 0 & 15 & 3 \\ -12 & 24 & 18 \\ 6 & -3 & 3 \end{bmatrix} = \begin{bmatrix} 6 & 15 & 10 \\ -7 & 23 & 20 \\ 9 & -1 & 6 \end{bmatrix}$

101. $\begin{bmatrix} 2 & 1 & 0 \\ 0 & 5 & -4 \end{bmatrix} - 3\begin{bmatrix} 5 & 3 & -6 \\ 0 & -2 & 5 \end{bmatrix} = \begin{bmatrix} 2 & 1 & 0 \\ 0 & 5 & -4 \end{bmatrix} - \begin{bmatrix} 15 & 9 & -18 \\ 0 & -6 & 15 \end{bmatrix}$

$$= \begin{bmatrix} -13 & -8 & 18 \\ 0 & 11 & -19 \end{bmatrix}$$

103. $-\begin{bmatrix} 8 & -1 \\ -2 & 4 \end{bmatrix} - 5\begin{bmatrix} -2 & 0 \\ 3 & -1 \end{bmatrix} + \begin{bmatrix} 7 & -8 \\ 4 & 3 \end{bmatrix} = \begin{bmatrix} -8 & 1 \\ 2 & -4 \end{bmatrix} - \begin{bmatrix} -10 & 0 \\ 15 & -5 \end{bmatrix} + \begin{bmatrix} 7 & -8 \\ 4 & 3 \end{bmatrix}$

$$= \begin{bmatrix} 9 & -7 \\ -9 & 4 \end{bmatrix}$$

105. $\begin{bmatrix} 32 & -\frac{17}{2} & -\frac{3}{2} \\ 6 & 46 & 33 \end{bmatrix}$

107. $X = 3A - 2B = 3\begin{bmatrix} -4 & 0 \\ 1 & -5 \\ -3 & 2 \end{bmatrix} - 2\begin{bmatrix} 1 & 2 \\ -2 & 1 \\ 4 & 4 \end{bmatrix} = \begin{bmatrix} -14 & -4 \\ 7 & -17 \\ -17 & -2 \end{bmatrix}$

109. $X = \frac{1}{3}[B - 2A] = \frac{1}{3}\left(\begin{bmatrix} 1 & 2 \\ -2 & 1 \\ 4 & 4 \end{bmatrix} - 2\begin{bmatrix} -4 & 0 \\ 1 & -5 \\ -3 & 2 \end{bmatrix}\right) = \frac{1}{3}\begin{bmatrix} 9 & 2 \\ -4 & 11 \\ 10 & 0 \end{bmatrix}$

111. $\begin{bmatrix} 1 & 2 \\ 5 & -4 \\ 6 & 0 \end{bmatrix}\begin{bmatrix} 6 & -2 & 8 \\ 4 & 0 & 0 \end{bmatrix} = \begin{bmatrix} 1(6) + 2(4) & 1(-2) + 2(0) & 1(8) + 2(0) \\ 5(6) + (-4)(4) & 5(-2) + (-4)(0) & 5(8) + (-4)(0) \\ 6(6) + (0)(4) & 6(-2) + (0)(0) & 6(8) + (0)(0) \end{bmatrix} = \begin{bmatrix} 14 & -2 & 8 \\ 14 & -10 & 40 \\ 36 & -12 & 48 \end{bmatrix}$

113. $AB = \begin{bmatrix} 3 & -2 & 0 \\ 1 & 4 & 9 \end{bmatrix}\begin{bmatrix} 7 & 0 \\ 5 & 3 \\ -1 & 3 \end{bmatrix} = \begin{bmatrix} 11 & -6 \\ 18 & 39 \end{bmatrix}$ **115.** $\begin{bmatrix} 4 & 1 \\ 11 & -7 \\ 12 & 3 \end{bmatrix}\begin{bmatrix} 3 & -5 & 6 \\ 2 & -2 & -2 \end{bmatrix} = \begin{bmatrix} 14 & -22 & 22 \\ 19 & -41 & 80 \\ 42 & -66 & 66 \end{bmatrix}$

117. $\begin{bmatrix} 2 & 1 \\ 6 & 0 \end{bmatrix}\left(\begin{bmatrix} 4 & 2 \\ -3 & 1 \end{bmatrix} + \begin{bmatrix} -2 & 4 \\ 0 & 4 \end{bmatrix}\right) = \begin{bmatrix} 2 & 1 \\ 6 & 0 \end{bmatrix}\begin{bmatrix} 2 & 6 \\ -3 & 5 \end{bmatrix}$

$$= \begin{bmatrix} 2(2) + 1(-3) & 2(6) + 1(5) \\ 6(2) + 0 & 6(6) + 0 \end{bmatrix}$$

$$= \begin{bmatrix} 1 & 17 \\ 12 & 36 \end{bmatrix}$$

119. (a) $AB = \begin{bmatrix} 40 & 64 & 52 \\ 60 & 82 & 76 \\ 76 & 96 & 84 \end{bmatrix}\begin{bmatrix} 2.65 & 0.25 \\ 2.81 & 0.30 \\ 2.93 & 0.35 \end{bmatrix}$

$$= \begin{bmatrix} 438.2 & 47.4 \\ 612.1 & 66.2 \\ 717.28 & 77.2 \end{bmatrix}$$

This is the sales and profit for milk on Friday, Saturday and Sunday.

(b) Profit = 47.4 + 66.2 + 77.2 = \$190.80

121. $AB = \begin{bmatrix} -4 & -1 \\ 7 & 2 \end{bmatrix} \begin{bmatrix} -2 & -1 \\ 7 & 4 \end{bmatrix} = \begin{bmatrix} 1 & 0 \\ 0 & 1 \end{bmatrix}$; $BA = I_2$

123. $\begin{bmatrix} -6 & 5 & \vdots & 1 & 0 \\ -5 & 4 & \vdots & 0 & 1 \end{bmatrix}$ row reduces to $\begin{bmatrix} 1 & 0 & \vdots & 4 & -5 \\ 0 & 1 & \vdots & 5 & -6 \end{bmatrix}$.

$\begin{bmatrix} -6 & 5 \\ -5 & 4 \end{bmatrix}^{-1} = \begin{bmatrix} 4 & -5 \\ 5 & -6 \end{bmatrix}$

125. $\begin{bmatrix} -1 & -2 & -2 & \vdots & 1 & 0 & 0 \\ 3 & 7 & 9 & \vdots & 0 & 1 & 0 \\ 1 & 4 & 7 & \vdots & 0 & 0 & 1 \end{bmatrix}$ row reduces to $\begin{bmatrix} 1 & 0 & 0 & \vdots & 13 & 6 & -4 \\ 0 & 1 & 0 & \vdots & -12 & -5 & 3 \\ 0 & 0 & 1 & \vdots & 5 & 2 & -1 \end{bmatrix}$.

$\begin{bmatrix} -1 & -2 & -2 \\ 3 & 7 & 9 \\ 1 & 4 & 7 \end{bmatrix}^{-1} = \begin{bmatrix} 13 & 6 & -4 \\ -12 & -5 & 3 \\ 5 & 2 & -1 \end{bmatrix}$

127. $\begin{bmatrix} 2 & 6 \\ 3 & -6 \end{bmatrix}^{-1} = \begin{bmatrix} \frac{1}{5} & \frac{1}{5} \\ \frac{1}{10} & -\frac{1}{15} \end{bmatrix}$

129. $\begin{bmatrix} 1 & 2 & 0 \\ -1 & 1 & 1 \\ 0 & -1 & 0 \end{bmatrix}^{-1} = \begin{bmatrix} 1 & 0 & 2 \\ 0 & 0 & -1 \\ 1 & 1 & 3 \end{bmatrix}$

131. $\begin{bmatrix} -7 & 2 \\ -8 & 2 \end{bmatrix}^{-1} = \frac{1}{(-7)(2) - (2)(-8)} \begin{bmatrix} 2 & -2 \\ 8 & -7 \end{bmatrix} = \begin{bmatrix} 1 & -1 \\ 4 & -\frac{7}{2} \end{bmatrix}$

133. $\begin{bmatrix} -1 & 10 \\ 2 & 20 \end{bmatrix}^{-1} = \frac{1}{(-1)(20) - 10(2)} \begin{bmatrix} 20 & -10 \\ -2 & -1 \end{bmatrix} = \frac{1}{-40} \begin{bmatrix} 20 & -10 \\ -2 & -1 \end{bmatrix} = \begin{bmatrix} -\frac{1}{2} & \frac{1}{4} \\ \frac{1}{20} & \frac{1}{40} \end{bmatrix}$

135. $x + 5y = -1$
$3x - 5y = 5$

$\begin{bmatrix} 1 & 5 \\ 3 & -5 \end{bmatrix}^{-1} = \begin{bmatrix} \frac{1}{4} & \frac{1}{4} \\ \frac{3}{20} & -\frac{1}{20} \end{bmatrix}$

$\begin{bmatrix} x \\ y \end{bmatrix} = \begin{bmatrix} \frac{1}{4} & \frac{1}{4} \\ \frac{3}{20} & -\frac{1}{20} \end{bmatrix} \begin{bmatrix} -1 \\ 5 \end{bmatrix} = \begin{bmatrix} 1 \\ -\frac{2}{5} \end{bmatrix}$

Answer: $\left(1, -\frac{2}{5}\right)$

137. $\begin{bmatrix} 3 & 2 & -1 \\ 1 & -1 & 2 \\ 5 & 1 & 1 \end{bmatrix}^{-1} = \begin{bmatrix} -1 & -1 & 1 \\ 3 & \frac{8}{3} & -\frac{7}{3} \\ 2 & \frac{7}{3} & -\frac{5}{3} \end{bmatrix}$

$\begin{bmatrix} x \\ y \\ z \end{bmatrix} = \begin{bmatrix} -1 & -1 & 1 \\ 3 & \frac{8}{3} & -\frac{7}{3} \\ 2 & \frac{7}{3} & -\frac{5}{3} \end{bmatrix} \begin{bmatrix} 6 \\ -1 \\ 7 \end{bmatrix} = \begin{bmatrix} 2 \\ -1 \\ -2 \end{bmatrix}$

Answer: $(2, -1, -2)$

139. $\begin{bmatrix} 1 & 2 & 1 & -1 \\ 2 & 1 & 1 & 1 \\ 1 & -1 & -3 & 0 \\ 0 & 0 & 1 & 1 \end{bmatrix}^{-1} = \begin{bmatrix} -1 & \frac{4}{3} & -\frac{2}{3} & -\frac{7}{3} \\ 2 & -\frac{5}{3} & \frac{4}{3} & \frac{11}{3} \\ -1 & 1 & -1 & -2 \\ 1 & -1 & 1 & 3 \end{bmatrix}^{-1}$

$\begin{bmatrix} x \\ y \\ z \\ w \end{bmatrix} = \begin{bmatrix} -1 & \frac{4}{3} & -\frac{2}{3} & -\frac{7}{3} \\ 2 & -\frac{5}{3} & \frac{4}{3} & \frac{11}{3} \\ -1 & 1 & -1 & -2 \\ 1 & -1 & 1 & 3 \end{bmatrix} \begin{bmatrix} -2 \\ 1 \\ 0 \\ 1 \end{bmatrix} = \begin{bmatrix} 1 \\ -2 \\ 1 \\ 0 \end{bmatrix}$

Answer: $(1, -2, 1, 0)$

141. $\begin{cases} x + 2y = -1 \\ 3x + 4y = -5 \end{cases}$

$\begin{bmatrix} 1 & 2 \\ 3 & 4 \end{bmatrix}^{-1} = \begin{bmatrix} -2 & 1 \\ \frac{3}{2} & -\frac{1}{2} \end{bmatrix} \Rightarrow \begin{bmatrix} x \\ y \end{bmatrix} = \begin{bmatrix} -2 & 1 \\ \frac{3}{2} & -\frac{1}{2} \end{bmatrix} \begin{bmatrix} -1 \\ -5 \end{bmatrix} = \begin{bmatrix} -3 \\ 1 \end{bmatrix}$

$x = -3, y = 1$

Answer: $(-3, 1)$

143. $\begin{cases} -3x - 3y - 4z = 2 \\ y + z = -1 \\ 4x + 3y + 4z = -1 \end{cases}$

$\begin{bmatrix} -3 & -3 & -4 \\ 0 & 1 & 1 \\ 4 & 3 & 4 \end{bmatrix}^{-1} = \begin{bmatrix} 1 & 0 & 1 \\ 4 & 4 & 3 \\ 4 & -3 & -3 \end{bmatrix} \Rightarrow \begin{bmatrix} x \\ y \\ z \end{bmatrix} = \begin{bmatrix} 1 & 0 & 1 \\ 4 & 4 & 3 \\ -4 & -3 & -3 \end{bmatrix} \begin{bmatrix} 2 \\ -1 \\ -1 \end{bmatrix} = \begin{bmatrix} 1 \\ 1 \\ -2 \end{bmatrix}$

$x = 1, y = 1, z = -2$

Answer: $(1, 1, -2)$

145. $\begin{vmatrix} 8 & 5 \\ 2 & -4 \end{vmatrix} = 8(-4) - 2(5) = -42$

147. $\begin{vmatrix} 50 & -30 \\ 10 & 5 \end{vmatrix} = 50(5) - (-30)(10) = 550$

149. $A = \begin{bmatrix} 2 & -1 \\ 7 & 4 \end{bmatrix}$

Minors: $\quad M_{11} = 4, \qquad M_{21} = -1$

$\qquad\qquad M_{12} = 7, \qquad M_{22} = 2$

Cofactors: $C_{11} = 4, \qquad C_{21} = 1$

$\qquad\qquad C_{12} = -7, \qquad C_{22} = 2$

151. $A = \begin{bmatrix} 3 & 2 & -1 \\ -2 & 5 & 0 \\ 1 & 8 & 6 \end{bmatrix}$

Minors: $M_{11} = \begin{vmatrix} 5 & 0 \\ 8 & 6 \end{vmatrix} = 30, M_{12} = \begin{vmatrix} -2 & 0 \\ 1 & 6 \end{vmatrix} = -12, M_{13} = \begin{vmatrix} -2 & 5 \\ 1 & 8 \end{vmatrix} = -21$

$\qquad M_{21} = \begin{vmatrix} 2 & -1 \\ 8 & 6 \end{vmatrix} = 20, M_{22} = \begin{vmatrix} 3 & -1 \\ 1 & 6 \end{vmatrix} = 19, M_{23} = \begin{vmatrix} 3 & 2 \\ 1 & 8 \end{vmatrix} = 22$

$\qquad M_{31} = \begin{vmatrix} 2 & -1 \\ 5 & 0 \end{vmatrix} = 5, M_{32} = \begin{vmatrix} 3 & -1 \\ -2 & 0 \end{vmatrix} = -2, M_{33} = \begin{vmatrix} 3 & 2 \\ -2 & 5 \end{vmatrix} = 19$

Cofactors: $C_{11} = 30, C_{12} = 12, C_{13} = -21$

$\qquad\qquad C_{21} = -20, C_{22} = 19, C_{23} = -22$

$\qquad\qquad C_{31} = 5, C_{32} = 2, C_{33} = 19$

153. $\begin{vmatrix} -2 & 4 & 1 \\ -6 & 0 & 2 \\ 5 & 3 & 4 \end{vmatrix} = 6 \begin{vmatrix} 4 & 1 \\ 3 & 4 \end{vmatrix} - 2 \begin{vmatrix} -2 & 4 \\ 5 & 3 \end{vmatrix} = 6(13) - 2(-26) = 130$

155. $\begin{vmatrix} 1 & 0 & -2 \\ 0 & 1 & 0 \\ -2 & 0 & 1 \end{vmatrix} = 1 \begin{vmatrix} 1 & -2 \\ -2 & 1 \end{vmatrix} = 1(1) - (-2)(-2) = 1 - 4 = -3$

157. $\begin{vmatrix} 3 & 0 & -4 & 0 \\ 0 & 8 & 1 & 2 \\ 6 & 1 & 8 & 2 \\ 0 & 3 & -4 & 1 \end{vmatrix} = 3 \begin{vmatrix} 8 & 1 & 2 \\ 1 & 8 & 2 \\ 3 & -4 & 1 \end{vmatrix} + (-4) \begin{vmatrix} 0 & 8 & 2 \\ 6 & 1 & 2 \\ 0 & 3 & 1 \end{vmatrix}$ (Expansion along Row 1)

$$= 3[8(8 - (-8)) - 1(1 - 6) + 2(-4 - 24)] - 4[0 - 6(8 - 6) + 0]$$

$$= 3[128 + 5 - 56] - 4[-12]$$

$$= 279$$

159. $\det(A) = 8(-1)(4)(3) = -96$, (Upper Triangular)

161. $(1, 0),\ (5, 0),\ (5, 8)$

$\frac{1}{2} \begin{vmatrix} 1 & 0 & 1 \\ 5 & 0 & 1 \\ 5 & 8 & 1 \end{vmatrix} = \frac{1}{2}(32) = 16$

Area $= 16$ square units

163. $\frac{1}{2} \begin{vmatrix} \frac{1}{2} & 1 & 1 \\ 2 & -\frac{5}{2} & 1 \\ \frac{3}{2} & 1 & 1 \end{vmatrix} = \frac{1}{2}\left(\frac{7}{2}\right) = \frac{7}{4}$

Area $= \frac{7}{4}$ square units

165. $\frac{1}{2} \begin{vmatrix} 2 & 4 & 1 \\ 5 & 6 & 1 \\ 4 & 1 & 1 \end{vmatrix} = \frac{1}{2}(-13)$

Area $= \frac{13}{2}$ square units

167. The figure is a rhombus.

$\begin{vmatrix} -2 & -1 & 1 \\ 4 & 9 & 1 \\ -2 & -9 & 1 \end{vmatrix} = -48$

Area $= 48$ square units

169. $\begin{vmatrix} -1 & 7 & 1 \\ 2 & 5 & 1 \\ 4 & 1 & 1 \end{vmatrix} = -8 \neq 0$

Not collinear

171. $x = \dfrac{\begin{vmatrix} 5 & 2 \\ 1 & 1 \end{vmatrix}}{\begin{vmatrix} 1 & 2 \\ -1 & 1 \end{vmatrix}} = \dfrac{3}{3} = 1$

$y = \dfrac{\begin{vmatrix} 1 & 5 \\ -1 & 1 \end{vmatrix}}{\begin{vmatrix} 1 & 2 \\ -1 & 1 \end{vmatrix}} = \dfrac{6}{3} = 2$

Answer: $(1, 2)$

173. $x = \dfrac{\begin{vmatrix} 6 & -2 \\ -23 & 3 \end{vmatrix}}{\begin{vmatrix} 5 & -2 \\ -11 & 3 \end{vmatrix}} = \dfrac{-28}{-7} = 4$

$y = \dfrac{\begin{vmatrix} 5 & 6 \\ -11 & -23 \end{vmatrix}}{\begin{vmatrix} 5 & -2 \\ -11 & 3 \end{vmatrix}} = \dfrac{-49}{-7} = 7$

Answer: $(4, 7)$

175. $x = \dfrac{\begin{vmatrix} -11 & 3 & -5 \\ -3 & -1 & 1 \\ 15 & -4 & 6 \end{vmatrix}}{\begin{vmatrix} -2 & 3 & -5 \\ 4 & -1 & 1 \\ -1 & -4 & 6 \end{vmatrix}} = \dfrac{-14}{14} = -1$

$y = \dfrac{\begin{vmatrix} -2 & -11 & -5 \\ 4 & -3 & 1 \\ -1 & 15 & 6 \end{vmatrix}}{14} = \dfrac{56}{14} = 4$

$z = \dfrac{\begin{vmatrix} -2 & 3 & -11 \\ 4 & -1 & -3 \\ -1 & -4 & 15 \end{vmatrix}}{14} = \dfrac{70}{14} = 5$

Answer: $(-1, 4, 5)$

177. $\begin{vmatrix} 1 & -3 & 2 \\ 2 & 2 & -3 \\ 1 & -7 & 8 \end{vmatrix} = 20$

$|A_1| = 0$

$|A_2| = -48$

$|A_3| = -52$

$x = 0,\ y = -\dfrac{48}{20} = -2.4,\ z = -\dfrac{52}{20} = -2.6$

Answer: $(0, -2.4, -2.6)$

179. (a) $\begin{cases} x - 3y + 2z = 5 \\ 2x + y - 4z = -1 \\ 2x + 4y + 2z = 3 \end{cases}$

$\begin{cases} x - 3y + 2z = 5 \\ 7y - 8z = -11 \\ 10y - 2z = -7 \end{cases}$

$\begin{cases} x - 3y + 2z = 5 \\ 7y - 8z = -11 \\ \frac{66}{7}z = \frac{61}{7} \end{cases}$

$z = \dfrac{61}{66}$

$7y - 8\left(\dfrac{61}{66}\right) = -11 \Rightarrow y = -\dfrac{17}{33}$

$x - 3\left(-\dfrac{17}{33}\right) + 2\left(\dfrac{61}{66}\right) = 5 \Rightarrow x = \dfrac{53}{33}$

Answer: $\left(\dfrac{53}{33}, -\dfrac{17}{33}, \dfrac{61}{66}\right)$

(b) $D = \begin{vmatrix} 1 & -3 & 2 \\ 2 & 1 & -4 \\ 2 & 4 & 2 \end{vmatrix} = 66$

$x = \dfrac{\begin{vmatrix} 5 & -3 & 2 \\ -1 & 1 & -4 \\ 3 & 4 & 2 \end{vmatrix}}{66} = \dfrac{106}{66} = \dfrac{53}{33}$

$y = \dfrac{\begin{vmatrix} 1 & 5 & 2 \\ 2 & -1 & -4 \\ 2 & 3 & 2 \end{vmatrix}}{66} = \dfrac{-34}{66} = \dfrac{-17}{33}$

$z = \dfrac{\begin{vmatrix} 1 & -3 & 5 \\ 2 & 1 & -1 \\ 2 & 4 & 3 \end{vmatrix}}{66} = \dfrac{61}{66}$

Answer: $\left(\dfrac{53}{33}, -\dfrac{17}{33}, \dfrac{61}{66}\right)$

181. I _ H A V E _ A _ D R E A M

[9 0 8] [1 22 5] [0 1 0] [4 18 5] [1 13 0]

$[9\ 0\ 8]A = [-30\ -2\ 24]$

$[1\ 22\ 5]A = [38\ 8\ -51]$

$[0\ 1\ 0]A = [3\ 0\ -3]$

$[4\ 18\ 5]A = [32\ 2\ -39]$

$[1\ 13\ 0]A = [41\ -2\ -39]$

Cryptogram: $-30\ -2\ 24\ 38\ 8\ -51\ 3\ 0\ -3\ 32\ 2\ -39\ 41\ -2\ -39$

183. $A^{-1} = \begin{bmatrix} \frac{1}{2} & -\frac{1}{4} & \frac{1}{4} \\ -\frac{1}{4} & -\frac{3}{8} & -\frac{1}{8} \\ -\frac{1}{2} & -\frac{1}{4} & \frac{1}{4} \end{bmatrix}$

$\begin{bmatrix} 32 & -46 & 37 \\ 9 & -48 & 15 \\ 3 & -14 & 10 \\ -1 & -6 & 2 \\ -8 & -22 & -3 \end{bmatrix} \begin{bmatrix} \frac{1}{2} & -\frac{1}{4} & \frac{1}{4} \\ -\frac{1}{4} & -\frac{3}{8} & -\frac{1}{8} \\ -\frac{1}{2} & -\frac{1}{4} & \frac{1}{4} \end{bmatrix} = \begin{bmatrix} 9 & 0 & 23 \\ 9 & 12 & 12 \\ 0 & 2 & 5 \\ 0 & 2 & 1 \\ 3 & 11 & 0 \end{bmatrix} = \begin{matrix} \text{I} & - & \text{W} \\ \text{I} & \text{L} & \text{L} \\ - & \text{B} & \text{E} \\ - & \text{B} & \text{A} \\ \text{C} & \text{K} & - \end{matrix}$

Message: I WILL BE BACK

185. $A^{-1} = \begin{bmatrix} \frac{2}{3} & \frac{1}{3} & \frac{1}{3} \\ -\frac{1}{3} & \frac{1}{3} & \frac{1}{3} \\ \frac{1}{6} & \frac{1}{3} & -\frac{1}{6} \end{bmatrix}$

$\begin{bmatrix} 21 & -11 & 14 \\ 29 & -11 & -18 \\ 32 & -6 & -26 \\ 31 & -19 & -12 \\ 10 & 6 & 26 \\ 13 & -11 & -2 \\ 37 & 28 & -8 \\ 5 & 13 & 36 \end{bmatrix} \begin{bmatrix} \frac{2}{3} & \frac{1}{3} & \frac{1}{3} \\ -\frac{1}{3} & \frac{1}{3} & \frac{1}{3} \\ \frac{1}{6} & \frac{1}{3} & -\frac{1}{6} \end{bmatrix} = \begin{bmatrix} 20 & 8 & 1 \\ 20 & 0 & 9 \\ 19 & 0 & 13 \\ 25 & 0 & 6 \\ 9 & 14 & 1 \\ 12 & 0 & 1 \\ 14 & 19 & 23 \\ 5 & 18 & 0 \end{bmatrix} \begin{matrix} \text{T} & \text{H} & \text{A} \\ \text{T} & - & \text{I} \\ \text{S} & - & \text{M} \\ \text{Y} & - & \text{F} \\ \text{I} & \text{N} & \text{A} \\ \text{L} & - & \text{A} \\ \text{N} & \text{S} & \text{W} \\ \text{E} & \text{R} & - \end{matrix}$

Message: THAT IS MY FINAL ANSWER

187. (a) $\begin{cases} 4b + 44a = 65 \\ 44b + 504a = 723.2 \end{cases}$

Using Cramer's Rule, you obtain

$a = 0.41$ and $b = 11.74$

$y = 0.41t + 11.74.$

(c) $y = 20$ when $t \approx 20.15$, or 2010.

(b)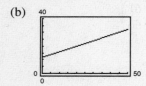

(d) $0.41t + 11.74 > 20$

$0.41t > 8.26$

$t > 20.15$

That is, year 2010.

189. True. Expansion by Row 3 gives

$$\begin{vmatrix} a_{11} & a_{12} & a_{13} \\ a_{21} & a_{22} & a_{23} \\ a_{31} + c_1 & a_{32} + c_2 & a_{33} + c_3 \end{vmatrix} = (a_{31} + c_1) \begin{vmatrix} a_{12} & a_{13} \\ a_{22} & a_{23} \end{vmatrix} - (a_{32} + c_2) \begin{vmatrix} a_{11} & a_{13} \\ a_{21} & a_{23} \end{vmatrix} + (a_{33} + c_3) \begin{vmatrix} a_{11} & a_{12} \\ a_{21} & a_{22} \end{vmatrix}$$

$$= a_{31} \begin{vmatrix} a_{12} & a_{13} \\ a_{22} & a_{23} \end{vmatrix} - a_{32} \begin{vmatrix} a_{11} & a_{13} \\ a_{21} & a_{23} \end{vmatrix} + a_{33} \begin{vmatrix} a_{11} & a_{12} \\ a_{21} & a_{22} \end{vmatrix}$$

$$+ c_1 \begin{vmatrix} a_{12} & a_{13} \\ a_{22} & a_{23} \end{vmatrix} - c_2 \begin{vmatrix} a_{11} & a_{13} \\ a_{21} & a_{23} \end{vmatrix} + c_3 \begin{vmatrix} a_{11} & a_{12} \\ a_{21} & a_{22} \end{vmatrix}$$

$$= \begin{vmatrix} a_{11} & a_{12} & a_{13} \\ a_{21} & a_{22} & a_{23} \\ a_{31} & a_{32} & a_{33} \end{vmatrix} + \begin{vmatrix} a_{11} & a_{12} & a_{13} \\ a_{21} & a_{22} & a_{23} \\ c_1 & c_2 & c_3 \end{vmatrix}.$$

Note: Expand each of these matrices by Row 3 to see the previous step.

191. A square $n \times n$ matrix has an inverse if $\det(A) \neq 0$.

Chapter 5   Practice Test

For Exercises 1–3, solve the given system by the method of substitution.

1. $x + y = 1$
$3x - y = 15$

2. $x - 3y = -3$
$x^2 + 6y = 5$

3. $x + y + z = 6$
$2x - y + 3z = 0$
$5x + 2y - z = -3$

4. Find the two numbers whose sum is 110 and product is 2800.

5. Find the dimensions of a rectangle if its perimeter is 170 feet and its area is 2800 square feet.

For Exercises 6–7, solve the linear system by elimination.

6. $2x + 15y = 4$
$x - 3y = 23$

7. $x + y = 2$
$38x - 19y = 7$

8. Use a graphing utility to graph the two equations. Use the graph to approximate the solution of the system. Verify your answer analytically.

$0.4x + 0.5y = 0.112$

$0.3x - 0.7y = -0.131$

9. Herbert invests $17,000 in two funds that pay 11% and 13% simple interest, respectively. If he receives $2080 in yearly interest, how much is invested in each fund?

10. Find the least squares regression line for the points $(4, 3)$, $(1, 1)$, $(-1, -2)$, and $(-2, -1)$.

For Exercises 11–13, solve the system of equations.

11. $x + y = -2$
$2x - y + z = 11$
$4y - 3z = -20$

12. $4x - y + 5z = 4$
$2x + y - z = 0$
$2x + 4y + 8z = 0$

13. $3x + 2y - z = 5$
$6x - y + 5z = 2$

14. Find the equation of the parabola $y = ax^2 + bx + c$ passing through the points $(0, -1)$, $(1, 4)$ and $(2, 13)$.

15. Find the position equation $s = \frac{1}{2}at^2 + v_0t + s_0$ given that $s = 12$ feet after 1 second, $s = 5$ feet after 2 seconds, and $s = 4$ feet after 3 seconds.

16. Write the matrix in reduced row-echelon form.

$$\begin{bmatrix} 1 & -2 & 4 \\ 3 & -5 & 9 \end{bmatrix}$$

For Exercises 17–19, use matrices to solve the system of equations.

17. $3x + 5y = 3$
$2x - y = -11$

18. $2x + 3y = -3$
$3x + 2y = 8$
$x + y = 1$

19. $x + 3z = -5$
$2x + y = 0$
$3x + y - z = 3$

20. Multiply $\begin{bmatrix} 1 & 4 & 5 \\ 2 & 0 & -3 \end{bmatrix} \begin{bmatrix} 1 & 6 \\ 0 & -7 \\ -1 & 2 \end{bmatrix}$

21. Given $A = \begin{bmatrix} 9 & 1 \\ -4 & 8 \end{bmatrix}$ and $B = \begin{bmatrix} 6 & -2 \\ 3 & 5 \end{bmatrix}$, find $3A - 5B$.

22. Find $f(A)$:

$$f(x) = x^2 - 7x + 8, \quad A = \begin{bmatrix} 3 & 0 \\ 7 & 1 \end{bmatrix}$$

23. True or false:

$(A + B)(A + 3B) = A^2 + 4AB + 3B^2$ where A and B are matrices.

(Assume that A^2, AB, and B^2 exist.)

For Exercises 24 and 25, find the inverse of the matrix, if it exists.

24. $\begin{bmatrix} 1 & 2 \\ 3 & 5 \end{bmatrix}$

25. $\begin{bmatrix} 1 & 1 & 1 \\ 3 & 6 & 5 \\ 6 & 10 & 8 \end{bmatrix}$

26. Use an inverse matrix to solve the systems.

(a) $\begin{aligned} x + 2y &= 4 \\ 3x + 5y &= 1 \end{aligned}$

(b) $\begin{aligned} x + 2y &= 3 \\ 3x + 5y &= -2 \end{aligned}$

For Exercises 27 and 28, find the determinant of the matrix.

27. $\begin{bmatrix} 6 & -1 \\ 3 & 4 \end{bmatrix}$

28. $\begin{bmatrix} 1 & 3 & -1 \\ 5 & 9 & 0 \\ 6 & 2 & -5 \end{bmatrix}$

29. Use a graphing utility to find the determinant of the matrix.

$$\begin{bmatrix} 1 & 4 & 2 & 3 \\ 0 & 1 & -2 & 0 \\ 3 & 5 & -1 & 1 \\ 2 & 0 & 6 & 1 \end{bmatrix}$$

30. Evaluate $\begin{vmatrix} 6 & 4 & 3 & 0 & 6 \\ 0 & 5 & 1 & 4 & 8 \\ 0 & 0 & 2 & 7 & 3 \\ 0 & 0 & 0 & 9 & 2 \\ 0 & 0 & 0 & 0 & 1 \end{vmatrix}$.

31. Use a determinant to find the area of the triangle with vertices $(0, 7)$, $(5, 0)$, and $(3, 9)$.

32. Use a determinant to find the equation of the line through $(2, 7)$ and $(-1, 4)$.

For Exercises 33–35, use Cramer's Rule to find the indicated value.

33. Find x.

$$\begin{aligned} 6x - 7y &= 4 \\ 2x + 5y &= 11 \end{aligned}$$

34. Find z.

$$\begin{aligned} 3x \qquad + z &= 1 \\ y + 4z &= 3 \\ x - y \qquad &= 2 \end{aligned}$$

35. Find y.

$$\begin{aligned} 721.4x - 29.1y &= 33.77 \\ 45.9x + 105.6y &= 19.85 \end{aligned}$$

C H A P T E R   6
Sequences, Series, and Probability

CHAPTER 6
Sequences, Series, and Probability

Section 6.1   Sequences and Series

■ Given the general nth term in a sequence, you should be able to find, or list, some of the terms.

■ You should be able to find an expression for the nth term of a sequence.

■ You should be able to use and evaluate factorials.

■ You should be able to use sigma notation for a sum.

Vocabulary Check

1. infinite sequence

2. terms

3. finite

4. recursively

5. factorial

6. summation notation

7. index, upper limit, lower limit

8. series

9. nth partial sum

---

1. $a_n = 2n + 5$

$a_1 = 2(1) + 5 = 7$

$a_2 = 2(2) + 5 = 9$

$a_3 = 2(3) + 5 = 11$

$a_4 = 2(4) + 5 = 13$

$a_5 = 2(5) + 5 = 15$

3. $a_n = 2^n$

$a_1 = 2^1 = 2$

$a_2 = 2^2 = 4$

$a_3 = 2^3 = 8$

$a_4 = 2^4 = 16$

$a_5 = 2^5 = 32$

5. $a_n = \left(-\frac{1}{2}\right)^n$

$a_1 = \left(-\frac{1}{2}\right)^1 = -\frac{1}{2}$

$a_2 = \left(-\frac{1}{2}\right)^2 = \frac{1}{4}$

$a_3 = \left(-\frac{1}{2}\right)^3 = -\frac{1}{8}$

$a_4 = \left(-\frac{1}{2}\right)^4 = \frac{1}{16}$

$a_5 = \left(-\frac{1}{2}\right)^5 = -\frac{1}{32}$

7. $a_n = \dfrac{n+1}{n}$

$a_1 = \dfrac{1+1}{1} = 2$

$a_2 = \dfrac{3}{2}$

$a_3 = \dfrac{4}{3}$

$a_4 = \dfrac{5}{4}$

$a_5 = \dfrac{6}{5}$

9. $a_n = \dfrac{n}{n^2 + 1}$

$a_1 = \dfrac{1}{1^2 + 1} = \dfrac{1}{2}$

$a_2 = \dfrac{2}{2^2 + 1} = \dfrac{2}{5}$

$a_3 = \dfrac{3}{3^2 + 1} = \dfrac{3}{10}$

$a_4 = \dfrac{4}{4^2 + 1} = \dfrac{4}{17}$

$a_5 = \dfrac{5}{5^2 + 1} = \dfrac{5}{26}$

11. $a_n = \dfrac{1 + (-1)^n}{n}$

$a_1 = 0$

$a_2 = \dfrac{2}{2} = 1$

$a_3 = 0$

$a_4 = \dfrac{2}{4} = \dfrac{1}{2}$

$a_5 = 0$

13. $a_n = 1 - \dfrac{1}{2^n}$

$a_1 = 1 - \dfrac{1}{2^1} = \dfrac{1}{2}$

$a_2 = 1 - \dfrac{1}{2^2} = 1 - \dfrac{1}{4} = \dfrac{3}{4}$

$a_3 = 1 - \dfrac{1}{2^3} = \dfrac{7}{8}$

$a_4 = 1 - \dfrac{1}{2^4} = \dfrac{15}{16}$

$a_5 = 1 - \dfrac{1}{2^5} = \dfrac{31}{32}$

15. $a_n = \dfrac{1}{n^{3/2}}$

$a_1 = \dfrac{1}{1} = 1$

$a_2 = \dfrac{1}{2^{3/2}}$

$a_3 = \dfrac{1}{3^{3/2}}$

$a_4 = \dfrac{1}{4^{3/2}} = \dfrac{1}{8}$

$a_5 = \dfrac{1}{5^{3/2}}$

17. $a_n = \dfrac{(-1)^n}{n^2}$

$a_1 = \dfrac{-1}{1} = -1$

$a_2 = \dfrac{1}{4}$

$a_3 = \dfrac{-1}{9}$

$a_4 = \dfrac{1}{16}$

$a_5 = \dfrac{-1}{25}$

19. $a_n = (2n - 1)(2n + 1)$

$a_1 = (1)(3) = 3$

$a_2 = (3)(5) = 15$

$a_3 = (5)(7) = 35$

$a_4 = (7)(9) = 63$

$a_5 = (9)(11) = 99$

21. $a_{25} = (-1)^{25}[3(25) - 2]$

$\qquad = -73$

23. $a_{10} = \dfrac{10^2}{10^2 + 1} = \dfrac{100}{101}$

25. $a_6 = \dfrac{2^6}{2^6 + 1} = \dfrac{64}{65}$

27. $a_n = \dfrac{2}{3}n$

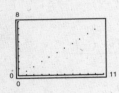

29. $a_n = 16(-0.5)^{n-1}$

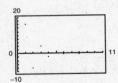

31. $a_n = \dfrac{2n}{n + 1}$

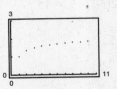

33. $a_n = 2(3n - 1) + 5$

n	1	2	3	4	5	6	7	8	9	10
a_n	9	15	21	27	33	39	45	51	57	63

35. $a_n = 1 + \dfrac{n + 1}{n}$

n	1	2	3	4	5	6	7	8	9	10
a_n	3	2.5	2.33	2.25	2.2	2.17	2.14	2.13	2.11	2.1

37. $a_n = (-1)^n + 1$

n	1	2	3	4	5	6	7	8	9	10
a_n	0	2	0	2	0	2	0	2	0	2

39. $a_n = \dfrac{8}{n+1}$

$a_n \to 0$ as $n \to \infty$

$a_1 = 4$, $a_{10} = \dfrac{8}{11}$

Matches graph (c).

41. $a_n = 4(0.5)^{n-1}$

$a_n \to 0$ as $n \to \infty$

$a_1 = 4$, $a_{10} \approx 0.008$

Matches graph (d).

43. $1, 4, 7, 10, 13, \ldots$

$a_n = 1 + (n-1)3 = 3n - 2$

45. $0, 3, 8, 15, 24, \ldots$

$a_n = n^2 - 1$

47. $\dfrac{2}{3}, \dfrac{3}{4}, \dfrac{4}{5}, \dfrac{5}{6}, \dfrac{6}{7}, \ldots$

$a_n = \dfrac{n+1}{n+2}$

49. $\dfrac{1}{2}, \dfrac{-1}{4}, \dfrac{1}{8}, \dfrac{-1}{16}, \ldots$

$a_n = \dfrac{(-1)^{n+1}}{2^n}$

51. $1 + \dfrac{1}{1}, 1 + \dfrac{1}{2}, 1 + \dfrac{1}{3}, 1 + \dfrac{1}{4}, 1 + \dfrac{1}{5}, \ldots$

$a_n = 1 + \dfrac{1}{n}$

53. $1, \dfrac{1}{2}, \dfrac{1}{6}, \dfrac{1}{24}, \dfrac{1}{120}, \ldots$

$a_n = \dfrac{1}{n!}$

55. $1, 3, 1, 3, 1, 3, \ldots$

$a_n = 2 + (-1)^n$

57. $a_1 = 28$ and $a_{k+1} = a_k - 4$

$a_1 = 28$

$a_2 = a_1 - 4 = 28 - 4 = 24$

$a_3 = a_2 - 4 = 24 - 4 = 20$

$a_4 = a_3 - 4 = 20 - 4 = 16$

$a_5 = a_4 - 4 = 16 - 4 = 12$

59. $a_1 = 3$ and $a_{k+1} = 2(a_k - 1)$

$a_1 = 3$

$a_2 = 2(a_1 - 1) = 2(3 - 1) = 4$

$a_3 = 2(a_2 - 1) = 2(4 - 1) = 6$

$a_4 = 2(a_3 - 1) = 2(6 - 1) = 10$

$a_5 = 2(a_4 - 1) = 2(10 - 1) = 18$

61. $a_1 = 6$ and $a_{k+1} = a_k + 2$

$a_1 = 6$

$a_2 = a_1 + 2 = 6 + 2 = 8$

$a_3 = a_2 + 2 = 8 + 2 = 10$

$a_4 = a_3 + 2 = 10 + 2 = 12$

$a_5 = a_4 + 2 = 12 + 2 = 14$

In general, $a_n = 2n + 4$.

63. $a_1 = 81$ and $a_{k+1} = \frac{1}{3}a_k$

$a_1 = 81$

$a_2 = \frac{1}{3}a_1 = \frac{1}{3}(81) = 27$

$a_3 = \frac{1}{3}a_2 = \frac{1}{3}(27) = 9$

$a_4 = \frac{1}{3}a_3 = \frac{1}{3}(9) = 3$

$a_5 = \frac{1}{3}a_4 = \frac{1}{3}(3) = 1$

In general, $a_n = 81\left(\frac{1}{3}\right)^{n-1} = 81(3)\left(\frac{1}{3}\right)^n = \frac{243}{3^n}$.

65. $a_n = \frac{1}{n!}$

$a_0 = \frac{1}{0!} = 1$

$a_1 = \frac{1}{1!} = 1$

$a_2 = \frac{1}{2}$

$a_3 = \frac{1}{3!} = \frac{1}{6}$

$a_4 = \frac{1}{4!} = \frac{1}{24}$

67. $a_n = \frac{n!}{2n+1}$

$a_0 = \frac{0!}{1} = 1$

$a_1 = \frac{1!}{2+1} = \frac{1}{3}$

$a_2 = \frac{2!}{4+1} = \frac{2}{5}$

$a_3 = \frac{3!}{6+1} = \frac{6}{7}$

$a_4 = \frac{4!}{8+1} = \frac{24}{9} = \frac{8}{3}$

69. $a_n = \frac{(-1)^{2n}}{(2n)!}$

$a_0 = \frac{(-1)^0}{0!} = 1$

$a_1 = \frac{(-1)^2}{2!} = \frac{1}{2}$

$a_2 = \frac{(-1)^4}{4!} = \frac{1}{24}$

$a_3 = \frac{(-1)^6}{6!} = \frac{1}{720}$

$a_4 = \frac{(-1)^8}{8!} = \frac{1}{40,320}$

71. $\dfrac{2!}{4!} = \dfrac{2!}{4 \cdot 3 \cdot 2!} = \dfrac{1}{12}$

73. $\dfrac{12!}{4!8!} = \dfrac{12 \cdot 11 \cdot 10 \cdot 9 \cdot 8!}{4!8!}$

$= \dfrac{12 \cdot 11 \cdot 10 \cdot 9}{4 \cdot 3 \cdot 2} = 495$

75. $\dfrac{(n+1)!}{n!} = \dfrac{(n+1)n!}{n!} = n + 1$

77. $\dfrac{(2n-1)!}{(2n+1)!} = \dfrac{(2n-1)!}{(2n+1)(2n)(2n-1)!}$

$= \dfrac{1}{2n(2n+1)}$

79. $\displaystyle\sum_{i=1}^{5}(2i+1) = (2+1) + (4+1) + (6+1) + (8+1) + (10+1) = 35$

81. $\displaystyle\sum_{k=1}^{4} 10 = 10 + 10 + 10 + 10 = 40$

83. $\displaystyle\sum_{i=0}^{4} i^2 = 0^2 + 1^2 + 2^2 + 3^2 + 4^2 = 30$

85. $\displaystyle\sum_{k=0}^{3} \frac{1}{k^2+1} = \frac{1}{1} + \frac{1}{1+1} + \frac{1}{4+1} + \frac{1}{9+1} = \frac{9}{5}$

87. $\displaystyle\sum_{i=1}^{4}[(i-1)^2 + (i+1)^3] = [(0)^2 + (2)^3] + [(1)^2 + (3)^3] + [(2)^2 + (4)^3] + [(3)^2 + (5)^3] = 238$

89. $\displaystyle\sum_{i=1}^{4} 2^i = 2^1 + 2^2 + 2^3 + 2^4$

$= 30$

91. $\displaystyle\sum_{j=1}^{6}(24 - 3j) = 81$

93. $\displaystyle\sum_{k=0}^{4}\frac{(-1)^k}{k+1} = \frac{47}{60}$

95. $\displaystyle\frac{1}{3(1)} + \frac{1}{3(2)} + \frac{1}{3(3)} + \cdots + \frac{1}{3(9)} = \sum_{i=1}^{9}\frac{1}{3i} \approx 0.94299$

97. $\displaystyle\left[2\left(\frac{1}{8}\right)+3\right] + \left[2\left(\frac{2}{8}\right)+3\right] + \left[2\left(\frac{3}{8}\right)+3\right] + \cdots + \left[2\left(\frac{8}{8}\right)+3\right] = \sum_{i=1}^{8}\left[2\left(\frac{i}{8}\right)+3\right] = 33$

99. $\displaystyle 3 - 9 + 27 - 81 + 243 - 729 = \sum_{i=1}^{6}(-1)^{i+1}3^i = -546$

101. $\displaystyle\frac{1}{1^2} - \frac{1}{2^2} + \frac{1}{3^2} - \frac{1}{4^2} + \cdots - \frac{1}{20^2} = \sum_{i=1}^{20}\frac{(-1)^{i+1}}{i^2} \approx 0.82128$

103. $\displaystyle\frac{1}{4} + \frac{3}{8} + \frac{7}{16} + \frac{15}{32} + \frac{31}{64} = \sum_{i=1}^{5}\frac{2^i - 1}{2^{i+1}} = \frac{129}{64} = 2.015625$

105. $\displaystyle\sum_{i=1}^{4} 5\left(\frac{1}{2}\right)^i = 4.6875 = \frac{75}{16}$

107. $\displaystyle\sum_{n=1}^{3} 4\left(-\frac{1}{2}\right)^n = -1.5 = -\frac{3}{2}$

109. (a) $\displaystyle\sum_{i=1}^{4} 6\left(\frac{1}{10}\right)^i = 6\left(\frac{1}{10}\right) + 6\left(\frac{1}{10}\right)^2 + 6\left(\frac{1}{10}\right)^3 + 6\left(\frac{1}{10}\right)^4$

$= 0.6666$

$= \dfrac{3333}{5000}$

(b) $\displaystyle\sum_{i=1}^{\infty} 6\left(\frac{1}{10}\right)^i = 6[0.1 + 0.01 + 0.001 + \ldots]$

$= 6[0.111\ldots]$

$= 0.666\ldots$

$= \dfrac{2}{3}$

111. (a) $\displaystyle\sum_{k=1}^{4}\left(\frac{1}{10}\right)^k = \frac{1}{10} + \frac{1}{100} + \frac{1}{1000} + \frac{1}{10{,}000}$

$= 0.1111$

$= \dfrac{1111}{10{,}000}$

(b) $\displaystyle\sum_{k=1}^{\infty}\left(\frac{1}{10}\right)^k = 0.1 + 0.01 + 0.001 + \ldots$

$= 0.111\ldots$

$= \dfrac{1}{9}$

113. $A_n = 5000\left(1 + \dfrac{0.03}{4}\right)^n$, $n = 1, 2, 3, \ldots$

   (a) $A_1 = 5000\left(1 + \dfrac{0.03}{4}\right)^1 = \$5037.50$

      $A_2 \approx \$5075.28$ $A_3 \approx \$5113.35$

      $A_4 \approx \$5151.70$ $A_5 \approx \$5190.33$

      $A_6 \approx \$5229.26$ $A_7 \approx \$5268.48$

      $A_8 \approx \$5307.99$

   (b) $A_{40} \approx \$6741.74$

115. (a) $p_0 = 5500$ (year 2008)

      $p_n = 0.75p_{n-1} + 500$

   (b) $p_1 = 0.75p_0 + 500 = 4625$ (2009)

      $p_2 = 0.75p_1 + 500 \approx 3969$ (2010)

      $p_3 = 0.75p_2 + 500 \approx 3477$ (2011)

      (Answers will vary slightly.)

   (c) The population approaches 2000 trout because $0.75(2000) + 500 = 2000$.

117. (a) $a_0 = 50$ (end of January)

      $a_n = \left(1 + \dfrac{0.06}{12}\right)a_{n-1} + 50 = 1.005a_{n-1} + 50$

   (b) $a_1 = 1.005\, a_0 + 50 = 100.25$ (end of February)

      $a_2 = 1.005\, a_1 + 50 = 150.75$

      $a_3 = 201.51$ $a_4 = 252.51$

      $a_5 = 303.78$ $a_6 = 355.29$

      $a_7 = 407.07$ $a_8 = 459.11$

      $a_9 = 511.40$ $a_{10} = 563.96$

      $a_{11} = 616.78$ (end of December)

      After one year, the IRA has $\$616.78$.

   (c) After 50 deposits, $a_{49} \approx \$2832.26$.

119. (a)

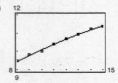

   (b) For 2010, $n = 20$ and $r_{20} \approx \$12.25$.

      For 2015, $n = 25$ and $r_{25} \approx \$12.64$.

   (c) Answers will vary.

   (d) $r_{18} = 11.97$ and $r_{19} = 12.12$. So, the average hourly wage reaches $12 in 2008.

121. (a)

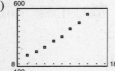

   (b) Linear: $R_n = 54.58n - 336.3$

      Quadratic: $R_n = 3.088n^2 - 22.62n + 130.0$

      Coefficient of determination for linear model: 0.98656

      Coefficient of determination for quadratic model: 0.99919.

   (c)

 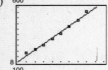 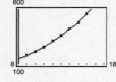

   (d) The quadratic model is better.

      The quadratic model is better because its coefficient of determination is closer to 1.

   (e) For 2010, $n = 20$ and $R_{20} \approx 912.8$ million.

      For 2015, $n = 25$ and $R_{25} \approx 1494.5$ million.

   (f) $R_n = 1000$ when $n \approx 20.8$, or in 2010.

123. True

125. $a_0 = 1, a_1 = 1, a_{k+2} = a_{k+1} + a_k$

$a_0 = 1$	$b_0 = \frac{1}{1} = 1$	$a_6 = 8 + 5 = 13$	$b_6 = \frac{21}{13}$
$a_1 = 1$	$b_1 = \frac{2}{1} = 2$	$a_7 = 13 + 8 = 21$	$b_7 = \frac{34}{21}$
$a_2 = 1 + 1 = 2$	$b_2 = \frac{3}{2}$	$a_8 = 21 + 13 = 34$	$b_8 = \frac{55}{34}$
$a_3 = 2 + 1 = 3$	$b_3 = \frac{5}{3}$	$a_9 = 34 + 21 = 55$	$b_9 = \frac{89}{55}$
$a_4 = 3 + 2 = 5$	$b_4 = \frac{8}{5}$	$a_{10} = 55 + 34 = 89$	
$a_5 = 5 + 3 = 8$	$b_5 = \frac{13}{8}$	$a_{11} = 89 + 55 = 144$	

127. $a_n = \dfrac{\left(1 + \sqrt{5}\right)^n - \left(1 - \sqrt{5}\right)^n}{2^n \sqrt{5}}$

$a_1 = \dfrac{\left(1 + \sqrt{5}\right)^1 - \left(1 - \sqrt{5}\right)^1}{2^1 \sqrt{5}} = 1$

$a_2 = 1, \quad a_3 = 2$

$a_4 = 3, \quad a_5 = 5$

129. $a_{n+1} = \dfrac{\left(1 + \sqrt{5}\right)^{n+1} - \left(1 - \sqrt{5}\right)^{n+1}}{2^{n+1} \sqrt{5}}$

$a_{n+2} = \dfrac{\left(1 + \sqrt{5}\right)^{n+2} - \left(1 - \sqrt{5}\right)^{n+2}}{2^{n+2} \sqrt{5}}$

131. $a_n = \dfrac{x^n}{n!}$

$a_1 = \dfrac{x}{1} = x$

$a_2 = \dfrac{x^2}{2!} = \dfrac{x^2}{2}$

$a_3 = \dfrac{x^3}{3!} = \dfrac{x^3}{6}$

$a_4 = \dfrac{x^4}{4!} = \dfrac{x^4}{24}$

$a_5 = \dfrac{x^5}{5!} = \dfrac{x^5}{120}$

133. $a_n = \dfrac{(-1)^n x^{2n+1}}{2n + 1}$

$a_1 = \dfrac{-x^3}{3}$

$a_2 = \dfrac{x^5}{5}$

$a_3 = -\dfrac{x^7}{7}$

$a_4 = \dfrac{x^9}{9}$

$a_5 = \dfrac{-x^{11}}{11}$

135. $a_n = \dfrac{(-1)^n x^{2n}}{(2n)!}$

$a_1 = \dfrac{-x^2}{2}$

$a_2 = \dfrac{x^4}{4!} = \dfrac{x^4}{24}$

$a_3 = \dfrac{-x^6}{6!} = \dfrac{-x^6}{720}$

$a_4 = \dfrac{x^8}{8!} = \dfrac{x^8}{40,320}$

$a_5 = \dfrac{-x^{10}}{10!} = \dfrac{-x^{10}}{3,628,800}$

137. $a_n = \dfrac{(-1)^n x^n}{n!}$

$a_1 = -x$

$a_2 = \dfrac{x^2}{2}$

$a_3 = \dfrac{-x^3}{3!} = \dfrac{-x^3}{6}$

$a_4 = \dfrac{x^4}{4!} = \dfrac{x^4}{24}$

$a_5 = \dfrac{-x^5}{5!} = -\dfrac{x^5}{120}$

139. $a_n = \dfrac{(-1)^{n+1}(x + 1)^n}{n!}$

$a_1 = x + 1$

$a_2 = \dfrac{-(x + 1)^2}{2}$

$a_3 = \dfrac{(x + 1)^3}{6}$

$a_4 = -\dfrac{(x + 1)^4}{24}$

$a_5 = \dfrac{(x + 1)^5}{120}$

141. $a_n = \dfrac{1}{2n} - \dfrac{1}{2n+2}$

$a_1 = \dfrac{1}{2} - \dfrac{1}{4} = \dfrac{1}{4}$

$a_2 = \dfrac{1}{4} - \dfrac{1}{6} = \dfrac{1}{12}$

$a_3 = \dfrac{1}{6} - \dfrac{1}{8} = \dfrac{1}{24}$

$a_4 = \dfrac{1}{8} - \dfrac{1}{10} = \dfrac{1}{40}$

$a_5 = \dfrac{1}{10} - \dfrac{1}{12} = \dfrac{1}{60}$

nth partial sum $= \left(\dfrac{1}{2} - \dfrac{1}{4}\right) + \left(\dfrac{1}{4} - \dfrac{1}{6}\right) + \cdots + \left(\dfrac{1}{2n} - \dfrac{1}{2n+2}\right) = \dfrac{1}{2} - \dfrac{1}{2n+2}$

143. $a_n = \dfrac{1}{n+1} - \dfrac{1}{n+2}$

$a_1 = \dfrac{1}{2} - \dfrac{1}{3} = \dfrac{1}{6}$

$a_2 = \dfrac{1}{3} - \dfrac{1}{4} = \dfrac{1}{12}$

$a_3 = \dfrac{1}{4} - \dfrac{1}{5} = \dfrac{1}{20}$

$a_4 = \dfrac{1}{5} - \dfrac{1}{6} = \dfrac{1}{30}$

$a_5 = \dfrac{1}{6} - \dfrac{1}{7} = \dfrac{1}{42}$

nth partial sum $= \left(\dfrac{1}{2} - \dfrac{1}{3}\right) + \left(\dfrac{1}{3} - \dfrac{1}{4}\right) + \cdots + \left(\dfrac{1}{n+1} - \dfrac{1}{n+2}\right) = \dfrac{1}{2} - \dfrac{1}{n+2}$

145. $a_n = \ln n$

$a_1 = \ln 1 = 0$

$a_2 = \ln 2$

$a_3 = \ln 3$

$a_4 = \ln 4$

$a_5 = \ln 5$

nth partial sum $= \ln 2 + \ln 3 + \cdots + \ln n$

$\qquad\qquad\quad = \ln(2 \cdot 3 \cdots n)$

$\qquad\qquad\quad = \ln(n!)$

147. (a) $A - B = \begin{bmatrix} 8 & 1 \\ -3 & 7 \end{bmatrix}$

(b) $2B - 3A = \begin{bmatrix} -22 & -7 \\ 3 & -18 \end{bmatrix}$

(c) $AB = \begin{bmatrix} 18 & 9 \\ 18 & 0 \end{bmatrix}$

(d) $BA = \begin{bmatrix} 0 & 6 \\ 27 & 18 \end{bmatrix}$

149. (a) $A - B = \begin{bmatrix} -3 & -7 & 4 \\ 4 & 4 & 1 \\ 1 & 4 & 3 \end{bmatrix}$ (b) $2B - 3A = \begin{bmatrix} 8 & 17 & -14 \\ -12 & -13 & -9 \\ -3 & -15 & -10 \end{bmatrix}$

(c) $AB = \begin{bmatrix} -2 & 7 & -16 \\ 4 & 42 & 45 \\ 1 & 23 & 48 \end{bmatrix}$ (d) $BA = \begin{bmatrix} 16 & 31 & 42 \\ 10 & 47 & 31 \\ 13 & 22 & 25 \end{bmatrix}$

Section 6.2   Arithmetic Sequences and Partial Sums

- ■ You should be able to recognize an arithmetic sequence, find its common difference, and find its nth term.
- ■ You should be able to find the nth partial sum of an arithmetic sequence with common difference d using the formula

$$S_n = \frac{n}{2}(a_1 + a_n).$$

Vocabulary Check

1. arithmetic, common   **2.** $a_n = dn + c$ **3.** nth partial sum

1. 10, 8, 6, 4, 2, . . .

Arithmetic sequence, $d = -2$

3. $3, \frac{5}{2}, 2, \frac{3}{2}, 1, . . .$

Arithmetic sequence, $d = -\frac{1}{2}$

5. $-24, -16, -8, 0, 8$

Arithmetic sequence, $d = 8$

7. 3.7, 4.3, 4.9, 5.5, 6.1, . . .

Arithmetic sequence, $d = 0.6$

9. $a_n = 8 + 13n$

21, 34, 47, 60, 73

Arithmetic sequence, $d = 13$

11. $a_n = \dfrac{1}{n + 1}$

$\dfrac{1}{2}, \dfrac{1}{3}, \dfrac{1}{4}, \dfrac{1}{5}, \dfrac{1}{6}$

Not an arithmetic sequence

13. $a_n = 150 - 7n$

143, 136, 129, 122, 115

Arithmetic sequence, $d = -7$

15. $a_n = 3 + 2(-1)^n$

1, 5, 1, 5, 1

Not an arithmetic sequence

17. $a_1 = 1, \ d = 3$

$a_n = a_1 + (n - 1)d = 1 + (n - 1)(3) = 3n - 2$

19. $a_1 = 100, \ d = -8$

$a_n = a_1 + (n - 1)d$

$= 100 + (n - 1)(-8) = 108 - 8n$

21. $4, \frac{3}{2}, -1, -\frac{7}{2}, . . . , d = -\frac{5}{2}$

$a_n = a_1 + (n - 1)d = 4 + (n - 1)\left(-\frac{5}{2}\right) = \frac{13}{2} - \frac{5}{2}n$

23. $a_1 = 5, \ a_4 = 15$

$a_4 = a_1 + 3d \Rightarrow 15 = 5 + 3d \Rightarrow d = \frac{10}{3}$

$a_n = a_1 + (n - 1)d = 5 + (n - 1)\left(\frac{10}{3}\right) = \frac{10}{3}n + \frac{5}{3}$

25. $a_3 = 94,\ a_6 = 85$

$a_6 = a_3 + 3d \implies 85 = 94 + 3d \implies d = -3$

$a_1 = a_3 - 2d \implies a_1 = 94 - 2(-3) = 100$

$a_n = a_1 + (n-1)d$

$\quad = 100 + (n-1)(-3) = 103 - 3n$

27. $a_1 = 5,\ d = 6$

$a_1 = 5$

$a_2 = 5 + 6 = 11$

$a_3 = 11 + 6 = 17$

$a_4 = 17 + 6 = 23$

$a_5 = 23 + 6 = 29$

29. $a_1 = -10,\ \ d = -12$

$a_1 = -10$

$a_2 = -10 - 12 = -22$

$a_3 = -22 - 12 = -34$

$a_4 = -34 - 12 = -46$

$a_5 = -46 - 12 = -58$

31. $a_8 = 26,\ a_{12} = 42$

$26 = a_8 = a_1 + (n-1)d = a_1 + 7d$

$42 = a_{12} = a_1 + (n-1)d = a_1 + 11d$

Answer: $d = 4,\ a_1 = -2$

$a_1 = -2$

$a_2 = -2 + 4 = 2$

$a_3 = 2 + 4 = 6$

$a_4 = 6 + 4 = 10$

$a_5 = 10 + 4 = 14$

33. $a_3 = 19,\ a_{15} = -1.7$

$a_{15} = a_3 + 12d$

$-1.7 = 19 + 12d \implies d = -1.725$

$a_3 = a_1 + 2d \implies 19 = a_1 + 2(-1.725)$

$\qquad\qquad \implies a_1 = 22.45$

$a_2 = a_1 - 1.725 = 20.725$

$a_3 = 19$

$a_4 = 19 - 1.725 = 17.275$

$a_5 = 17.275 - 1.725 = 15.55$

35. $a_1 = 15,\ \ a_{k+1} = a_k + 4$

$a_2 = a_1 + 4 = 15 + 4 = 19$

$a_3 = 19 + 4 = 23$

$a_4 = 23 + 4 = 27$

$a_5 = 27 + 4 = 31$

$d = 4,\ \ a_n = 11 + 4n$

37. $a_1 = \frac{3}{5},\ a_{k+1} = -\frac{1}{10} + a_k$

$a_2 = -\frac{1}{10} + \frac{3}{5} = \frac{5}{10} = \frac{1}{2}$

$a_3 = -\frac{1}{10} + \frac{1}{2} = \frac{4}{10} = \frac{2}{5}$

$a_4 = -\frac{1}{10} + \frac{2}{5} = \frac{3}{10}$

$a_5 = -\frac{1}{10} + \frac{3}{10} = \frac{1}{5}$

$d = -\frac{1}{10}$

$a_n = \frac{7}{10} - \frac{1}{10}n$

39. $a_1 = 5,\ a_2 = 11 \implies d = 6$

$a_{10} = a_1 + 9d = 5 + 9(6) = 59$

41. $a_1 = 4.2,\ a_2 = 6.6 \implies d = 2.4$

$a_7 = a_1 + 6d = 4.2 + 6(2.4) = 18.6$

43. $a_n = 15 - \frac{3}{2}n$

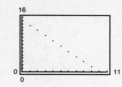

45. $a_n = 0.5n + 4$

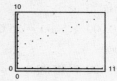

47. $a_n = 4n - 5$

n	1	2	3	4	5	6	7	8	9	10
a_n	-1	3	7	11	15	19	23	27	31	35

49. $a_n = 20 - \frac{3}{4}n$

n	1	2	3	4	5	6	7	8	9	10
a_n	19.25	18.5	17.75	17	16.25	15.5	14.75	14	13.25	12.5

51. $a_n = 1.5 + 0.05n$

n	1	2	3	4	5	6	7	8	9	10
a_n	1.55	1.6	1.65	1.7	1.75	1.8	1.85	1.9	1.95	2.0

53. $S_{10} = \frac{10}{2}(2 + 20) = 110$

55. $S_5 = \frac{5}{2}(-1 + (-9)) = -25$

57. $S_{50} = \frac{50}{2}(2 + 100) = 2550$

59. $S_{131} = \frac{131}{2}(-100 + 30) = -4585$

61. $8, 20, 32, 44, \ldots n = 10$

$a_1 = 8, a_2 = 20 \implies d = 12$

$a_{10} = a_1 + 9d = 8 + 9(12) = 116$

$S_{10} = \frac{n}{2}[a_1 + a_{10}] = \frac{10}{2}[8 + 116] = 620$

63. $a_1 = 0.5, a_2 = 1.3 \implies d = 0.8$

$a_{10} = a_1 + 9d = 0.5 + 9(0.8) = 7.7$

$S_{10} = \frac{10}{2}(a_1 + a_{10}) = 5(0.5 + 7.7) = 41$

65. $a_1 = 100, a_{25} = 220$

$S_{25} = \frac{25}{2}(a_1 + a_{25}) = 12.5(100 + 220) = 4000$

67. $a_1 = 1, \; a_{50} = 50, \; n = 50$

$\sum_{n=1}^{50} n = \frac{50}{2}(1 + 50) = 1275$

69. $a_1 = 5, \; a_{100} = 500, n = 100$

$\sum_{n=1}^{100} 5n = \frac{100}{2}(5 + 500) = 25{,}250$

71. $\sum_{n=11}^{30} n - \sum_{n=1}^{10} n = \frac{20}{2}(11 + 30) - \frac{10}{2}(1 + 10)$

$= 410 - 55 = 355$

73. $\sum_{n=1}^{500} (n + 8) = \frac{500}{2}[9 + 508] = 129{,}250$

75. $\sum_{n=1}^{20} (2n + 1) = 440$

77. $\sum_{n=1}^{100} \frac{n + 1}{2} = 2575$

79. $\sum_{i=1}^{60} \left(250 - \frac{2}{5}i\right) = 14{,}268$

81. $a_1 = 14, \; a_{18} = 31$

$S_{18} = \frac{18}{2}(14 + 31) = 405 \text{ bricks}$

83. $a_1 = 20{,}000$

$a_2 = 20{,}000 + 5000 = 25{,}000$

$d = 5000$

$a_5 = 20{,}000 + 4(5000) = 40{,}000$

$S_5 = \frac{5}{2}(20{,}000 + 40{,}000) = 150{,}000$

85. (a) $S_n = 0.91n + 5.7$

(b)

Year	1997	1998	1999	2000	2001	2002	2003	2004
Sales (Billions of $)	12.1	13.0	13.9	14.8	15.7	16.6	17.5	18.4

The model is a good fit.

(c) Total $= \frac{8}{2}(12.1 + 18.4) = \$122$ billion

(d) For 2005, $n = 15$ and $S_{15} = 19.35$.

For 2012, $n = 22$ and $S_{22} = 25.72$.

Total $= \frac{8}{2}(19.35 + 25.72) = \$180.3$ billion

Answers will vary.

87. True. Given a_1 and a_2, you know $d = a_2 - a_1$. Thus, $a_n = a_1 + (n-1)d$.

89. $a_1 = x$ $a_6 = 11x$

$a_2 = x + 2x = 3x$ $a_7 = 13x$

$a_3 = 3x + 2x = 5x$ $a_8 = 15x$

$a_4 = 7x$ $a_9 = 17x$

$a_5 = 9x$ $a_{10} = 19x$

91. $a_{20} = a_1 + 19(3) = a_1 + 57$

$$S = \frac{n}{2}(a_1 + a_{20})$$

$$= \frac{20}{2}(a_1 + (a_1 + 57)) = 650$$

$10(2a_1 + 57) = 650$

$20a_1 = 80$

$a_1 = 4$

93. (a) $-7, -4, -1, 2, 5, 8, 11$

$a_{n+1} = a_n + 3, \; a_1 = -7$

(b) $17, 23, 29, 35, 41, 47, 53, 57$

$a_{n+1} = a_n + 6, \; a_1 = 17$

(c) Not arithmetic

(d) $4, 7.5, 11, 14.5, 18, 21.5, 25, 28.5$

$a_{n+1} = a_n + 3.5, \; a_1 = 4$

(e) Not arithmetic

95. $S = \dfrac{n(n+1)}{2} = \dfrac{200(201)}{2} = 20{,}100$

97. $S = 1 + 3 + 5 + \cdots + 101$

$= (1 + 2 + 3 + \cdots + 101) - (2 + 4 + \cdots + 100)$

$= \dfrac{101(102)}{2} - 2\left(\dfrac{50(51)}{2}\right)$

$= 5151 - 2550 = 2601$

99. $\begin{bmatrix} 2 & -1 & 7 & \vdots & -10 \\ 3 & 2 & -4 & \vdots & 17 \\ 6 & -5 & 1 & \vdots & -20 \end{bmatrix}$ row reduces to $\begin{bmatrix} 1 & 0 & 0 & \vdots & 1 \\ 0 & 1 & 0 & \vdots & 5 \\ 0 & 0 & 1 & \vdots & -1 \end{bmatrix}$.

Answer: $(1, 5, -1)$

101. $\begin{vmatrix} 0 & 0 & 1 \\ 4 & -3 & 1 \\ 2 & 6 & 1 \end{vmatrix} = 30$

Area $= \frac{1}{2}(30) = 15$ square units

103. Answers will vary.

Section 6.3   Geometric Sequences and Series

- ■ You should be able to identify a geometric sequence, find its common ratio, and find the *n*th term.
- ■ You should be able to find the *n*th partial sum of a geometric sequence with common ratio *r* using the formula.

$$S_n = a_1\left(\frac{1 - r^n}{1 - r}\right)$$

- ■ You should know that if $|r| < 1$, then

$$\sum_{n=1}^{\infty} a_1 r^{n-1} = \frac{a_1}{1 - r}.$$

Vocabulary Check

1. geometric, common

2. $a_n = a_1 r^{n-1}$

3. $S_n = \sum_{i=1}^{n} a_1 r^{i-1} = a_1\left(\frac{1 - r^n}{1 - r}\right)$

4. geometric series

5. $S = \sum_{i=0}^{\infty} a_1 r^i = \frac{a_1}{1 - r}$

---

1. 5, 15, 45, 135, . . .

Geometric sequence

$r = 3$

3. 6, 18, 30, 42, . . .

Not a geometric sequence

(**Note:** It is an arithmetic sequence with $d = 12$.)

5. $1, -\frac{1}{2}, \frac{1}{4}, -\frac{1}{8}, . . .$

Geometric sequence

$r = -\frac{1}{2}$

7. $\frac{1}{8}, \frac{1}{4}, \frac{1}{2}, 1, . . .$

Geometric sequence

$r = 2$

9. $1, \frac{1}{2}, \frac{1}{3}, \frac{1}{4}, . . .$

Not a geometric sequence

11. $a_1 = 6, r = 3$

$a_2 = 6(3) = 18$

$a_3 = 18(3) = 54$

$a_4 = 54(3) = 162$

$a_5 = 162(3) = 486$

13. $a_1 = 1$, $r = \frac{1}{2}$

$a_1 = 1$

$a_2 = 1\left(\frac{1}{2}\right) = \frac{1}{2}$

$a_3 = \frac{1}{2}\left(\frac{1}{2}\right) = \frac{1}{4}$

$a_4 = \frac{1}{4}\left(\frac{1}{2}\right) = \frac{1}{8}$

$a_5 = \frac{1}{8}\left(\frac{1}{2}\right) = \frac{1}{16}$

15. $a_1 = 5$, $r = -\frac{1}{10}$

$a_1 = 5$

$a_2 = 5\left(-\frac{1}{10}\right) = -\frac{1}{2}$

$a_3 = \left(-\frac{1}{2}\right)\left(-\frac{1}{10}\right) = \frac{1}{20}$

$a_4 = \frac{1}{20}\left(-\frac{1}{10}\right) = -\frac{1}{200}$

$a_5 = \left(-\frac{1}{200}\right)\left(-\frac{1}{10}\right) = \frac{1}{2000}$

17. $a_1 = 1$, $r = e$

$a_1 = 1$

$a_2 = 1(e) = e$

$a_3 = (e)(e) = e^2$

$a_4 = (e^2)(e) = e^3$

$a_5 = (e^3)(e) = e^4$

19. $a_1 = 64$, $a_{k+1} = \frac{1}{2}a_k$

$a_1 = 64$

$a_2 = \frac{1}{2}(64) = 32$

$a_3 = \frac{1}{2}(32) = 16$

$a_4 = \frac{1}{2}(16) = 8$

$a_5 = \frac{1}{2}(8) = 4$

$r = \frac{1}{2}$, $a_n = 64\left(\frac{1}{2}\right)^{n-1} = 128\left(\frac{1}{2}\right)^n$

21. $a_1 = 9$, $a_{k+1} = 2a_k$

$a_2 = 2(9) = 18$

$a_3 = 2(18) = 36$

$a_4 = 2(36) = 72$

$a_5 = 2(72) = 144$

$r = 2$

$a_n = \left(\frac{9}{2}\right)2^n = 9(2^{n-1})$

23. $a_1 = 6$, $a_{k+1} = -\frac{3}{2}a_k$

$a_1 = 6$

$a_2 = -\frac{3}{2}(6) = -9$

$a_3 = -\frac{3}{2}(-9) = \frac{27}{2}$

$a_4 = -\frac{3}{2}\left(\frac{27}{2}\right) = -\frac{81}{4}$

$a_5 = -\frac{3}{2}\left(-\frac{81}{4}\right) = \frac{243}{8}$

$r = -\frac{3}{2}$, $a_n = 6\left(-\frac{3}{2}\right)^{n-1} = -4\left(-\frac{3}{2}\right)^n$

25. $a_1 = 4$, $a_4 = \frac{1}{2}$, $n = 10$

$a_1 r^3 = a_4$

$4r^3 = \frac{1}{2}$

$r^3 = \frac{1}{8}$

$r = \frac{1}{2}$

$a_{10} = a_4 r^6 = \frac{1}{2}\left(\frac{1}{2}\right)^6 = \frac{1}{2^7} = \frac{1}{128}$

27. $a_1 = 6$, $r = -\frac{1}{3}$, $n = 12$

$a_n = a_1 r^{n-1}$

$a_{12} = 6\left(-\frac{1}{3}\right)^{11} = \frac{-2}{3^{10}}$

29. $a_1 = 500$, $r = 1.02$, $n = 14$

$a_n = a_1 r^{n-1}$

$a_{14} = 500(1.02)^{13} \approx 646.8$

31. $a_2 = a_1 r = -18 \implies a_1 = \frac{-18}{r}$

$a_5 = a_1 r^4 = (a_1 r)r^3 = -18r^3 = \frac{2}{3} \implies r = -\frac{1}{3}$

$a_1 = \frac{-18}{r} = \frac{-18}{-1/3} = 54$

$a_6 = a_1 r^5 = 54\left(\frac{-1}{3}\right)^5 = \frac{-54}{243} = -\frac{2}{9}$

33. 7, 21, 63

$r = 3$

$a_n = 7(3)^{n-1}$

$a_9 = 7(3)^{9-1} = 45{,}927$

35. 5, 30, 180

$r = \frac{30}{5} = 6$

$a_n = 5(6)^{n-1}$

$a_{10} = 5(6)^{10-1} = 50{,}388{,}480$

37. $a_n = 12(-0.75)^{n-1}$

39. $a_n = 2(1.3)^{n-1}$

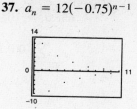

41. 8, -4, 2, -1, $\frac{1}{2}$

$S_1 = 8$

$S_2 = 8 + (-4) = 4$

$S_3 = 8 + (-4) + 2 = 6$

$S_4 = 8 + (-4) + 2 + (-1) = 5$

43. $\displaystyle\sum_{n=1}^{\infty} 16\left(-\frac{1}{2}\right)^{n-1}$

n	1	2	3	4	5	6	7	8	9	10
S_n	16	24	28	30	31	31.5	31.75	31.875	31.9375	31.96875

45. $\displaystyle\sum_{n=1}^{9} 2^{n-1} \Rightarrow a_1 = 1,\ r = 2$

$S_9 = \dfrac{1(1 - 2^9)}{1 - 2} = 511$

47. $\displaystyle\sum_{i=1}^{7} 64\left(-\frac{1}{2}\right)^{i-1} \Rightarrow a_1 = 64,\ r = -\frac{1}{2}$

$S_7 = 64\left[\dfrac{1 - (-1/2)^7}{1 - (-1/2)}\right] = \dfrac{128}{3}\left[1 - \left(-\dfrac{1}{2}\right)^7\right] = 43$

49. $\displaystyle\sum_{n=0}^{20} 3\left(\frac{3}{2}\right)^n = \sum_{n=1}^{21} 3\left(\frac{3}{2}\right)^{n-1} \Rightarrow a_1 = 3,\ r = \frac{3}{2}$

$S_{21} = 3\left[\dfrac{1 - (3/2)^{21}}{1 - (3/2)}\right]$

$= -6\left[1 - \left(\dfrac{3}{2}\right)^{21}\right] \approx 29{,}921.31$

51. $\displaystyle\sum_{i=1}^{10} 8\left(-\frac{1}{4}\right)^{i-1} \Rightarrow a_1 = 8,\ r = -\frac{1}{4}$

$S_{10} = 8\left[\dfrac{1 - (-1/4)^{10}}{1 - (-1/4)}\right] = \dfrac{32}{5}\left[1 - \left(-\dfrac{1}{4}\right)^{10}\right] \approx 6.4$

53. $\displaystyle\sum_{n=0}^{5} 300(1.06)^n = \sum_{n=1}^{6} 300(1.06)^{n-1} \Rightarrow a_1 = 300,\ r = 1.06$

$S_6 = 300\left[\dfrac{1 - (1.06)^6}{1 - 1.06}\right] \approx 2092.60$

55. $5 + 15 + 45 + \cdots + 3645$

$r = 3$ and $3645 = 5(3)^{n-1} \Rightarrow n = 7$

Thus, the sum can be written as $\displaystyle\sum_{n=1}^{7} 5(3)^{n-1}$.

57. $2 - \frac{1}{2} + \frac{1}{8} - \cdots + \frac{1}{2048}$

$r = -\frac{1}{4}$ and $\frac{1}{2048} = 2\left(-\frac{1}{4}\right)^{n-1} \Rightarrow n = 7$

$\displaystyle\sum_{n=1}^{7} 2\left(-\frac{1}{4}\right)^{n-1}$

59. $a_1 = 10,\ r = \dfrac{4}{5}$

$\displaystyle\sum_{n=0}^{\infty} 10\left(\frac{4}{5}\right)^n = \frac{a_1}{1 - r} = \frac{10}{1 - \frac{4}{5}} = 50$

61. $a_1 = 5,\ r = -\dfrac{1}{2}$

$\displaystyle\sum_{n=0}^{\infty} 5\left(-\frac{1}{2}\right)^n = \frac{a_1}{1 - r} = \frac{5}{1 - \left(-\frac{1}{2}\right)} = \frac{5}{\left(\frac{3}{2}\right)} = \frac{10}{3}$

63. $\sum\limits_{n=1}^{\infty} 2\left(\frac{7}{3}\right)^{n-1}$ does not have a finite sum $\left(\frac{7}{3} > 1\right)$.

65. $a_1 = 10, r = 0.11$

$$\sum_{n=0}^{\infty} 10(0.11)^n = \frac{a_1}{1-r} = \frac{10}{1-0.11} = \frac{10}{0.89}$$

$$= \frac{1000}{89} \approx 11.236$$

67. $a_1 = -3, r = -0.9$

$$\sum_{n=0}^{\infty} -3(-0.9)^n = \frac{a_1}{1-r} = \frac{-3}{1-(-0.9)}$$

$$= \frac{-3}{1.9} = \frac{-30}{19} \approx -1.579$$

69. $8 + 6 + \frac{9}{2} + \frac{27}{8} + \cdots = \sum\limits_{n=0}^{\infty} 8\left(\frac{3}{4}\right)^n$

$$= \frac{8}{1-3/4} = 32$$

71. $3 - 1 + \frac{1}{3} - \frac{1}{9} + \cdots = \sum\limits_{n=0}^{\infty} 3\left(-\frac{1}{3}\right)^n = \frac{a_1}{1-r} = \frac{3}{1-(-1/3)} = 3\left(\frac{3}{4}\right) = \frac{9}{4}$

73. $0.\overline{36} = \sum\limits_{n=0}^{\infty} 0.36(0.01)^n$

$$= \frac{0.36}{1-0.01} = \frac{0.36}{0.99} = \frac{36}{99} = \frac{4}{11}$$

75. $1.2\overline{5} = 1.2 + \sum\limits_{n=0}^{\infty} 0.05(0.1)^n$

$$= \frac{6}{5} + \frac{0.05}{1-0.1}$$

$$= \frac{6}{5} + \frac{0.05}{0.9}$$

$$= \frac{6}{5} + \frac{5}{90} = \frac{113}{90}$$

77. $A = P\left(1 + \frac{r}{n}\right)^{nt} = 1000\left(1 + \frac{0.03}{n}\right)^{n(10)}$

   (a) $n = 1$: $A = 1000(1 + 0.03)^{10} \approx 1343.92$

   (b) $n = 2$: $A = 1000\left(1 + \frac{0.03}{2}\right)^{2(10)} \approx 1346.86$

   (c) $n = 4$: $A = 1000\left(1 + \frac{0.03}{4}\right)^{4(10)} \approx 1348.35$

   (d) $n = 12$: $A = 1000\left(1 + \frac{0.03}{12}\right)^{12(10)} \approx 1349.35$

   (e) $n = 365$: $A = 1000\left(1 + \frac{0.03}{365}\right)^{365(10)} \approx 1349.84$

79. $A = \sum\limits_{n=1}^{60} 100\left(1 + \frac{0.03}{12}\right)^n$

$$= 100\left(1 + \frac{0.03}{12}\right) \cdot \frac{[1 - (1 + 0.03/12)^{60}]}{[1 - (1 + 0.03/12)]}$$

$$= 100(1.0025) \cdot \left[\frac{1 - 1.0025^{60}}{1 - 1.0025}\right]$$

$$\approx \$6480.83$$

81. Let $N = 12t$ be the total number of deposits.

$$A = P\left(1 + \frac{r}{12}\right) + P\left(1 + \frac{r}{12}\right)^2 + \cdots + P\left(1 + \frac{r}{12}\right)^N$$

$$= \left(1 + \frac{r}{12}\right)\left[P + P\left(1 + \frac{r}{12}\right) + \cdots + P\left(1 + \frac{r}{12}\right)^{N-1}\right]$$

$$= P\left(1 + \frac{r}{12}\right)\sum_{n=1}^{N}\left(1 + \frac{r}{12}\right)^{n-1}$$

$$= P\left(1 + \frac{r}{12}\right)\frac{1 - \left(1 + \frac{r}{12}\right)^N}{1 - \left(1 + \frac{r}{12}\right)}$$

$$= P\left(1 + \frac{r}{12}\right)\left(-\frac{12}{r}\right)\left[1 - \left(1 + \frac{r}{12}\right)^N\right]$$

$$= P\left(\frac{12}{r} + 1\right)\left[-1 + \left(1 + \frac{r}{12}\right)^N\right]$$

$$= P\left[\left(1 + \frac{r}{12}\right)^N - 1\right]\left(1 + \frac{12}{r}\right)$$

$$= P\left[\left(1 + \frac{r}{12}\right)^{12t} - 1\right]\left(1 + \frac{12}{r}\right)$$

83. $P = \$50,\ r = 7\%,\ t = 20$ years

(a) Compounded monthly: $A = 50\left[\left(1 + \frac{0.07}{12}\right)^{12(20)} - 1\right]\left(1 + \frac{12}{0.07}\right) \approx \$26{,}198.27$

(b) Compounded continuously: $A = \dfrac{50e^{0.07/12}(e^{0.07(20)} - 1)}{e^{0.07/12} - 1} \approx \$26{,}263.88$

85. $P = 100,\ r = 5\% = 0.05,\ t = 40$

(a) Compounded monthly: $A = 100\left[\left(1 + \frac{0.05}{12}\right)^{12(40)} - 1\right]\left(1 + \frac{12}{0.05}\right) \approx \$153{,}237.86$

(b) Compounded continuously: $A = \dfrac{100e^{0.05/12}(e^{0.05(40)} - 1)}{e^{0.05/12} - 1} \approx \$153{,}657.02$

87. First shaded area: $\dfrac{16^2}{4}$

Second shaded area: $\dfrac{16^2}{4} + \dfrac{1}{2} \cdot \dfrac{16^2}{4}$

Third shaded area: $\dfrac{16^2}{4} + \dfrac{1}{2}\dfrac{16^2}{4} + \dfrac{1}{4}\dfrac{16^2}{4}$, etc

Total area of shaded region: $\dfrac{16^2}{4}\sum_{n=0}^{5}\left(\dfrac{1}{2}\right)^n = 64\left[\dfrac{1 - (1/2)^6}{1 - 1/2}\right] = 128\left(1 - \left(\dfrac{1}{2}\right)^6\right) = 126$ square units

89. (a) $a_0 = 70$ degrees

$a_1 = 0.8(70) = 56$ degrees

$\vdots$

$a_n = (0.8)^n(70)$

(b) $a_6 = (0.8)^6(70) \approx 18.35$ degrees

$a_{12} = (0.8)^{12}(70) \approx 4.81$ degrees

(c)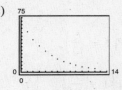

$a_3 \approx 35.8$

$a_4 \approx 28.7$

Thus, the water freezes between 3 and 4 hours, about 3.5 hours.

91. $400 + 0.75(400) + (0.75)^2(400) + \cdots = \displaystyle\sum_{n=0}^{\infty} 400(0.75)^n$

$$= \frac{400}{1 - 0.75} = \$1600$$

93. $250 + 0.80(250) + (0.80)^2(250) + \cdots = \displaystyle\sum_{n=0}^{\infty} 250(0.80)^2$

$$= \frac{250}{1 - 0.80} = \$1250$$

95. $600 + 0.725(600) + (0.725)^2(600) + \cdots = \displaystyle\sum_{n=0}^{\infty} 600(0.725)^n$

$$= \frac{600}{1 - 0.725} \approx \$2181.82$$

97. (a) Option 1: $30{,}000 + 1.025(30{,}000) + \cdots + (1.025)^4(30{,}000) = \displaystyle\sum_{n=0}^{4} 30{,}000(1.025)^n$

$$\approx \$157{,}689.86$$

Option 2: $32{,}500 + 1.02(32{,}500) + \cdots + (1.02)^4(32{,}500) = \displaystyle\sum_{n=0}^{4} 32{,}500(1.02)^n$

$$\approx \$169{,}131.31$$

Option 2 has the larger cumulative amount.

(b) Option 1: $(1.025)^4(30{,}000) \approx \$33{,}114.39$

Option 2: $(1.02)^4(32{,}500) \approx \$35{,}179.05$

Option 2 has the larger amount.

99. (a) Downward: $850 + 0.75(850) + (0.75)^2(850) + \cdots + (0.75)^9(850) = \displaystyle\sum_{n=0}^{9} 850(0.75)^n$

$$\approx 3208.53 \text{ feet}$$

Upward: $0.75(850) + (0.75)^2(850) + \cdots + (0.75)^{10}(850) = \displaystyle\sum_{n=0}^{9} (0.75)(850)(0.75)^n$

$$= \displaystyle\sum_{n=0}^{9} 637.5(0.75)^n \approx 2406.4 \text{ feet}$$

Total distance: $3208.53 + 2406.4 = 5614.93$ feet

(b) $\displaystyle\sum_{n=0}^{\infty} 850(0.75)^n + \displaystyle\sum_{n=0}^{\infty} 637.5(0.75)^n = \frac{850}{1 - 0.75} + \frac{637.5}{1 - 0.75} = 5950$ feet

101. False. See definition page 535.

103. $a_1 = 3,\ r = \dfrac{x}{2}$

$$a_2 = 3\left(\frac{x}{2}\right) = \frac{3x}{2}$$

$$a_3 = \frac{3x}{2}\left(\frac{x}{2}\right) = \frac{3x^2}{4}$$

$$a_4 = \frac{3x^2}{4}\left(\frac{x}{2}\right) = \frac{3x^3}{8}$$

$$a_5 = \frac{3x^3}{8}\left(\frac{x}{2}\right) = \frac{3x^4}{16}$$

105. $a_1 = 100,\ r = e^x,\ n = 9$

$$a_n = a_1 r^{n-1}$$

$$a_9 = 100(e^x)^8 = 100e^{8x}$$

107. (a) $f(x) = 6\left[\dfrac{1 - 0.5^x}{1 - 0.5}\right]$

$$\sum_{n=0}^{\infty} 6\left(\frac{1}{2}\right)^n = \frac{6}{1 - 1/2} = 12$$

The horizontal asymptote of $f(x)$ is $y = 12$.
This corresponds to the sum of the series.

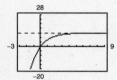

(b) $f(x) = 2\left[\dfrac{1 - 0.8^x}{1 - 0.8}\right]$

$$\sum_{n=0}^{\infty} 2\left(\frac{4}{5}\right)^n = \frac{2}{1 - 4/5} = 10$$

The horizontal asymptote of $f(x)$ is $y = 10$.
This corresponds to the sum of the series.

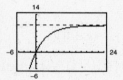

109. To use the first two terms of a geometric series to find the nth term, first divide the second term by the first term to obtain the constant ratio. The nth term is the first term multiplied by the common ratio raised to the $(n - 1)$ power.

$$r = \frac{a_2}{a_1},\ a_n = a_1 r^{n-1}$$

111. $\text{Time} = \dfrac{\text{Distance}}{\text{Speed}} = \dfrac{200}{50} + \dfrac{200}{42} = 200\left[\dfrac{92}{2100}\right]$ hours

$\text{Speed} = \dfrac{\text{Distance}}{\text{Time}} = \dfrac{400}{200\left[92/2100\right]} = \dfrac{2(2100)}{92} \approx 45.65$ mph

113. $\det\begin{bmatrix} -1 & 3 & 4 \\ -2 & 8 & 0 \\ 2 & 5 & -1 \end{bmatrix} = 4(-10 - 16) - 1(-8 + 6)$

$$= -104 + 2 = -102$$

115. Answers will vary.

Section 6.4   Mathematical Induction

- ■ You should be sure that you understand the principle of mathematical induction. If P_n is a statement involving the positive integer n, where P_1 is true and the truth of P_k implies the truth of P_{k+1}, then P_n is true for all positive integers n.
- ■ You should be able to verify (by induction) the formulas for the sums of powers of integers and be able to use these formulas.
- ■ You should be able to work with finite differences.

Vocabulary Check

1. mathematical induction

2. first

3. arithmetic

4. second

---

1. $P_k = \dfrac{5}{k(k+1)}$

$P_{k+1} = \dfrac{5}{(k+1)[(k+1)+1]} = \dfrac{5}{(k+1)(k+2)}$

3. $P_k = \dfrac{2^k}{(k+1)!}$

$P_{k+1} = \dfrac{2^{k+1}}{((k+1)+1)!} = \dfrac{2^{k+1}}{(k+2)!}$

5. $P_k = 1 + 6 + 11 + \cdots + [5(k-1) - 4] + [5k - 4]$

$P_{k+1} = 1 + 6 + 11 + \cdots + [5k - 4] + [5(k+1) - 4]$

$= 1 + 6 + 11 + \cdots + [5k - 4] + [5k + 1]$

7. 1. When $n = 1$, $S_1 = 2 = 1(1 + 1)$.

   2. Assume that

   $S_k = 2 + 4 + 6 + 8 + \cdots + 2k = k(k+1)$.

   Then,

   $S_{k+1} = 2 + 4 + 6 + 8 + \cdots + 2k + 2(k+1)$

   $= S_k + 2(k+1) = k(k+1) + 2(k+1) = (k+1)(k+2)$.

   Therefore, by mathematical induction, the formula is valid for all positive integer values of n.

9. 1. When $n = 1$, $S_1 = 3 = \dfrac{1}{2}(5(1) + 1)$

   2. Assume that $S_k = 3 + 8 + 13 + \cdots + (5k - 2) = \dfrac{k}{2}(5k + 1)$.

   Then, $S_{k+1} = 3 + 8 + 13 + \cdots + (5k - 2) + [5(k+1) - 2]$

   $= S_k + [5k + 3] = \dfrac{k}{2}(5k + 1) + 5k + 3$

   $= \dfrac{1}{2}[5k^2 + 11k + 6] = \dfrac{1}{2}(k+1)(5k + 6)$

   $= \dfrac{1}{2}(k+1)(5(k+1) + 1)$.

   Therefore, by mathematical induction, the formula is valid for all positive integer values of n.

11. 1. When $n = 1$, $S_1 = 1 = 2^1 - 1$.

2. Assume that

$$S_k = 1 + 2 + 2^2 + 2^3 + \cdots + 2^{k-1} = 2^k - 1.$$

Then,

$$S_{k+1} = 1 + 2 + 2^2 + 2^3 + \cdots + 2^{k-1} + 2^k$$
$$= S_k + 2^k = 2^k - 1 + 2^k = 2(2^k) - 1 = 2^{k+1} - 1.$$

Therefore, by mathematical induction, the formula is valid for all positive integer values of n.

13. 1. When $n = 1$, $S_1 = 1 = \dfrac{1(1 + 1)}{2}$.

2. Assume that

$$S_k = 1 + 2 + 3 + 4 + \cdots + k = \frac{k(k + 1)}{2}.$$

Then,

$$S_{k+1} = 1 + 2 + 3 + 4 + \cdots + k + (k + 1)$$
$$= S_k + (k + 1) = \frac{k(k + 1)}{2} + \frac{2(k + 1)}{2} = \frac{(k + 1)(k + 2)}{2}.$$

Therefore, the formula is valid for all positive integer values of n.

15. 1. When $n = 1$,

$$S_1 = 1^4 = \frac{1(1 + 1)(2 \cdot 1 + 1)(3 \cdot 1^2 + 3 \cdot 1 - 1)}{30}.$$

2. Assume that $S_k = \displaystyle\sum_{i=1}^{k} i^4 = \dfrac{k(k + 1)(2k + 1)(3k^2 + 3k - 1)}{30}.$

Then, $S_{k+1} = S_k + (k + 1)^4$

$$= \frac{k(k + 1)(2k + 1)(3k^2 + 3k - 1)}{30} + (k + 1)^4 = \frac{k(k + 1)(2k + 1)(3k^2 + 3k - 1) + 30(k + 1)^4}{30}$$

$$= \frac{(k + 1)[k(2k + 1)(3k^2 + 3k - 1) + 30(k + 1)^3]}{30} = \frac{(k + 1)(6k^4 + 39k^3 + 91k^2 + 89k + 30)}{30}$$

$$= \frac{(k + 1)(k + 2)(2k + 3)(3k^2 + 9k + 5)}{30} = \frac{(k + 1)(k + 2)(2(k + 1) + 1)(3(k + 1)^2 + 3(k + 1) - 1)}{30}.$$

Therefore, the formula is valid for all positive integer values of n.

17. 1. When $n = 1$, $S_1 = 2 = \dfrac{1(2)(3)}{3}$.

2. Assume that $S_k = 1(2) + 2(3) + 3(4) + \cdots + k(k + 1) = \dfrac{k(k + 1)(k + 2)}{3}$.

Then,

$$S_{k+1} = 1(2) + 2(3) + 3(4) + \cdots + k(k + 1) + (k + 1)(k + 2)$$

$$= S_k + (k + 1)(k + 2)$$

$$= \frac{k(k + 1)(k + 2)}{3} + \frac{3(k + 1)(k + 2)}{3}$$

$$= \frac{(k + 1)(k + 2)(k + 3)}{3}.$$

Therefore, the formula is valid for all positive integer values of n.

19. 1. When $n = 1$, $S_1 = \dfrac{1}{1(1 + 1)} = \dfrac{1}{2}$

2. Assume $S_k = \displaystyle\sum_{i=1}^{k} \frac{1}{i(i + 1)} = \frac{k}{k + 1}$.

Thus,

$$S_{k+1} = \sum_{i=1}^{k+1} \frac{1}{i(i + 1)}$$

$$= \frac{1}{1(2)} + \frac{1}{2(3)} + \cdots + \frac{1}{k(k + 1)} + \frac{1}{(k + 1)(k + 2)}$$

$$= \frac{k}{k + 1} + \frac{1}{(k + 1)(k + 2)}$$

$$= \frac{k(k + 2) + 1}{(k + 2)(k + 2)}$$

$$= \frac{(k + 1)^2}{(k + 1)(k + 2)}$$

$$= \frac{k + 1}{k + 2}$$

$$= \frac{k + 1}{(k + 1) + 1}.$$

Therefore, the formula is valid for all positive integer values of n.

21. $\displaystyle\sum_{n=1}^{50} n^3 = \frac{50^2(50 + 1)^2}{4} = 1{,}625{,}625$

23. $\displaystyle\sum_{n=1}^{12} (n^2 - n) = \sum_{n=1}^{12} n^2 - \sum_{n=1}^{12} n$

$$= \frac{12(12 + 1)(2 \cdot 12 + 1)}{6} - \frac{12(12 + 1)}{2}$$

$$= 650 - 78 = 572$$

25. 1. When $n = 4$, $4! = 24$ and $2^4 = 16$, thus $4! > 2^4$.

2. Assume $k! > 2^k$, $k > 4$. Then, $(k + 1)! = k!(k + 1) > 2^k(2)$ since $k + 1 > 2$. Thus, $(k + 1)! > 2^{k+1}$.

Therefore, by mathematical induction, the formula is valid for all integers n such that $n \geq 4$.

27. 1. When $n = 2$, $\dfrac{1}{\sqrt{1}} + \dfrac{1}{\sqrt{2}} \approx 1.707$ and $\sqrt{2} \approx 1.414$, thus $\dfrac{1}{\sqrt{1}} + \dfrac{1}{\sqrt{2}} > \sqrt{2}$.

2. Assume $\dfrac{1}{\sqrt{1}} + \dfrac{1}{\sqrt{2}} + \dfrac{1}{\sqrt{3}} + \cdots + \dfrac{1}{\sqrt{k}} > \sqrt{k}$, $k > 2$.

Then, $\dfrac{1}{\sqrt{1}} + \dfrac{1}{\sqrt{2}} + \dfrac{1}{\sqrt{3}} + \cdots + \dfrac{1}{\sqrt{k}} + \dfrac{1}{\sqrt{k+1}} > \sqrt{k} + \dfrac{1}{\sqrt{k+1}}$.

Now we need to show that $\sqrt{k} + \dfrac{1}{\sqrt{k+1}} > \sqrt{k+1}$, $k > 2$.

This is true because $\sqrt{k(k+1)} > k$

$$\sqrt{k(k+1)} + 1 > k + 1$$

$$\frac{\sqrt{k(k+1)} + 1}{\sqrt{k+1}} > \frac{k+1}{\sqrt{k+1}}$$

$$\sqrt{k} + \frac{1}{\sqrt{k+1}} > \sqrt{k+1}.$$

Therefore, $\dfrac{1}{\sqrt{1}} + \dfrac{1}{\sqrt{2}} + \dfrac{1}{\sqrt{3}} + \cdots + \dfrac{1}{\sqrt{k}} + \dfrac{1}{\sqrt{k+1}} > \sqrt{k+1}$.

Therefore, by mathematical induction, the formula is valid for all integers n such that $n \geq 2$.

29. 1. When $n = 1$, $1 + a \geq a$ since $1 > 0$.

2. Assume $(1 + a)^k \geq ka$.

Then, $(1 + a)^{k+1} = (1 + a)^k(1 + a) \geq ka(1 + a)$

$$= ka + ka^2 \geq ka + a \quad \text{(because } a > 1\text{)}$$

$$= (k + 1)a.$$

Therefore, by mathematical induction, the inequality is valid for all integers $n \geq 1$.

31. 1. When $n = 1$, $(ab)^1 = a^1 b^1 = ab$.

2. Assume that $(ab)^k = a^k b^k$.

Then, $(ab)^{k+1} = (ab)^k(ab)$

$$= a^k b^k ab$$

$$= a^{k+1} b^{k+1}.$$

Thus, $(ab)^n = a^n b^n$.

33. 1. When $n = 1$, $(x_1)^{-1} = x_1^{-1}$.

    2. Assume that

$$(x_1 x_2 x_3 \cdots x_k)^{-1} = x_1^{-1} x_2^{-1} x_3^{-1} \cdots x_k^{-1}.$$

    Then,

$$\begin{aligned}
(x_1 x_2 x_3 \cdots x_k x_{k+1})^{-1} &= [(x_1 x_2 x_3 \cdots x_k) x_{k+1}]^{-1} \\
&= (x_1 x_2 x_3 \cdots x_k)^{-1} x_{k+1}^{-1} \\
&= x_1^{-1} x_2^{-1} x_3^{-1} \cdots x_k^{-1} x_{k+1}^{-1}.
\end{aligned}$$

Thus, the formula is valid.

35. 1. When $n = 1$, $x(y_1) = xy_1$.

    2. Assume that $x(y_1 + y_2 + \cdots + y_k) = xy_1 + xy_2 + \cdots + xy_k$.

    Then,

$$\begin{aligned}
xy_1 + xy_2 + \cdots + xy_k + xy_{k+1} &= x(y_1 + y_2 + \cdots + y_k) + xy_{k+1} \\
&= x[(y_1 + y_2 + \cdots + y_k) + y_{k+1}] \\
&= x(y_1 + y_2 + \cdots + y_k + y_{k+1}).
\end{aligned}$$

Hence, the formula holds.

37. 1. When $n = 1$, $[1^3 + 3(1)^2 + 2(1)] = 6$ and 3 is a factor.

    2. Assume that 3 is a factor of $(k^3 + 3k^2 + 2k)$.

    Then,

$$\begin{aligned}
[(k + 1)^3 + 3(k + 1)^2 + 2(k + 1)] &= k^3 + 3k^2 + 3k + 1 + 3k^2 + 6k + 3 + 2k + 2 \\
&= (k^3 + 3k^2 + 2k) + (3k^2 + 9k + 6) \\
&= (k^3 + 3k^2 + 2k) + 3(k^2 + 3k + 2).
\end{aligned}$$

Since 3 is a factor of $(k^3 + 3k^2 + 2k)$ by our assumption, and 3 is a factor of $3(k^2 + 3k + 2)$ then 3 is a factor of the whole sum.

Thus, 3 is a factor of $(n^3 + 3n^2 + 2n)$ for every positive integer n.

39. 1. When $n = 1$, $[1^3 - 1 + 3] = 3$, and 3 is a factor.

    2. Assume that 3 is a factor of $k^3 - k + 3$. Then,

$$\begin{aligned}
[(k + 1)^3 - (k + 1) + 3] &= k^3 + 3k^2 + 3k + 1 - k - 1 + 3 \\
&= k^3 + 3k^2 + 2k + 3 \\
&= (k^3 - k + 3) + 3k^2 + 3k \\
&= (k^3 - k + 3) + 3(k^2 + k).
\end{aligned}$$

Since 3 is a factor of $k^3 - k + 3$ by our assumption, and 3 is a factor of $3(k^2 + k)$, then 3 is a factor of the whole sum.

Thus, 3 is a factor of $n^3 - n + 3$ for every positive integer n.

41. 1. When $n = 1$, $2^{2+1} + 1 = 9$, and 3 is a factor.

2. Assume that 3 is a factor of $2^{2k+1} + 1$.

Then,

$$2^{2(k+1)+1} + 1 = 2^{2k+3} + 1$$
$$= 4 \cdot 2^{2k+1} + 1$$
$$= (3 + 1)2^{2k+1} + 1$$
$$= (2^{2k+1} + 1) + 3 \cdot 2^{2k+1}.$$

Since 3 is a factor of $2^{2k+1} + 1$ by our assumption, and 3 is a factor of $3 \cdot 2^{2k+1}$, then 3 is a factor of the whole sum.

Thus, 3 is a factor of $2^{2n+1} + 1$ for every positive integer n.

43. $a_1 = 0$, $a_n = a_{n-1} + 3$

$a_1 = 0$

$a_2 = a_1 + 3 = 0 + 3 = 3$

$a_3 = a_2 + 3 = 3 + 3 = 6$

$a_4 = a_3 + 3 = 6 + 3 = 9$

$a_5 = a_4 + 3 = 9 + 3 = 12$

a_n: 0   3   6   9   12

First differences: 3   3   3   3

Second differences: 0   0   0

Since the first differences are equal, the sequence has a linear model.

45. $a_1 = 3$, $a_n = a_{n-1} - n$

$a_1 = 3$

$a_2 = a_1 - 2 = 3 - 2 = 1$

$a_3 = a_2 - 3 = 1 - 3 = -2$

$a_4 = a_3 - 4 = -2 - 4 = -6$

$a_5 = a_4 - 5 = -6 - 5 = -11$

a_n: 3   1   -2   -6   -11

First differences: -2   -3   -4   -5

Second differences: -1   -1   -1

Since the second differences are all the same, the sequence has a quadratic model.

47. $a_0 = 0$, $a_n = a_{n-1} + n$

$a_0 = 0$

$a_1 = a_0 + 1 = 0 + 1 = 1$

$a_2 = a_1 + 2 = 1 + 2 = 3$

$a_3 = a_2 + 3 = 3 + 3 = 6$

$a_4 = a_3 + 4 = 6 + 4 = 10$

a_n: 0   1   3   6   10

First differences: 1   2   3   4

Second differences: 1   1   1

Since the second differences are equal, the sequence has a quadratic model.

49. $a_1 = 2$, $a_n = a_{n-1} + 2$

$a_1 = 2$

$a_2 = a_1 + 2 = 2 + 2 = 4$

$a_3 = a_2 + 2 = 4 + 2 = 6$

$a_4 = a_3 + 2 = 6 + 2 = 8$

$a_5 = a_4 + 2 = 8 + 2 = 10$

a_n: 2   4   6   8   10

First differences: 2   2   2   2

Second differences: 0   0   0

Since the first differences are equal, the sequence has a linear model.

51. $a_1 = 3$, $a_2 = 3$, $a_3 = 5$

Let $a_n = an^2 + bn + c$.

$a_1 = a(1)^2 + b(1) + c = 3 \implies a + b + c = 3$

$a_2 = a(2)^2 + b(2) + c = 3 \implies 4a + 2b + c = 3$

$a_3 = a(3)^2 + b(3) + c = 5 \implies 9a + 3b + c = 5$

Solving the system, $a = 1$, $b = -3$, $c = 5$,
$a_n = n^2 - 3n + 5$, $n \geq 1$.

53. $a_0 = -3, a_2 = 1, a_4 = 9$

Let $a_n = an^2 + bn + c$. Then:

$a_0 = a(0)^2 + b(0) + c = -3 \implies c = -3$

$a_2 = a(2)^2 + b(2) + c = 1 \implies 4a + 2b + c = 1$

$$4a + 2b \qquad = 4$$

$$2a + b \qquad = 2$$

$a_4 = a(4)^2 + b(4) + c = 9 \implies 16a + 4b + c = 9$

$$16a + 4b \qquad = 12$$

$$4a + b \qquad = 3$$

By elimination: $-2a - b = -2$

$$\underline{4a + b = 3}$$

$$2a \qquad = 1$$

$$a = \tfrac{1}{2} \implies b = 1$$

Thus, $a_n = \tfrac{1}{2}n^2 + n - 3$.

55. (a) $n = 1$: 3 sides

$n = 2$: $3 \cdot 4 = 12$ sides

$n = 3$: $3 \cdot 4^2 = 48$ sides

nth Koch snowflake: $3(4)^{n-1}$ sides

To prove this, use mathematical induction.

1. For $n - 1$, the number of sides is $3 \cdot 4^{1-1} = 3$.

2. Assume that the number of sides of the kth Koch snowflake is $3 \cdot 4^{k-1}$. When the $(k + 1)^{st}$ Koch snowflake is created, each side is replaced with 4 sides. That is, the number of sides is increased by a factor of 4:

Number sides $= 4(3 \cdot 4^{k-1}) = 3 \cdot 4^k$.

Hence, the formula is valid for all positive integers n.

(b) $n = 1$: $A_1 = \dfrac{\sqrt{3}}{4}(1)^2 = \dfrac{\sqrt{3}}{4}$

$n = 2$: $A_2 = \dfrac{\sqrt{3}}{4}\left[1 + \dfrac{1}{3}\right]$

$n = 3$: $A_3 = \dfrac{\sqrt{3}}{4}\left[1 + \dfrac{1}{3} + \dfrac{1}{3}\left(\dfrac{4}{9}\right)\right]$

$n = 4$: $A_4 = \dfrac{\sqrt{3}}{4}\left[1 + \dfrac{1}{3} + \dfrac{1}{3}\left(\dfrac{4}{9}\right) + \dfrac{1}{3}\left(\dfrac{4}{9}\right)^2\right]$

$A_n = \dfrac{\sqrt{3}}{4}\left[1 + \displaystyle\sum_{k=2}^{n} \dfrac{1}{3}\left(\dfrac{4}{9}\right)^{k-2}\right], n > 1$

(c) For the nth Koch snowflake, the length of a single side is $(1/3)^{n-1}$, and the number of sides is $3 \cdot 4^{n-1}$. Hence, the perimeter is

$$\left(\dfrac{1}{3}\right)^{n-1} 3 \cdot 4^{n-1} = 3\left(\dfrac{4}{3}\right)^{n-1}.$$

57. False. P_1 might not even be defined.

59. False. It has $n - 2$ second differences.

61. $(2x^2 - 1)^2 = 4x^4 - 4x^2 + 1$

63. $(5 - 4x)^3 = -64x^3 + 240x^2 - 300x + 125$

65. $3\sqrt{-27} - \sqrt{-12} = 3\sqrt{3 \cdot 3 \cdot (-3)} - \sqrt{2 \cdot 2(-3)} = 9\sqrt{3}i - 2\sqrt{3}i = 7\sqrt{3}i$

67. $10\left(\sqrt[3]{64} - 2\sqrt[3]{-16}\right) = 10\left(4 - 2^2\sqrt[3]{-2}\right)$

$$= 40 - 40\sqrt[3]{-2}$$

$$= 40\left(1 + \sqrt[3]{2}\right)$$

Section 6.5   The Binomial Theorem

■ You should be able to use the Binomial Theorem

$$(x + y)^n = x^n + nx^{n-1}y + \frac{n(n-1)}{2!}x^{n-2}y^2 + \cdots + {}_nC_r x^{n-r}y^r + \cdots + y^n$$

where ${}_nC_r = \frac{n!}{(n-r)!r!}$, to expand $(x + y)^n$.

■ You should be able to use Pascal's Triangle.

Vocabulary Check

1. binomial coefficients

2. Binomial Theorem, Pascal's Triangle

3. ${}_nC_r$ or $\binom{n}{r}$

4. expanding, binomial

---

1. ${}_7C_5 = \frac{7!}{2!5!} = \frac{7 \cdot 6 \cdot 5!}{2 \cdot 5!} = \frac{42}{2} = 21$

3. $\binom{12}{0} = {}_{12}C_0 = \frac{12!}{0!12!} = 1$

5. ${}_{20}C_{15} = \frac{20!}{15!5!} = \frac{20 \cdot 19 \cdot 18 \cdot 17 \cdot 16}{5 \cdot 4 \cdot 3 \cdot 2 \cdot 1} = 15{,}504$

7. ${}_{14}C_1 = \frac{14!}{13!1!} = \frac{14 \cdot 13!}{13!} = 14$

9. $\binom{100}{98} = {}_{100}C_{98} = \frac{100!}{98!2!} = \frac{100 \cdot 99}{2 \cdot 1} = 4950$

11. ${}_{41}C_{36} = 749{,}398$

13. ${}_{100}C_{98} = 4950$

15. ${}_{250}C_2 = 31{,}125$

17. $(x + 2)^4 = {}_4C_0 x^4 + {}_4C_1 x^3(2) + {}_4C_2 x^2(2)^2 + {}_4C_3 x(2)^3 + {}_4C_4(2)^4$
$$= x^4 + 8x^3 + 24x^2 + 32x + 16$$

19. $(a + 3)^3 = {}_3C_0 a^3 + {}_3C_1 a^2(3) + {}_3C_2 a(3)^2 + {}_3C_3(3)^3$
$$= a^3 + 3a^2(3) + 3a(3)^2 + (3)^3 = a^3 + 9a^2 + 27a + 27$$

21. $(y - 2)^4 = {}_4C_0 y^4 - {}_4C_1 y^3(2) + {}_4C_2 y^2(2)^2 - {}_4C_3 y(2)^3 + {}_4C_4(2)^4$
$$= y^4 - 4y^3(2) + 6y^2(4) - 4y(8) + 16$$
$$= y^4 - 8y^3 + 24y^2 - 32y + 16$$

23. $(x + y)^5 = {}_5C_0 x^5 + {}_5C_1 x^4 y + {}_5C_2 x^3 y^2 + {}_5C_3 x^2 y^3 + {}_5C_4 xy^4 + {}_5C_5 y^5$
$$= x^5 + 5x^4 y + 10x^3 y^2 + 10x^2 y^3 + 5xy^4 + y^5$$

25. $(3r + 2s)^6 = {}_6C_0(3r)^6 + {}_6C_1(3r)^5(2s) + {}_6C_2(3r)^4(2s)^2 + {}_6C_3(3r)^3(2s)^3 + {}_6C_4(3r)^2(2s)^4 + {}_6C_5(3r)(2s)^5 + {}_6C_6(2s)^6$
$$= 729r^6 + 2916r^5 s + 4860r^4 s^2 + 4320r^3 s^3 + 2160r^2 s^4 + 576rs^5 + 64s^6$$

27. $(x - y)^5 = {}_5C_0x^5 - {}_5C_1x^4y + {}_5C_2x^3y^2 - {}_5C_3x^2y^3 + {}_5C_4xy^4 - {}_5C_5y^5$

$\qquad = x^5 - 5x^4y + 10x^3y^2 - 10x^2y^3 + 5xy^4 - y^5$

29. $(1 - 4x)^3 = {}_3C_0 1^3 - {}_3C_1 1^2(4x) + {}_3C_2 1(4x)^2 - {}_3C_3(4x)^3$

$\qquad = 1 - 3(4x) + 3(4x)^2 - (4x)^3$

$\qquad = 1 - 12x + 48x^2 - 64x^3$

31. $(x^2 + 2)^4 = {}_4C_0(x^2)^4 + {}_4C_1(x^2)^3(2) + {}_4C_2(x^2)^2 2^2 + {}_4C_3(x^2)2^3 + {}_4C_4(2)^4$

$\qquad = x^8 + 8x^6 + 24x^4 + 32x^2 + 16$

33. $(x^2 - 5)^5 = {}_5C_0(x^2)^5 - {}_5C_1(x^2)^4(5) + {}_5C_2(x^2)^3(5^2) - {}_5C_3(x^2)^2(5^3) + {}_5C_4(x^2)(5^4) - {}_5C_5(5^5)$

$\qquad = x^{10} - 25x^8 + 250x^6 - 1250x^4 + 3125x^2 - 3125$

35. $(x^2 + y^2)^4 = {}_4C_0(x^2)^4 + {}_4C_1(x^2)^3(y^2) + {}_4C_2(x^2)^2(y^2)^2 + {}_4C_3(x^2)(y^2)^3 + {}_4C_4(y^2)^4$

$\qquad = x^8 + 4x^6y^2 + 6x^4y^4 + 4x^2y^6 + y^8$

37. $(x^3 - y)^6 = {}_6C_0(x^3)^6 - {}_6C_1(x^3)^5 y + {}_6C_2(x^3)^4y^2 - {}_6C_3(x^3)^3 y^3 + {}_6C_4(x^3)^2y^4 - {}_6C_5(x^3)y^5 + {}_6C_6 y^6$

$\qquad = x^{18} - 6x^{15}y + 15x^{12}y^2 - 20x^9y^3 + 15x^6y^4 - 6x^3y^5 + y^6$

39. $\left(\dfrac{1}{x} + y\right)^5 = {}_5C_0\left(\dfrac{1}{x}\right)^5 + {}_5C_1\left(\dfrac{1}{x}\right)^4 y + {}_5C_2\left(\dfrac{1}{x}\right)^3 y^2 + {}_5C_3\left(\dfrac{1}{x}\right)^2 y^3 + {}_5C_4\left(\dfrac{1}{x}\right)y^4 + {}_5C_5 y^5$

$\qquad = \dfrac{1}{x^5} + \dfrac{5y}{x^4} + \dfrac{10y^2}{x^3} + \dfrac{10y^3}{x^2} + \dfrac{5y^4}{x} + y^5$

41. $\left(\dfrac{2}{x} - y\right)^4 = {}_4C_0\left(\dfrac{2}{x}\right)^4 - {}_4C_1\left(\dfrac{2}{x}\right)^3 y + {}_4C_2\left(\dfrac{2}{x}\right)^2 y^2 - {}_4C_3\left(\dfrac{2}{x}\right)y^3 + {}_4C_4 y^4$

$\qquad = \dfrac{16}{x^4} - \dfrac{32}{x^3}y + \dfrac{24}{x^2}y^2 - \dfrac{8}{x}y^3 + y^4$

43. $(4x - 1)^3 - 2(4x - 1)^4 = (64x^3 - 48x^2 + 12x - 1) - 2(256x^4 - 256x^3 + 96x^2 - 16x + 1)$

$\qquad = -512x^4 + 576x^3 - 240x^2 + 44x - 3$

45. $2(x - 3)^4 + 5(x - 3)^2 = 2[x^4 - 4(x^3)(3) + 6(x^2)(3^2) - 4(x)(3^3) + 3^4] + 5[x^2 - 2(x)(3) + 3^2]$

$\qquad = 2(x^4 - 12x^3 + 54x^2 - 108x + 81) + 5(x^2 - 6x + 9)$

$\qquad = 2x^4 - 24x^3 + 113x^2 - 246x + 207$

47. $-3(x - 2)^3 - 4(x + 1)^6 = [-3x^3 + 18x^2 - 36x + 24] - [4x^6 + 24x^5 + 60x^4 + 80x^3 + 60x^2 + 24x + 4]$

$\qquad = -4x^6 - 24x^5 - 60x^4 - 83x^3 - 42x^2 - 60x + 20$

49. $(x + 8)^{10}, n = 4$

$\qquad {}_{10}C_3 x^{10-3}(8)^3 = 120x^7(512) = 61{,}440x^7$

51. $(x - 6y)^5, n = 3$

$\qquad {}_5C_2 x^{5-2}(-6y)^2 = 10x^3(36)y^2 = 360x^3y^2$

53. $(4x + 3y)^9$, $n = 8$

$$_9C_7(4x)^{9-7}(3y)^7 = 36(16)x^2(3^7)y^7$$
$$= 1,259,712x^2y^7$$

55. $(10x - 3y)^{12}$, $n = 9$

$$_{12}C_8(10x)^{12-8}(-3y)^8 = 495(10^4)(3^8)x^4y^8$$
$$= 32,476,950,000x^4y^8$$

57. The term involving x^4 in the expansion of $(x + 3)^{12}$ is $_{12}C_8x^4(3)^8 = 495x^4(3)^8 = 3,247,695x^4$. The coefficient is 3,247,695.

59. The term involving x^8y^2 in the expansion of $(x - 2y)^{10}$ is

$$_{10}C_2x^8(-2y)^2 = \frac{10!}{2!8!} \cdot 4x^8y^2 = 180x^8y^2.$$

The coefficient is 180.

61. The term involving x^6y^3 in $(3x - 2y)^9$ is

$$_9C_3(3x)^6(-2y)^3 = 84(3)^6(-2)^3x^6y^3$$
$$= -489,888x^6y^3.$$

The coefficient is $-489,888$.

63. The coefficient of $x^8y^6 = (x^2)^4y^6$ in the expansion of $(x^2 + y)^{10}$ is $_{10}C_6 = 210$.

65. 5th entry of 7th row: $_7C_5 = 21$

67. 5th entry of 6th row: $_6C_5 = 6$

69. 4th row of Pascal's Triangle: 1   4   6   4   1

$$(3t - 2v)^4 = 1(3t)^4 - 4(3t)^3(2v) + 6(3t)^2(2v)^2 - 4(3t)(2v)^3 + 1(2v)^4$$
$$= 81t^4 - 216t^3v + 216t^2v^2 - 96tv^3 + 16v^4$$

71. 5th row of Pascal's Triangle: 1   5   10   10   5   1

$$(2x - 3y)^5 = 1(2x)^5 - 5(2x)^4(3y) + 10(2x)^3(3y)^2 - 10(2x)^2(3y)^3 + 5(2x)(3y)^4 - (3y)^5$$
$$= 32x^5 - 240x^4y + 720x^3y^2 - 1080x^2y^3 + 810xy^4 - 243y^5$$

73. $(\sqrt{x} + 5)^4 = (\sqrt{x})^4 + 4(\sqrt{x})^3(5) + 6(\sqrt{x})^2(5)^2 + 4(\sqrt{x})(5^3) + 5^4$
$$= x^2 + 20x\sqrt{x} + 150x + 500\sqrt{x} + 625$$
$$= x^2 + 20x^{3/2} + 150x + 500x^{1/2} + 625$$

75. $(x^{2/3} - y^{1/3})^3 = (x^{2/3})^3 - 3(x^{2/3})^2(y^{1/3}) + 3(x^{2/3})(y^{1/3})^2 - (y^{1/3})^3$
$$= x^2 - 3x^{4/3}y^{1/3} + 3x^{2/3}y^{2/3} - y$$

77. $\dfrac{f(x + h) - f(x)}{h} = \dfrac{(x + h)^3 - x^3}{h}$

$$= \frac{x^3 + 3x^2h + 3xh^2 + h^3 - x^3}{h}$$

$$= \frac{h(3x^2 + 3xh + h^2)}{h}$$

$$= 3x^2 + 3xh + h^2, \ h \neq 0$$

79. $\dfrac{f(x + h) - f(x)}{h} = \dfrac{(x + h)^6 - x^6}{h}$

$$= \dfrac{(x^6 + 6x^5h + 15x^4h^2 + 20x^3h^3 + 15x^2h^4 + 6xh^5 + h^6) - x^6}{h}$$

$$= \dfrac{h(6x^5 + 15x^4h + 20x^3h^2 + 15x^2h^3 + 6xh^4 + h^5)}{h}$$

$$= 6x^5 + 15x^4h + 20x^3h^2 + 15x^2h^3 + 6xh^4 + h^5, h \neq 0$$

81. $\dfrac{f(x + h) - f(x)}{h} = \dfrac{\sqrt{x + h} - \sqrt{x}}{h}$

$$= \dfrac{\sqrt{x + h} - \sqrt{x}}{h} \cdot \dfrac{\sqrt{x + h} + \sqrt{x}}{\sqrt{x + h} + \sqrt{x}}$$

$$= \dfrac{(x + h) - x}{h\left[\sqrt{x + h} + \sqrt{x}\right]}$$

$$= \dfrac{1}{\sqrt{x + h} + \sqrt{x}}, h \neq 0$$

83. $(1 + i)^4 = {}_4C_0 1^4 + {}_4C_1(1)^3 i + {}_4C_2(1)^2 i^2 + {}_4C_3 1 \cdot i^3 + {}_4C_4 i^4$

$$= 1 + 4i - 6 - 4i + 1$$

$$= -4$$

85. $(4 + i)^4 = {}_4C_0(4)^4 + {}_4C_1(4^3)i + {}_4C_2(4^2)(i^2) + {}_4C_3(4)(i^3) + {}_4C_4 i^4$

$$= 256 + 256i - 96 - 16i + 1$$

$$= 161 + 240i$$

87. $(2 - 3i)^6 = {}_6C_0 2^6 - {}_6C_1 2^5(3i) + {}_6C_2 2^4(3i)^2 - {}_6C_3 2^3(3i)^3 + {}_6C_4 2^2(3i)^4 - {}_6C_5 2(3i)^5 + {}_6C_6(3i)^6$

$$= 64 - 576i - 2160 + 4320i + 4860 - 2916i - 729$$

$$= 2035 + 828i$$

89. $\left(5 + \sqrt{-16}\right)^3 = (5 + 4i)^3$

$$= 5^3 + 3(5^2)(4i) + 3(5)(4i)^2 + (4i)^3$$

$$= 125 + 300i - 240 - 64i$$

$$= -115 + 236i$$

91. $\left(4 + \sqrt{3}i\right)^4 = 4^4 + 4(4^3)\left(\sqrt{3}i\right) + 6(4^2)\left(\sqrt{3}i\right)^2 + 4(4)\left(\sqrt{3}i\right)^3 + \left(\sqrt{3}\right)^4$

$$= 256 + 256\sqrt{3}i - 288 - 48\sqrt{3}i + 9$$

$$= -23 + 208\sqrt{3}i$$

93. $\left(-\dfrac{1}{2} + \dfrac{\sqrt{3}}{2}i\right)^3 = \dfrac{1}{8}(-1 + \sqrt{3}i)^3$

$$= \dfrac{1}{8}\left[(-1)^3 + 3(-1)^2(\sqrt{3}i) + 3(-1)(\sqrt{3}i)^2 + (\sqrt{3}i)^3\right]$$

$$= \dfrac{1}{8}\left[-1 + 3\sqrt{3}i + 9 - 3\sqrt{3}i\right]$$

$$= 1$$

95. $\left(\dfrac{1}{4} - \dfrac{\sqrt{3}}{4}i\right)^3 = \left(\dfrac{1}{4}\right)^3 - 3\left(\dfrac{1}{4}\right)^2\left(\dfrac{\sqrt{3}}{4}i\right) + 3\left(\dfrac{1}{4}\right)\left(\dfrac{-\sqrt{3}}{4}i\right)^2 - \left(\dfrac{\sqrt{3}}{4}i\right)^3$

$$= \left[\dfrac{1}{64} - \dfrac{3}{4}\left(\dfrac{3}{16}\right)\right] + \left[\dfrac{-3}{16}\dfrac{\sqrt{3}}{4} + \dfrac{3\sqrt{3}}{64}\right]i$$

$$= -\dfrac{1}{8}$$

97. $(1.02)^8 = (1 + 0.02)^8 = 1 + 8(0.02) + 28(0.02)^2 + 56(0.02)^3 + 70(0.02)^4 + 56(0.02)^5$

$$+ 28(0.02)^6 + 8(0.02)^7 + (0.02)^8$$

$$= 1 + 0.16 + 0.0112 + 0.000448 + \cdots \approx 1.172$$

99. $(2.99)^{12} = (3 - 0.01)^{12}$

$$= 3^{12} - 12(3)^{11}(0.01) + 66(3)^{10}(0.01)^2 - 220(3)^9(0.01)^3 + 495(3)^8(0.01)^4$$

$$- 792(3)^7(0.01)^5 + 924(3)^6(0.01)^6 - 792(3)^5(0.01)^7 + 495(3)^4(0.01)^8$$

$$- 220(3)^3(0.01)^9 + 66(3)^2(0.01)^{10} - 12(3)(0.01)^{11} + (0.01)^{12}$$

$$\approx 510{,}568.785$$

101. $f(x) = x^3 - 4x$

$g(x) = f(x + 3)$

$\qquad = (x + 3)^3 - 4(x + 3)$

$\qquad = x^3 + 9x^2 + 27x + 27 - 4x - 12$

$\qquad = x^3 + 9x^2 + 23x + 15$

g is shifted three units to the left.

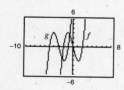

103. $f(x) = (1 - x)^3$

$g(x) = 1 - 3x$

$h(x) = 1 - 3x + 3x^2$

$p(x) = 1 - 3x + 3x^2 - x^3$

Since $p(x)$ is the expansion of $f(x)$, they have the same graph.

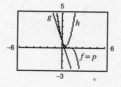

105. $_7C_4\left(\dfrac{1}{2}\right)^4\left(\dfrac{1}{2}\right)^3 = 35\left(\dfrac{1}{16}\right)\left(\dfrac{1}{8}\right) \approx 0.273$

107. $_8C_4\left(\dfrac{1}{3}\right)^4\left(\dfrac{2}{3}\right)^4 = 70\left(\dfrac{1}{81}\right)\left(\dfrac{16}{81}\right) \approx 0.171$

109. $f(t) = 0.064t^2 - 9.30t + 416.5, 5 \le t \le 23$

   (a) $g(t) = f(t + 20)$

$$= 0.064(t + 20)^2 - 9.30(t + 20) + 416.5$$

$$= 0.064t^2 - 6.74t + 256.1, -15 \le t \le 3$$

   (b)

111. False. The x^4y^8 term is

$$_{12}C_4 x^4(-2y)^8 = 495x^4(-2)^8 y^8 = 126,720x^4 y^8.$$

[**Note:** 7920 is the coefficient of $x^8 y^4$.]

113. Answers will vary. See page 557.

115. The expansions of $(x + y)^n$ and $(x - y)^n$ are almost the same except that the signs of the terms in the expansion of $(x - y)^n$ alternate from positive to negative.

117. $_nC_{n-r} = \dfrac{n!}{[n - (n - r)]!(n - r)!}$

$$= \dfrac{n!}{r!(n - r)!} = \dfrac{n!}{(n - r)!r!} = {_nC_r}$$

119. $_nC_r + {_nC_{r-1}} = \dfrac{n!}{(n - r)!r!} + \dfrac{n!}{(n - r + 1)!(r - 1)!}$

$$= \dfrac{n!(n - r + 1)}{(n - r)!r!(n - r + 1)} + \dfrac{n!}{(n - r + 1)!(r - 1)!r}$$

$$= \dfrac{n!(n - r + 1)}{(n - r + 1)!r!} + \dfrac{n!r}{(n - r + 1)!r!}$$

$$= \dfrac{n!(n - r + 1 + r)}{(n - r + 1)!r!}$$

$$= \dfrac{n!(n + 1)}{(n - r + 1)!r!}$$

$$= \dfrac{(n + 1)!}{(n + 1 - r)!r!} = {_{n+1}C_r}$$

121. $g(x) = f(x) + 8$

   $g(x)$ is shifted eight units up from $f(x)$.

123. $g(x) = f(-x)$

   $g(x)$ is the reflection of $f(x)$ in the y-axis.

125. $\begin{bmatrix} -6 & 5 \\ -5 & 4 \end{bmatrix}^{-1} = \dfrac{1}{-24 + 25}\begin{bmatrix} 4 & -5 \\ 5 & -6 \end{bmatrix} = \begin{bmatrix} 4 & -5 \\ 5 & -6 \end{bmatrix}$

Section 6.6   Counting Principles

- You should know The Fundamental Counting Principle.

- $_nP_r = \dfrac{n!}{(n-r)!}$ is the number of permutations of n elements taken r at a time.

- Given a set of n objects that has n_1 of one kind, n_2 of a second kind, and so on, the number of distinguishable permutations is

$$\frac{n!}{n_1!n_2!\cdots n_k!}.$$

- $_nC_r = \dfrac{n!}{(n-r)!r!}$ is the number of combinations of n elements taken r at a time.

Vocabulary Check

1. Fundamental Counting Principle

2. permutation

3. $_nP_r = \dfrac{n!}{(n-r)!}$

4. distinguishable permutations

5. combinations

---

1. Odd integers: 1, 3, 5, 7, 9, 11

   6 ways

3. Prime integers: 2, 3, 5, 7, 11

   5 ways

5. Divisible by 4: 4, 8, 12

   3 ways

7. Sum is 8:

   $1 + 7, 2 + 6, 3 + 5, 4 + 4, 5 + 3, 6 + 2, 7 + 1$

   7 ways

9. Amplifiers: 4 choices

   Compact disc players: 6 choices

   Speakers: 5 choices

   Total: $4 \cdot 6 \cdot 5 = 120$ ways

11. $2^{10} = 1024$ ways

13. (a) $9 \cdot 10 \cdot 10 = 900$

   (b) $9 \cdot 9 \cdot 8 = 648$

15. $2(8 \cdot 10 \cdot 10)(10 \cdot 10 \cdot 10 \cdot 10) = 16{,}000{,}000$ numbers

17. (a) $26^3 + 26^3 = 35{,}152$

   (b) There are $2 \cdot 25^3$ possibilities that don't have Q. Hence, $2 \cdot 26^3 - 2 \cdot 25^3 = 3902$ have at least one Q.

19. (a) $10^5 = 100{,}000$ zip codes

   (b) $2 \cdot 10^4 = 20{,}000$ zip codes beginning with a one or a two

21. (a) $6 \cdot 5 \cdot 4 \cdot 3 \cdot 2 \cdot 1 = 720$

   (b) $6 \cdot 1 \cdot 4 \cdot 1 \cdot 2 \cdot 1 = 48$

23. $_nP_r = \dfrac{n!}{(n-r)!}$

   So, $_4P_4 = \dfrac{4!}{0!} = 4! = 24$.

25. $_8P_3 = \dfrac{8!}{5!} = 8 \cdot 7 \cdot 6 = 336$

27. $_5P_4 = \dfrac{5!}{1!} = 120$

29. $_{20}P_6 = 27,907,200$

31. $_{120}P_4 = 197,149,680$

33. $5! = 120$ ways

35. $9! = 362,880$ ways

37. $_{12}P_4 = \dfrac{12!}{8!}$

$\qquad = 12 \cdot 11 \cdot 10 \cdot 9$

$\qquad = 11,880$ ways

39. $37 \cdot 37 \cdot 37 = 50,653$

41.

ABCD	BACD	CABD	DABC
ABDC	BADC	CADB	DACB
ACBD	BCAD	CBAD	DBAC
ACDB	BCDA	CBDA	DBCA
ADBC	BDAC	CDAB	DCAB
ADCB	BDCA	CDBA	DCBA

43. $\dfrac{7!}{2!1!3!1!} = \dfrac{7!}{2!3!} = 420$

45. $\dfrac{7!}{2!1!1!1!1!1!1!} = \dfrac{7!}{2!}$

$\qquad = 7 \cdot 6 \cdot 5 \cdot 4 \cdot 3$

$\qquad = 2520$

47. $_5C_2 = \dfrac{5!}{2!3!} = \dfrac{5 \cdot 4}{2} = 10$

49. $_4C_1 = \dfrac{4!}{1!3!} = 4$

51. $_{25}C_0 = \dfrac{25!}{0!25!} = 1$

53. $_{20}C_4 = 4845$

55. $_{42}C_5 = 850,668$

57. AB, AC, AD, AE, AF,
BC, BD, BE, BF, CD
CE, CF, DE, DF, EF

$_6C_2 = 15$ ways

59. $_{100}C_{14} = \dfrac{100!}{14!86!} \approx 4.42 \times 10^{16}$ ways

61. $_{49}C_6 = 13,983,816$ ways

63. $_9C_2 = 36$ lines

65. Select type of card for three of a kind: $_{13}C_1$

Select three of four cards for three of a kind: $_4C_3$

Select type of card for pair: $_{12}C_1$

Select two of four cards for pair: $_4C_2$

$_{13}C_1 \cdot {_4C_3} \cdot {_{12}C_1} \cdot {_4C_2} = 13 \cdot 4 \cdot 12 \cdot 6 = 3744$ ways to get a full house

67. (a) $_{12}C_4 = 495$ ways

(b) $(_5C_2)(_7C_2) = (10)(21) = 210$ ways

69. $(_7C_1)(_{12}C_3)(_{20}C_2) = 7 \cdot 220 \cdot 190$

$\qquad = 292,600$ ways

71. $_5C_2 - 5 = 10 - 5 = 5$ diagonals

73. $_8C_2 - 8 = 28 - 8 = 20$ diagonals

75. $\qquad 14 \cdot {}_nP_3 = {}_{n+2}P_4$

Note: $n \geq 3$ for this to be defined.

$$14\left[\frac{n!}{(n-3)!}\right] = \frac{(n+2)!}{(n-2)!}$$

$$14n(n-1)(n-2) = (n+2)(n+1)n(n-1) \quad \text{(We can divide here by } n(n-1) \text{ since } n \neq 0, n \neq 1.\text{)}$$

$$14n - 28 = n^2 + 3n + 2$$

$$0 = n^2 - 11n + 30$$

$$0 = (n-5)(n-6)$$

$$n = 5 \quad \text{or} \quad n = 6$$

77. $\qquad {}_nP_4 = 10 \cdot {}_{n-1}P_3$

$$\frac{n!}{(n-4)!} = 10\frac{(n-1)!}{(n-4)!}$$

$$n! = 10(n-1)!$$

$$n = 10$$

79. $\qquad {}_{n+1}P_3 = 4 \cdot {}_nP_2$

$$\frac{(n+1)!}{(n-2)!} = 4\frac{n!}{(n-2)!}$$

$$(n+1)! = 4n!$$

$$n = 3$$

81. $\quad 4 \cdot {}_{n+1}P_2 = {}_{n+2}P_3$

$$4\frac{(n+1)!}{(n-1)!} = \frac{(n+2)!}{(n-1)!}$$

$$4(n+1)! = (n+2)!$$

$$n = 2$$

83. False

85. ${}_{100}P_{80} \approx 3.836 \times 10^{139}$.

This number is too large for some calculators to evaluate.

87. ${}_nC_r = {}_nC_{n-r} = \dfrac{n!}{r!(n-r)!}$

89. ${}_nP_{n-1} = \dfrac{n!}{(n-(n-1))!} = \dfrac{n!}{1!} = \dfrac{n!}{0!} = {}_nP_n$

91. ${}_nC_{n-1} = \dfrac{n!}{[n-(n-1)]!(n-1)!}$

$$= \frac{n!}{(1)!(n-1)!}$$

$$= \frac{n!}{(n-1)!1!} = {}_nC_1$$

93. From the graph of $y = \sqrt{x-3} - x + 6$, you see that there is one zero, $x \approx 8.303$. Analytically,

$$\sqrt{x-3} = x - 6$$

$$x - 3 = x^2 - 12x + 36$$

$$0 = x^2 - 13x + 39.$$

By the Quadratic Formula, $x = \dfrac{13 \pm \sqrt{(-13)^2 - 4(39)}}{2} = \dfrac{13 \pm \sqrt{13}}{2}$.

Selecting the larger solution, $x = \dfrac{13 + \sqrt{13}}{2} \approx 8.303$. (The other solution is extraneous.)

95. $\log_2(x-3) = 5$

$$2^5 = x - 3$$

$$2^5 + 3 = x$$

$$x = 35$$

97. $x = \dfrac{\begin{vmatrix} -14 & 3 \\ 2 & -2 \end{vmatrix}}{\begin{vmatrix} -5 & 3 \\ 7 & -2 \end{vmatrix}} = \dfrac{22}{-11} = -2$

$y = \dfrac{\begin{vmatrix} -5 & -14 \\ 7 & 2 \end{vmatrix}}{\begin{vmatrix} -5 & 3 \\ 7 & -2 \end{vmatrix}} = \dfrac{88}{-11} = -8$

Answer: $(-2, -8)$

99. $x = \dfrac{\begin{vmatrix} -1 & -4 \\ -4 & 5 \end{vmatrix}}{\begin{vmatrix} -3 & -4 \\ 9 & 5 \end{vmatrix}} = \dfrac{-21}{21} = -1$

$y = \dfrac{\begin{vmatrix} -3 & -1 \\ 9 & -4 \end{vmatrix}}{\begin{vmatrix} -3 & -4 \\ 9 & 5 \end{vmatrix}} = \dfrac{21}{21} = 1$

Answer: $(-1, 1)$

Section 6.7   Probability

You should know the following basic principles of probability.

■ If an event E has $n(E)$ equally likely outcomes and its sample space has $n(S)$ equally likely outcomes, then the probability of event E is

$$P(E) = \frac{n(E)}{n(S)}, \text{ where } 0 \le P(E) \le 1.$$

■ If A and B are mutually exclusive events, then $P(A \cup B) = P(A) + P(B)$.

If A and B are not mutually exclusive events, then $P(A \cup B) = P(A) + P(B) - P(A \cap B)$.

■ If A and B are independent events, then the probability that both A and B will occur is $P(A)P(B)$.

■ The probability of the complement of an event A is $P(A') = 1 - P(A)$.

Vocabulary Check

1. experiment, outcomes

2. sample space

3. probability

4. impossible, certain

5. mutually exclusive

6. independent

7. complement

8. (a) iii   (b) i   (c) iv   (d) ii

1. $\{(H, 1), (H, 2), (H, 3), (H, 4), (H, 5), (H, 6),$
$(T, 1), (T, 2), (T, 3), (T, 4), (T, 5), (T, 6)\}$

3. $\{ABC, ACB, BAC, BCA, CAB, CBA\}$

5. $\{(A, B), (A, C), (A, D), (A, E), (B, C),$
$(B, D), (B, E), (C, D), (C, E), (D, E)\}$

7. $E = \{HTT, THT, TTH\}$

$P(E) = \dfrac{n(E)}{n(S)} = \dfrac{3}{8}$

9. $E = \{HHH, HHT, HTH, HTT, THH, THT, TTH\}$

$P(E) = \dfrac{n(E)}{n(S)} = \dfrac{7}{8}$

11. $E = \{K, K, K, K, Q, Q, Q, Q, J, J, J, J\}$

$P(E) = \dfrac{n(E)}{n(S)} = \dfrac{12}{52} = \dfrac{3}{13}$

13. $E = \{A, A, A, A, K, K, K, K, Q, Q, Q, Q, J, J, J, J\}$

$P(E) = \dfrac{n(E)}{n(S)} = \dfrac{16}{52} = \dfrac{4}{13}$

15. $E = \{(1, 5), (2, 4), (3, 3), (4, 2), (5, 1)\}$

$P(E) = \dfrac{n(E)}{n(S)} = \dfrac{5}{36}$

17. not $E = \{(5, 6), (6, 5), (6, 6)\}$

$n(E) = n(S) - n(\text{not } E) = 36 - 3 = 33$

$P(E) = \dfrac{n(E)}{n(S)} = \dfrac{33}{36} = \dfrac{11}{12}$

19. $P(E) = \dfrac{{}_3C_2}{{}_6C_2} = \dfrac{3}{15} = \dfrac{1}{5}$

21. $P(E) = \dfrac{{}_4C_2}{{}_6C_2} = \dfrac{6}{15} = \dfrac{2}{5}$

23. $P(E') = 1 - P(E) = 1 - 0.75 = 0.25$

25. $P(E') = 1 - P(E) = 1 - \dfrac{2}{3} = \dfrac{1}{3}$

27. $P(E) = 1 - P(E') = 1 - p = 1 - 0.12 = 0.88$

29. $P(E) = 1 - P(E') = 1 - \dfrac{13}{20} = \dfrac{7}{20}$

31. (a) $0.15(8.15) \approx 1.22$ million

(b) $\dfrac{0.41}{1.0} = 0.41$

(c) $\dfrac{0.24}{1.0} = 0.24$

(d) $\dfrac{0.24 + 0.02}{1.0} = 0.26$

33. (a) $(0.128)(293.66) \approx 37.6$ million

(b) $\dfrac{0.01}{1.0} = 0.01$

(c) $\dfrac{0.01 + 0.002}{1.0} = 0.012$

35. (a) $\dfrac{34}{100} = 0.34$

(b) $\dfrac{45}{100} = 0.45$

(c) $\dfrac{23}{100} = 0.23$

37. (a) $\dfrac{672}{1254}$

(b) $\dfrac{582}{1254}$

(c) $\dfrac{672 - 124}{1254} = \dfrac{548}{1254}$

39. $p + p + 2p = 1$

$p = 0.25$

Taylor: $0.50 = \dfrac{1}{2}$

Perez: $0.25 = \dfrac{1}{4}$

Moore: $0.25 = \dfrac{1}{4}$

41. (a) $\dfrac{{}_{15}C_{10}}{{}_{20}C_{10}} = \dfrac{3003}{184{,}756} = \dfrac{21}{1292} \approx 0.016$

(b) $\dfrac{{}_{15}C_8 \cdot {}_5C_2}{{}_{20}C_{10}} = \dfrac{64{,}350}{184{,}756} = \dfrac{225}{646} \approx 0.348$

(c) $\dfrac{{}_{15}C_9 \cdot {}_5C_1}{{}_{20}C_{10}} + \dfrac{{}_{15}C_{10}}{{}_{20}C_{10}} = \dfrac{25{,}025 + 3003}{184{,}756}$

$= \dfrac{28{,}028}{184{,}756} = \dfrac{49}{323} \approx 0.152$

43. (a) $\dfrac{1}{{}_5P_5} = \dfrac{1}{120}$

(b) $\dfrac{1}{{}_4P_4} = \dfrac{1}{24}$

45. (a) There are three letters to be selected, and two must be Q and Y.

QY__, YQ__, Q__Y, Y__Q, __YQ, __QY

Thus, the probability is

$\dfrac{6(26)}{26^3} = \dfrac{6}{26^2} \approx 0.008876.$

(b) The three letters must be Q, Y, and X.

QYX, QXY, YQX, YXQ, XQY, XYQ

Thus, the probability is $\dfrac{6}{26^3} = \dfrac{3}{8778}$.

47. (a) $\dfrac{100}{(_{55}C_5)(_{42}C_1)} = \dfrac{100}{(3{,}478{,}761)(42)}$

(b) $\dfrac{1000}{(_{55}C_5)(_{42}C_1)} = \dfrac{1000}{(3{,}478{,}761)(42)}$

49. (a) $\dfrac{20}{52} = \dfrac{5}{13}$

(b) $\dfrac{13 + 13}{52} = \dfrac{1}{2}$

(c) $\dfrac{4 + 12}{52} = \dfrac{4}{13}$

51. (a) $\dfrac{_9C_4}{_{12}C_4} = \dfrac{126}{495} = \dfrac{14}{55}$ (4 good units)

(b) $\dfrac{(_9C_2)\,(_3C_2)}{_{12}C_4} = \dfrac{108}{495} = \dfrac{12}{55}$ (2 good units)

(c) $\dfrac{(_9C_3)(_3C_1)}{_{12}C_4} = \dfrac{252}{495} = \dfrac{28}{55}$ (3 good units)

At least 2 good units: $\dfrac{12}{55} + \dfrac{28}{55} + \dfrac{14}{55} = \dfrac{54}{55}$

53. $(0.32)^2 = 0.1024$

55. (a) $P(SS) = (0.985)^2 \approx 0.9702$

(b) $P(S) = 1 - P(FF) = 1 - (0.015)^2 \approx 0.9998$

(c) $P(FF) = (0.015)^2 \approx 0.0002$

57. (a) $\left(\dfrac{1}{5}\right)^6 = \dfrac{1}{15{,}625}$

(b) $\left(\dfrac{4}{5}\right)^6 = \dfrac{4096}{15{,}625} = 0.262144$

(c) $1 - 0.262144 = 0.737856 = \dfrac{11{,}529}{15{,}625}$

59. (a) If the *center* of the coin falls within the circle of radius $d/2$ around a vertex, the coin will cover the vertex.

$$P(\text{coin covers a vertex}) = \dfrac{\begin{array}{c}\text{Area in which coin may fall}\\ \text{so that it covers a vertex}\\ \hline \text{Total area}\end{array}}{} = \dfrac{n\left[\pi\left(\dfrac{d}{2}\right)^2\right]}{nd^2} = \dfrac{1}{4}\pi$$

(b) Experimental results will vary.

61. True. $P(E) + P(E') = 1$

63. (a) As you consider successive people with distinct birthdays, the probabilities must decrease to take into account the birth dates already used. Since the birth dates of people are independent events, multiply the respective probabilities of distinct birthdays.

(b) $\dfrac{365}{365} \cdot \dfrac{364}{365} \cdot \dfrac{363}{365} \cdot \dfrac{362}{365}$

(c) $P_1 = \dfrac{365}{365} = 1$

$P_2 = \dfrac{365}{365} \cdot \dfrac{364}{365} = \dfrac{364}{365}P_1 = \dfrac{365 - (2 - 1)}{365}P_1$

$P_3 = \dfrac{365}{365} \cdot \dfrac{364}{365} \cdot \dfrac{363}{365} = \dfrac{363}{365}P_2 = \dfrac{365 - (3 - 1)}{365}P_2$

$P_n = \dfrac{365}{365} \cdot \dfrac{364}{365} \cdot \dfrac{363}{365} \cdot \ldots \cdot \dfrac{365 - (n - 1)}{365} = \dfrac{365 - (n - 1)}{365}P_{n-1}$

—CONTINUED—

63. —CONTINUED—

(d) Q_n is the probability that the birthdays are *not* distinct which is equivalent to at least 2 people having the same birthday.

(e)

n	10	15	20	23	30	40	50
P_n	0.88	0.75	0.59	0.49	0.29	0.11	0.03
Q_n	0.12	0.25	0.41	(0.51)	0.71	0.89	0.97

(f) 23, See the chart above.

65. $\dfrac{2}{x-5} = 4$

$2 = 4(x-5) = 4x - 20$

$4x = 22$

$x = \dfrac{11}{2}$

67. $\dfrac{3}{x-2} + \dfrac{x}{x+2} = 1$

$3(x+2) + x(x-2) = (x-2)(x+2)$

$3x + 6 + x^2 - 2x = x^2 - 4$

$x = -10$

69. $e^x + 7 = 35$

$e^x = 28$

$x = \ln(28) \approx 3.332$

71. $4 \ln 6x = 16$

$\ln 6x = 4$

$e^4 = 6x$

$x = \frac{1}{6}e^4 \approx 9.10$

73. $_5P_3 = \dfrac{5!}{(5-3)!} = \dfrac{120}{2} = 60$

75. $_{11}P_8 = \dfrac{11!}{(11-8)!} = \dfrac{11!}{3!} = 6{,}652{,}800$

77. $_6C_2 = \dfrac{6!}{4!2!} = \dfrac{6 \cdot 5 \cdot 4!}{4!2} = 15$

79. $_{11}C_8 = \dfrac{11!}{8!3!} = \dfrac{11 \cdot 10 \cdot 9 \cdot 8!}{8!6} = 165$

Review Exercises for Chapter 6

1. $a_n = \dfrac{2^n}{2^n + 1}$

$a_1 = \dfrac{2^1}{2^1 + 1} = \dfrac{2}{3}$

$a_2 = \dfrac{2^2}{2^2 + 1} = \dfrac{4}{5}$

$a_3 = \dfrac{2^3}{2^3 + 1} = \dfrac{8}{9}$

$a_4 = \dfrac{2^4}{2^4 + 1} = \dfrac{16}{17}$

$a_5 = \dfrac{2^5}{2^5 + 1} = \dfrac{32}{33}$

3. $a_n = \dfrac{(-1)^n}{n!}$

$a_1 = \dfrac{(-1)^1}{1!} = -1$

$a_2 = \dfrac{(-1)^2}{2!} = \dfrac{1}{2}$

$a_3 = \dfrac{(-1)^3}{3!} = -\dfrac{1}{6}$

$a_4 = \dfrac{(-1)^4}{4!} = \dfrac{1}{24}$

$a_5 = \dfrac{(-1)^5}{5!} = -\dfrac{1}{120}$

5. Common difference is 5.

$a_n = 5n, \ n = 1, 2, \ldots$

7. Denominators are successive odd numbers.

$$a_n = \frac{2}{2n-1}, n = 1, 2, 3, \ldots$$

9. $a_1 = 9, a_{k+1} = a_k - 4$

$a_2 = a_1 - 4 = 9 - 4 = 5$

$a_3 = 5 - 4 = 1$

$a_4 = 1 - 4 = -3$

$a_5 = -3 - 4 = -7$

11. $\dfrac{18!}{20!} = \dfrac{18!}{20 \cdot 19 \cdot 18!}$

$= \dfrac{1}{20 \cdot 19} = \dfrac{1}{380}$

13. $\dfrac{(n+1)!}{(n-1)!} = \dfrac{(n+1)n(n-1)!}{(n-1)!} = n(n+1)$

15. $\displaystyle\sum_{i=1}^{6} 5 = 6(5) = 30$

17. $\displaystyle\sum_{j=1}^{4} \frac{6}{j^2} = \frac{6}{1^2} + \frac{6}{2^2} + \frac{6}{3^2} + \frac{6}{4^2}$

$= 6 + \dfrac{3}{2} + \dfrac{2}{3} + \dfrac{3}{8} = \dfrac{205}{24}$

19. $\displaystyle\sum_{k=1}^{100} 2k^3 = 2 \cdot \frac{100^2(101)^2}{4} = 51{,}005{,}000$

21. $\displaystyle\sum_{n=0}^{50} (n^2 + 3) = \frac{50(51)(101)}{6} + 3(51) = 43{,}078$

23. $\dfrac{1}{2(1)} + \dfrac{1}{2(2)} + \dfrac{1}{2(3)} + \cdots + \dfrac{1}{2(20)} = \displaystyle\sum_{k=1}^{20} \frac{1}{2k}$

≈ 1.799

25. $\dfrac{1}{2} + \dfrac{2}{3} + \dfrac{3}{4} + \cdots + \dfrac{9}{10} = \displaystyle\sum_{k=1}^{9} \frac{k}{k+1} \approx 7.071$

27. (a) $\displaystyle\sum_{k=1}^{4} \frac{5}{10^k} = \frac{5}{10} + \frac{5}{100} + \frac{5}{1000} + \frac{5}{10{,}000} = 0.5 + 0.05 + 0.005 + 0.0005 = 0.5555 = \frac{1111}{2000}$

(b) $\displaystyle\sum_{k=1}^{\infty} \frac{5}{10^k} = \frac{5}{10} \sum_{k=0}^{\infty} \frac{1}{10^k} = \frac{5}{10} \cdot \frac{1}{1 - 1/10} = \frac{5}{10} \cdot \frac{10}{9} = \frac{5}{9}$

29. $\displaystyle\sum_{k=1}^{\infty} 2(0.5)^k$

(a) $\displaystyle\sum_{k=1}^{4} 2(0.5)^k = 2(0.5) + 2(0.5)^2 + 2(0.5)^3 + 2(0.5)^4$

$= 1.875 = \dfrac{15}{8}$

(b) $\displaystyle\sum_{k=1}^{\infty} 2(0.5)^k = 2(0.5)\frac{1}{1 - 0.5} = 2$

31. $a_n = 2500\left(1 + \dfrac{0.02}{4}\right)^n, n = 1, 2, 3$

(a) $a_1 = 2500\left(1 + \dfrac{0.02}{4}\right)^1 = 2512.5$

$a_2 = 2525.06 \qquad a_3 = 2537.69$

$a_4 = 2550.38 \qquad a_5 = 2563.13$

$a_6 = 2575.94 \qquad a_7 = 2588.82$

$a_8 = 2601.77$

(b) $a_{40} = 2500\left(1 + \dfrac{0.02}{4}\right)^{40} = \$3051.99$

33. Yes

$d = 3 - 5 = -2$

35. Yes

$d = 1 - \tfrac{1}{2} = \tfrac{1}{2}$

37. $a_1 = 3, d = 4$

$a_1 = 3$

$a_2 = 3 + 4 = 7$

$a_3 = 7 + 4 = 11$

$a_4 = 11 + 4 = 15$

$a_5 = 15 + 4 = 19$

39. $a_4 = 10, \ a_{10} = 28$

$a_{10} = a_4 + 6d$

$28 = 10 + 6d$

$18 = 6d$

$3 = d$

$a_1 = a_4 - 3d$

$a_1 = 10 - 3(3)$

$a_1 = 1$

$a_2 = 1 + 3 = 4$

$a_3 = 4 + 3 = 7$

$a_4 = 7 + 3 = 10$

$a_5 = 10 + 3 = 13$

41. $a_1 = 35, a_{k+1} = a_k - 3$

$a_1 = 35$

$a_2 = a_1 - 3 = 35 - 3 = 32$

$a_3 = a_2 - 3 = 32 - 3 = 29$

$a_4 = a_3 - 3 = 29 - 3 = 26$

$a_5 = a_4 - 3 = 26 - 3 = 23$

$a_n = 35 + (n - 1)(-3) = 38 - 3n, d = -3$

43. $a_1 = 9, a_{k+1} = a_k + 7$

$a_1 = 9$

$a_2 = a_1 + 7 = 9 + 7 = 16$

$a_3 = a_2 + 7 = 16 + 7 = 23$

$a_4 = a_3 + 7 = 23 + 7 = 30$

$a_5 = a_4 + 7 = 30 + 7 = 37$

$a_n = 9 + (n - 1)(7) = 2 + 7n, d = 7$

45. $a_n = 100 + (n - 1)(-3) = 103 - 3n$

$$\sum_{n=1}^{20} (103 - 3n) = \sum_{n=1}^{20} 103 - 3 \sum_{n=1}^{20} n = 20(103) - 3\left[\frac{(20)(21)}{2}\right] = 1430$$

47. $\displaystyle\sum_{j=1}^{10} (2j - 3) = 2\sum_{j=1}^{10} j - \sum_{j=1}^{10} 3$

$$= 2\left[\frac{10(11)}{2}\right] - 10(3) = 80$$

49. $\displaystyle\sum_{k=1}^{11} \left(\frac{2}{3}k + 4\right) = \frac{2}{3}\sum_{k=1}^{11} k + \sum_{k=1}^{11} 4$

$$= \frac{2}{3} \cdot \frac{(11)(12)}{2} + 11(4) = 88$$

51. $\displaystyle\sum_{k=1}^{100} 5k = 5\left[\frac{(100)(101)}{2}\right] = 25{,}250$

53. (a) $34{,}000 + 4(2250) = \$43{,}000$

(b) $\displaystyle\sum_{k=1}^{5} [34{,}000 + (k - 1)(2250)]$

$$= \sum_{k=1}^{5} (31{,}750 + 2250k)$$

$$= \$192{,}500$$

55. 5, 10, 20, 40

Geometric: $r = 2$

57. Geometric:

$r = -\frac{1}{3}$

59. $a_1 = 4, \; r = -\frac{1}{4}$

$a_1 = 4$

$a_2 = 4\left(-\frac{1}{4}\right) = -1$

$a_3 = -1\left(-\frac{1}{4}\right) = \frac{1}{4}$

$a_4 = \frac{1}{4}\left(-\frac{1}{4}\right) = -\frac{1}{16}$

$a_5 = -\frac{1}{16}\left(-\frac{1}{4}\right) = \frac{1}{64}$

61. $a_1 = 9, \; a_3 = 4$

$a_3 = a_1 r^2$ $a_1 = 9$ $a_1 = 9$

$4 = 9r^2$ $a_2 = 9\left(\frac{2}{3}\right) = 6$ $a_2 = 9\left(-\frac{2}{3}\right) = -6$

$\frac{4}{9} = r^2 \implies r = \pm\frac{2}{3}$ $a_3 = 6\left(\frac{2}{3}\right) = 4$ or $a_3 = -6\left(-\frac{2}{3}\right) = 4$

                       $a_4 = 4\left(\frac{2}{3}\right) = \frac{8}{3}$ $a_4 = 4\left(-\frac{2}{3}\right) = -\frac{8}{3}$

                       $a_5 = \frac{8}{3}\left(\frac{2}{3}\right) = \frac{16}{9}$ $a_5 = -\frac{8}{3}\left(-\frac{2}{3}\right) = \frac{16}{9}$

63. $a_1 = 120, \; a_{k+1} = \frac{1}{3}a_k$

$a_1 = 120$

$a_2 = \frac{1}{3}(120) = 40$

$a_3 = \frac{1}{3}(40) = \frac{40}{3}$

$a_4 = \frac{1}{3}\left(\frac{40}{3}\right) = \frac{40}{9}$

$a_5 = \frac{1}{3}\left(\frac{40}{9}\right) = \frac{40}{27}$

$a_n = 120\left(\frac{1}{3}\right)^{n-1}, \; r = \frac{1}{3}$

65. $a_1 = 25, \; a_{k+1} = -\frac{3}{5}a_k$

$a_1 = 25$

$a_2 = -\frac{3}{5}(25) = -15$

$a_3 = -\frac{3}{5}(-15) = 9$

$a_4 = -\frac{3}{5}(9) = -\frac{27}{3}$

$a_5 = -\frac{3}{5}\left(-\frac{27}{5}\right) = \frac{81}{25}$

$a_n = 25\left(-\frac{3}{5}\right)^{n-1}, \; r = -\frac{3}{5}$

67. $a_2 = a_1 r$

$-8 = 16r$

$-\frac{1}{2} = r$

$a_n = 16\left(-\frac{1}{2}\right)^{n-1}$

$\sum_{n=1}^{20} 16\left(-\frac{1}{2}\right)^{n-1} = 16\left[\frac{1 - (-1/2)^{20}}{1 - (-1/2)}\right] \approx 10.67$

69. $a_1 = 100, \; r = 1.05$

$a_n = 100(1.05)^{n-1}$

$\sum_{n=1}^{20} 100(1.05)^{n-1} = 100\left[\frac{1 - (1.05)^{20}}{1 - 1.05}\right] \approx 3306.60$

71. $\displaystyle\sum_{i=1}^{7} 2^{i-1} = \frac{1 - 2^7}{1 - 2} = 127$

73. $\displaystyle\sum_{n=1}^{7} (-4)^{n-1} = \frac{1 - (-4)^7}{1 - (-4)} = 3277$

75. $\displaystyle\sum_{n=0}^{4} 250(1.02)^n = 250\left(\frac{1 - 1.02^5}{1 - 1.02}\right) = 1301.01004$

77. $\displaystyle\sum_{i=1}^{10} 10\left(\frac{3}{5}\right)^{i-1} \approx 24.849$

79. $\sum\limits_{i=1}^{\infty} 4\left(\dfrac{7}{8}\right)^{i-1} = \sum\limits_{i=0}^{\infty} 4\left(\dfrac{7}{8}\right)^{i} = \dfrac{4}{1 - 7/8} = 32$

81. $\sum\limits_{k=1}^{\infty} 4\left(\dfrac{2}{3}\right)^{k-1} = \dfrac{4}{1 - 2/3} = 12$

83. (a) $a_t = 120{,}000(0.7)^t$

(b) $a_5 = 120{,}000(0.7)^5 = \$20{,}168.40$

85. 1. When $n = 1$, $2 = \dfrac{1}{2}(5(1) - 1)$.

2. Assume that $S_k = 2 + 7 + \cdots + (5k - 3) = \dfrac{k}{2}(5k - 1)$. Then,

$$S_{k+1} = 2 + 7 + \cdots + (5k - 3) + [5(k + 1) - 3]$$

$$= S_k + 5k + 2$$

$$= \dfrac{k}{2}(5k - 1) + 5k + 2$$

$$= \dfrac{1}{2}[5k^2 + 9k + 4]$$

$$= \dfrac{1}{2}[(5k + 4)(k + 1)]$$

$$= \dfrac{k + 1}{2}(5(k + 1) - 1).$$

Therefore, by mathematical induction, the formula is true for all positive integers n.

87. 1. When $n = 1$, $a = a\left(\dfrac{1 - r}{1 - r}\right)$.

2. Assume that

$$S_k = \sum\limits_{i=0}^{k-1} ar^i = \dfrac{a(1 - r^k)}{1 - r}.$$

Then,

$$S_{k+1} = \sum\limits_{i=0}^{k} ar^i = \sum\limits_{i=0}^{k-1} ar^i + ar^k = \dfrac{a(1 - r^k)}{1 - r} + ar^k$$

$$= \dfrac{a(1 - r^k + r^k - r^{k+1})}{1 - r} = \dfrac{a(1 - r^{k+1})}{1 - r}.$$

Therefore, by mathematical induction, the formula is valid for all positive integer values of n.

89. $\sum\limits_{n=1}^{30} n = \dfrac{30(31)}{2} = 465$

91. $\sum\limits_{n=1}^{7} (n^4 - n) = \sum\limits_{n=1}^{7} n^4 - \sum\limits_{n=1}^{7} n$

$$= \dfrac{7(8)(15)[3(7)^2 + 3(7) - 1]}{30} - \dfrac{7(8)}{2}$$

$$= \dfrac{840(167)}{30} - 28 = 4676 - 28 = 4648$$

93. $a_1 = f(1) = 5$

$a_2 = a_1 + 5 = 5 + 5 = 10$

$a_3 = a_2 + 5 = 15$

$a_4 = a_3 + 5 = 20$

$a_5 = a_4 + 5 = 25$

n:	1	2	3	4	5
a_n:	5	10	15	20	25

First differences: $\quad 5 \quad 5 \quad 5 \quad 5$

Second difference: $\quad 0 \quad 0 \quad 0$

Linear model: $a_n = 5n$

95. $a_1 = f(1) = 16$

$a_2 = a_1 - 1 = 16 - 1 = 15$

$a_3 = a_2 - 1 = 15 - 1 = 14$

$a_4 = 14 - 1 = 13$

$a_5 = 13 - 1 = 12$

n:	1	2	3	4	5
a_n:	16	15	14	13	12

First differences: $\quad -1 \quad -1 \quad -1 \quad -1$

Second difference: $\quad 0 \quad 0 \quad 0$

Linear model: $a_n = 17 - n$

97. $_{10}C_8 = 45$

99. $\binom{9}{4} = {}_9C_4 = 126$

101. 4th number in 6th row is $_6C_3 = 20$.

103. 5th number in 8th row is $\binom{8}{4} = {}_8C_4 = 70$.

105. $(x + 5)^4 = x^4 + 4x^3(5) + 6x^2(5^2) + 4x(5^3) + 5^4$

$\qquad = x^4 + 20x^3 + 150x^2 + 500x + 625$

107. $(a - 4b)^5 = a^5 - 5a^4(4b) + 10a^3(4b)^2 - 10a^2(4b)^3 + 5a(4b)^4 - (4b)^5$

$\qquad = a^5 - 20a^4b + 160a^3b^2 - 640a^2b^3 + 1280ab^4 - 1024b^5$

109. $(7 + 2i)^4 = 7^4 + 4(7)^3(2i) + 6(7)^2(2i)^2 + 4(7)(2i)^3 + (2i)^4$

$\qquad = 2401 + 2744i - 1176 - 224i + 16$

$\qquad = 1241 + 2520i$

111. $E = \{(1, 11), (2, 10), (3, 9), (4, 8), (5, 7), (7, 5), (8, 4), (9, 3), (10, 2), (11, 1)\}$

$n(E) = 10$

113. (a) $(4)(3)(6)(3) = 216$ schedules

(b) $(2)(3)(6)(3) = 108$ schedules

(c) $(2)(3)(2)(3) = 36$ schedules

115. $_{10}C_8 = \dfrac{10!}{2!8!} = \dfrac{10 \cdot 9}{2} = 45$

117. $_{12}P_{10} = \dfrac{12!}{2!} = 239{,}500{,}800$

119. $_{100}C_{98} = \dfrac{100!}{2!98!} = \dfrac{100 \cdot 99}{2} = 4950$

121. $_{1000}P_2 = \dfrac{1000!}{998!} = 1000(999) = 999{,}000$

123. $\dfrac{8!}{2!2!2!1!1!} = \dfrac{8!}{8} = 7! = 5040$ permutations

125. $10! = 3{,}628{,}800$ ways

127. $_{20}C_{15} = 15{,}504$ ways

129. $_{n+1}P_2 = 4 \cdot {}_nP_1$

$\dfrac{(n + 1)!}{(n - 1)!} = 4 \cdot \dfrac{n!}{(n - 1)!}$

$(n + 1)! = 4 \cdot n!$

$n = 3$

131. $\frac{10}{10} \cdot \frac{1}{9} = \frac{1}{9}$

133. (a) $\frac{208}{500} = 0.416$

     (b) $\frac{400}{500} = 0.8$

     (c) $\frac{37}{500} = 0.074$

135. $P(2 \text{ pairs}) = \dfrac{(_{13}C_2)(_4C_2)(_4C_2)(_{44}C_1)}{(_{52}C_5)} = 0.0475$

137. True

$$\frac{(n + 2)!}{n!} = \frac{(n + 2)(n + 1)n!}{n!} = (n + 2)(n + 1)$$

139. Answers will vary. See pages 526 and 535.

141. (a) Arithmetic-linear model

     (b) Geometric model

143. Answers will vary. See page 528. To define a sequence recursively, you need to be given one or more of the first few terms. All other terms are defined using previous terms.

145. If n is even, the expansion are the same. If n is odd, the expansion of $(-x + y)^n$ is the negative of that of $(x - y)^n$.

Chapter 6   Practice Test

1. Write out the first five terms of the sequence $a_n = \dfrac{2n}{(n + 2)!}$.

2. Write an expression for the nth term of the sequence $\left\{\dfrac{4}{3}, \dfrac{5}{9}, \dfrac{6}{27}, \dfrac{7}{81}, \dfrac{8}{243}, \ldots\right\}$.

3. Find the sum $\displaystyle\sum_{i=1}^{6}(2i - 1)$.

4. Write out the first five terms of the arithmetic sequence where $a_1 = 23$ and $d = -2$.

5. Find a_{50} for the arithmetic sequence with $a_1 = 12$, $d = 3$, and $n = 50$.

6. Find the sum of the first 200 positive integers.

7. Write out the first five terms of the geometric sequence with $a_1 = 7$ and $r = 2$.

8. Evaluate $\displaystyle\sum_{n=0}^{9}6\left(\dfrac{2}{3}\right)^n$.

9. Evaluate $\displaystyle\sum_{n=0}^{\infty}(0.03)^n$.

10. Use mathematical induction to prove that $1 + 2 + 3 + 4 + \cdots + n = \dfrac{n(n + 1)}{2}$.

11. Use mathematical induction to prove that $n! > 2^n$, $n \geq 4$.

12. Evaluate $_{13}C_4$. Verify with a graphing utility.

13. Expand $(x + 3)^5$.

14. Find the term involving x^7 in $(x - 2)^{12}$.

15. Evaluate $_{30}P_4$.

16. How many ways can six people sit at a table with six chairs?

17. Twelve cars run in a race. How many different ways can they come in first, second, and third place? (Assume that there are no ties.)

18. Two six-sided dice are tossed. Find the probability that the total of the two dice is less than 5.

19. Two cards are selected at random from a deck of 52 playing cards without replacement. Find the probability that the first card is a King and the second card is a black ten.

20. A manufacturer has determined that for every 1000 units it produces, 3 will be faulty. What is the probability that an order of 50 units will have one or more faulty units?

C H A P T E R 7
Conics and Parametric Equations

CHAPTER 7
Conics and Parametric Equations

Section 7.1   Circles and Parabolas

- ■ A **parabola** is the set of all points (x, y) that are equidistant from a fixed line (**directrix**) and a fixed point (**focus**) not on the line.
- ■ The standard equation of a parabola with vertex (h, k) and
  - (a) Vertical axis $x = h$ and directrix $y = k - p$ is
    $$(x - h)^2 = 4p(y - k), \ p \neq 0.$$
  - (b) Horizontal axis $y = k$ and directrix $x = h - p$ is
    $$(y - k)^2 = 4p(x - h), \ p \neq 0.$$
- ■ The tangent line to a parabola at a point P makes **equal angles** with
  - (a) the line through P and the focus.
  - (b) the axis of the parabola.

Vocabulary Check

1. conic section
2. locus
3. circle, center
4. parabola, directrix, focus
5. vertex
6. axis
7. tangent

---

1. $x^2 + y^2 = \left(\sqrt{18}\right)^2$

$x^2 + y^2 = 18$

3. Radius $= \sqrt{(3 - 1)^2 + (7 - 0)^2}$

$\qquad = \sqrt{4 + 49} = \sqrt{53}$

$(x - h)^2 + (y - k)^2 = r^2$

$(x - 3)^2 + (y - 7)^2 = 53$

5. Diameter $= 2\sqrt{7} \implies$ radius $= \sqrt{7}$

$(x - h)^2 + (y - k)^2 = r^2$

$(x + 3)^2 + (y + 1)^2 = 7$

7. $x^2 + y^2 = 49$

Center: $(0, 0)$

Radius: 7

9. $(x + 2)^2 + (y - 7)^2 = 16$

Center: $(-2, 7)$

Radius: 4

11. $(x - 1)^2 + y^2 = 15$

Center: $(1, 0)$

Radius: $\sqrt{15}$

13. $\frac{1}{4}x^2 + \frac{1}{4}y^2 = 1$

$x^2 + y^2 = 4$

Center: $(0, 0)$

Radius: 2

15. $\dfrac{4}{3}x^2 + \dfrac{4}{3}y^2 = 1$

$\qquad x^2 + y^2 = \dfrac{3}{4}$

Center: $(0, 0)$

Radius: $\dfrac{\sqrt{3}}{2}$

17. $(x^2 - 2x + 1) + (y^2 + 6y + 9) = -9 + 1 + 9$

$\qquad\qquad (x - 1)^2 + (y + 3)^2 = 1$

Center: $(1, -3)$

Radius: 1

19. $4\left(x^2 + 3x + \dfrac{9}{4}\right) + 4(y^2 - 6y + 9) = -41 + 9 + 36$

$\qquad\qquad 4\left(x + \dfrac{3}{2}\right)^2 + 4(y - 3)^2 = 4$

$\qquad\qquad\quad \left(x + \dfrac{3}{2}\right)^2 + (y - 3)^2 = 1$

Center: $\left(-\dfrac{3}{2}, 3\right)$

Radius: 1

21. $\qquad x^2 = 16 - y^2$

$\qquad x^2 + y^2 = 16$

Center: $(0, 0)$

Radius: 4

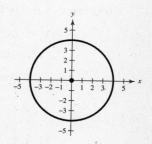

23. $\qquad x^2 + 4x + y^2 + 4y - 1 = 0$

$(x^2 + 4x + 4) + (y^2 + 4y + 4) = 1 + 4 + 4$

$\qquad\qquad (x + 2)^2 + (y + 2)^2 = 9$

Center: $(-2, -2)$

Radius: 3

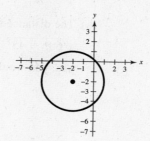

25. $\qquad x^2 - 14x + y^2 + 8y + 40 = 0$

$(x^2 - 14x + 4) + (y^2 + 8y + 16) = -40 + 49 + 16$

$\qquad\qquad\quad (x - 7)^2 + (y + 4)^2 = 25$

Center: $(7, -4)$

Radius: 5

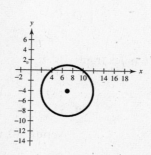

27. $x^2 + 2x + y^2 - 35 = 0$

$(x^2 + 2x + 1) + y^2 = 35 + 1$

$\qquad (x + 1)^2 + y^2 = 36$

Center: $(-1, 0)$

Radius: 6

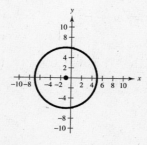

29. y-intercepts: $(0 - 2)^2 + (y + 3)^2 = 9$

$\qquad\qquad\qquad\quad 4 + (y + 3)^2 = 9$

$\qquad\qquad\qquad\qquad (y + 3)^2 = 5$

$\qquad\qquad\qquad\qquad\qquad y = -3 \pm \sqrt{5}$

$\qquad\qquad \left(0, -3 \pm \sqrt{5}\right)$

x-intercepts: $(x - 2)^2 + (0 + 3)^2 = 9$

$\qquad\qquad\qquad\quad (x - 2)^2 = 0$

$\qquad\qquad\qquad\qquad\quad x = 2$

$\qquad\qquad (2, 0)$

31. *y*-intercepts: Let $x = 0$.

$$y^2 - 6y - 27 = 0$$
$$y^2 - 6y + 9 = 27 + 9$$
$$(y - 3)^2 = 36$$
$$y - 3 = \pm 6$$
$$y = 9, -3$$
$$(0, 9), (0, -3)$$

x-intercepts: Let $y = 0$.

$$x^2 - 2x - 27 = 0$$
$$x^2 - 2x + 1 = 27 + 1$$
$$(x - 1)^2 = 28$$
$$x - 1 = \pm \sqrt{28}$$
$$x = 1 \pm 2\sqrt{7}$$
$$\left(1 \pm 2\sqrt{7}, 0\right)$$

33. *y*-intercepts: $(0 - 6)^2 + (y + 3)^2 = 16$

$$(y + 3)^2 = 16 - 36$$
$$= -20$$

No solution

No *y*-intercepts

x-intercepts: $(x - 6)^2 + (0 + 3)^2 = 16$

$$(x - 6)^2 = 7$$
$$x - 6 = \pm\sqrt{7}$$
$$x = 6 \pm \sqrt{7}$$
$$\left(6 \pm \sqrt{7}, 0\right)$$

35. (a) Radius: 81; Center: $(0, 0)$

$$x^2 + y^2 = 81^2 = 6561$$

(c)

You were $81 - 75 = 6$ miles from the outer boundary.

(b) The distance from $(60, 45)$ to $(0, 0)$ is

$$\sqrt{60^2 + 45^2} = \sqrt{5625} = 75 \text{ miles.}$$

Yes, you would feel the earthquake.

37. $y^2 = -4x$

Vertex: $(0, 0)$

Opens to the left since *p* is negative.

Matches graph (e).

39. $x^2 = -8y$

Vertex: $(0, 0)$

Opens downward since *p* is negative.

Matches graph (d).

41. $(y - 1)^2 = 4(x - 3)$

Vertex: $(3, 1)$

Opens to the right since *p* is positive.

Matches graph (a).

43. Vertex: $(0, 0) \Rightarrow h = 0, k = 0$

Graph opens upward.

$$x^2 = 4py$$

Point on graph: $(3, 6)$

$$3^2 = 4p(6)$$
$$9 = 24p$$
$$\tfrac{3}{8} = p$$

Thus, $x^2 = 4\left(\tfrac{3}{8}\right)y \Rightarrow y = \tfrac{2}{3}x^2$

$$\Rightarrow x^2 = \tfrac{3}{2}y.$$

45. Vertex: $(0, 0) \Rightarrow h = 0, k = 0$

Focus: $\left(0, -\tfrac{3}{2}\right) \Rightarrow p = -\tfrac{3}{2}$

$$(x - h)^2 = 4p(y - k)$$
$$x^2 = 4\left(-\tfrac{3}{2}\right)y$$
$$x^2 = -6y$$

47. Vertex: $(0, 0) \Rightarrow h = 0,$
$$k = 0$$

Focus: $(-2, 0) \Rightarrow p = -2$

$$(y - k)^2 = 4p(x - h)$$
$$y^2 = 4(-2)x$$
$$y^2 = -8x$$

49. Vertex: $(0, 0) \implies h = 0, k = 0$

Directrix: $y = -1 \implies p = 1$

$(x - h)^2 = 4p(y - k)$

$(x - 0)^2 = 4(1)(y - 0)$

$x^2 = 4y$ or $y = \frac{1}{4}x^2$

51. Vertex: $(0, 0) \implies h = 0, k = 0$

Directrix: $x = 2 \implies p = -2$

$y^2 = 4px$

$y^2 = -8x$

53. Vertex: $(0, 0) \implies h = 0, k = 0$

Horizontal axis and passes through the point $(4, 6)$

$(y - k)^2 = 4p(x - h)$

$(y - 0)^2 = 4p(x - 0)$

$y^2 = 4px$

$6^2 = 4p(4)$

$36 = 16p \implies p = \frac{9}{4}$

$y^2 = 4\left(\frac{9}{4}\right)x$

$y^2 = 9x$

55. $y = \frac{1}{2}x^2$

$x^2 = 2y = 4\left(\frac{1}{2}\right)y; \ p = \frac{1}{2}$

Vertex: $(0, 0)$

Focus: $\left(0, \frac{1}{2}\right)$

Directrix: $y = -\frac{1}{2}$

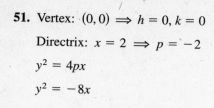

57. $y^2 = -6x$

$y^2 = 4\left(-\frac{3}{2}\right)x; \ p = -\frac{3}{2}$

Vertex: $(0, 0)$

Focus: $\left(-\frac{3}{2}, 0\right)$

Directrix: $x = \frac{3}{2}$

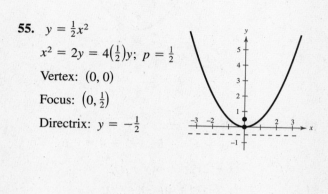

59. $x^2 + 8y = 0$

$x^2 = 4(-2)y;$

$p = -2$

Vertex: $(0, 0)$

Focus: $(0, -2)$

Directrix: $y = 2$

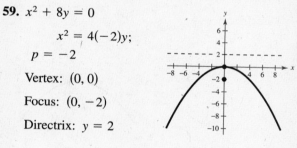

61. $(x + 1)^2 + 8(y + 3) = 0$

$(x + 1)^2 = 4(-2)(y + 3)$

$h = -1, k = -3, p = -2$

Vertex: $(-1, -3)$

Focus: $(-1, -5)$

Directrix: $y = -1$

63. $y^2 + 6y + 8x + 25 = 0$

$(y + 3)^2 = 4(-2)(x + 2); \ p = -2$

Vertex: $(-2, -3)$

Focus: $(-4, -3)$

Directrix: $x = 0$

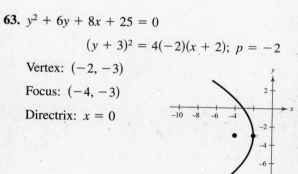

65. $\left(x + \frac{3}{2}\right)^2 = 4(y - 2) \implies h = -\frac{3}{2}, k = 2, p = 1$

Vertex: $\left(-\frac{3}{2}, 2\right)$

Focus: $\left(-\frac{3}{2}, 2 + 1\right) = \left(-\frac{3}{2}, 3\right)$

Directrix: $y = 1$

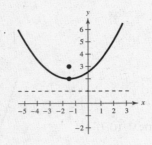

67. $y = \frac{1}{4}(x^2 - 2x + 5)$

$4y - 4 = (x - 1)^2$

$(x - 1)^2 = 4(1)(y - 1)$

$h = 1, k = 1, p = 1$

Vertex: $(1, 1)$

Focus: $(1, 2)$

Directrix: $y = 0$

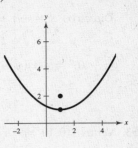

69. $x^2 + 4x + 6y - 2 = 0$

$x^2 + 4x + 4 = -6y + 2 + 4 = -6y + 6$

$(x + 2)^2 = -6(y - 1)$

$(x + 2)^2 = 4\left(-\frac{3}{2}\right)(y - 1)$

Vertex: $(-2, 1)$

Focus: $\left(-2, 1 - \frac{3}{2}\right) = \left(-2, -\frac{1}{2}\right)$

Directrix: $y = \frac{5}{2}$

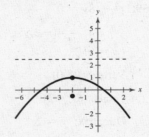

71. $y^2 + x + y = 0$

$y^2 + y + \frac{1}{4} = -x + \frac{1}{4}$

$\left(y + \frac{1}{2}\right)^2 = 4\left(-\frac{1}{4}\right)\left(x - \frac{1}{4}\right)$

$h = \frac{1}{4}, k = -\frac{1}{2}, p = -\frac{1}{4}$

Vertex: $\left(\frac{1}{4}, -\frac{1}{2}\right)$

Focus: $\left(0, -\frac{1}{2}\right)$

Directrix: $x = \frac{1}{2}$

To use a graphing calculator, enter:

$y_1 = -\frac{1}{2} + \sqrt{\frac{1}{4} - x}$

$y_2 = -\frac{1}{2} - \sqrt{\frac{1}{4} - x}$

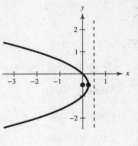

73. Vertex: $(3, 1)$,
          opens downward

Passes through: $(2, 0), (4, 0)$

$\quad y = -(x - 2)(x - 4)$

$\quad\quad = -x^2 + 6x - 8$

$\quad\quad = -(x - 3)^2 + 1$

$(x - 3)^2 = -(y - 1)$

75. Vertex: $(-2, 0)$,
          opens to the right

Focus: $\left(-\frac{3}{2}, 0\right)$

$\frac{1}{2} = p$

$y^2 = 4\left(\frac{1}{2}\right)(x + 2)$

$y^2 = 2(x + 2)$

77. Vertex: $(5, 2)$

Focus: $(3, 2)$

Horizontal axis: $p = 3 - 5 = -2$

$(y - 2)^2 = 4(-2)(x - 5)$

$(y - 2)^2 = -8(x - 5)$

79. Vertex: $(0, 4)$

Directrix: $y = 2$

Vertical axis

$p = 4 - 2 = 2$

$(x - 0)^2 = 4(2)(y - 4)$

$\quad x^2 = 8(y - 4)$

81. Focus: $(2, 2)$

Directrix: $x = -2$

Horizontal axis

Vertex: $(0, 2)$

$p = 2 - 0 = 2$

$(y - 2)^2 = 4(2)(x - 0)$

$(y - 2)^2 = 8x$

83. $y^2 - 8x = 0$ and $x - y + 2 = 0$

$$y^2 = 8x \qquad\qquad y_3 = x + 2$$

$$y_1 = \sqrt{8x}$$

$$y_2 = -\sqrt{8x}$$

The point of tangency is $(2, 4)$.

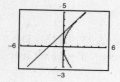

85. $x^2 = 2y$, $(4, 8)$, $p = \dfrac{1}{2}$, focus: $\left(0, \dfrac{1}{2}\right)$

Following Example 4, we find the y-intercept $(0, b)$.

$$d_1 = \frac{1}{2} - b$$

$$d_2 = \sqrt{(4 - 0)^2 + \left(8 - \frac{1}{2}\right)^2} = \frac{17}{2}$$

$$d_1 = d_2 \implies \frac{1}{2} - b = \frac{17}{2} \implies b = -8$$

$$m = \frac{8 - (-8)}{4 - 0} = 4$$

$$y = 4x - 8, \quad \text{Tangent line}$$

Let $y = 0 \implies x = 2 \implies$ x-intercept $(2, 0)$.

87. $y = -2x^2 \implies x^2 = -\dfrac{1}{2}y = 4\left(-\dfrac{1}{8}\right)y$

$$\implies p = -\frac{1}{8}$$

Focus: $\left(0, -\dfrac{1}{8}\right)$

Following Example 4, we find the y-intercept $(0, b)$.

$$d_1 = \frac{1}{8} + b$$

$$d_2 = \sqrt{(-1 - 0)^2 + \left(-2 + \frac{1}{8}\right)^2} = \frac{17}{8}$$

$$d_1 = d_2 \implies \frac{1}{8} + b = \frac{17}{8} \implies b = 2$$

$$m = \frac{-2 - 2}{-1 - 0} = 4$$

$$y = 4x + 2$$

Let $y = 0 \implies x = -\dfrac{1}{2} \implies$ x-intercept $\left(-\dfrac{1}{2}, 0\right)$.

89. $R = 375x - \dfrac{3}{2}x^2$

R is a maximum of \$23,437.50 when $x = 125$ televisions.

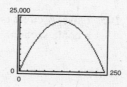

91. (a) $x^2 = 4py$, $p = \frac{3}{2}$

$$x^2 = 4\left(\tfrac{3}{2}\right)y = 6y$$

(or $y^2 = 6x$)

(b) When $x = 4$,

$$6y = 16$$

$$y = \frac{16}{6} = \frac{8}{3}.$$

Depth: $\frac{8}{3}$ inches

93. (a)

(b) $x^2 = 4py$

$$640^2 = 4p(152)$$

$$p = \frac{12,800}{19}$$

$$y = \frac{19}{51,200}x^2$$

(c)

x	0	200	400	500	600
y	0	14.84	59.38	92.77	133.59

95. Vertex: $(0, 0]$

$y^2 = 4px$

Point: $(1000, 800)$

$800^2 = 4p(1000] \implies p = 160$

$y^2 = 4(160)x$

$y^2 = 640x$

97. $-12.5(y - 7.125) = (x - 6.25)^2$

$-12.5y + 89.0625 = x^2 - 12.5x + 39.0625$

$y = -0.08x^2 + x + 4$

(a)

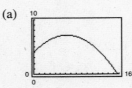

(b) The highest point is at $(6.25, 7.125)$. The distance is the x-intercept of ≈ 15.69 feet.

99. The slope of the line joining $(3, -4)$ and the center is $-\frac{4}{3}$. The slope of the tangent line at $(3, -4)$ is $\frac{3}{4}$. Thus,

$$y + 4 = \frac{3}{4}(x - 3)$$

$$4y + 16 = 3x - 9$$

$$3x - 4y = 25, \quad \text{tangent line.}$$

101. The slope of the line joining $\left(2, -2\sqrt{2}\right)$ and the center is $\left(-2\sqrt{2}\right)/2 = -\sqrt{2}$. The slope of the tangent line is $1/\sqrt{2} = \sqrt{2}/2$. Thus,

$$y + 2\sqrt{2} = \frac{\sqrt{2}}{2}(x - 2)$$

$$2y + 4\sqrt{2} = \sqrt{2}x - 2\sqrt{2}$$

$$\sqrt{2}x - 2y = 6\sqrt{2}, \quad \text{tangent line.}$$

103. False. The center is $(0, -5)$.

105. False. A circle is a conic section.

107. True

109. Answers will vary. See the reflective property of parabolas, page 599.

111. $(y - 3)^2 = 6(x + 1)$

For the upper half of the parabola,

$$y - 3 = \sqrt{6(x + 1)}$$

$$y = \sqrt{6(x + 1)} + 3.$$

113. $f(x) = 3x^3 - 4x + 2$

Relative maximum: $(-0.67, 3.78)$

Relative minimum: $(0.67, 0.22)$

115. $f(x) = x^4 + 2x + 2$

Relative minimum: $(-0.79, 0.81)$

Section 7.2   Ellipses

■ An **ellipse** is the set of all points (x, y) the sum of whose distances from two distinct fixed points (**foci**) is constant.

■ The standard equation of an ellipse with center (h, k) and major and minor axes of lengths $2a$ and $2b$ is

(a) $\dfrac{(x - h)^2}{a^2} + \dfrac{(y - k)^2}{b^2} = 1$ if the major axis is horizontal.

(b) $\dfrac{(x - h)^2}{b^2} + \dfrac{(y - k)^2}{a^2} = 1$ if the major axis is vertical.

■ $c^2 = a^2 - b^2$ where c is the distance from the center to a focus.

■ The eccentricity of an ellipse is $e = \dfrac{c}{a}$.

Vocabulary Check

1. ellipse

2. major axis, center

3. minor axis

4. eccentricity

---

1. $\dfrac{x^2}{4} + \dfrac{y^2}{9} = 1$

Center: $(0, 0)$

$a = 3, b = 2$

Vertical major axis

Matches graph (b).

3. $\dfrac{x^2}{4} + \dfrac{y^2}{25} = 1$

Center: $(0, 0)$

$a = 5, b = 2$

Vertical major axis

Matches graph (d).

5. $\dfrac{(x - 2)^2}{16} + (y + 1)^2 = 1$

Center: $(2, -1)$

$a = 4, b = 1$

Horizontal major axis

Matches graph (a).

7. $\dfrac{x^2}{64} + \dfrac{y^2}{9} = 1$

Center: $(0, 0)$

$a = 8, b = 3,$

$c = \sqrt{64 - 9} = \sqrt{55}$

Vertices: $(\pm 8, 0)$

Foci: $\left(\pm \sqrt{55}, 0\right)$

$e = \dfrac{c}{a} = \dfrac{\sqrt{55}}{8}$

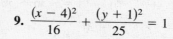

9. $\dfrac{(x - 4)^2}{16} + \dfrac{(y + 1)^2}{25} = 1$

Center: $(4, -1)$

$a = 5, b = 4, c = 3$

Vertices: $(4, -1 \pm 5);\ (4, -6), (4, 4)$

Foci: $(4, -1 \pm 3);\ (4, -4), (4, 2)$

$e = \dfrac{c}{a} = \dfrac{3}{5}$

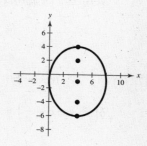

11. $\dfrac{(x + 5)^2}{9/4} + (y - 1)^2 = 1$

Center: $(-5, 1)$

$a = \dfrac{3}{2}, b = 1, c = \sqrt{\dfrac{9}{4} - 1} = \dfrac{\sqrt{5}}{2}$

Foci: $\left(-5 + \dfrac{\sqrt{5}}{2}, 1\right), \left(-5 - \dfrac{\sqrt{5}}{2}, 1\right)$

Vertices: $\left(-5 + \dfrac{3}{2}, 1\right) = \left(-\dfrac{7}{2}, 1\right), \left(-5 - \dfrac{3}{2}, 1\right) = \left(-\dfrac{13}{2}, 1\right)$

$e = \dfrac{\sqrt{5}/2}{3/2} = \dfrac{\sqrt{5}}{3}$

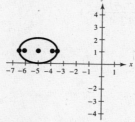

13. (a) $x^2 + 9y^2 = 36$

$$\frac{x^2}{36} + \frac{y^2}{4} = 1$$

(c)

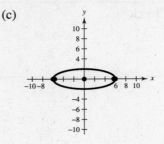

(b) $a = 6, b = 2, c = \sqrt{36 - 4} = \sqrt{32} = 4\sqrt{2}$

Center: $(0, 0)$

Vertices: $(\pm 6, 0)$

Foci: $\left(\pm 4\sqrt{2}, 0\right)$

$$e = \frac{c}{a} = \frac{4\sqrt{2}}{6} = \frac{2\sqrt{2}}{3}$$

15. (a) $9x^2 + 4y^2 + 36x - 24y + 36 = 0$

$$9(x^2 + 4x + 4) + 4(y^2 - 6y + 9) = -36 + 36 + 36$$

$$\frac{(x + 2)^2}{4} + \frac{(y - 3)^2}{9} = 1$$

(c)

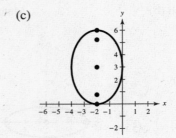

(b) $a = 3, b = 2, c = \sqrt{5}$

Center: $(-2, 3)$

Foci: $\left(-2, 3 \pm \sqrt{5}\right)$

Vertices: $(-2, 6), (-2, 0)$

$$e = \frac{\sqrt{5}}{3}$$

17. (a) $6x^2 + 2y^2 + 18x - 10y + 2 = 0$

$$6\left(x^2 + 3x + \frac{9}{4}\right) + 2\left(y^2 - 5y + \frac{25}{4}\right) = -2 + \frac{27}{2} + \frac{25}{2}$$

$$6\left(x + \frac{3}{2}\right)^2 + 2\left(y - \frac{5}{2}\right)^2 = 24$$

$$\frac{\left(x + \frac{3}{2}\right)^2}{4} + \frac{\left(y - \frac{5}{2}\right)^2}{12} = 1$$

(c)

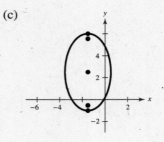

(b) $a = 2\sqrt{3}, b = 2, c = 2\sqrt{2}$

Center: $\left(-\frac{3}{2}, \frac{5}{2}\right)$

Foci: $\left(-\frac{3}{2}, \frac{5}{2} \pm 2\sqrt{2}\right)$

Vertices: $\left(-\frac{3}{2}, \frac{5}{2} \pm 2\sqrt{3}\right)$

$$e = \frac{\sqrt{2}}{\sqrt{3}} = \frac{\sqrt{6}}{3}$$

19. (a) $16x^2 + 25y^2 - 32x + 50y + 16 = 0$

$16(x^2 - 2x + 1) + 25(y^2 + 2y + 1) = -16 + 16 + 25$

$$\frac{(x-1)^2}{25/16} + (y+1)^2 = 1$$

(b) $a = \dfrac{5}{4}, b = 1, c = \dfrac{3}{4}$

Center: $(1, -1)$

Foci: $\left(\dfrac{7}{4}, -1\right), \left(\dfrac{1}{4}, -1\right)$

Vertices: $\left(\dfrac{9}{4}, -1\right), \left(-\dfrac{1}{4}, -1\right)$

$e = \dfrac{3}{5}$

(c)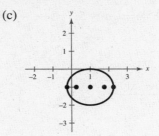

21. (a) $12x^2 + 20y^2 - 12x + 40y - 37 = 0$

$12\left(x^2 - 1 + \dfrac{1}{4}\right) + 20(y^2 + 2y + 1) = 37 + 3 + 20$

$12\left(x - \dfrac{1}{2}\right)^2 + 20(y+1)^2 = 60$

$$\frac{\left(x - \frac{1}{2}\right)^2}{5} + \frac{(y+1)^2}{3} = 1$$

(b) $a = \sqrt{5}, b = \sqrt{3}, c = \sqrt{5-3} = \sqrt{2}$

Center: $\left(\dfrac{1}{2}, -1\right)$

Vertices: $\left(\dfrac{1}{2} \pm \sqrt{5}, -1\right)$

Foci: $\left(\dfrac{1}{2} \pm \sqrt{2}, -1\right)$

Eccentricity: $\dfrac{c}{a} = \dfrac{\sqrt{2}}{\sqrt{5}} = \dfrac{\sqrt{10}}{5}$

(c)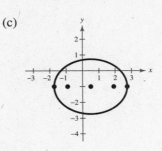

23. Center: $(0, 0)$

$a = 4, b = 2$

Vertical major axis

$$\frac{x^2}{4} + \frac{y^2}{16} = 1$$

25. Center: $(0, 0)$

$a = 3,$

$c = 2 \Rightarrow b = \sqrt{9-4} = \sqrt{5}$

Horizontal major axis

$$\frac{x^2}{9} + \frac{y^2}{5} = 1$$

27. Center: $(0, 0)$

$c = 3$

$a = 4 \Rightarrow b = \sqrt{16-9} = \sqrt{7}$

Horizontal major axis

$$\frac{x^2}{16} + \frac{y^2}{7} = 1$$

29. Vertices: $(0, \pm 5) \implies a = 5$

Center: $(0, 0)$

Vertical major axis

$$\frac{(x - h)^2}{b^2} + \frac{(y - k)^2}{a^2} = 1$$

$$\frac{x^2}{b^2} + \frac{y^2}{25} = 1$$

Point: $(4, 2)$

$$\frac{4^2}{b^2} + \frac{2^2}{25} = 1$$

$$\frac{16}{b^2} = 1 - \frac{4}{25} = \frac{21}{25}$$

$$400 = 21b^2$$

$$\frac{400}{21} = b^2$$

$$\frac{x^2}{400/21} + \frac{y^2}{25} = 1$$

$$\frac{21x^2}{400} + \frac{y^2}{25} = 1$$

31. Center: $(2, 3)$

$a = 3, b = 1$

Vertical major axis

$$\frac{(x - h)^2}{b^2} + \frac{(y - k)^2}{a^2} = 1$$

$$\frac{(x - 2)^2}{1} + \frac{(y - 3)^2}{9} = 1$$

33. Center: $(4, 2)$

$a = 4, b = 1 \implies c = \sqrt{16 - 1} = \sqrt{15}$

Horizontal major axis

$$\frac{(x - 4)^2}{16} + \frac{(y - 2)^2}{1} = 1$$

35. Center: $(0, 4)$

$c = 4,$

$a = 18 \implies b^2 = a^2 - c^2 = 324 - 16 = 308$

Vertical major axis

$$\frac{x^2}{308} + \frac{(y - 4)^2}{324} = 1$$

37. Vertices: $(3, 1), (3, 9) \implies a = 4$

Center: $(3, 5)$

Minor axis of length $6 \implies b = 3$

Vertical major axis

$$\frac{(x - h)^2}{b^2} + \frac{(y - k)^2}{a^2} = 1$$

$$\frac{(x - 3)^2}{9} + \frac{(y - 5)^2}{16} = 1$$

39. Center: $(0, 4)$

Vertices: $(-4, 4), (4, 4) \implies a = 4$

$a = 2c \implies 4 = 2c \implies c = 2$

$2^2 = 4^2 - b^2 \implies b^2 = 12$

Horizontal major axis

$$\frac{(x - h)^2}{a^2} + \frac{(y - k)^2}{b^2} = 1$$

$$\frac{x^2}{16} + \frac{(y - 4)^2}{12} = 1$$

41. $\dfrac{x^2}{4} + \dfrac{y^2}{9} = 1$

$a = 3, b = 2,$

$c = \sqrt{9 - 4} = \sqrt{5}$

$e = \dfrac{c}{a} = \dfrac{\sqrt{5}}{3}$

43. $\qquad x^2 + 9y^2 - 10x + 36y + 52 = 0$

$(x^2 - 10x + 25) + 9(y^2 + 4y + 4) = -52 + 25 + 36$

$\qquad (x - 5)^2 + 9(y + 2)^2 = 9$

$\qquad \dfrac{(x - 5)^2}{9} + \dfrac{(y + 2)^2}{1} = 1$

$a = 3, b = 1, c = \sqrt{9 - 1} = 2\sqrt{2}$

$e = \dfrac{c}{a} = \dfrac{2\sqrt{2}}{3}$

45. Vertices: $(\pm 5, 0) \implies a = 5$

Eccentricity: $\dfrac{4}{5} = \dfrac{c}{a} \implies c = \dfrac{4}{5}a = 4$

$b^2 = a^2 - c^2 = 25 - 16 = 9$

Center: $(0, 0)$

Horizontal major axis

$\dfrac{x^2}{25} + \dfrac{y^2}{9} = 1$

47. (a)

(b) Vertices: $(\pm 50, 0) \implies a = 50$

Height at center:

$40 \implies b = 40$

Horizontal major axis

$\dfrac{x^2}{a^2} + \dfrac{y^2}{b^2} = 1$

$\dfrac{x^2}{2500} + \dfrac{y^2}{1600} = 1, \ y \geq 0$

(c) For $x = 45, \dfrac{45^2}{2500} + \dfrac{y^2}{1600} = 1.$

$y^2 = 1600\left(1 - \dfrac{45^2}{2500}\right)$

$y^2 = 304$

$y \approx 17.44$

The height five feet from the edge of the tunnel is approximately 17.44 feet.

49. Let $\dfrac{x^2}{a^2} + \dfrac{y^2}{b^2} = 1$ be the equation of the ellipse. Then $b = 2$ and

$a = 3 \implies c^2 = a^2 - b^2 = 9 - 4 = 5.$ Thus, the tacks are placed

at $\left(\pm \sqrt{5}, 0\right).$ The string has a length of $2a = 6$ feet.

51. Area of ellipse = 2(area of circle)

$$\pi ab = 2\pi r^2$$

$$\pi a(10) = 2\pi(10)^2$$

$$\pi a(10) = 200$$

$$a = 20$$

Length of major axis: $2a = 2(20) = 40$ units

53. $a + c = 4.08$

$a - c = 0.34$

$2a = 4.42 \Rightarrow a = 2.21 \Rightarrow c = 1.87$

$b^2 = a^2 - c^2 \Rightarrow b^2 = 1.3872$

$$\frac{x^2}{4.8841} + \frac{y^2}{1.3872} = 1$$

55. For $\dfrac{x^2}{a^2} + \dfrac{y^2}{b^2} = 1$, we have $c^2 = a^2 - b^2$.

When $x = c$,

$$\frac{c^2}{a^2} + \frac{y^2}{b^2} = 1 \Rightarrow y^2 = b^2\left(1 - \frac{a^2 - b^2}{a^2}\right)$$

$$\Rightarrow y^2 = \frac{b^4}{a^2}$$

$$\Rightarrow 2y = \frac{2b^2}{a}.$$

57. $\dfrac{x^2}{9} + \dfrac{y^2}{16} = 1$

$a = 4, b = 3, c = \sqrt{7}$

Points on the ellipse:
$(\pm 3, 0), (0, \pm 4)$

Length of latus recta:

$$\frac{2b^2}{a} = \frac{2(3)^2}{4} = \frac{9}{2}$$

Additional points: $\left(\pm\dfrac{9}{4}, -\sqrt{7}\right), \left(\pm\dfrac{9}{4}, \sqrt{7}\right)$

59. $5x^2 + 3y^2 = 15$

$$\frac{x^2}{3} + \frac{y^2}{5} = 1$$

$a = \sqrt{5}, b = \sqrt{3},$

$c = \sqrt{2}$

Points on the ellipse:
$\left(\pm\sqrt{3}, 0\right), \left(0, \pm\sqrt{5}\right)$

Length of latus recta: $\dfrac{2b^2}{a} = \dfrac{2 \cdot 3}{\sqrt{5}} = \dfrac{6\sqrt{5}}{5}$

Additional points: $\left(\pm\dfrac{3\sqrt{5}}{5}, -\sqrt{2}\right), \left(\pm\dfrac{3\sqrt{5}}{5}, \sqrt{2}\right)$

61. True. If $e \approx 1$ then the ellipse is elongated, not circular.

63. (a) The length of the string is $2a$.

(b) The path is an ellipse because the sum of the distances from the two thumbtacks is always the length of the string, that is, it is constant.

65. Center: $(6, 2)$

Foci: $(2, 2), (10, 2) \Rightarrow c = 4$

$(a + c) + (a - c) = 2a = 36 \Rightarrow a = 18$

$b^2 = a^2 - b^2 \Rightarrow b = \sqrt{18^2 - 16} = \sqrt{308}$

Horizontal major axis

$$\frac{(x - 6)^2}{324} + \frac{(y - 2)^2}{308} = 1$$

67. Arithmetic: $d = -11$ **69.** Geometric: $r = 2$ **71.** $\displaystyle\sum_{n=0}^{6} 3^n = 1093$ **73.** $\displaystyle\sum_{n=1}^{10} 4\left(\frac{3}{4}\right)^{n-1} \approx 15.099$

Section 7.3   Hyperbolas

■ A **hyperbola** is the set of all points (x, y) the difference of whose distances from two distinct fixed points (**foci**) is constant.

■ The standard equation of a hyperbola with center (h, k) and transverse and conjugate axes of lengths $2a$ and $2b$ is:

   (a) $\dfrac{(x - h)^2}{a^2} - \dfrac{(y - k)^2}{b^2} = 1$ if the transverse axis is horizontal.

   (b) $\dfrac{(y - k)^2}{a^2} - \dfrac{(x - h)^2}{b^2} = 1$ if the transverse axis is vertical.

■ $c^2 = a^2 + b^2$ where c is the distance from the center to a focus.

■ The asymptotes of a hyperbola are:

   (a) $y = k \pm \dfrac{b}{a}(x - h)$ if the transverse axis is horizontal.

   (b) $y = k \pm \dfrac{a}{b}(x - h)$ the transverse axis is vertical.

■ The eccentricity of a hyperbola is $e = \dfrac{c}{a}$.

■ To classify a nondegenerate conic from its general equation $Ax^2 + Cy^2 + Dx + Ey + F = 0$:
   (a) If $A = C\ (A \neq 0, C \neq 0)$, then it is a circle.
   (b) If $AC = 0\ (A = 0$ or $C = 0$, but not both), then it is a parabola.
   (c) If $AC > 0$, then it is an ellipse.
   (d) If $AC < 0$, then it is a hyperbola.

Vocabulary Check

1. hyperbola

2. branches

3. transverse axis, center

4. asymptotes

5. $Ax^2 + Cy^2 + Dx + Ey + F = 0$

1. Center: $(0, 0)$

   $a = 3, b = 5, c = \sqrt{34}$

   Vertical transverse axis

   Matches graph (b).

3. Center: $(1, 0)$

   $a = 4, b = 2$

   Horizontal transverse axis

   Matches graph (a).

5. $x^2 - y^2 = 1$

   $a = 1, b = 1, c = \sqrt{2}$

   Center: $(0, 0)$

   Vertices: $(\pm 1, 0)$

   Foci: $\left(\pm\sqrt{2}, 0\right)$

   Asymptotes: $y = \pm x$

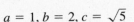

7. $\dfrac{y^2}{1} - \dfrac{x^2}{4} = 1$

   $a = 1, b = 2, c = \sqrt{5}$

   Center: $(0, 0)$

   Vertices: $(0, \pm 1)$

   Foci: $\left(0, \pm\sqrt{5}\right)$

   Asymptotes: $y = \pm\dfrac{1}{2}x$

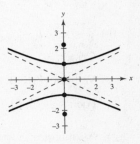

9. $\dfrac{y^2}{25} - \dfrac{x^2}{81} = 1$

$a = 5, b = 9, c = \sqrt{a^2 + b^2} = \sqrt{106}$

Center: $(0, 0)$

Vertices: $(0, \pm 5)$

Foci: $\left(0, \pm\sqrt{106}\right)$

Asymptotes:

$y = \pm\dfrac{a}{b}x = \pm\dfrac{5}{9}x$

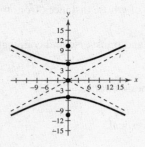

11. $\dfrac{(x-1)^2}{4} - \dfrac{(y+2)^2}{1} = 1$

$a = 2, b = 1, c = \sqrt{5}$

Center: $(1, -2)$

Vertices:

$(-1, -2), (3, -2)$

Foci: $\left(1 \pm \sqrt{5}, -2\right)$

Asymptotes: $y = -2 \pm \dfrac{1}{2}(x - 1)$

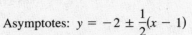

13. $\dfrac{(y+5)^2}{1/9} - \dfrac{(x-1)^2}{1/4} = 1$

$a = \dfrac{1}{3}, b = \dfrac{1}{2}, c = \sqrt{\dfrac{1}{9} + \dfrac{1}{4}} = \dfrac{\sqrt{13}}{6}$

Center: $(1, -5)$

Vertices: $\left(1, -5 \pm \dfrac{1}{3}\right): \left(1, -\dfrac{16}{3}\right), \left(1, -\dfrac{14}{3}\right)$

Foci: $\left(1, -5 \pm \dfrac{\sqrt{13}}{6}\right)$

Asymptotes: $y = k \pm \dfrac{a}{b}(x - h)$

$$y = -5 \pm \dfrac{2}{3}(x - 1)$$

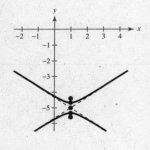

15. (a) $4x^2 - 9y^2 = 36$

$$\dfrac{x^2}{9} - \dfrac{y^2}{4} = 1$$

(b) Center: $(0, 0)$

$a = 3, b = 2, c = \sqrt{9 + 4} = \sqrt{13}$

Vertices: $(\pm 3, 0)$

Foci: $\left(\pm\sqrt{13}, 0\right)$

Asymptotes: $y = \pm\dfrac{b}{a}x = \pm\dfrac{2}{3}x$

(c)

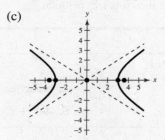

17. (a) $2x^2 - 3y^2 = 6$

$$\dfrac{x^2}{3} - \dfrac{y^2}{2} = 1$$

(b) $a = \sqrt{3}, b = \sqrt{2}, c = \sqrt{5}$

Center: $(0, 0)$

Vertices: $\left(\pm\sqrt{3}, 0\right)$

Foci: $\left(\pm\sqrt{5}, 0\right)$

Asymptotes: $y = \pm\sqrt{\dfrac{2}{3}}x$

$$= \pm\dfrac{\sqrt{6}}{3}x$$

(c) To use a graphing calculator, solve first for y.

$$y^2 = \dfrac{2x^2 - 6}{3}$$

$\left. \begin{array}{l} y_1 = \sqrt{\dfrac{2x^2 - 6}{3}} \\[3mm] y_2 = -\sqrt{\dfrac{2x^2 - 6}{3}} \end{array} \right\}$ Hyperbola

$\left. \begin{array}{l} y_3 = \sqrt{\dfrac{2}{3}}x \\[3mm] y_4 = -\sqrt{\dfrac{2}{3}}x \end{array} \right\}$ Asymptotes

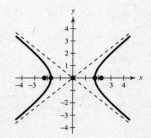

19. (a) $9x^2 - y^2 - 36x - 6y + 18 = 0$

$9(x^2 - 4x + 4) - (y^2 + 6y + 9) = -18 + 36 - 9$

$$\frac{(x-2)^2}{1} - \frac{(y+3)^2}{9} = 1$$

(b) $a = 1, b = 3, c = \sqrt{10}$

Center: $(2, -3)$

Vertices: $(1, -3), (3, -3)$

Foci: $\left(2 \pm \sqrt{10}, -3\right)$

Asymptotes: $y = -3 \pm 3(x - 2)$

(c)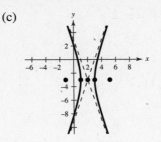

21. (a) $x^2 - 9y^2 + 2x - 54y - 80 = 0$

$(x^2 + 2x + 1) - 9(y^2 + 6y + 9) = 80 + 1 - 81$

$(x + 1)^2 - 9(y + 3)^2 = 0$

$y + 3 = \pm \frac{1}{3}(x + 1)$

(b) Degenerate hyperbola is two lines intersecting at $(-1, -3)$.

(c)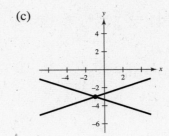

23. (a) $9y^2 - x^2 + 2x + 54y + 62 = 0$

$9(y^2 + 6y + 9) - (x^2 - 2x + 1) = -62 - 1 + 81$

$$\frac{(y+3)^2}{2} - \frac{(x-1)^2}{18} = 1$$

(b) $a = \sqrt{2}, b = 3\sqrt{2}, c = 2\sqrt{5}$

Center: $(1, -3)$

Vertices: $\left(1, -3 \pm \sqrt{2}\right)$

Foci: $\left(1, -3 \pm 2\sqrt{5}\right)$

Asymptotes: $y = -3 \pm \frac{1}{3}(x - 1)$

(c) To use a graphing calculator, solve for y first.

$9(y + 3)^2 = 18 + (x - 1)^2$

$$y = -3 \pm \sqrt{\frac{18 + (x-1)^2}{9}}$$

$y_1 = -3 + \frac{1}{3}\sqrt{18 + (x-1)^2}$

$\left. \begin{array}{l} y_2 = -3 - \frac{1}{3}\sqrt{18 + (x-1)^2} \end{array} \right\}$ Hyperbola

$\left. \begin{array}{l} y_3 = -3 + \frac{1}{3}(x - 1) \\[2mm] y_4 = -3 - \frac{1}{3}(x - 1) \end{array} \right\}$ Asymptotes

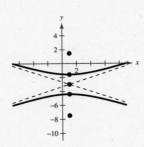

25. Vertices: $(0, \pm 2) \implies a = 2$

Foci: $(0, \pm 4) \implies c = 4$

$b^2 = c^2 - a^2 = 16 - 4 = 12$

Center: $(0, 0) = (h, k)$

$$\frac{(y - k)^2}{a^2} - \frac{(x - h)^2}{b^2} = 1$$

$$\frac{y^2}{4} - \frac{x^2}{12} = 1$$

27. Vertices: $(\pm 1, 0) \implies a = 1$

Asymptotes:

$$y = \pm 5x \implies \frac{b}{a} = 5$$

$$\implies b = 5$$

Center: $(0, 0)$

$$\frac{x^2}{1} - \frac{y^2}{25} = 1$$

29. Foci: $(0, \pm 8) \implies c = 8$

Asymptotes: $y = \pm 4x \implies \dfrac{a}{b} = 4 \implies a = 4b$

Center: $(0, 0) = (h, k)$

$c^2 = a^2 + b^2 \implies 64 = 16b^2 + b^2$

$$\frac{64}{17} = b^2 \implies a^2 = \frac{1024}{17}$$

$$\frac{(y - k)^2}{a^2} - \frac{(x - h)^2}{b^2} = 1$$

$$\frac{y^2}{1024/17} - \frac{x^2}{64/17} = 1$$

$$\frac{17y^2}{1024} - \frac{17x^2}{64} = 1$$

31. Vertices: $(2, 0), (6, 0) \implies a = 2$

Foci: $(0, 0), (8, 0) \implies c = 4$

$b^2 = c^2 - a^2 = 16 - 4 = 12$

Center: $(4, 0) = (h, k)$

$$\frac{(x - h)^2}{a^2} - \frac{(y - k)^2}{b^2} = 1$$

$$\frac{(x - 4)^2}{4} - \frac{y^2}{12} = 1$$

33. Vertices: $(4, 1), (4, 9) \implies a = 4$

Foci: $(4, 0), (4, 10) \implies c = 5$

$b^2 = c^2 - a^2 = 25 - 16 = 9$

Center: $(4, 5) = (h, k)$

$$\frac{(y - k)^2}{a^2} - \frac{(x - h)^2}{b^2} = 1$$

$$\frac{(y - 5)^2}{16} - \frac{(x - 4)^2}{9} = 1$$

35. Vertices: $(2, 3), (2, -3) \implies a = 3$

Solution point: $(0, 5)$

Center: $(2, 0) = (h, k)$

$$\frac{(y - k)^2}{a^2} - \frac{(x - h)^2}{b^2} = 1$$

$$\frac{y^2}{9} - \frac{(x - 2)^2}{b^2} = 1 \implies$$

$$b^2 = \frac{9(x - 2)^2}{y^2 - 9}$$

$$= \frac{9(-2)^2}{25 - 9} = \frac{36}{16} = \frac{9}{4}$$

$$\frac{y^2}{9} - \frac{(x - 2)^2}{9/4} = 1$$

37. Vertices: $(0, 4), (0, 0)$

Center: $(0, 2), a = 2$

$$\frac{(y - 2)^2}{4} - \frac{x^2}{b^2} = 1$$

Passes through $\left(\sqrt{5}, -1\right)$

$$\frac{(-1 - 2)^2}{4} - \frac{5}{b^2} = 1$$

$$\frac{9}{4} - 1 = \frac{5}{b^2}$$

$$b^2 = 4 \implies b = 2$$

$$\frac{(y - 2)^2}{4} - \frac{x^2}{4} = 1$$

39. Vertices:

$(1, 2), (3, 2) \implies a = 1$

Center: $(2, 2)$

Asymptotes:

$y = x, y = 4 - x$

$$\frac{b}{a} = 1 \implies b = 1$$

$$\frac{(x - 2)^2}{1} - \frac{(y - 2)^2}{1} = 1$$

41. Vertices: $(0, 2), (6, 2) \implies a = 3$

Asymptotes: $y = \frac{2}{3}x, y = 4 - \frac{2}{3}x$

$$\frac{b}{a} = \frac{2}{3} \implies b = 2$$

Center: $(3, 2) = (h, k)$

$$\frac{(x - h)^2}{a^2} - \frac{(y - k)^2}{b^2} = 1$$

$$\frac{(x - 3)^2}{9} - \frac{(y - 2)^2}{4} = 1$$

43. F_1: Friend's location $(-10{,}560, 0)$

F_2: Your location $(10{,}560, 0)$

$P(x, y)$: Location of lightning strike

$(1100)(18) = 19{,}800$

$$\frac{x^2}{a^2} - \frac{y^2}{b^2} = 1$$

$c = 10{,}560,$

$$a = \frac{19{,}800}{2} = 9900 \implies a^2 = 98{,}010{,}000$$

$$b^2 = c^2 - a^2 = 13{,}503{,}600$$

$$\frac{x^2}{98{,}010{,}000} - \frac{y^2}{13{,}503{,}600}$$

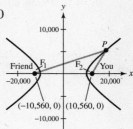

45. (a) $\dfrac{x^2}{a^2} - \dfrac{y^2}{b^2} = 1$

$a = 1$; $(2, 9)$ is on the curve, so

$$\frac{4}{1} - \frac{81}{b^2} = 1 \implies \frac{81}{b^2} = 3$$

$$\implies b^2 = \frac{81}{3} \implies b = 3\sqrt{3}.$$

$$\frac{x^2}{1} - \frac{y^2}{27} = 1, \quad -9 \le y \le 9$$

(b) Because each unit is $\frac{1}{2}$ foot, 4 inches is $\frac{2}{3}$ of a unit. The base is 9 units from the origin, so

$$y = 9 - \frac{2}{3} = 8\frac{1}{3}.$$

When $y = \dfrac{25}{3}$,

$$x^2 = 1 + \frac{(25/3)^2}{27} \implies x \approx 1.88998.$$

So the width is $2x \approx 3.779956$ units, or 22.68 inches, or 1.88998 feet.

47. Center: $(0, 0)$

Focus: $(24, 0)$

$b^2 = c^2 - a^2 = 24^2 - a^2 = 576 - a^2$

$$\frac{x^2}{a^2} - \frac{y^2}{576 - a^2} = 1$$

$$\frac{24^2}{a^2} - \frac{24^2}{576 - a^2} = 1$$

$$\frac{576}{a^2} - \frac{576}{576 - a^2} = 1$$

$576(576 - a^2) - 576a^2 = a^2(576 - a^2)$

$a^4 - 1728a^2 + 331,776 = 0$

$$a \approx \pm 38.83 \quad \text{or} \quad a \approx \pm 14.83$$

Since $a < c$ and $c = 24$, we choose $a = 14.83$. The vertex is approximate at $(14.83, 0)$.
[**Note:** By the Quadratic Formula, the exact value of a is $a = 12(\sqrt{5} - 1)$.]

49. $9x^2 + 4y^2 - 18x + 16y - 119 = 0$

$A = 9, C = 4$

$AC = 36 > 0$, Ellipse

51. $16x^2 - 9y^2 + 32x + 54y - 209 = 0$

$A = 16, C = -9$

$AC = 16(-9) < 0$, Hyperbola

53. $y^2 + 12x + 4y + 28 = 0$

$C = 1, A = 0$

$AC = 0$, Parabola

55. $x^2 + y^2 + 2x - 6y = 0$

$A = C = 1$, Circle

57. $x^2 - 6x - 2y + 7 = 0$

$A = 1, C = 0, D = -6,$

$E = -2, F = 7$

$AC = 0 \implies$ Parabola

59. True. $e = \dfrac{c}{a} = \dfrac{\sqrt{a^2 + b^2}}{a}$

61. False. For example,

$$x^2 - y^2 - 2x + 2y = 0$$

$$(x - 1)^2 - (y - 1)^2 = 0$$

is the graph of two intersecting lines.

63. Let (x, y) be such that the difference of the distances from $(c, 0)$ and $(-c, 0)$ is $2a$ (again only deriving one of the forms).

$$2a = \left| \sqrt{(x + c)^2 + y^2} - \sqrt{(x - c) + y^2} \right|$$

$$2a + \sqrt{(x - c)^2 + y^2} = \sqrt{(x + c)^2 + y^2}$$

$$4a^2 + 4a\sqrt{(x - c)^2 + y^2} + (x - c)^2 + y^2 = (x + c)^2 + y^2$$

$$4a\sqrt{(x - c)^2 + y^2} = 4cx - 4a^2$$

$$a\sqrt{(x - c)^2 + y^2} = cx - a^2$$

$$a^2(x^2 - 2cx + c^2 + y^2) = c^2x^2 - 2a^2cx + a^4$$

$$a^2(c^2 - a^2) = (c^2 - a^2)x^2 - a^2y^2$$

Let $b^2 = c^2 - a^2$. Then $a^2b^2 = b^2x^2 - a^2y^2 \implies 1 = \dfrac{x^2}{a^2} - \dfrac{y^2}{b^2}$.

65. $|d_2 - d_1| = $ constant by definition of hyperbola

At the point $(a, 0)$, $|d_2 - d_1| = |(a + c) - (c - a)| = 2a$.

67. At the point $(a, 0)$, the difference of the distances to the foci $(\pm c, 0)$ is $(c + a) - (c - a) = 2a$. Let (x, y) be a point on the hyperbola.

$$2a = \sqrt{(x + c)^2 + y^2} - \sqrt{(x - c)^2 + y^2}$$
$$2a + \sqrt{(x - c)^2 + y^2} = \sqrt{(x + c)^2 + y^2}$$
$$4a^2 + 4a\sqrt{(x - c)^2 + y^2} + (x - c)^2 + y^2 = (x + c)^2 + y^2$$
$$4a\sqrt{(x - c)^2 + y^2} = 4cx - 4a^2$$
$$a\sqrt{(x - c)^2 + y^2} = cx - a^2$$
$$a^2(x^2 - 2cx + c^2 + y^2) = c^2x^2 - 2a^2cx + a^4$$
$$a^2(c^2 - a^2) = (c^2 - a^2)x^2 - a^2y^2$$
$$1 = \frac{x^2}{a^2} - \frac{y^2}{c^2 - a^2}$$

Thus, $c^2 - a^2 = b^2$, as desired.

69. $(x^3 - 3x^2) - (6 - 2x - 4x^2) = x^3 + x^2 + 2x - 6$

71.

$$-2 \,\begin{array}{|rrrr} 1 & 0 & -3 & 4 \\ & -2 & 4 & -2 \\ \hline 1 & -2 & 1 & 2 \end{array}$$

$$\frac{x^3 - 3x + 4}{x + 2} = x^2 - 2x + 1 + \frac{2}{x + 2}$$

73. $x^3 - 16x = x(x^2 - 16) = x(x - 4)(x + 4)$

75. $2x^3 - 24x^2 + 72x = 2x(x^2 - 12x + 36)$
$$= 2x(x - 6)^2$$

77. $16x^3 + 54 = 2(8x^3 + 27)$
$$= 2(2x + 3)(4x^2 - 6x + 9)$$

Section 7.4   Parametric Equations

- If f and g are continuous functions of t on an interval I, then the set of ordered pairs $(f(t), g(t))$ is a *plane curve C*. The equations $x = f(t)$ and $y = g(t)$ are *parametric equations* for C and t is the *parameter*.
- You should be able to graph plane curves with your graphing utility.
- To eliminate the parameter:
    Solve for t in one equation and substitute into the second equation.
- You should be able to find the parametric equations for a graph.

Vocabulary Check

1. plane curve, parametric equations, parameter

2. orientation

3. eliminating, parameter

1. $x = t$

$y = t + 2$

$y = x + 2$, Line

Matches (c).

3. $x = \sqrt{t}$

$y = t$

$y = x^2$, parabola, $x \geq 0$

Matches (b).

5. $x = \ln t \iff t = e^x$

$y = \dfrac{1}{2}t - 2$

$y = \dfrac{1}{2}e^x - 2$

Matches (f).

7. $x = \sqrt{t}, \ y = 2 - t$

(a)

t	0	1	2	3	4
x	0	1	$\sqrt{2}$	$\sqrt{3}$	2
y	2	1	0	-1	-2

(b) Graph by hand.

Note: $x \geq 0$

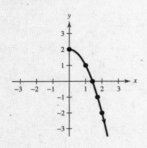

(c)

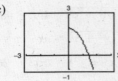

(d) $y = 2 - t = 2 - x^2$, Parabola

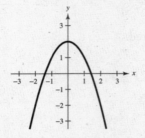

In part (c), $x \geq 0$.

9. $x = t, \ y = -4t$

$y = -4x$

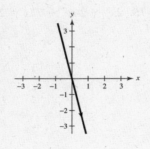

11. $x = 3t - 3, \ y = 2t + 1$

$y = 2\left[\dfrac{x + 3}{3}\right] + 1$

$= \dfrac{2}{3}x + 3$, Line

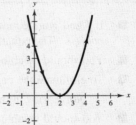

13. $x = \frac{1}{4}t, \ y = t^2$

$y = (4x)^2$

$y = 16x^2$

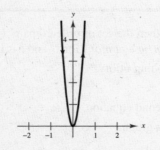

15. $x = t + 2, \ y = t^2$

$y = (x - 2)^2$, Parabola

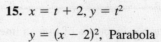

17. $x = t + 1, y = \dfrac{t}{t + 1}$

$y = \dfrac{x - 1}{x} = 1 - \dfrac{1}{x}$

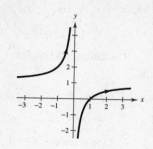

19. $x = 2t$

$y = |t - 2|$

$t = \dfrac{x}{2} \implies y = |t - 2|$

$\qquad = \left|\dfrac{x}{2} - 2\right|$

$\qquad = \dfrac{1}{2}|x - 4|$

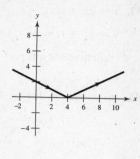

21. $x = e^{-t} \implies \dfrac{1}{x} = e^t$

$y = e^{3t} \implies y = (e^t)^3$

$y = \left(\dfrac{1}{x}\right)^3$

$y = \dfrac{1}{x^3}, \; x > 0, \; y > 0$

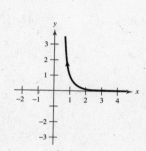

23. $x = t^3 \implies x^{1/3} = t$

$y = 3 \ln t \implies y = \ln t^3$

$y = \ln(x^{1/3})^3$

$y = \ln x$

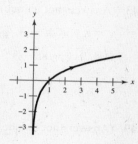

25. $x = 12t$

$y = -8t^2 + 32t$

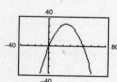

27. $x = \dfrac{3t}{1 + t^3}, \; y = \dfrac{3t^2}{1 + t^3}$

Undefined when $t = -1$

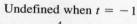
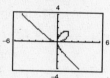

29. $x = \dfrac{t}{2}$

$y = \ln(t^2 + 1)$

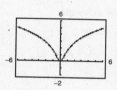

31. $x = 10 - 0.01e^t$

$y = 0.4t^2$

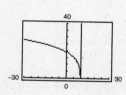

33. Each curve represents a portion of the line $y = 2x + 1$.

(a) Domain: $(-\infty, \infty)$; orientation: left to right

(b) Domain: all $x \neq 0$; orientation: oscillates, right to left

(c) Domain: $x > 0$; orientation: right to left

(d) Domain: $x > 0$; orientation: left to right

35. Each curve represents a portion of the graph of $y = x^3 - 1$. The portions are as follows.

(a) $0 \le x \le 1$ (b) $0 \le x \le 3$ (c) $-2 \le x \le 3$ (d) $-3 \le x \le 3$

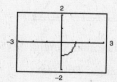
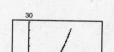
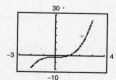
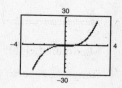

37. $x = x_1 + t(x_2 - x_1) = 0 + t(5 - 0) = 5t$

$y = y_1 + t(y_2 - y_1) = 0 + t(-2 - 0) = -2t$

(Solution not unique)

39. $x = x_1 + t(x_2 - x_1) = -2 + t[3 - (-2)] = 5t - 2$

$y = y_1 + t(y_2 - y_1) = 3 + t(10 - 3) = 7t + 3$

(Solution not unique)

41. (a) $x = x_1 + t(x_2 - x_1) = 3 + t(3 - 3) = 3$

$y = y_1 + t(y_2 - y_1) = -1 + t[5 - (-1)] = 6t - 1$

One answer: $x = 3$, $y = 6t - 1$, $0 \le t \le 1$

Alternative: $x = 3$, $y = t$, $-1 \le t \le 5$

(b) $x = x_1 + t(x_2 - x_1) = 3 + t(3 - 3) = 3$

$y = y_1 + t(y_2 - y_1) = 5 + t(-1 - 5) = 5 - 6t$

One answer: $x = 3$, $y = 5 - 6t$, $0 \le t \le 1$

Alternative: $x = 3$, $y = -t$, $-5 \le t \le 1$

43. (Answers not unique)

$x = t, y = 4t - 3$

$x = 2t, y = 8t - 3$

45. (Answers not unique)

$x = t, y = \dfrac{1}{t}$

$x = \dfrac{1}{t}, y = t$

47. (Answers not unique)

$x = t, y = t^2 + 4$

$x = -t, y = t^2 + 4$

49. (Answers not unique)

$x = t, y = t^3 + 2t$

$x = 2t, y = 8t^3 + 4t$

51. $y = e^{-x} + 1$

Sample answers:

$x = t, y = e^{-t} + 1$

$x = -t, y = e^t + 1$

53. $y = \ln(x^2 + 1)$

Sample answers:

$x = t, y = \ln(t^2 + 1)$

$x = 2t, y = \ln(4t^2 + 1)$

55. Sample answers:

$x = 4 - t^2, y = t$

$x = 4 - 4t^2, y = 2t$

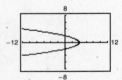

57. Sample answers:

$x = t$

$y = \pm\sqrt{4 - (t + 2)^2}$

$x = 2t$

$y = \pm\sqrt{4 - (2t + 2)^2}$

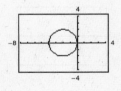

59. Sample answers:

$x = \pm\sqrt{36 - 4t^2}$

$y = t$

$x = \pm\sqrt{36 - 16t^2}$

$y = 2t$

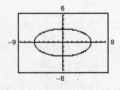

61. Sample answers:

$x = t$

$y = \pm\sqrt{1 + t^2}$

$x = 2t$

$y = \pm\sqrt{1 + 4t^2}$

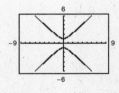

63. (a)

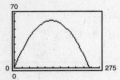

Maximum height: 60.5 feet

Range: 242 feet

(b)

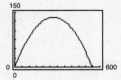

Maximum height: 136.1 feet

Range: 544.5 feet

(c)

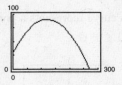

Maximum height: 90.5 feet

Range: 269 feet

65. (a) $x = 145t$

$y = 3 + 39t - 16t^2$

When $x = 400 = 145t$, $t = \frac{400}{145}$ and $y = 3 + 39\left(\frac{400}{145}\right) - 16\left(\frac{400}{145}\right)^2 \approx -11$.

Hence, it is not a home run. You can verify this graphically by tracing along the curve.

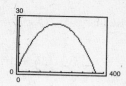

(b) $x = 138t$

$y = 3 + 59t - 16t^2$

When $x = 400 = 138t$, $t = \frac{400}{138}$ and $y \approx 39.6$.

Yes, it is a home run.

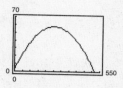

67. True

$x = t$ first set

$y = t^2 + 1 = x^2 + 1$

$x = 3t$ second set

$y = 9t^2 + 1$

$= (3t)^2 + 1 = x^2 + 1$

69. False. For example, let $x = t^2$ and $y = t$. Then $x = y^2$ and y is not a function of x.

71. The graph is the same, but the orientation is reversed.

73. $5x^2 + 8 = 0$

$x^2 = -\frac{8}{5}$

$x = \pm\sqrt{\frac{8}{5}}i = \pm\frac{2}{5}\sqrt{10}i$

75. $4x^2 + 4x - 11 = 0$

$x = \frac{-4 \pm \sqrt{16 + 4(4)(11)}}{8}$

$= \frac{-1 \pm \sqrt{12}}{2} = -\frac{1}{2} \pm \sqrt{3}$

77. $f(-x) = \frac{4(-x)^2}{(-x)^2 + 1} = \frac{4x^2}{x^2 + 1} = f(x)$

Symmetric about the y-axis

Even function

79. $y = e^x \neq e^{-x}$; $e^{-x} \neq -e^x$

No symmetry

Neither even nor odd

81. $\sum_{n=1}^{50} 8n = 8\frac{50(51)}{2} = 10,200$

83. $\sum_{n=1}^{40} \left(300 - \frac{1}{2}n\right) = 300(40) - \frac{1}{2}\frac{40(41)}{2} = 11,590$

Review Exercises for Chapter 7

1. Radius $= \sqrt{(-3 - 0)^2 + (-4 - 0)^2}$

$= \sqrt{9 + 16} = \sqrt{25} = 5$

$x^2 + y^2 = 25$

3. Radius $= \sqrt{(-4 - (-1))^2 + (5 - 1)^2}$

$= \sqrt{9 + 16} = \sqrt{25} = 5$

$(x + 1)^2 + (y - 1)^2 = 25$

5. Radius $= \frac{1}{2}\sqrt{(5-(-1))^2 + (6-2)^2}$

$$= \frac{1}{2}\sqrt{36+16} = \frac{1}{2}\sqrt{52} = \sqrt{13}$$

Center $= \left(\frac{5+(-1)}{2}, \frac{6+2}{2}\right) = (2,4)$

$(x-2)^2 + (y-4)^2 = 13$

7. $\frac{1}{2}x^2 + \frac{1}{2}y^2 = 18$

$x^2 + y^2 = 36$

Center: $(0,0)$

Radius: 6

9. $16x^2 + 16y^2 - 16x + 24y - 3 = 0$

$16\left(x^2 - x + \frac{1}{4}\right) + 16\left(y^2 + \frac{3}{2}y + \frac{9}{16}\right) = 3 + 4 + 9$

$16\left(x - \frac{1}{2}\right)^2 + 16\left(y + \frac{3}{4}\right)^2 = 16$

$\left(x - \frac{1}{2}\right)^2 + \left(y + \frac{3}{4}\right)^2 = 1$

Center: $\left(\frac{1}{2}, -\frac{3}{4}\right)$

Radius: 1

11. $(x^2 + 4x + 4) + (y^2 + 6y + 9) = 3 + 4 + 9$

$(x+2)^2 + (y+3)^2 = 16$

Center: $(-2, -3)$

Radius: 4

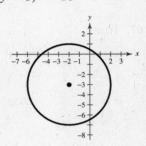

13. x-intercepts: $(x-3)^2 + (0+1)^2 = 7$

$(x-3)^2 = 6$

$x - 3 = \pm\sqrt{6}$

$x = 3 \pm \sqrt{6}$

$\left(3 \pm \sqrt{6}, 0\right)$

y-intercepts: $(0-3)^2 + (y+1)^2 = 7$

$(y+1)^2 = -2$, impossible

No y-intercepts

15. $4x - y^2 = 0$

$y^2 = 4(1)x, \; p = 1$

Vertex: $(0,0)$

Focus: $(1,0)$

Directrix: $x = -1$

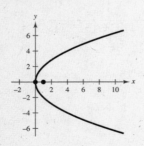

17. $\frac{1}{2}y^2 + 18x = 0$

$\frac{1}{2}y^2 = -18x$

$y^2 = -36x = 4(-9)x, \; p = -9$

Vertex: $(0,0)$

Focus: $(-9, 0)$

Directrix: $x = 9$

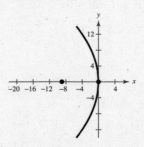

19. Vertex: $(0, 0)$

Focus: $(-6, 0)$

Parabola opens to left.

$y^2 = 4px$

$y^2 = 4(-6)x$

$y^2 = -24x$

21. Vertex: $(-6, 4)$

Passes through $(0, 0)$

Vertical axis

$(x + 6)^2 = 4p(y - 4)$

$(0 + 6)^2 = 4p(0 - 4)$

$36 = -16p$

$-\frac{9}{4} = p$

$(x + 6)^2 = 4\left(-\frac{9}{4}\right)(y - 4)$

$(x + 6)^2 = -9(y - 4)$

23. Vertex: $(0, 2) = (h, k)$

Directrix: $x = -3 \Rightarrow p = 3$

$(y - k)^2 = 4p(x - h)$

$(y - 2)^2 = 12x$

25. $x^2 = -2y = 4\left(-\frac{1}{2}\right)y, \ p = -\frac{1}{2}$

Focus: $\left(0, -\frac{1}{2}\right)$

$d_1 = \frac{1}{2} + b$

$d_2 = \sqrt{(2 - 0)^2 + \left(-2 + \frac{1}{2}\right)^2} = \frac{5}{2}$

$d_1 = d_2 \Rightarrow \frac{1}{2} + b = \frac{5}{2} \Rightarrow b = 2$

Slope of tangent line: $\dfrac{b + 2}{0 - 2} = \dfrac{4}{-2} = -2$

Equation: $y + 2 = -2(x - 2)$

$y = -2x + 2$

x-intercept: $(1, 0)$

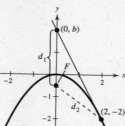

27. $y = \dfrac{x^2}{200}, \ -100 \leq x \leq 100$

Vertex: $(0, 0)$

$x^2 = 200y$

$4p = 200$

$p = 50$

Focus: $(0, 50)$

29. $x^2 = 4p(y - 12)$

$(4, 10)$ on curve:

$16 = 4p(10 - 12) = -8p \Rightarrow p = -2$

$x^2 = 4(-2)(y - 12) = -8y + 96$

$y = \dfrac{-x^2 + 96}{8}$

$y = 0$ if $x^2 = 96 \Rightarrow x = 4\sqrt{6} \Rightarrow$ width

is $8\sqrt{6}$ meters.

31. $\dfrac{x^2}{4} + \dfrac{y^2}{16} = 1$

$a = 4, b = 2, c = \sqrt{16 - 4} = \sqrt{12} = 2\sqrt{3}$

Center: $(0, 0)$

Vertices: $(0, \pm 4)$

Foci: $\left(0, \pm 2\sqrt{3}\right)$

Eccentricity $= \dfrac{c}{a} = \dfrac{2\sqrt{3}}{4} = \dfrac{\sqrt{3}}{2}$

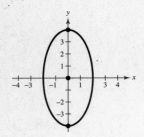

33. $\dfrac{(x-4)^2}{6} + \dfrac{(y+4)^2}{9} = 1$

$a = 3, b = \sqrt{6}, c = \sqrt{9 - 6} = \sqrt{3}$

Center: $(4, -4)$

Vertices: $(4, -1), (4, -7)$

Foci: $\left(4, -4 \pm \sqrt{3}\right)$

Eccentricity $= \dfrac{c}{a} = \dfrac{\sqrt{3}}{3}$

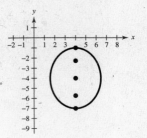

35. (a) $16(x^2 - 2x + 1) + 9(y^2 + 8y + 16) = -16 + 16 + 144$

$16(x - 1)^2 + 9(y + 4)^2 = 144$

$\dfrac{(x-1)^2}{9} + \dfrac{(y+4)^2}{16} = 1$

(b) Center: $(1, -4)$

$a = 4, b = 3, c = \sqrt{16 - 9} = \sqrt{7}$

Vertices: $(1, 0), (1, -8)$

Foci: $\left(1, -4 \pm \sqrt{7}\right)$

$e = \dfrac{c}{a} = \dfrac{\sqrt{7}}{4}$

(c)

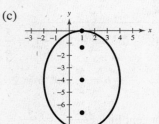

37. (a) $3(x^2 + 4x + 4) + 8(y^2 - 14y + 49) = -403 + 12 + 392$

$3(x + 2)^2 + 8(y - 7)^2 = 1$

$\dfrac{(x+2)^2}{1/3} + \dfrac{(y-7)^2}{1/8} = 1$

(b) Center: $(-2, 7)$

$a = \dfrac{\sqrt{3}}{3}, b = \dfrac{\sqrt{2}}{4}$

$c^2 = a^2 - b^2 = \dfrac{1}{3} - \dfrac{1}{8} = \dfrac{5}{24} \Rightarrow c = \dfrac{\sqrt{30}}{12}$

Vertices: $\left(-2 \pm \dfrac{\sqrt{3}}{3}, 7\right)$

Foci: $\left(-2 \pm \dfrac{\sqrt{30}}{12}, 7\right)$

Eccentricity: $\dfrac{c}{a} = \dfrac{\sqrt{30}/12}{\sqrt{3}/3} = \dfrac{\sqrt{10}}{4}$

(c)

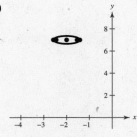

39. Vertices: $(\pm 5, 0)$

Foci: $(\pm 4, 0)$

$a = 5, c = 4 \implies b = 3$

$\dfrac{x^2}{25} + \dfrac{y^2}{9} = 1$

41. Vertices: $(0, 3), (10, 3)$

Passes through $(5, 0)$

Center: $(5, 3)$

Horizontal major axis

$a = 5, \ b = 3$

$\dfrac{(x-5)^2}{25} + \dfrac{(y-3)^2}{9} = 1$

43. Vertices: $(-3, 0), (7, 0)$

Foci: $(0, 0), (4, 0)$

Horizontal major axis

Center: $(2, 0)$

$a = 5, c = 2,$

$b = \sqrt{25 - 4} = \sqrt{21}$

$\dfrac{(x-h)^2}{a^2} + \dfrac{(y-k)^2}{b^2} = 1$

$\dfrac{(x-2)^2}{25} + \dfrac{y^2}{21} = 1$

45. Vertices: $(0, 1), (4, 1)$

Endpoints of minor axis: $(2, 0), (2, 2)$

Horizontal major axis

Center: $(2, 1)$

$a = 2, b = 1$

$\dfrac{(x-h)^2}{a^2} + \dfrac{(y-k)^2}{b^2} = 1$

$\dfrac{(x-2)^2}{4} + \dfrac{(y-1)^2}{1} = 1$

47. $a = 5, \ b = 4, c = \sqrt{a^2 - b^2} = \sqrt{25 - 16} = 3$

The foci should be placed 3 feet on either side of the center and have the same height as the pillars.

49. $a - c = 1.3495 \times 10^9$

$a + c = 1.5045 \times 10^9$

Adding, $2a = 2.854 \times 10^9 \implies a = 1.427 \times 10^9$. Then

$c = 1.5045 \times 10^9 - 1.427 \times 10^9 = 0.0775 \times 10^9$.

$e = \dfrac{c}{a} \approx 0.0543$

51. $\dfrac{x^2}{36} - \dfrac{y^2}{25} = 1$

$a = 6, b = 5,$

$c = \sqrt{36 + 25} = \sqrt{61}$

Center: $(0, 0)$

Vertices: $(\pm 6, 0)$

Foci: $\left(\pm \sqrt{61}, 0\right)$

Asymptotes: $y = \pm \dfrac{b}{a}x = \pm \dfrac{5}{6}x$

53. $\dfrac{(y-2)^2}{9} - \dfrac{(x-3)^2}{4} = 1$

$a = 3, b = 2,$

$c = \sqrt{9 + 4} = \sqrt{13}$

Center: $(3, 2)$

Vertices: $(3, 5), (3, -1)$

Foci: $\left(3, 2 \pm \sqrt{13}\right)$

Asymptotes: $y - 2 = \pm \dfrac{a}{b}(x - 3)$

$$y = 2 \pm \dfrac{3}{2}(x - 3)$$

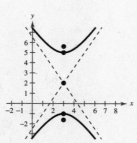

55. (a) $5y^2 - 4x^2 = 20$

$$\frac{y^2}{4} - \frac{x^2}{5} = 1$$

(b) $a = 2, b = \sqrt{5},$

$c = \sqrt{4 + 5} = 3$

Center: $(0, 0)$

Vertices: $(0, \pm 2)$

Foci: $(0, \pm 3)$

Eccentricity $= \dfrac{c}{a} = \dfrac{3}{2}$

(c)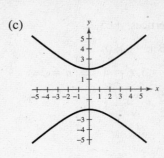

57. (a) $9(x^2 - 2x + 1) - 16(y^2 + 2y + 1) = 151 + 9 - 16$

$$9(x - 1)^2 - 16(y + 1)^2 = 144$$

$$\frac{(x - 1)^2}{16} - \frac{(y + 1)^2}{9} = 1$$

(b) Center: $(1, -1)$, $a = 4, b = 3, c = 5$

Vertices: $(5, -1), (-3, -1)$

Foci: $(6, -1), (-4, -1)$

Eccentricity: $\dfrac{5}{4}$

Asymptotes: $y = -1 \pm \dfrac{3}{4}(x - 1)$

(c)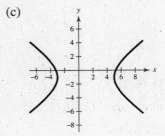

59. (a) $(y^2 - 2y + 1) - 4(x^2 + 12x + 36) = -59 + 1 - 144$

$$(y - 1)^2 - 4(x + 6)^2 = -202$$

$$\frac{(x + 6)^2}{(101/2)} - \frac{(y - 1)^2}{202} = 1$$

(b) Center: $(-6, 1)$

$a^2 = \dfrac{101}{2}, b^2 = 202, c^2 = \dfrac{101}{2} + 202 = \dfrac{505}{2}$

Vertices: $\left(-6 \pm \sqrt{\dfrac{101}{2}}, 1\right)$

Foci: $\left(-6 \pm \sqrt{\dfrac{505}{2}}, 1\right)$

Asymptotes: $y = 1 \pm 2(x + 6)$

Eccentricity: $e = \dfrac{c}{a} = \dfrac{\sqrt{505}}{\sqrt{101}} = \sqrt{5}$

(c)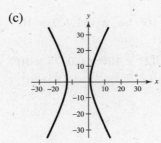

61. Vertices: $(\pm 1, 0)$

Horizontal transverse axis

Center: $(0, 0)$

$a = 1$

$\pm \dfrac{b}{a} = \pm 2 \implies b = 2$

$$\frac{x^2}{1} - \frac{y^2}{4} = 1$$

63. $\dfrac{x^2}{a^2} - \dfrac{y^2}{b^2} = 1$

$a = 4$

$c^2 = a^2 + b^2 \implies 36 = 16 + b^2$

$\implies b = \sqrt{20} = 2\sqrt{5}$

$$\frac{x^2}{16} - \frac{y^2}{20} = 1$$

65. Foci: $(0, 0)$, $(8, 0)$ $\Rightarrow$ $c = 4$

Center: $(4, 0)$

Asymptotes:

$y = \pm 2(x - 4) \Rightarrow \dfrac{b}{a} = 2 \Rightarrow b = 2a$

$c^2 = a^2 + b^2$

$16 = a^2 + (2a)^2 = 5a^2 \Rightarrow a = \dfrac{4}{\sqrt{5}}, b = \dfrac{8}{\sqrt{5}}$

$\dfrac{(x - 4)^2}{16/5} - \dfrac{y^2}{64/5} = 1$

67. $d_2 - d_1 = 186{,}000(0.0005)$

$2a = 93$

$a = 46.5$

$c = 100$

$b = \sqrt{c^2 - a^2}$

$\dfrac{x^2}{a^2} - \dfrac{y^2}{b^2} = 1$

$x = 60 \Rightarrow y^2 = b^2\left(\dfrac{x^2}{a^2} - 1\right) = (100^2 - 46.5^2)\left(\dfrac{60^2}{46.5^2} - 1\right) \approx 5211.57 \Rightarrow y \approx 72.2$

72.2 miles north

69. $3x^2 + 2y^2 - 12x + 12y + 29 = 0$

$3(x^2 - 4x + 4) + 2(y^2 + 6y + 9) = -29 + 12 + 18$

$3(x - 2)^2 + 2(y + 3)^2 = 1$

Ellipse

71. $5x^2 - 2y^2 + 10x - 4y + 17 = 0$

$5(x^2 + 2x + 1) - 2(y^2 + 2y + 1) = -17 + 5 - 2$

$5(x + 1)^2 - 2(y + 1)^2 = -14$

$\dfrac{(y + 1)^2}{7} - \dfrac{(x + 1)^2}{(14/5)} = 1$

Hyperbola

73.

t	-2	-1	0	1	2	3
x	-8	-5	-2	1	4	7
y	15	11	7	3	-1	-5

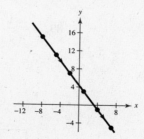

75.

t	-2	-1	1	2	3	4
x	-3	-6	6	3	2	$\frac{3}{2}$
y	2	3	5	6	7	8

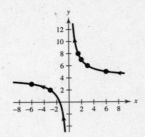

77. $x = t^3, y = t^2 + 1$

$t = x^{1/3} \implies y = x^{2/3} + 1, -1 \le x \le 1$

Matches (a).

79. $x = (2t)^3 = 8t^3, y = (2t)^2 + 1 = 4t^2 + 1$

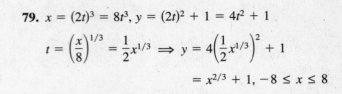

$$= x^{2/3} + 1, -8 \le x \le 8$$

Matches (d).

81. $x = 5t - 1, y = 2t + 5$

$t = \frac{1}{5}(x + 1) \implies$

$y = \frac{2}{5}(x + 1) + 5 = \frac{2}{5}x + \frac{27}{5}$, line

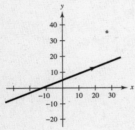

83. $x = t^2 + 2, y = 4t^2 - 3$

$t^2 = x - 2 \implies$

$y = 4(x - 2) - 3 = 4x - 11, \quad x \ge 2$

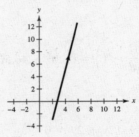

85. $x = t^3, y = \dfrac{1}{2}t^2$

$t = x^{1/3} \implies y = \dfrac{1}{2}x^{2/3}$

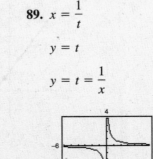

87. $x = \sqrt[3]{t}$

$y = t$

$t = x^3 \implies y = t = x^3$

$y = x^3$

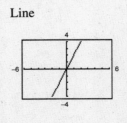

89. $x = \dfrac{1}{t}$

$y = t$

$y = t = \dfrac{1}{x}$

91. $x = 2t$

$y = 4t$

$y = 2(2t) = 2x$

Line

93. $x = 1 + 4t$

$y = 2 - 3t$

$t = \dfrac{x - 1}{4} \implies y = 2 - 3\left(\dfrac{x - 1}{4}\right) = 2 - \dfrac{3}{4}x + \dfrac{3}{4}$

$y = \dfrac{11}{4} - \dfrac{3}{4}x$

$3x + 4y - 11 = 0$

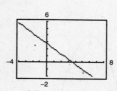

95. $x = \dfrac{1}{t}$

$y = t^2$

$y = \left(\dfrac{1}{x}\right)^2$

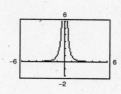

97. $x = 3$

$y = t$

Vertical line: $x = 3$

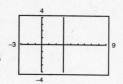

99. $y = 6x + 2$

$x = t, y = 6t + 2$

$x = -t, y = -6t + 2$

Other answers possible

101. $y = x^2 + 2$

$x = t, y = t^2 + 2$

$x = t + 1, y = (t + 1)^2 + 2 = t^2 + 2t + 3$

Other answers possible

103. $x = x_1 + t(x_2 - x_1) = 3 + t(8 - 3) = 5t + 3$

$y = y_1 + t(y_2 - y_1) = 5 + t(0 - 0) = 5$

or $x = t, \ y = 5$

105. $x = x_1 + t(x_2 - x_1)$

$\quad = -1 + t[10 - (-1)] = 11t - 1$

$y = y_1 + t(y_2 - y_1) = 6 + t(0 - 6) = -6t + 6$

107. $(90, 4)$ is on the curve:

$90 = 0.82v_0 t \implies v_0 = \dfrac{90}{0.82t}$

$4 = 7 + 0.57\left[\dfrac{90}{0.82t}\right]t - 16t^2 \implies$

$16t^2 = 3 + \dfrac{0.57(90)}{0.82} \implies t \approx 2.024$

Hence, $v_0 \approx \dfrac{90}{0.82(2.024)} \approx 54.2$ ft/sec.

109. From Exercise 108:

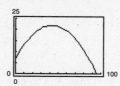

The maximum height is approximately 21.9 feet for $t \approx 0.97$.

111. False. $\dfrac{x^2}{4} - y^4 = 1$ is not of degree 2 in y.

113. False. There are an infinite number of parametric equations. Two examples are:

$x = t, y = 3 - 2t$

$x = t + 1, y = 3 - 2(t + 1) = 1 - 2t$

115. $x^2 - 6x + 2y + 9 = 0$

$$(x - 3)^2 = -2y$$

Parabola

Vertex: $(3, 0)$

Focus: $\left(3, -\frac{1}{2}\right)$

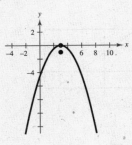

117. $x^2 + 9y^2 + 10x - 18y + 25 = 0$

$$(x + 5)^2 + 9(y - 1)^2 = 9$$

$$\frac{(x + 5)^2}{9} + (y - 1)^2 = 1$$

Ellipse

Center: $(-5, 1)$

Vertices: $(-8, 1), (-2, 1)$

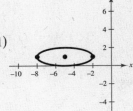

119. $$4x^2 - 4x - 4y^2 + 8y = 11$$

$$4\left(x^2 - x + \frac{1}{4}\right) - 4(y^2 - 2y + 1) = 11 + 1 - 4$$

$$4\left(x - \frac{1}{2}\right)^2 - 4(y - 1)^2 = 8$$

$$\frac{\left(x - \frac{1}{2}\right)^2}{2} - \frac{(y - 1)^2}{2} = 1$$

Hyperbola

121. $4x^2 + y^2 - 16x + 15 = 0$

$$4(x - 2)^2 + y^2 = 1$$

Ellipse

Center: $(2, 0)$

Vertices: $(2, -1), (2, 1)$

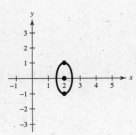

123. The number b must be less than 5. The ellipse becomes more circular and approaches a circle of radius 5.

Chapter 7   Practice Test

1. Find the vertex, focus, and directrix of the parabola $x^2 = 20y$.

2. Find the equation of the parabola with vertex $(0, 0)$ and focus $(7, 0)$.

3. Find the center, foci, and vertices of the ellipse $\dfrac{x^2}{144} + \dfrac{y^2}{25} = 1$.

4. Find the equation of the ellipse with foci $(\pm 4, 0)$ and minor axis of length 6.

5. Find the center, vertices, foci, and asymptotes of the hyperbola $\dfrac{y^2}{144} - \dfrac{x^2}{169} = 1$.

6. Find the equation of the hyperbola with vertices $(\pm 4, 0)$ and asymptotes $y = \pm \frac{1}{2}x$.
   Use a graphing utility to graph the curve.

7. Find the equation of the parabola with vertex $(6, -1)$ and focus $(6, 3)$.
   Use a graphing utility to graph the curve.

8. Find the center, foci, and vertices of the ellipse $16x^2 + 9y^2 - 96x + 36y + 36 = 0$.

9. Find the equation of the ellipse with vertices $(-1, 1)$ and $(7, 1)$ and minor axis of length 2.

10. Find the center, vertices, foci, and asymptotes of the hyperbola $4(x + 3)^2 - 9(y - 1)^2 = 1$.

11. Find the equation of the hyperbola with vertices $(3, 4)$ and $(3, -4)$ and foci $(3, 7)$ and $(3, -7)$.

12. Use a graphing utility to sketch the curve represented by the parametric equations.

    $x = 2t + 1$

    $y = -1 - 3t$

    Then eliminate the parameter and write the corresponding rectangular equation.

13. Use a graphing utility to sketch the curve represented by the parametric equations

    $x = 2 \ln t$

    $y = t^3$

14. Describe how the plane curves $x = t$, $y = t^2$ and $x = t^2$, $y = t^4$ differ from each other.

A P P E N D I C E S

APPENDIX B
Concepts in Statistics

Appendix B.1   Measures of Central Tendency and Dispersion

Vocabulary Check

1. measure, central tendency

2. modes, bimodal

3. variance, standard deviation

4. Quartiles

---

1. Mean $= \dfrac{5 + 12 + 7 + 14 + 8 + 9 + 7}{7}$

   $= \dfrac{62}{7} \approx 8.86$

   Median: 8

   Mode: 7

3. Mean $= \dfrac{5 + 12 + 7 + 24 + 8 + 9 + 7}{7}$

   $= \dfrac{72}{7} \approx 10.29$

   Median: 8

   Mode: 7

5. Mean $= \dfrac{5 + 12 + 7 + 14 + 9 + 7}{6} = \dfrac{54}{6} = 9$

   Median: $\dfrac{7 + 9}{2} = 8$

   Mode: 7

7. (a) The mean is sensitive to extreme values

   (b) Mean: 14.86

   Median: 14

   Mode: 13

   Each is increased by 6.

   (c) Each will increase by k.

9. Mean $= \dfrac{410 + 260 + 320 + 320 + 460 + 150}{6}$

   $= \dfrac{1920}{6} = 320$

   Median: 320

   Mode: 320

11. (a) Jay: $\dfrac{181 + 222 + 196}{3} = 199\dfrac{2}{3}$

   Hank: $\dfrac{199 + 195 + 205}{3} = 199\dfrac{2}{3}$

   Buck: $\dfrac{202 + 251 + 235}{3} = 229\dfrac{1}{3}$

   (b) Adding all nine numbers, you obtain

   Mean $= \dfrac{1886}{9} = 209\dfrac{5}{9}$.

   (c) Median $= 202$ (four scores below 202 and four scores above 202)

13. There are many possible answers. For example: $\{4, 4, 10\}$

15. The mean is 76.55 and the median is 82. The median is the best description.

17. (a) Mean = 12, $\sigma \approx 2.83$ (b) Mean = 20, $\sigma \approx 2.83$

     (c) Mean = 12, $\sigma \approx 1.41$ (d) Mean = 9, $\sigma \approx 1.41$

19. $\bar{x} = 6$ **21.** $\bar{x} = 2$ **23.** $\bar{x} = 4$ **25.** $\bar{x} = 47$

     $v = 10$ $v = \frac{4}{3}$ $v = 4$ $v = 226$

     $\sigma \approx 3.16$ $\sigma \approx 1.15$ $\sigma \approx 2$ $\sigma \approx 15.03$

27. $\bar{x} = 6$

$$\sigma = \sqrt{\frac{2^2 + 4^2 + 6^2 + 6^2 + 13^2 + 5^2}{6} - 6^2} = \sqrt{\frac{286}{6} - 36} = \sqrt{\frac{35}{3}} \approx 3.42$$

29. $\bar{x} = 5.8$

$$\sigma = \sqrt{\frac{8.1^2 + 6.9^2 + 3.7^2 + 4.2^2 + 6.1^2}{5} - 5.8^2} = \sqrt{2.712} \approx 1.65$$

31. $\bar{x} = 12$ and $|x_i - 12| = 8$ for all x_i. Hence, $\sigma = 8$.

33. The mean will increase by 5. The standard deviation will not change.

35. $\bar{x} = 235$

     $\sigma = 28$

     $n = 600$

     $1 - \dfrac{1}{2^2} = \dfrac{3}{4}$ lies within two standard deviations:

         $[235 - 2(28), 235 + 2(28)] = [179, 291]$.

     $1 - \dfrac{1}{3^2} = \dfrac{8}{9}$ lies within three standard deviations:

         $[235 - 3(28), 235 + 3(28)] = [151, 319]$.

     If $\sigma = 16$, then

         $[235 - 2(16), 235 + 2(16)] = [203, 267]$

         $[235 - 3(16), 235 + 3(16)] = [187, 283]$.

37. (a) 12, 13, 13, 14, 14, 15, 20, 23, 23

        Median: 14

        Lower quartile is median of $\{12, 13, 13\} = 13$.

        Upper quartile is median of $\{15, 20, 23, 23\} = 21.5$.

     (b)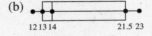

39. (a) 46, 47, 47, 48, 48, 49, 50, 51, 52, 53

        Median: $\dfrac{48 + 49}{2} = 48.5$

        Lower quartile is median of $\{46, 47, 47, 48, 48\} = 47$.

        Upper quartile is median of $\{49, 50, 51, 52, 53\} = 51$.

     (b)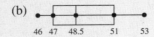

41.

43.

45.

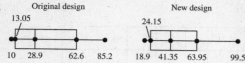

Original design

13.05

10   28.9   62.6   85.2

New design

24.15

18.9   41.35   63.95   99.5

From the plots, you can see that the lifetimes of the units in the new design are greater than the original design. The median increased by over 12 months.

Appendix B.2   Least Squares Regression

1.

x	y	xy	x^2
−4	1	−4	16
−3	3	−9	9
−2	4	−8	4
−1	6	−6	1
Total	−10	14	−27	30

$n = 4$

$4b + (-10)a = 14$

$(-10)b + 30a = -27$

Solving this system, $a = 1.6$ and $b = 7.5$.

Answer: $y = 1.6x + 7.5$

3.

x	y	xy	x^2
−3	1	−3	9
−1	2	−2	1
1	2	2	1
4	3	12	16
Total	1	8	9	27

$n = 4$

$4b + a = 8$

$b + 27a = 9$

Solving this system, $a \approx 0.262$ and $b \approx 1.93$.

Answer: $y = 0.262x + 1.93$

A P P E N D I X   C
Variation

Vocabulary Check

1. directly proportional

2. constant, variation

3. directly proportional

4. inverse

5. combined

6. jointly proportional

1. $y = kx$

$12 = k(5)$

$\frac{12}{5} = k$

$y = \frac{12}{5}x$

3. $y = kx$

$2050 = k(10)$

$205 = k$

$y = 205x$

5. $y = kx$

$33 = k(13)$

$\frac{33}{13} = k$

$y = \frac{33}{13}x$

When $x = 10$ inches, $y \approx 25.4$ centimeters.

When $x = 20$ inches, $y \approx 50.8$ centimeters.

7. $y = kx$

$5520 = k(150,000)$

$0.0368 = k$

$y = 0.0368x$

$y = 0.0368(200,000)$

$= \$7360$

The property tax is $7360.

9. $d = kF$

$0.15 = k(265)$

$\frac{3}{5300} = k$

$d = \frac{3}{5300}F$

(a) $d = \frac{3}{5300}(90) \approx 0.05$ meter

(b) $0.1 = \frac{3}{5300}F$

$\frac{530}{3} = F$

$F = 176\frac{2}{3}$ newtons

11. $k = 1$

x	2	4	6	8	10
$y = kx^2$	4	16	36	64	100

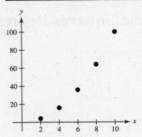

13. $k = \frac{1}{2}$

x	2	4	6	8	10
$y = kx^2$	2	8	18	32	50

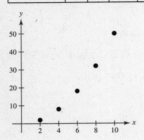

15. $d = kv^2$

$0.02 = k\left(\frac{1}{4}\right)^2$

$k = 0.32$

$d = 0.32v^2$

$0.12 = 0.32v^2$

$v^2 = \frac{0.12}{0.32} = \frac{3}{8}$

$v = \frac{\sqrt{3}}{2\sqrt{2}} = \frac{\sqrt{6}}{4} \approx 0.61$ mi/hr

17. $k = 2$

x	2	4	6	8	10
$y = \dfrac{k}{x^2}$	$\frac{1}{2}$	$\frac{1}{8}$	$\frac{1}{18}$	$\frac{1}{32}$	$\frac{1}{50}$

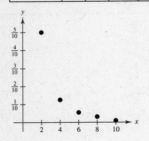

19. $k = 10$

x	2	4	6	8	10
$y = \dfrac{k}{x^2}$	$\frac{5}{2}$	$\frac{5}{8}$	$\frac{5}{18}$	$\frac{5}{32}$	$\frac{1}{10}$

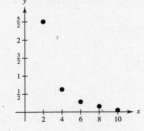

21. The table represents the equation $y = 5/x$.

23. $y = kx$

$-7 = k(10)$

$-\frac{7}{10} = k$

$y = -\frac{7}{10}x$

This equation checks with the other points given in the table.

25. $A = kr^2$

27. $y = \dfrac{k}{x^2}$

29. $F = \dfrac{kg}{r^2}$

31. $P = \dfrac{k}{V}$

33. $R = k(T - T_e)$

35. $A = \dfrac{1}{2}bh$

The area of a triangle is jointly proportional to its base and height.

37. $V = \dfrac{4}{3}\pi r^3$

The volume of a sphere varies directly as the cube of its radius.

39. $r = \dfrac{d}{t}$

Average speed is directly proportional to the distance and inversely proportional to the time.

41. $A = kr^2$

$9\pi = k(3)^2$

$\pi = k$

$A = \pi r^2$

43. $y = \dfrac{k}{x}$

$7 = \dfrac{k}{4}$

$28 = k$

$y = \dfrac{28}{x}$

45. $F = krs^3$

$4158 = k(11)(3)^3$

$k = 14$

$F = 14rs^3$

47. $z = \dfrac{kx^2}{y}$

$6 = \dfrac{k(6)^2}{4}$

$\dfrac{24}{36} = k$

$\dfrac{2}{3} = k$

$z = \dfrac{2/3x^2}{y} = \dfrac{2x^2}{3y}$

49. $r = \dfrac{kl}{A}, \; A = \pi r^2 = \dfrac{\pi d^2}{4}$

$r = \dfrac{4kl}{\pi d^2}$

$66.17 = \dfrac{4(1000)k}{\pi\left(\frac{0.0126}{12}\right)^2}$

$k \approx 5.73 \times 10^{-8}$

$r = \dfrac{4(5.73 \times 10^{-8})l}{\pi\left(\frac{0.0126}{12}\right)^2}$

$33.5 = \dfrac{4(5.73 \times 10^{-8})l}{\pi\left(\frac{0.0126}{12}\right)^2}$

$\dfrac{33.5\pi\left(\frac{0.0126}{12}\right)^2}{4(5.73 \times 10^{-8})} = l$

$l \approx 506$ feet

51. $W = kmh$

$2116.8 = k(120)(1.8)$

$k = \dfrac{2116.8}{(120)(1.8)} = 9.8$

$W = 9.8mh$

When $m = 100$ kilograms and $h = 1.5$ meters, we have $W = 9.8(100)(1.5) = 1470$ joules.

53. $v = \dfrac{k}{A}$

$v = \dfrac{k}{0.75A} = \dfrac{4}{3}\left(\dfrac{k}{A}\right)$

The velocity is increased by one-third.

55. (a)

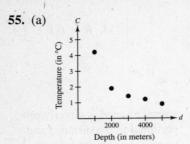

(b) Yes, the data appears to be modeled (approximately) by the inverse proportion model.

$$4.2 = \frac{k_1}{1000} \qquad 1.9 = \frac{k_2}{2000} \qquad 1.4 = \frac{k_3}{3000} \qquad 1.2 = \frac{k_4}{4000} \qquad 0.9 = \frac{k_5}{5000}$$

$$4200 = k_1 \qquad 3800 = k_2 \qquad 4200 = k_3 \qquad 4800 = k_4 \qquad 4500 = k_5$$

(c) Mean: $k = \dfrac{4200 + 3800 + 4200 + 4800 + 4500}{5} = 4300,$ Model: $C = \dfrac{4300}{d}$

(d)

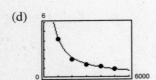

(e) $3 = \dfrac{4300}{d}$

$$d = \frac{4300}{3} = 1433\frac{1}{3} \text{ meters}$$

57. False. y will increase if k is positive and y will decrease if k is negative.

59. The graph appears to represent $y = 4/x$, so y varies inversely as x.

A P P E N D I X   D
Solving Linear Equations and Inequalities

Vocabulary Check

1. linear

2. equivalent inequalities

1. $x + 11 = 15$

$x = 15 - 11$

$x = 4$

3. $x - 2 = 5$

$x = 5 + 2$

$x = 7$

5. $3x = 12$

$x = \dfrac{12}{3}$

$x = 4$

7. $\dfrac{x}{5} = 4$

$x = 4(5)$

$x = 20$

9. $8x + 7 = 39$

$8x = 32$

$x = 4$

11. $24 - 7x = 3$

$-7x = -21$

$x = 3$

13. $8x - 5 = 3x + 20$

$5x = 25$

$x = 5$

15. $-2(x + 5) = 10$

$\qquad -2x - 10 = 10$

$\qquad -2x = 20$

$\qquad x = -10$

17. $2x + 3 = 2x - 2$

$\qquad 3 = -2$

No solution

19. $\frac{3}{2}(x + 5) - \frac{1}{4}(x + 24) = 0$

$\qquad \frac{3}{2}(x + 5) = \frac{1}{4}(x + 24)$

$\qquad 12(x + 5) = 2(x + 24)$

$\qquad 12x + 60 = 2x + 48$

$\qquad 10x = -12$

$\qquad x = -\frac{12}{10}$

$\qquad x = -\frac{6}{5}$

21. $0.25x + 0.75(10 - x) = 3$

$\qquad 25x + 75(10 - x) = 300$

$\qquad 25x + 750 - 75x = 300$

$\qquad -50x = -450$

$\qquad x = 9$

23. $x + 6 < 8$

$\qquad x < 8 - 6$

$\qquad x < 2$

25. $-x - 8 > -17$

$\qquad 17 - 8 > x$

$\qquad 9 > x$

$\qquad x < 9$

27. $6 + x \leq -8$

$\qquad x \leq -8 - 6$

$\qquad x \leq -14$

29. $\frac{4}{5}x > 8$

$\qquad x > \frac{5}{4}(8)$

$\qquad x > 10$

31. $-\frac{3}{4}x > -3$

$\qquad \frac{3}{4}x < 3$

$\qquad x < 4$

33. $4x < 12$

$\qquad x < 3$

35. $-11x \leq -22$

$\qquad 11x \geq 22$

$\qquad x \geq 2$

37. $x - 3(x + 1) \geq 7$

$\qquad x - 3x - 3 \geq 7$

$\qquad -2x \geq 10$

$\qquad x \leq -5$

39. $7x - 12 < 4x + 6$

$\qquad 3x < 18$

$\qquad x < 6$

41. $\frac{3}{4}x - 6 \leq x - 7$

$\qquad 1 \leq \frac{1}{4}x$

$\qquad 4 \leq x$

$\qquad x \geq 4$

43. $3.6x + 11 \geq -3.4$

$\qquad 3.6x \geq -14.4$

$\qquad x \geq \frac{-14.4}{3.6}$

$\qquad x \geq -4$

APPENDIX E
Systems of Inequalities

Appendix E.1   Solving Systems of Inequalities

Vocabulary Check

1. solution

2. graph

3. linear

4. point, equilibrium

1. $x < 2$

Vertical boundary

Matches graph (g).

3. $2x + 3y \geq 6$

$\qquad y \geq -\frac{2}{3}x + 2$

Line with negative slope

Matches (a).

5. $x^2 + y^2 < 9$

Circular boundary

Matches (e).

7. $xy > 1$ or $y > \dfrac{1}{x}$

Matches (f).

9. $y < 2 - x^2$

Graph the parabola $y = 2 - x^2$. The region lies below the parabola.

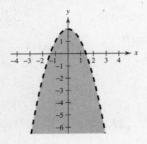

11. $y^2 + 1 \geq x$

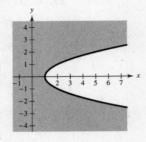

13. $x \geq 4$

Using a solid line, graph the vertical line $x = 4$ and shade to the right of this line.

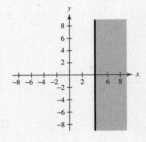

15. $y \geq -1$

Using a solid line, graph the horizontal line $y = -1$ and shade above this line.

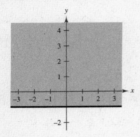

17. $2y - x \geq 4$

Using a solid line, graph $2y - x = 4$, and then shade above the line. (Use $(0, 0)$ as a test point.)

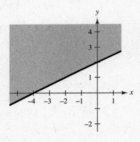

19. $2x + 3y < 6$

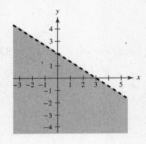

21. $4x - 3y \leq 24$

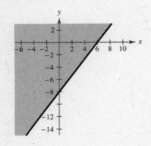

23. $y > 3x^2 + 1$

Sketch the parabola $y = 3x^2 + 1$. The region lies above the parabola.

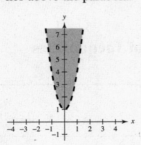

25. $2x - y^2 > 0$

$2x > y^2$

27. $(x + 1)^2 + y^2 < 9$

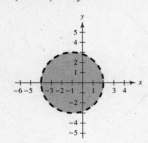

29. $y \geq \frac{2}{3}x - 1$

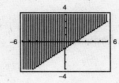

31. $y < -3.8x + 1.1$

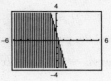

33. $x^2 + 5y - 10 \leq 0$

$$y \leq 2 - \frac{x^2}{5}$$

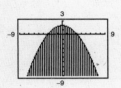

35. $y \leq \dfrac{1}{1 + x^2}$

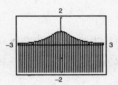

37. $y < \ln x$

Using a dashed line, graph $y = \ln x$, and shade to the right of the curve. (Use $(2, 0)$ as a test point.)

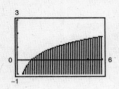

39. $y > 3^{-x-4}$

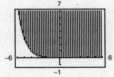

41. The line through $(0, 2)$ and $(3, 0)$ is $y = -\frac{2}{3}x + 2$. For the shaded region above the line, we have:

$$y > -\frac{2}{3}x + 2$$

$$3y > -2x + 6$$

$$2x + 3y > 6$$

$$\frac{x}{3} + \frac{y}{2} > 1$$

43. The circle shown is $x^2 + y^2 = 9$. For the shaded region inside the circle, we have $x^2 + y^2 \leq 9$.

45. (a) $(0, 2)$ is a solution: $-2(0) + 5(2) \geq 3$

$$2 < 4$$

$$-4(0) + 2(2) < 7$$

(b) $(-6, 4)$ is not a solution: $4 \not< 4$

(c) $(-8, -2)$ is not a solution: $-4(-8) + 2(-2) \not< 7$

(d) $(-3, 2)$ is not a solution: $-4(-3) + 2(2) \not< 7$

47. $\begin{cases} x + y \le 1 \\ -x + y \le 1 \\ \quad\quad y \ge 0 \end{cases}$

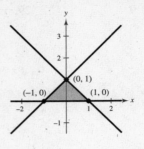

First, find the points of intersection of each pair of equations.

Vertex A	**Vertex B**	**Vertex C**
$\begin{cases} x + y = 1 \\ -x + y = 1 \end{cases}$	$\begin{cases} x + y = 1 \\ \quad\quad y = 0 \end{cases}$	$\begin{cases} -x + y = 1 \\ \quad\quad y = 0 \end{cases}$
$(0, 1)$	$(1, 0)$	$(-1, 0)$

49. $\begin{cases} -3x + 2y < 6 \\ \quad x - 4y > -2 \\ \quad 2x + y < 3 \end{cases}$

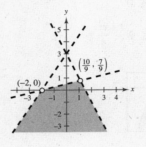

First, find the points of intersection of each pair of equations.

Vertex A	**Vertex B**	**Vertex C**
$\begin{cases} -3x + 2y = 6 \\ \quad x - 4y = -2 \end{cases}$	$\begin{cases} -3x + 2y = 6 \\ \quad 2x + y = 3 \end{cases}$	$\begin{cases} x - 4y = -2 \\ 2x + y = \cdot 3 \end{cases}$
$(-2, 0)$	$(0, 3)$	$\left(\frac{10}{9}, \frac{7}{9}\right)$

51. $3x + y \le y^2$

$x - y > 0$

The curves given by $3x + y = y^2$ and $x - y = 0$ intersect as follows:

$3x + x = x^2$

$4x = x^2$

$x = 0, 4$

Intersection points:
$(0, 0), (4, 4)$

53. $2x + y < 2 \implies y < 2 - 2x$

$x + 3y > 2 \implies y > \frac{1}{3}(2 - x)$

$2 - 2x = \frac{1}{3}(2 - x)$

$6 - 6x = (2 - x)$

$4 = 5x$

$x = \frac{4}{5}$

Intersection: $\left(\frac{4}{5}, \frac{2}{5}\right)$

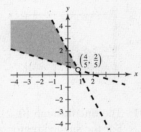

55. $\begin{cases} x < y^2 \\ x > y + 2 \end{cases}$

Points of intersection:

$y^2 = y + 2$

$y^2 - y - 2 = 0$

$(y + 1)(y - 2) = 0$

$y = -1, 2$

$(1, -1), (4, 2)$

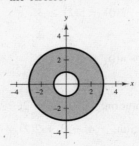

57. $\begin{cases} x^2 + y^2 \le 9 \\ x^2 + y^2 \ge 1 \end{cases}$

There are no points of intersection. The region in common to both inequalities is the region between the circles.

59. $\begin{cases} y \le \sqrt{3x} + 1 \\ y \ge x^2 + 1 \end{cases}$

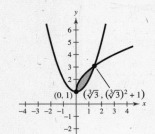

61. $\begin{cases} y < x^3 - 2x + 1 \\ y > -2x \\ x \le 1 \end{cases}$

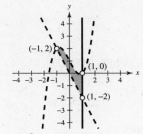

63. $\begin{cases} x^2 y \ge 1 \\ 0 < x \le 4 \\ y \le 4 \end{cases}$

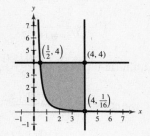

65. $\begin{cases} y < -x + 4 \implies \dfrac{x}{4} + \dfrac{y}{4} < 1 \\ x \ge 0 \qquad\qquad\quad x \ge 0 \\ y \ge 0 \qquad\qquad\quad y \ge 0 \end{cases}$

67. $(0, 4), (4, 0)$

Line: $y \le 4 - x$

$(0, 2), (8, 0)$

Line: $y \le -\dfrac{1}{4}x + 2$

$x \ge 0,\ y \ge 0$

69. Circle of radius 2 and center $(0, 2)$

$x^2 + (y - 2)^2 \le 4$

71. $\begin{cases} x \ge 2 \\ x \le 5 \\ y \ge 1 \\ y \le 7 \end{cases}$

Thus,
$2 \le x \le 5,\ 1 \le y \le 7$.

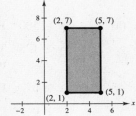

73. $(0, 0), (5, 0)$

Line: $y \ge 0$

$(0, 0), (2, 3)$

Line: $y \le \dfrac{3}{2}x$

$(2, 3), (5, 0)$

Line: $y \le -x + 5$

75.   Demand $=$ Supply

$50 - 0.5x = 0.125x$

$50 = 0.625x$

$x = 80$

$p = 10$

Point of equilibrium: $(80, 10)$

Consumer surplus $= \dfrac{1}{2}(40)(80) = 1600$

Producer surplus $= \dfrac{1}{2}(10)(80) = 400$

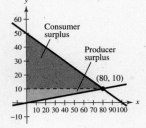

77.
$$\text{Demand} = \text{Supply}$$
$$300 - 0.0002x = 225 + 0.0005x$$
$$75 = 0.0007x$$
$$x = \frac{75}{0.0007} = \frac{750,000}{7}$$

Equilibrium point: $\left(\dfrac{750,000}{7}, \dfrac{1950}{7}\right) \approx (107,142.86, 278.57)$

Consumer surplus: $\dfrac{(107,142.86)(300 - 278.57)}{2} \approx 1,148,036$

Producer surplus: $\dfrac{(107,142.86)(278.57 - 225)}{2} \approx 2,869,822$

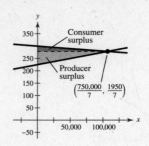

79. $x + y \le 30,000$

$\qquad x \ge 7500$

$\qquad y \ge 7500$

$\qquad x \ge 2y$

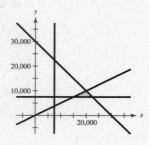

81. (a) Let x = number of ounces of food X.

Let y = number of ounces of food Y.

Calcium: $20x + 10y \ge 280$

Iron: $15x + 10y \ge 160$

Vitamin B: $10x + 20y \ge 180$

$\qquad\qquad\quad x \ge 0$

$\qquad\qquad\quad y \ge 0$

(b)

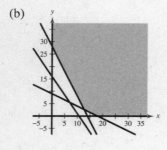

83. (a)

$\quad xy \ge 500\qquad$ Body-building space

$2x + \pi y \ge 125\qquad$ Track (two semi-circles and two lengths)

$\qquad x \ge 0\qquad$ Physical constraint

$\qquad y \ge 0\qquad$ Physical constraint

(b)

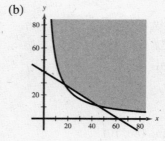

85. Area $= 9 \cdot 11 = 99$ square units

True

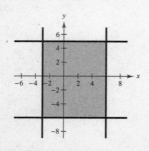

87. Test a point on either side of the boundary.

Appendix E.2   Linear Programming

Vocabulary Check

1. optimization

2. objective function

3. constraints, feasible solutions

---

1. $z = 3x + 5y$

At $(0, 6)$: $z = 3(0) + 5(6) = 30$

At $(0, 0)$: $z = 3(0) + 5(0) = 0$

At $(6, 0)$: $z = 3(6) + 5(0) = 18$

The minimum value is 0 at $(0, 0)$.

The maximum value is 30 at $(0, 6)$.

3. $z = 10x + 7y$

At $(0, 6)$: $z = 10(0) + 7(6) = 42$

At $(0, 0)$: $z = 10(0) + 7(0) = 0$

At $(6, 0)$: $z = 10(6) + 7(0) = 60$

The minimum value is 0 at $(0, 0)$.

The maximum value is 60 at $(6, 0)$.

5. $z = 3x + 2y$

$x + 3y = 15 \implies y = \frac{1}{3}(15 - x)$

$4x + y = 16 \implies y = (16 - 4x)$

$\frac{1}{3}(15 - x) = 16 - 4x$

$(15 - x) = 48 - 12x$

$11x = 33$

$x = 3$

$y = 4$

At $(0, 0)$: $z = 0$

At $(0, 5)$: $z = 10$

At $(4, 0)$: $z = 12$

At $(3, 4)$: $z = 17$

The minimum value is 0 at $(0, 0)$.

The maximum value is 17 at $(3, 4)$.

7. $z = 5x + 0.5y$

At $(0, 0)$: $z = 0$

At $(0, 5)$: $z = 2.5$

At $(4, 0)$: $z = 20$

At $(3, 4)$: $z = 17$

The minimum value is 0 at $(0, 0)$.

The maximum value is 20 at $(4, 0)$.

9. $z = 10x + 7y$

At $(0, 45)$: $z = 10(0) + 7(45) = 315$

At $(30, 45)$: $z = 10(30) + 7(45) = 615$

At $(60, 20)$: $z = 10(60) + 7(20) = 740$

At $(60, 0)$: $z = 10(60) + 7(0) = 600$

At $(0, 0)$: $z = 10(0) + 7(0) = 0$

The minimum value is 0 at $(0, 0)$.

The maximum value is 740 at $(60, 20)$.

11. $z = 25x + 30y$

At $(0, 45)$: $z = 25(0) + 30(45) = 1350$

At $(30, 45)$: $z = 25(30) + 30(45) = 2100$

At $(60, 20)$: $z = 25(60) + 30(20) = 2100$

At $(60, 0)$: $z = 25(60) + 30(0) = 1500$

At $(0, 0)$: $z = 25(0) + 30(0) = 0$

The minimum value is 0 at $(0, 0)$.

The maximum value is 2100 at any point along the line segment connecting $(30, 45)$ and $(60, 20)$.

13. $z = 6x + 10y$

At $(0, 2)$: $z = 6(0) + 10(2) = 20$

At $(5, 0)$: $z = 6(5) + 10(0) = 30$

At $(0, 0)$: $z = 6(0) + 10(0) = 0$

The minimum value is 0 at $(0, 0)$.

The maximum value is 30 at $(5, 0)$.

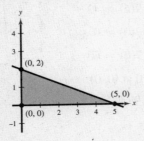

15. $z = 3x + 4y$

At $(0, 0)$: $z = 0$

At $(7, 0)$: $z = 21$

At $(0, 10)$: $z = 40$

At $(5, 8)$: $z = 47$

The minimum value is 0 at $(0, 0)$.

The maximum value is 47 at $(5, 8)$.

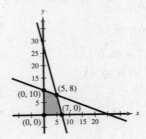

17. $z = x + 2y$

At $(0, 0)$: $z = 0 + 2(0) = 0$

At $(0, 10)$: $z = 0 + 2(10) = 20$

At $(5, 8)$: $z = 5 + 2(8) = 21$

At $(7, 0)$: $z = 7 + 2(0) = 7$

The minimum value is 0 at $(0, 0)$.

The maximum value is 21 at $(5, 8)$.

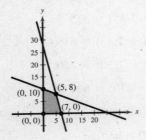

19. $z = 2x$

At $(0, 0)$: $z = 2(0) = 0$

At $(0, 10)$: $z = 2(0) = 0$

At $(5, 8)$: $z = 2(5) = 10$

At $(7, 0)$: $z = 2(7) = 14$

The maximum value is 14 at $(7, 0)$.

The minimum value is 0 along the line segment joining $(0, 0)$ and $(0, 10)$.

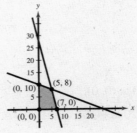

21. $z = 4x + y$

At $(36, 0)$: $z = 4(36) + 0 = 144$

At $(40, 0)$: $z = 4(40) + 0 = 160$

At $(24, 8)$: $z = 4(24) + 8 = 104$

The minimum value is 104 at $(24, 8)$.

The maximum value is 160 at $(40, 0)$.

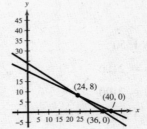

23. $z = x + 4y$

At $(36, 0)$: $z = 36 + 4(0) = 36$

At $(40, 0)$: $z = 40 + 4(0) = 40$

At $(24, 8)$: $z = 24 + 4(8) = 56$

The minimum value is 36 at $(36, 0)$.

The maximum value is 56 at $(24, 8)$.

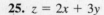

25. $z = 2x + 3y$

At $(36, 0)$: $z = 2(36) + 3(0) = 72$

At $(40, 0)$: $z = 2(40) + 3(0) = 80$

At $(24, 8)$: $z = 2(24) + 3(8) = 72$

The minimum value is 72 at any point on the line segment joining $(36, 0)$ and $(24, 8)$.

The maximum value is 80 at $(40, 0)$.

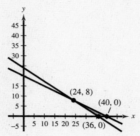

27. $z = 2x + y$

(a), (b)

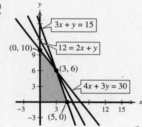

(c) At $(0, 10)$: $z = 2(0) + (10) = 10$

At $(3, 6)$: $z = 2(3) + (6) = 12$

At $(5, 0)$: $z = 2(5) + (0) = 10$

At $(0, 0)$: $z = 2(0) + (0) = 0$

The maximum value is 12 at $(3, 6)$.

29. $z = x + y$

(a), (b)

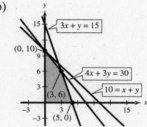

(c) At $(0, 10)$: $z = (0) + (10) = 10$

At $(3, 6)$: $z = (3) + (6) = 9$

At $(5, 0)$: $z = (5) + (0) = 5$

At $(0, 0)$: $z = (0) + (0) = 0$

The maximum value is 10 at $(0, 10)$.

31. $-x + y \leq 1 \implies y \leq x + 1$

$-x + 2y \leq 4 \implies y \leq \frac{1}{2}x + 2$

Intersection: $(2, 3)$

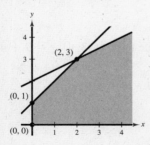

The constraints do not form a closed set of points.
Therefore, $z = x + y$ is unbounded.

33. $-x + y \leq 0 \implies y \leq x$

$-3x + y \geq 3 \implies y \geq 3x + 3$

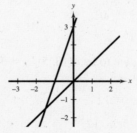

The feasible set is empty.

35. Let x = number of audits.

Let y = number of tax returns.

Constraints: $100x + 12.5y \leq 800$

$$8x + 2y \leq 96$$

$$x \geq 0$$

$$y \geq 0$$

Objective function: $R = 2000x + 300y$

Vertices of feasible region:

$(0, 0), (8, 0), (0, 48), (4, 32)$

At $(0, 0)$: $R = 0$

At $(8, 0)$: $R = 16{,}000$

At $(0, 48)$: $R = 14{,}400$

At $(4, 32)$: $R = 17{,}600$

4 audits, 32 tax returns yields maximum revenue of $17,600.

37. x = number of bags of Brand X

y = number of bags of Brand Y

Constraints: $2x + y \geq 12$

$$2x + 9y \geq 36$$

$$2x + 3y \geq 24$$

$$x \geq 0$$

$$y \geq 0$$

Objective function: $C = 25x + 20y$

Vertices: $(0, 12), (3, 6), (9, 2), (18, 0)$

At $(0, 12)$: $C = 25(0) + 20(12) = 240$

At $(3, 6)$: $C = 25(3) + 20(6) = 195$

At $(9, 2)$: $C = 25(9) + 20(2) = 265$

At $(18, 0)$: $C = 25(18) + 20(0) = 450$

To minimize cost, use three bags of Brand X and six bags of Brand Y for a total cost of $195.

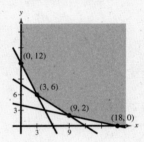

39. True, the maximum value is attained at all points in the segment joining these two vertices.

41. There are an infinite number of objective functions that would have a maximum at $(0, 4)$. One such objective function is $z = x + 5y$.

43. There are an infinite number of objective functions that would have a maximum at $(5, 0)$. One such objective function is $z = 4x + y$.

45. Constraints: $x \geq 0, y \geq 0, x + 3y \leq 15, 4x + y \leq 16$

Vertex	Value of $z = 3x + ty$
$(0, 0)$	$z = 0$
$(0, 5)$	$z = 5t$
$(3, 4)$	$z = 9 + 4t$
$(4, 0)$	$z = 12$

(a) For the maximum value to be at $(0, 5)$, $z = 5t$ must be greater than

$z = 9 + 4t$ and $z = 12$.

$5t > 9 + 4t$ and $5t > 12$

$t > 9$ $t > \frac{12}{5}$

Thus, $t > 9$.

(b) For the maximum value to be at $(3, 4)$, $z = 9 + 4t$ must be greater than $z = 5t$ and $z = 12$.

$9 + 4t > 5t$ and $9 + 4t > 12$

$9 > t$ $t > 3$

$t > \frac{3}{4}$

Thus, $\frac{3}{4} < t < 9$.

Chapter 1   Practice Test Solutions

1.

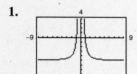

x-intercepts: ± 0.894

2.

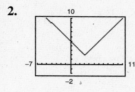

No *x*-intercepts

3. $3x - 5y = 15$

Line

x-intercept: $(5, 0)$

y-intercept: $(0, -3)$

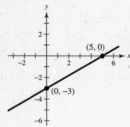

4. $y = \sqrt{9 - x}$

Domain: $(-\infty, 9]$

x-intercept: $(9, 0)$

y-intercept: $(0, 3)$

5. $5x + 4 = 7x - 8$

$4 + 8 = 7x - 5x$

$12 = 2x$

$x = 6$

6.
$$\frac{x}{3} - 5 = \frac{x}{5} + 1$$
$$15\left(\frac{x}{3} - 5\right) = 15\left(\frac{x}{5} + 1\right)$$
$$5x - 75 = 3x + 15$$
$$2x = 90$$
$$x = 45$$

7.
$$\frac{3x + 1}{6x - 7} = \frac{2}{5}$$
$$5(3x + 1) = 2(6x - 7)$$
$$15x + 5 = 12x - 14$$
$$3x = -19$$
$$x = -\frac{19}{3}$$

8.
$$(x - 3)^2 + 4 = (x + 1)^2$$
$$x^2 - 6x + 9 + 4 = x^2 + 2x + 1$$
$$-8x = -12$$
$$x = \frac{-12}{-8}$$
$$x = \frac{3}{2}$$

9. Slope $= \dfrac{-2 - (-5)}{3 - 4} = \dfrac{3}{-1} = -3$

$y + 2 = -3(x - 3)$

$y + 2 = -3x + 9$

$y + 3x = 7$ or $y = -3x + 7$

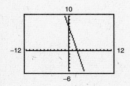

10. $y - 5 = -3(x + 1)$

$y - 5 = -3x - 3$

$y + 3x = 2$ or $y = -3x + 2$

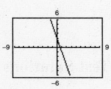

11. No, y is not a function of x. For example, $(0, 2)$ and $(0, -2)$ both satisfy the equation.

12. $f(0) = \dfrac{|0 - 2|}{(0 - 2)} = \dfrac{2}{-2} = -1$

$f(2)$ is not defined.

$f(4) = \dfrac{|4 - 2|}{(4 - 2)} = \dfrac{2}{2} = 1$

13. The domain of $f(x) = \dfrac{5}{x^2 - 16}$ is all $x \neq \pm 4$.

14. The domain of $g(t) = \sqrt{4 - t}$ consists of all t satisfying $4 - t \geq 0$ or $t \leq 4$.

15.

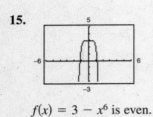

$f(x) = 3 - x^6$ is even.

16.

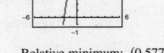

Relative minimum: $(0.577, 3.615)$

Relative maximum: $(-0.577, 4.385)$

17. $f(x) = x^3 - 3$ is a vertical shift of three units downward of $y = x^3$.

18. $f(x) = \sqrt{x - 6}$ is a horizontal shift six units to the right of $y = \sqrt{x}$.

19. $(g \circ f)(x) = g(f(x))$

$= g(\sqrt{x}) = (\sqrt{x})^2 - 2 = x - 2$

Domain: $x \geq 0$

20. $\left(\dfrac{f}{g}\right)(x) = \dfrac{f(x)}{g(x)} = \dfrac{3x^2}{16 - x^4}$

The domain is all $x \neq \pm 2$.

21. $(f \circ g)(x) = f\left(\dfrac{x - 1}{3}\right)$

$= 3\left(\dfrac{x - 1}{3}\right) + 1 = (x - 1) + 1 = x$

$(g \circ f)(x) = g(3x + 1) = \dfrac{(3x + 1) - 1}{3} = \dfrac{3x}{3} = x$

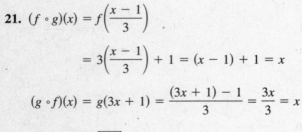

22. $y = \sqrt{9 - x^2}, \quad 0 \leq x \leq 3$

$x = \sqrt{9 - y^2}$

$x^2 = 9 - y^2$

$y^2 = 9 - x^2$

$y = \sqrt{9 - x^2}$

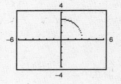

Chapter 2   Practice Test Solutions

1. $\frac{1}{2}x - \frac{1}{3}(x - 1) = 10$

$3x - 2(x - 1) = 60$

$3x - 2x + 2 = 60$

$x = 58$

2. $(x + 1)^2 - 6 = x^2 + 3x$

$x^2 + 2x + 1 - 6 = x^2 + 3x$

$2x - 5 = 3x$

$x = -5$

3. $A = \frac{1}{2}(a + b)h$

$2A = ah + bh$

$2A - bh = ah$

$\frac{2A - bh}{h} = a$

4. Percent $= \frac{301}{4300} = 0.07 = 7\%$

5. Let $x =$ number of quarters. Then $53 - x =$ number of nickels.

$25x + 5(53 - x) = 605$

$20x + 265 = 605$

$20x = 340$

$x = 17$ quarters

$53 - x = 36$ nickels

6. Let $x =$ amount in $9\frac{1}{2}\%$ fund. Then $15,000 - x =$ amount in 11% fund.

$0.095x + 0.11(15,000 - x) = 1582.50$

$-0.015x + 1650 = 1582.50$

$-0.015x = -67.5$

$x = \$4500$ at $9\frac{1}{2}\%$

$15,000 - x = \$10,500$ at 11%

7.   (1.257, 0.743), (−1.591, 3.591)

8.   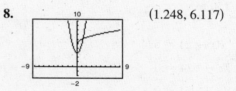   (1.248, 6.117)

9. $\frac{2}{1 + i} = \frac{2}{1 + i} \cdot \frac{1 - i}{1 - i} = \frac{2 - 2i}{1 + 1} = 1 - i$

10. $\frac{3 + i}{2} - \frac{i + 1}{4} = \frac{6 + 2i - i - 1}{4} = \frac{5}{4} + \frac{i}{4}$

11. $28 + 5x - 3x^2 = 0$

$(4 - x)(7 + 3x) = 0$

$4 - x = 0 \Rightarrow x = 4$

$7 + 3x = 0 \Rightarrow x = -\frac{7}{3}$

12. $(x - 2)^2 = 24$

$x - 2 = \pm\sqrt{24}$

$x - 2 = \pm 2\sqrt{6}$

$x = 2 \pm 2\sqrt{6}$

13. $x^2 - 4x - 9 = 0$

$x^2 - 4x + 2^2 = 9 + 2^2$

$(x - 2)^2 = 13$

$x - 2 = \pm\sqrt{13}$

$x = 2 \pm \sqrt{13}$

14. $x^2 + 5x - 1 = 0$

$a = 1, \; b = 5, \; c = -1$

$x = \frac{-5 \pm \sqrt{(5)^2 - 4(1)(-1)}}{2(1)}$

$= \frac{-5 \pm \sqrt{25 + 4}}{2} = \frac{-5 \pm \sqrt{29}}{2}$

15. $3x^2 - 2x + 4 = 0$

$a = 3, \; b = -2, \; c = 4$

$x = \frac{-(-2) \pm \sqrt{(-2)^2 - 4(3)(4)}}{2(3)}$

$= \frac{2 \pm \sqrt{4 - 48}}{6}$

$= \frac{2 \pm \sqrt{-44}}{6}$

$= \frac{2 \pm 2i\sqrt{11}}{6}$

$= \frac{1 \pm i\sqrt{11}}{3} = \frac{1}{3} \pm \frac{\sqrt{11}}{3}i$

16.
$$60,000 = xy$$
$$y = \frac{60,000}{x}$$
$$2x + 2y = 1100$$
$$2x + 2\left(\frac{60,000}{x}\right) = 1100$$
$$x + \frac{60,000}{x} = 550$$
$$x^2 + 60,000 = 550x$$
$$x^2 - 550x + 60,000 = 0$$
$$(x - 150)(x - 400) = 0$$
$$x = 150 \quad \text{or} \quad x = 400$$
$$y = 400 \qquad y = 150$$

Length: 400 feet

Width: 150 feet

17.
$$x(x + 2) = 624$$
$$x^2 + 2x - 624 = 0$$
$$(x - 24)(x + 26) = 0$$
$$x = 24 \quad \text{or} \quad x = -26, \text{ (extraneous solution)}$$
$$x + 2 = 26$$

18.
$$x^3 - 10x^2 + 24x = 0$$
$$x(x^2 - 10x + 24) = 0$$
$$x(x - 4)(x - 6) = 0$$
$$x = 0, \ x = 4, \ x = 6$$

19.
$$\sqrt[3]{6 - x} = 4$$
$$6 - x = 64$$
$$-x = 58$$
$$x = -58$$

20.
$$(x^2 - 8)^{2/5} = 4$$
$$x^2 - 8 = \pm 4^{5/2}$$
$$x^2 - 8 = 32 \quad \text{or} \quad x^2 - 8 = -32$$
$$x^2 = 40 \qquad\qquad x^2 = -24$$
$$x = \pm\sqrt{40} \qquad\quad x = \pm\sqrt{-24}$$
$$x = \pm 2\sqrt{10} \qquad\quad x = \pm 2\sqrt{6}\,i$$

21.
$$x^4 - x^2 - 12 = 0$$
$$(x^2 - 4)(x^2 + 3) = 0$$
$$x^2 = 4 \quad \text{or} \quad x^2 = -3$$
$$x^2 = \pm 2 \qquad x = \pm\sqrt{3}\,i$$

22.
$$4 - 3x > 16$$
$$-3x > 12$$
$$x < -4$$

23.
$$\left|\frac{x - 3}{2}\right| < 5$$
$$-5 < \frac{x - 3}{2} < 5$$
$$-10 < x - 3 < 10$$
$$-7 < x < 13$$

24.
$$\frac{x + 1}{x - 3} < 2$$

$$\frac{x + 1}{x - 3} - 2 < 0$$

$$\frac{x + 1 - 2(x - 3)}{x - 3} < 0$$

$$\frac{7 - x}{x - 3} < 0$$

Critical numbers: $x = 7$ and $x = 3$

Test intervals: $(-\infty, 3), (3, 7), (7, \infty)$

Solution intervals: $(-\infty, 3) \cup (7, \infty)$

25. $|3x - 4| \geq 9$

$3x - 4 \leq -9$ or $3x - 4 \geq 9$

$3x \leq -5$ $\qquad\qquad$ $3x \geq 13$

$x \leq -\dfrac{5}{3}$ $\qquad\qquad$ $x \geq \dfrac{13}{3}$

26. $y = 0.882 + 0.912x$

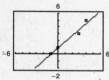

Chapter 3   Practice Test Solutions

1. x-intercepts: $(1, 0), (5, 0)$

y-intercept: $(0, 5)$

Vertex: $(3, -4)$

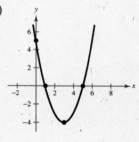

2. $a = 0.01, b = -90$

$$\frac{-b}{2a} = \frac{90}{2(0.01)} = 4500 \text{ units}$$

3. Vertex: $(1, 7)$

Opening downward through $(2, 5)$

$y = a(x - 1)^2 + 7$, Standard form

$5 = a(2 - 1)^2 + 7$

$5 = a + 7$

$a = -2$

$y = -2(x - 1)^2 + 7$

$\quad = -2(x^2 - 2x + 1) + 7$

$\quad = -2x^2 + 4x + 5$

4. $y = \pm a(x - 2)(3x - 4)$ where a is any real number.

$y = \pm(3x^2 - 10x + 8)$

5. Leading coefficient: -3

Degree: 5

Moves down to the right and up to the left.

6. $0 = x^5 - 5x^3 + 4x$

$\qquad = x(x^4 - 5x^2 + 4)$

$\qquad = x(x^2 - 1)(x^2 - 4)$

$\qquad = x(x + 1)(x - 1)(x + 2)(x - 2)$

$x = 0, x = \pm 1, x = \pm 2$

7. $f(x) = x(x - 3)(x + 2)$

$\qquad = x(x^2 - x - 6)$

$\qquad = x^3 - x^2 - 6x$

8. Intercepts: $(0, 0), (\pm 2\sqrt{3}, 0)$

Moves up to the right.

Moves down to the left.

x	-2	-1	0	1	2
y	16	11	0	-11	-16

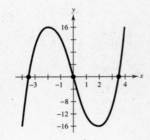

9.

$$3x^3 + 9x^2 + 20x + 62 + \frac{176}{x - 3}$$

$$x - 3 \overline{)3x^4 + 0x^3 - 7x^2 + 2x - 10}$$

$$\underline{3x^4 - 9x^3}$$

$$9x^3 - 7x^2$$

$$\underline{9x^3 - 27x^2}$$

$$20x^2 + 2x$$

$$\underline{20x^2 - 60x}$$

$$62x - 10$$

$$\underline{62x - 186}$$

$$176$$

10.

$$x - 2 + \frac{5x - 13}{x^2 + 2x - 1}$$

$$x^2 + 2x - 1 \overline{)x^3 + 0x^2 + 0x - 11}$$

$$\underline{x^3 + 2x^2 - x}$$

$$-2x^2 + x - 11$$

$$\underline{-2x^2 - 4x + 2}$$

$$5x - 13$$

11. $-5 \,\big|$

	3	13	0	0	12	-1
		-15	10	-50	250	-1310
	3	-2	10	-50	262	-1311

$$\frac{3x^5 + 13x^4 + 12x - 1}{x + 5} = 3x^4 - 2x^3 + 10x^2 - 50x + 262 - \frac{1311}{x + 5}$$

12. $-6 \,\big|$

	7	40	-12	15
		-42	12	0
	7	-2	0	15

$f(-6) = 15$

13. $0 = x^3 - 19x - 30$

Possible rational roots:

$\pm 1, \pm 2, \pm 3, \pm 5, \pm 6, \pm 10, \pm 15, \pm 30$

$-2 \,\big|$

	1	0	-19	-30
		-2	4	30
	1	-2	-15	0

-2 is a zero.

$0 = (x + 2)(x^2 - 2x - 15)$

$0 = (x + 2)(x + 3)(x - 5)$

Zeros: $x = -2, x = -3, x = 5$

14. $0 = x^4 + x^3 - 8x^2 - 9x - 9$

Possible rational roots: $\pm 1, \pm 3, \pm 9$

$$\begin{array}{r|rrrr} 3 & 1 & 1 & -8 & -9 & -9 \\ & & 3 & 12 & 12 & 9 \\ \hline & 1 & 4 & 4 & 3 & 0 \end{array}$$

$x = 3$ is a zero.

$0 = (x - 3)(x^3 + 4x^2 + 4x + 3)$

Possible rational roots of $x^3 + 4x^2 + 4x + 3$: $\pm 1, \pm 3$

$$\begin{array}{r|rrrr} -3 & 1 & 4 & 4 & 3 \\ & & -3 & -3 & -3 \\ \hline & 1 & 1 & 1 & 0 \end{array}$$

$x = -3$ is a zero.

$0 = (x - 3)(x + 3)(x^2 + x + 1)$

The zeros of $x^2 + x + 1$ are $x = \dfrac{-1 \pm \sqrt{3}\,i}{2}$.

Zeros: $x = 3, x = -3, x = -\dfrac{1}{2} + \dfrac{\sqrt{3}}{2}i, x = -\dfrac{1}{2} - \dfrac{\sqrt{3}}{2}i$

15. $0 = 6x^3 - 5x^2 + 4x - 15$

Possible rational roots: $\pm 1, \pm 3, \pm 5, \pm 15, \pm\frac{1}{2}, \pm\frac{3}{2}, \pm\frac{5}{2}, \pm\frac{15}{2}, \pm\frac{1}{3}, \pm\frac{5}{3}, \pm\frac{1}{6}, \pm\frac{5}{6}$

16. $0 = x^3 - \frac{20}{3}x^2 + 9x - \frac{10}{3}$

$0 = 3x^3 - 20x^2 + 27x - 10$

Possible rational roots:

$\pm 1, \pm 2, \pm 5, \pm 10, \pm\frac{1}{3}, \pm\frac{2}{3}, \pm\frac{5}{3}, \pm\frac{10}{3}$

$$\begin{array}{r|rrrr} 1 & 3 & -20 & 27 & -10 \\ & & 3 & -17 & 10 \\ \hline & 3 & -17 & 10 & 0 \end{array}$$

$x = 1$ is a zero.

$0 = (x - 1)(3x^2 - 17x + 10)$

$0 = (x - 1)(3x - 2)(x - 5)$

Zeros: $x = 1, x = \frac{2}{3}, x = 5$

17. $f(x) = x^4 + x^3 + 3x^2 + 5x - 10$

Possible rational roots: $\pm 1, \pm 2, \pm 5, \pm 10$

$$\begin{array}{r|rrrrr} 1 & 1 & 1 & 3 & 5 & -10 \\ & & 1 & 2 & 5 & 10 \\ \hline & 1 & 2 & 5 & 10 & 0 \end{array}$$

$x = 1$ is a zero.

$$\begin{array}{r|rrrr} -2 & 1 & 2 & 5 & 10 \\ & & -2 & 0 & -10 \\ \hline & 1 & 0 & 5 & 0 \end{array}$$

$x = -2$ is a zero.

$f(x) = (x - 1)(x + 2)(x^2 + 5)$

$\quad\;\; = (x - 1)(x + 2)(x + 5i)(x - 5i)$

18. $f(x) = (x - 2)[x - (3 + i)][x - (3 - i)][x - (3 - 2i)][x - (3 + 2i)]$

$\quad\;\; = (x - 2)[(x - 3)^2 + 1][(x - 3)^2 + 4]$

$\quad\;\; = (x - 2)(x^2 - 6x + 10)(x^2 - 6x + 13)$

$\quad\;\; = x^5 - 14x^4 + 83x^3 - 256x^2 + 406x - 260$

19. $3i$ | $\begin{array}{cccc} 1 & 4 & 9 & 36 \\ & 3i & 12i - 9 & -36 \\ \hline 1 & 4 + 3i & 12i & 0 \end{array}$

20. $z = \dfrac{kx^2}{\sqrt{y}}$

21. $f(x) = \dfrac{x - 1}{2x}$

Vertical asymptote: $x = 0$

Horizontal asymptote: $y = \dfrac{1}{2}$

x-intercept: $(1, 0)$

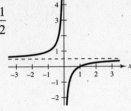

22. $f(x) = \dfrac{3x^2 - 4}{x}$

Vertical asymptote: $x = 0$

Slant asymptote: $y = 3x$

x-intercepts: $\left(\pm\dfrac{2}{\sqrt{3}}, 0\right)$

23. $y = 8$ is a horizontal asymptote since the degree of the numerator equals the degree of the denominator. There are no vertical asymptotes.

24. $x = 1$ is a vertical asymptote.

$$\dfrac{4x^2 - 2x + 7}{x - 1} = 4x + 2 + \dfrac{9}{x - 1}$$

so $y = 4x + 2$ is a slant asymptote.

25. $f(x) = \dfrac{x - 5}{(x - 5)^2} = \dfrac{1}{x - 5}$

Vertical asymptote: $x = 5$

Horizontal asymptote: $y = 0$

y-intercept: $\left(0, -\dfrac{1}{5}\right)$

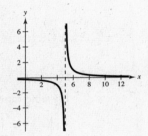

Chapter 4   Practice Test Solutions

1. $x^{3/5} = 8$

$x = 8^{5/3}$

$ = \left(\sqrt[3]{8}\right)^5 = 2^5 = 32$

2. $3^{x-1} = \frac{1}{81}$

$3^{x-1} = 3^{-4}$

$x - 1 = -4$

$x = -3$

3. $f(x) = 2^{-x} = \left(\frac{1}{2}\right)^x$

x	-2	-1	0	1	2
$f(x)$	4	2	1	$\frac{1}{2}$	$\frac{1}{4}$

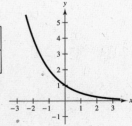

4. $g(x) = e^x + 1$

x	-2	-1	0	1	2
$g(x)$	1.14	1.37	2	3.72	8.39

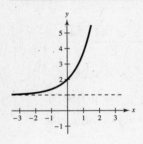

5. $A = P\left(1 + \dfrac{r}{n}\right)^{nt}$

(a) $A = 5000\left(1 + \dfrac{0.09}{12}\right)^{12(3)} \approx \$6543.23$

(b) $A = 5000\left(1 + \dfrac{0.09}{4}\right)^{4(3)} \approx \$6530.25$

(c) $A = 5000e^{(0.09)(3)} \approx \$6549.82$

6. $7^{-2} = \dfrac{1}{49}$

$\log_7 \dfrac{1}{49} = -2$

7. $x - 4 = \log_2 \dfrac{1}{64}$

$2^{x-4} = \dfrac{1}{64}$

$2^{x-4} = 2^{-6}$

$x - 4 = -6$

$x = -2$

8. $\log_b \sqrt[4]{\dfrac{8}{25}} = \dfrac{1}{4} \log_b \dfrac{8}{25}$

$= \dfrac{1}{4}[\log_b 8 - \log_b 25]$

$= \dfrac{1}{4}[\log_b 2^3 - \log_b 5^2]$

$= \dfrac{1}{4}[3 \log_b 2 - 2 \log_b 5]$

$= \dfrac{1}{4}[3(0.3562) - 2(0.8271)]$

$= -0.1464$

9. $5 \ln x - \dfrac{1}{2} \ln y + 6 \ln z = \ln x^5 - \ln \sqrt{y} + \ln z^6 = \ln\left(\dfrac{x^5 z^6}{\sqrt{y}}\right)$

10. $\log_9 28 = \dfrac{\log 28}{\log 9} \approx 1.5166$

11. $\log_{10} N = 0.6646$

$N = 10^{0.6646} \approx 4.62$

12.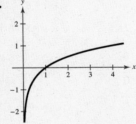

13. Domain:

$x^2 - 9 > 0$

$(x + 3)(x - 3) > 0$

$x < -3 \text{ or } x > 3$

14.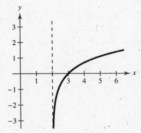

15. $\dfrac{\ln x}{\ln y} \neq \ln(x - y)$ since

$\dfrac{\ln x}{\ln y} = \log_y x.$

16. $5^x = 41$

$x = \log_5 41$

$= \dfrac{\ln 41}{\ln 5}$

≈ 2.3074

17. $x - x^2 = \log_5 \dfrac{1}{25}$

$5^{x-x^2} = \dfrac{1}{25}$

$5^{x-x^2} = 5^{-2}$

$x - x^2 = -2$

$0 = x^2 - x - 2$

$0 = (x + 1)(x - 2)$

$x = -1 \text{ or } x = 2$

18. $\log_2 x + \log_2(x - 3) = 2$

$\log_2[x(x - 3)] = 2$

$x(x - 3) = 2^2$

$x^2 - 3x = 4$

$x^2 - 3x - 4 = 0$

$(x + 1)(x - 4) = 0$

$x = 4$

$x = -1$ (extraneous solution)

19. $\dfrac{e^x + e^{-x}}{3} = 4$

$e^x(e^x + e^{-x}) = 12e^x$

$e^{2x} + 1 = 12e^x$

$e^{2x} - 12e^x + 1 = 0$

$e^x = \dfrac{12 \pm \sqrt{144 - 4}}{2}$

$e^x \approx 11.9161$ or $e^x \approx 0.0839$

$x \approx \ln 11.9161$ $x \approx \ln 0.0839$

$x \approx 2.4779$ $x \approx -2.4779$

20. $A = Pe^{rt}$

$12{,}000 = 6000e^{0.13t}$

$2 = e^{0.13t}$

$\ln 2 = 0.13t$

$\dfrac{\ln 2}{0.13} = t$

$t \approx 5.3319$ yr or 5 yr 4 mo

21. There are two points of intersection:

$(0.0169, -2.983), (1.731, 1.647)$

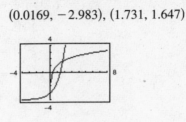

22. $y = 1.0597x^{1.9792}$

Chapter 5   Practice Test Solutions

1. $\begin{cases} x + y = 1 \\ 3x - y = 15 \end{cases} \Rightarrow y = 3x - 15$

$x + (3x - 15) = 1$

$4x = 16$

$x = 4$

$y = -3$

2. $\begin{cases} x - 3y = -3 \\ x^2 + 6y = 5 \end{cases} \Rightarrow x = 3y - 3$

$(3y - 3)^2 + 6y = 5$

$9y^2 - 18y + 9 + 6y = 5$

$9y^2 - 12y + 4 = 0$

$(3y - 2)^2 = 0$

$y = \tfrac{2}{3}$

$x = -1$

3. $\begin{cases} x + y + z = 6 \\ 2x - y + 3z = 0 \\ 5x + 2y - z = -3 \end{cases}$ $\Rightarrow z = 6 - x - y$

$2x - y + 3(6 - x - y) = 0 \Rightarrow -x - 4y = -18$

$5x + 2y - (6 - x - y) = -3 \Rightarrow 6x + 3y = 3$

$x = 18 - 4y$

$6(18 - 4y) + 3y = 3$

$-21y = -105$

$y = 5$

$x = 18 - 4y = -2$

$z = 6 - x - y = 3$

4. $\begin{cases} x + y = 110 \\ xy = 2800 \end{cases} \Rightarrow y = 110 - x$

$x(110 - x) = 2800$

$$0 = x^2 - 110x + 2800$$

$$0 = (x - 40)(x - 70)$$

$$x = 40 \quad \text{or} \quad x = 70$$

$$y = 70 \qquad y = 40$$

5. $\begin{cases} 2x + 2y = 170 \\ xy = 1500 \end{cases} \Rightarrow y = \dfrac{170 - 2x}{2} = 85 - x$

$x(85 - x) = 1500$

$$0 = x^2 - 85x + 1500$$

$$0 = (x - 25)(x - 60)$$

$$x = 25 \quad \text{or} \quad x = 60$$

$$y = 60 \qquad y = 25$$

Dimensions: $60' \times 25'$

6. $\begin{cases} 2x + 15y = 4 \\ x - 3y = 23 \end{cases} \Rightarrow \begin{array}{r} 2x + 15y = 4 \\ 5x - 15y = 115 \\ \hline 7x \qquad\quad = 119 \end{array}$

$$x = 17$$

$$y = \frac{x - 23}{3} = -2$$

7. $\begin{cases} x + y = 2 \\ 38x - 19y = 7 \end{cases} \Rightarrow \begin{array}{r} 19x + 19y = 38 \\ 38x - 19y = 7 \\ \hline 57x \qquad\quad = 45 \end{array}$

$$x = \frac{45}{57} = \frac{15}{19}$$

$$y = 2 - x$$

$$= \frac{38}{19} - \frac{15}{19}$$

$$= \frac{23}{19}$$

8. $y_1 = 2(0.112 - 0.4x)$

$$y_2 = \frac{(0.131 + 0.3x)}{0.7}$$

$\begin{cases} 0.4x + 0.5y = 0.112 \\ 0.3x - 0.7y = -0.131 \end{cases} \Rightarrow \begin{array}{r} 0.28x + 0.35y = 0.0784 \\ 0.15x - 0.35y = -0.0655 \\ \hline 0.43x \qquad\qquad = 0.0129 \end{array}$

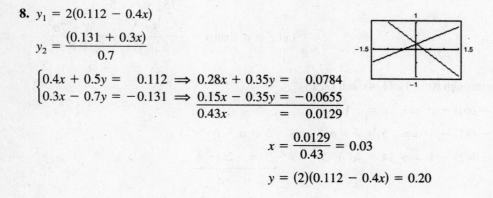

$$x = \frac{0.0129}{0.43} = 0.03$$

$$y = (2)(0.112 - 0.4x) = 0.20$$

9. Let x = amount in 11% fund and y = amount in 13% fund.

$\begin{cases} x + y = 17,000 \\ 0.11x + 0.13y = 2080 \end{cases} \Rightarrow y = 17,000 - x$

$$0.11x + 0.13(17,000 - x) = 2080$$

$$-0.02x = -130$$

$$x = \$6500$$

$$y = \$10,500$$

10. Using a graphing utility, you obtain
$y = 0.7857x - 0.1429$. Analytically, $(4, 3)$,
$(1, 1)$, $(-1, -2)$, $(-2, -1)$.

$$n = 4, \sum_{i=1}^{4} x_i = 2, \sum_{i=1}^{4} y_i = 1, \sum_{i=1}^{4} x_i^2 = 22, \sum_{i=1}^{4} x_i y_i = 17$$

$$4b + 2a = 1 \implies \quad 4b + 2a = 1$$
$$2b + 22a = 17 \implies \underline{-4b - 44a = -34}$$
$$-42a = -33$$

$a = \frac{33}{42} = \frac{11}{14}$

$b = \frac{1}{4}\left(1 - 2\left(\frac{33}{42}\right)\right) = -\frac{1}{7}$

$y = ax + b = \frac{11}{14}x - \frac{1}{7}$

11.
$$x + y \quad\quad = -2 \quad\quad \text{Equation 1}$$
$$2x - y + z = 11 \quad\quad \text{Equation 2}$$
$$4y - 3z = -20 \quad\quad \text{Equation 3}$$

$$\begin{cases} x + y \quad\quad = -2 \\ \quad -3y + z = 15 \quad -2\text{Eq.1} + \text{Eq.2} \\ \quad\quad 4y - 3z = -20 \end{cases}$$

$$\begin{cases} x + y \quad\quad = -2 \\ \quad -3y + z = 15 \\ \quad -5y \quad\quad = 25 \quad 3\text{Eq.2} + \text{Eq.3} \end{cases}$$

Answer: $y = -5$
$\quad\quad\quad\quad x = 3$
$\quad\quad\quad\quad z = 0$

12.
$$4x - y + 5z = 4 \quad\quad \text{Equation 1}$$
$$2x + y - z = 0 \quad\quad \text{Equation 2}$$
$$2x + 4y + 8z = 0 \quad\quad \text{Equation 3}$$

$$\begin{cases} 4x - y + 5z = 4 \\ \quad -3y + 7z = 4 \quad \text{Eq.1} - 2\text{Eq.2} \\ \quad\quad 3y + 9z = 0 \quad -\text{Eq.2} + \text{Eq.3} \end{cases}$$

$$\begin{cases} 4x - y + 5z = 4 \\ \quad -3y + 7z = 4 \\ \quad\quad 16z = 4 \quad \text{Eq.2} + \text{Eq.3} \end{cases}$$

Answer: $z = \frac{1}{4}$
$\quad\quad\quad\quad y = -\frac{3}{4}$
$\quad\quad\quad\quad x = \frac{1}{2}$

13.
$$\begin{cases} 3x + 2y - z = 5 \implies \quad 6x + 4y - 2z = 10 \\ 6x - y + 5z = 2 \implies \underline{-6x + y - 5z = -2} \\ \quad\quad\quad\quad\quad\quad\quad\quad\quad\quad 5y - 7z = 8 \end{cases}$$

$$y = \frac{8 + 7z}{5}$$

$$3x + 2y - z = 5$$
$$\underline{12x - 2y + 10z = 4}$$
$$15x \quad\quad + 9z = 9$$

$$x = \frac{9 - 9z}{15} = \frac{3 - 3z}{5}$$

Let $z = a$, then $x = \dfrac{3 - 3a}{5}$ and $y = \dfrac{8 + 7a}{5}$.

14. $y = ax^2 + bx + c$ passes through $(0, -1)$, $(1, 4)$, and $(2, 13)$.

At $(0, -1)$: $-1 = a(0)^2 + b(0) + c \implies c = -1$

At $(1, 4)$: $\quad 4 = a(1)^2 + b(1) - 1 \implies 5 = a + b \implies 5 = a + b$

At $(2, 13)$: $\quad 13 = a(2)^2 + b(2) - 1 \implies 14 = 4a + 2b \implies \underline{-7 = -2a - b}$
$$-2 = -a$$
$$a = 2$$
$$b = 3$$

Thus, $y = 2x^2 + 3x - 1$.

15. $s = \frac{1}{2}at^2 + v_0t + s_0$ passes through $(1, 12)$, $(2, 5)$, and $(3, 4)$.

At $(1, 12)$: $12 = \frac{1}{2}a + v_0 + s_0$ $\Rightarrow$ $\begin{cases} \frac{1}{2}a + v_0 + s_0 = 12 \\ -a \quad\;\; + s_0 = 19 \quad 2\text{Eq.1} - \text{Eq.2} \\ -3a \quad\;\; + s_0 = 7 \quad 3\text{Eq.2} - 2\text{Eq.3} \end{cases}$

At $(2, 5)$: $\;\;5 = 2a + 2v_0 + s_0$ $\Rightarrow$

At $(3, 4)$: $\;\;4 = \frac{9}{2}a + 3v_0 + s_0$ $\Rightarrow$

$a = 6$

$s_0 = 25$ $\qquad\begin{cases} \frac{1}{2}a + v_0 + s_0 = \;\; 12 \\ -a \quad\;\; + s_0 = \;\; 19 \\ -2a \quad\quad\quad = -12 \quad -\text{Eq.2} + \text{Eq.3} \end{cases}$

$v_0 = -16$

Thus,

$s = \frac{1}{2}(6)t^2 - 16t + 25 = 3t^2 - 16t + 25.$

---

16. $\begin{bmatrix} 1 & -2 & 4 \\ 3 & -5 & 9 \end{bmatrix}$

$-3R_1 + R_2 \rightarrow \begin{bmatrix} 1 & -2 & 4 \\ 0 & 1 & -3 \end{bmatrix}$

$2R_2 + R_1 \rightarrow \begin{bmatrix} 1 & 0 & -2 \\ 0 & 1 & -3 \end{bmatrix}$

17. $3x + 5y = \quad 3$

$\quad 2x - \;\; y = -11$

$\begin{bmatrix} 3 & 5 & \vdots & 3 \\ 2 & -1 & \vdots & -11 \end{bmatrix}$

$-R_2 + R_1 \rightarrow \begin{bmatrix} 1 & 6 & \vdots & 14 \\ 2 & -1 & \vdots & -11 \end{bmatrix}$

$-2R_1 + R_2 \rightarrow \begin{bmatrix} 1 & 6 & \vdots & 14 \\ 0 & -13 & \vdots & -39 \end{bmatrix}$

$-\frac{1}{13}R_2 \rightarrow \begin{bmatrix} 1 & 6 & \vdots & 14 \\ 0 & 1 & \vdots & 3 \end{bmatrix}$

$-6R_2 + R_1 \rightarrow \begin{bmatrix} 1 & 0 & \vdots & -4 \\ 0 & 1 & \vdots & 3 \end{bmatrix}$

Answer: $x = -4, y = 3$

18. $\begin{cases} 2x + 3y = -3 \\ 3x + 2y = \;\; 8 \\ x + \;\; y = \;\; 1 \end{cases}$

$\begin{bmatrix} 2 & 3 & \vdots & -3 \\ 3 & 2 & \vdots & 8 \\ 1 & 1 & \vdots & 1 \end{bmatrix}$

$\begin{array}{c} R_3 \\ \\ R_1 \end{array} \begin{bmatrix} 1 & 1 & \vdots & 1 \\ 3 & 2 & \vdots & 8 \\ 2 & 3 & \vdots & -3 \end{bmatrix}$

$\begin{array}{c} -3R_1 + R_2 \rightarrow \\ -2R_1 + R_3 \rightarrow \end{array} \begin{bmatrix} 1 & 1 & \vdots & 1 \\ 0 & -1 & \vdots & 5 \\ 0 & 1 & \vdots & -5 \end{bmatrix}$

$\begin{array}{c} R_2 + R_1 \rightarrow \\ -R_2 \rightarrow \\ -R_2 + R_3 \rightarrow \end{array} \begin{bmatrix} 1 & 0 & \vdots & 6 \\ 0 & 1 & \vdots & -5 \\ 0 & 0 & \vdots & 0 \end{bmatrix}$

Answer: $x = 6, y = -5$

19. $\begin{cases} x \quad\quad + 3z = -5 \\ 2x + y \quad\quad = \;\; 0 \\ 3x + y - \;\; z = \;\; 3 \end{cases}$

$\begin{bmatrix} 1 & 0 & 3 & \vdots & -5 \\ 2 & 1 & 0 & \vdots & 0 \\ 3 & 1 & -1 & \vdots & 3 \end{bmatrix}$

$\begin{array}{c} -2R_1 + R_2 \rightarrow \\ -3R_1 + R_3 \rightarrow \end{array} \begin{bmatrix} 1 & 0 & 3 & \vdots & -5 \\ 0 & 1 & -6 & \vdots & 10 \\ 0 & 1 & -10 & \vdots & 18 \end{bmatrix}$

$-R_2 + R_3 \rightarrow \begin{bmatrix} 1 & 0 & 3 & \vdots & -5 \\ 0 & 1 & -6 & \vdots & 10 \\ 0 & 0 & -4 & \vdots & 8 \end{bmatrix}$

$\begin{array}{c} -3R_3 + R_1 \rightarrow \\ 6R_3 + R_2 \rightarrow \\ -\frac{1}{4}R_3 \rightarrow \end{array} \begin{bmatrix} 1 & 0 & 0 & \vdots & 1 \\ 0 & 1 & 0 & \vdots & -2 \\ 0 & 0 & 1 & \vdots & -2 \end{bmatrix}$

Answer: $x = 1, y = -2, z = -2$

20. $\begin{bmatrix} 1 & 4 & 5 \\ 2 & 0 & -3 \end{bmatrix} \begin{bmatrix} 1 & 6 \\ 0 & -7 \\ -1 & 2 \end{bmatrix} = \begin{bmatrix} -4 & -12 \\ 5 & 6 \end{bmatrix}$

21. $3A - 5B = 3\begin{bmatrix} 9 & 1 \\ -4 & 8 \end{bmatrix} - 5\begin{bmatrix} 6 & -2 \\ 3 & 5 \end{bmatrix}$

$\qquad = \begin{bmatrix} -3 & 13 \\ -27 & -1 \end{bmatrix}$

22. $f(A) = \begin{bmatrix} 3 & 0 \\ 7 & 1 \end{bmatrix}^2 - 7\begin{bmatrix} 3 & 0 \\ 7 & 1 \end{bmatrix} + 8\begin{bmatrix} 1 & 0 \\ 0 & 1 \end{bmatrix}$

$\qquad = \begin{bmatrix} 3 & 0 \\ 7 & 1 \end{bmatrix}\begin{bmatrix} 3 & 0 \\ 7 & 1 \end{bmatrix} - \begin{bmatrix} 21 & 0 \\ 49 & 7 \end{bmatrix} + \begin{bmatrix} 8 & 0 \\ 0 & 8 \end{bmatrix}$

$\qquad = \begin{bmatrix} 9 & 0 \\ 28 & 1 \end{bmatrix} - \begin{bmatrix} 21 & 0 \\ 49 & 7 \end{bmatrix} + \begin{bmatrix} 8 & 0 \\ 0 & 8 \end{bmatrix}$

$\qquad = \begin{bmatrix} -4 & 0 \\ -21 & 2 \end{bmatrix}$

23. False

$(A + B)(A + 3B) = A(A + 3B) + B(A + 3B)$

$\qquad\qquad\qquad = A^2 + 3AB + BA + 3B^2$

24.

$\begin{bmatrix} 1 & 2 & \vdots & 1 & 0 \\ 3 & 5 & \vdots & 0 & 1 \end{bmatrix}$

$-3R_1 + R_2 \rightarrow \begin{bmatrix} 1 & 2 & \vdots & 1 & 0 \\ 0 & -1 & \vdots & -3 & 1 \end{bmatrix}$

$\begin{matrix} 2R_2 + R_1 \rightarrow \\ -R_2 \rightarrow \end{matrix} \begin{bmatrix} 1 & 0 & \vdots & -5 & 2 \\ 0 & 1 & \vdots & 3 & -1 \end{bmatrix}$

$A^{-1} = \begin{bmatrix} -5 & 2 \\ 3 & -1 \end{bmatrix}$

25. $\begin{bmatrix} 1 & 1 & 1 & \vdots & 1 & 0 & 0 \\ 3 & 6 & 5 & \vdots & 0 & 1 & 0 \\ 6 & 10 & 8 & \vdots & 0 & 0 & 1 \end{bmatrix}$

$\begin{matrix} -3R_1 + R_2 \rightarrow \\ -6R_1 + R_3 \rightarrow \end{matrix} \begin{bmatrix} 1 & 1 & 1 & \vdots & 1 & 0 & 0 \\ 0 & 3 & 2 & \vdots & -3 & 1 & 0 \\ 0 & 4 & 2 & \vdots & -6 & 0 & 1 \end{bmatrix}$

$\begin{matrix} -\frac{1}{3}R_2 + R_1 \rightarrow \\ \frac{1}{3}R_2 \rightarrow \\ -4R_2 + R_3 \rightarrow \end{matrix} \begin{bmatrix} 1 & 0 & \frac{1}{3} & \vdots & 2 & -\frac{1}{3} & 0 \\ 0 & 1 & \frac{2}{3} & \vdots & -1 & \frac{1}{3} & 0 \\ 0 & 0 & -\frac{2}{3} & \vdots & -2 & -\frac{4}{3} & 1 \end{bmatrix}$

$\begin{matrix} \frac{1}{2}R_3 + R_1 \rightarrow \\ R_3 + R_2 \rightarrow \\ -\frac{3}{2}R_3 \rightarrow \end{matrix} \begin{bmatrix} 1 & 0 & 0 & \vdots & 1 & -1 & \frac{1}{2} \\ 0 & 1 & 0 & \vdots & -3 & -1 & 1 \\ 0 & 0 & 1 & \vdots & 3 & 2 & -\frac{3}{2} \end{bmatrix}$

$A^{-1} = \begin{bmatrix} 1 & -1 & \frac{1}{2} \\ -3 & -1 & 1 \\ 3 & 2 & -\frac{3}{2} \end{bmatrix}$

26. (a) $x + 2y = 4$

$\qquad 3x + 5y = 1$

$\qquad \begin{bmatrix} 1 & 2 & \vdots & 1 & 0 \\ 3 & 5 & \vdots & 0 & 1 \end{bmatrix}$

$-3R_1 + R_2 \rightarrow \begin{bmatrix} 1 & 2 & \vdots & 1 & 0 \\ 0 & -1 & \vdots & -3 & 1 \end{bmatrix}$

$\begin{matrix} -2R_2 + R_1 \rightarrow \\ -R_2 \rightarrow \end{matrix} \begin{bmatrix} 1 & 0 & \vdots & -5 & 2 \\ 0 & 1 & \vdots & 3 & -1 \end{bmatrix}$

$X = A^{-1}B = \begin{bmatrix} -5 & 2 \\ 3 & -1 \end{bmatrix}\begin{bmatrix} 4 \\ 1 \end{bmatrix} = \begin{bmatrix} -18 \\ 11 \end{bmatrix}$

$x = -18, y = 11$

(b) $x + 2y = 3$

$\quad 3x + 5y = -2$

$X = A^{-1}B - \begin{bmatrix} -5 & 2 \\ 3 & -1 \end{bmatrix}\begin{bmatrix} 3 \\ -2 \end{bmatrix} = \begin{bmatrix} -19 \\ 11 \end{bmatrix}$

$x = -19, y = 11$

27. $\begin{vmatrix} 6 & -1 \\ 3 & 4 \end{vmatrix} = 24 - (-3) = 27$

28. $\begin{vmatrix} 1 & 3 & -1 \\ 5 & 9 & 0 \\ 6 & 2 & -5 \end{vmatrix} = 1(-45) + (-3)(-25) + (-1)(-44)$

$$= 74$$

29. $\begin{vmatrix} 1 & 4 & 2 & 3 \\ 0 & 1 & -2 & 0 \\ 3 & 5 & -1 & 1 \\ 2 & 0 & 6 & 1 \end{vmatrix} = -7$

30. $\begin{vmatrix} 6 & 4 & 3 & 0 & 6 \\ 0 & 5 & 1 & 4 & 8 \\ 0 & 0 & 2 & 7 & 3 \\ 0 & 0 & 0 & 9 & 2 \\ 0 & 0 & 0 & 0 & 1 \end{vmatrix} = 6(5)(2)(9)(1) = 540$

31. Area $= \dfrac{1}{2}\begin{vmatrix} 0 & 7 & 1 \\ 5 & 0 & 1 \\ 3 & 9 & 1 \end{vmatrix} = \dfrac{1}{2}(31)$

$= 15.5$ square units

32. $\begin{vmatrix} x & y & 1 \\ 2 & 7 & 1 \\ -1 & 4 & 1 \end{vmatrix} = 3x - 3y + 15 = 0$

or $x - y + 5 = 0$

33. $x = \dfrac{\begin{vmatrix} 4 & -7 \\ 11 & 5 \end{vmatrix}}{\begin{vmatrix} 6 & -7 \\ 2 & 5 \end{vmatrix}} = \dfrac{97}{44}$

34. $z = \dfrac{\begin{vmatrix} 3 & 0 & 1 \\ 0 & 1 & 3 \\ 1 & -1 & 2 \end{vmatrix}}{\begin{vmatrix} 3 & 0 & 1 \\ 0 & 1 & 4 \\ 1 & -1 & 0 \end{vmatrix}} = \dfrac{14}{11}$

35. $y = \dfrac{\begin{vmatrix} 721.4 & 33.77 \\ 45.9 & 19.85 \end{vmatrix}}{\begin{vmatrix} 721.4 & -29.1 \\ 45.9 & 105.6 \end{vmatrix}}$

$$= \dfrac{12,769.747}{77,515.530} \approx 0.1647$$

Chapter 6   Practice Test Solutions

1. $a_n = \dfrac{2n}{(n+2)!}$

$a_1 = \dfrac{2(1)}{3!} = \dfrac{2}{6} = \dfrac{1}{3}$

$a_2 = \dfrac{2(2)}{4!} = \dfrac{4}{24} = \dfrac{1}{6}$

$a_3 = \dfrac{2(3)}{5!} = \dfrac{6}{120} = \dfrac{1}{20}$

$a_4 = \dfrac{2(4)}{6!} = \dfrac{8}{720} = \dfrac{1}{90}$

$a_5 = \dfrac{2(5)}{7!} = \dfrac{10}{5040} = \dfrac{1}{504}$

Terms: $\dfrac{1}{3}, \dfrac{1}{6}, \dfrac{1}{20}, \dfrac{1}{90}, \dfrac{1}{504}$

2. $a_n = \dfrac{n+3}{3^n}$

3. $\displaystyle\sum_{i=1}^{6}(2i - 1) = 1 + 3 + 5 + 7 + 9 + 11 = 36$

4. $a_1 = 23, d = -2$

$a_2 = a_1 + d = 21$

$a_3 = a_2 + d = 19$

$a_4 = a_3 + d = 17$

$a_5 = a_4 + d = 15$

Terms: 23, 21, 19, 17, 15

5. $a_1 = 12, d = 3, n = 50$

$a_n = a_1 + (n - 1)d$

$a_{50} = 12 + (50 - 1)3 = 159$

6. $a_1 = 1$

$a_{200} = 200$

$S_n = \dfrac{n}{2}(a_1 + a_n)$

$S_{200} = \dfrac{200}{2}(1 + 200) = 20{,}100$

7. $a_1 = 7, r = 2$

$a_2 = a_1 r = 14$

$a_3 = a_1 r^2 = 28$

$a_4 = a_1 r^3 = 56$

$a_5 = a_1 r^4 = 112$

Terms: 7, 14, 28, 56, 112

8. $\displaystyle\sum_{n=0}^{9} 6\left(\dfrac{2}{3}\right)^n, a_1 = 6, r = \dfrac{2}{3}, n = 10$

$S_n = \dfrac{a_1(1 - r^n)}{1 - r} = \dfrac{6\left(1 - (2/3)^{10}\right)}{1 - (2/3)} \approx 17.6879$

9. $\displaystyle\sum_{n=0}^{\infty} (0.03)^n, a_1 = 1, r = 0.03$

$S = \dfrac{a_1}{1 - r} = \dfrac{1}{1 - 0.03} = \dfrac{1}{0.97} = \dfrac{100}{97} \approx 1.0309$

10. For $n = 1, 1 = \dfrac{1(1 + 1)}{2}$. Assume that $1 + 2 + 3 = 4 + \cdots + k = \dfrac{k(k + 1)}{2}$. Now for $n = k + 1$,

$1 + 2 + 3 + 4 + \cdots + k + (k + 1) = \dfrac{k(k + 1)}{2} + k + 1$

$= \dfrac{k(k + 1)}{2} + \dfrac{2(k + 1)}{2}$

$= \dfrac{(k + 1)(k + 2)}{2}.$

Thus, $1 + 2 + 3 + 4 + \cdots + n = \dfrac{n(n + 1)}{2}$ for all integers $n \geq 1$.

11. For $n = 4, 4! > 2^4$. Assume that $k! > 2^k$. Then

$(k + 1)! = (k + 1)(k!) > (k + 1)2^k > 2 \cdot 2^k$

$= 2^{k+1}.$

Thus, $n! > 2^n$ for all integers $n \geq 4$.

12. $_{13}C_4 = \dfrac{13!}{(13 - 4)!4!} = 715$

13. $(x + 3)^5 = x^5 + 5x^4(3) + 10x^3(3)^2 + 10x^2(3)^3 + 5x(3)^4 + (3)^5$

$= x^5 + 15x^4 + 90x^3 + 270x^2 + 405x + 243$

14. $_{12}C_5 x^7(-2)^5 = -25{,}344x^7$

15. $_{30}P_4 = \dfrac{30!}{(30 - 4)!} = 657{,}720$

16. $6! = 720$ ways

17. $_{12}P_3 = 1320$

18. $P(2) + P(3) + P(4) = \frac{1}{36} + \frac{2}{36} + \frac{3}{36}$
$$= \frac{6}{36} = \frac{1}{6}$$

19. $P(K, B10) = \frac{4}{52} \cdot \frac{2}{51} = \frac{2}{663}$

20. Let A = probability of no faulty units.

$P(A) = \left(\frac{997}{1000}\right)^{50} \approx 0.8605$

$P(A') = 1 - P(A) \approx 0.1395$

Chapter 7   Practice Test Solutions

1. $(x - 0)^2 = 4(5)(y - 0)$

Vertex: $(0, 0)$

Focus: $(0, 5)$

Directrix: $y = -5$

2. $(y - 0)^2 = 4(7)(x - 0)$
$$y^2 = 28x$$

3. $a = 12, b = 5, h = k = 0,$

$c = \sqrt{144 - 25} = \sqrt{119}$

Center: $(0, 0)$

Foci: $\left(\pm\sqrt{119}, 0\right)$

Vertices: $(\pm 12, 0)$

4. Center: $(0, 0)$

$c = 4, 2b = 6 \implies b = 3,$

$a = \sqrt{16 + 9} = 5$

$$\frac{x^2}{25} + \frac{y^2}{9} = 1$$

5. $a = 12, b = 13, c = \sqrt{144 + 169} = \sqrt{313}$

Center: $(0, 0)$

Foci: $\left(0, \pm\sqrt{313}\right)$

Vertices: $(0, \pm 12)$

Asymptotes: $y = \pm\frac{12}{13}x$

6. Center: $(0, 0)$

$a = 4, \pm\frac{1}{2} = \pm\frac{b}{4} \implies b = 2$

$$\frac{x^2}{16} - \frac{y^2}{4} = 1$$

$y_1 = 2\sqrt{\frac{x^2}{16} - 1}$

$y_2 = -2\sqrt{\frac{x^2}{16} - 1}$

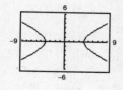

7. $p = 4$

$(x - 6)^2 = 4(4)(y + 1)$

$(x - 6)^2 = 16(y + 1)$

$y = \frac{(x - 6)^2}{16} - 1$

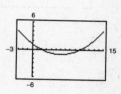

8.
$$16x^2 - 96x + 9y^2 + 36y = -36$$
$$16(x^2 - 6x + 9) + 9(y^2 + 4y + 4) = -36 + 144 + 36$$
$$16(x - 3)^2 + 9(y + 2)^2 = 144$$
$$\frac{(x - 3)^2}{9} + \frac{(y + 2)^2}{16} = 1$$

$a = 4, b = 3, c = \sqrt{16 - 9} = \sqrt{7}$

Center: $(3, -2)$

Foci: $\left(3, -2 \pm \sqrt{7}\right)$

Vertices: $(3, -2 \pm 4)$ or $(3, 2)$ and $(3, -6)$

9. Center: $(3, 1)$

$a = 4, 2b = 2 \implies b = 1$

$$\frac{(x - 3)^2}{16} + \frac{(y - 1)^2}{1} = 1$$

10. Center: $(-3, 1)$

Vertices: $\left(-3 \pm \frac{1}{2}, 1\right)$ or $\left(-\frac{5}{2}, 1\right)$ and $\left(-\frac{7}{2}, 1\right)$

Foci: $\left(-3 \pm \frac{\sqrt{13}}{6}, 1\right)$

Asymptotes: $y = \pm\frac{1/3}{1/2}(x + 3) + 1 = \pm\frac{2}{3}(x + 3) + 1$

$a = \frac{1}{2}, b = \frac{1}{3}, c = \sqrt{\frac{1}{4} + \frac{1}{9}} = \frac{\sqrt{13}}{6}$

$$\frac{(x + 3)^2}{1/4} - \frac{(y - 1)^2}{1/9} = 1$$

11. Center: $(3, 0)$

$a = 4, c = 7, b = \sqrt{49 - 16} = \sqrt{33}$

$$\frac{y^2}{16} - \frac{(x - 3)^2}{33} = 1$$

12. $x = 2t + 1 \implies t = \frac{x - 1}{2}$

$y = -1 - 3t = -1 - 3\left(\frac{x - 1}{2}\right)$

$= -1 - \frac{3}{2}x + \frac{3}{2} = -\frac{3}{2}x + \frac{1}{2}$

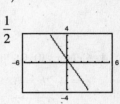

13.

14. The first curve consists of the entire parabola $y = x^2$, whereas the second consists of just the branch $x \geq 0$.

PART II

Chapter P   Chapter Test Solutions

1. $-\frac{10}{3} \approx -3.3$ and $-|-4| = -4$, hence $-\frac{10}{3} > -|-4|$.

2. $d(-17, 39) = |39 - (-17)| = |39 + 17| = |56| = 56$

3. $(5 - x) + 0 = 5 - x$ Additive Identity Property

4. (a) $27\left(\frac{-2}{3}\right) = -2\left(\frac{27}{3}\right) = -2(9) = -18$

   (b) $\frac{5}{18} \div \frac{15}{8} = \frac{5}{18} \cdot \frac{8}{15} = \frac{5 \cdot 2 \cdot 4}{2 \cdot 9 \cdot 3 \cdot 5} = \frac{4}{27}$

   (c) $\left(-\frac{2}{7}\right)^3 = -\frac{2^3}{7^3} = -\frac{8}{343}$

   (d) $\left(\frac{3^2}{2}\right)^{-3} = \left(\frac{2}{3^2}\right)^3 = \frac{2^3}{3^6} = \frac{8}{729}$

5. (a) $\sqrt{5} \cdot \sqrt{125} = \sqrt{5} \cdot 5\sqrt{5} = 25$

   (b) $\frac{\sqrt{72}}{\sqrt{2}} = \frac{6\sqrt{2}}{\sqrt{2}} = 6$

   (c) $\frac{5.4 \times 10^8}{3 \times 10^3} = \frac{5.4}{3} \times 10^5 = 1.8 \times 10^5$

   (d) $(3 \times 10^4)^3 = 3^3 \times (10^4)^3$

                   $= 27 \times 10^{12} = 2.7 \times 10^{13}$

6. (a) $3z^2(2z^3)^2 = 3z^2 4 \cdot z^6 = 12z^8$

   (b) $(u - 2)^{-4}(u - 2)^{-3} = (u - 2)^{-7} = \frac{1}{(u - 2)^7}$

   (c) $\left(\frac{x^{-2}y^2}{3}\right)^{-1} = \frac{3}{x^{-2}y^2} = \frac{3x^2}{y^2}$

7. (a) $9z\sqrt{8z} - 3\sqrt{2z^3} = 9z \cdot 2\sqrt{2z} - 3z\sqrt{2z}$

                  $= 15z\sqrt{2z}$

   (b) $-5\sqrt{16y} + 10\sqrt{y} = -5 \cdot 4\sqrt{y} + 10\sqrt{y}$

                  $= -10\sqrt{y}$

   (c) $\sqrt[3]{\frac{16}{v^5}} = \sqrt[3]{\frac{2^3 \cdot 2}{v^3 v^2}} = \frac{2}{v}\sqrt[3]{\frac{2}{v^2}}$

8. $3 - 2x^5 + 3x^3 - x^4 = -2x^5 - x^4 + 3x^3 + 3$

Degree: 5

Leading coefficient: -2

9. $(x^2 + 3) - [3x + (8 - x^2)] = x^2 + 3 - 3x - 8 + x^2 = 2x^2 - 3x - 5$

10. $(2x - 5)(4x^2 + 3) = 2x(4x^2) - 5(4x^2) + 2x(3) - 5(3)$

                    $= 8x^3 - 20x^2 + 6x - 15$

11. $\frac{8x}{x - 3} + \frac{24}{3 - x} = \frac{8x}{x - 3} - \frac{24}{x - 3} = \frac{8x - 24}{x - 3} = \frac{8(x - 3)}{x - 3} = 8, \; x \neq 3$

12. $\left(\frac{2}{x} - \frac{2}{x + 1}\right) \div \frac{4}{x^2 - 1} = \frac{2(x + 1) - 2x}{x(x + 1)} \cdot \frac{(x - 1)(x + 1)}{4} = \frac{2}{x} \cdot \frac{x - 1}{4} = \frac{x - 1}{2x}, \; x \neq \pm 1$

13. $(x + \sqrt{5})(x - \sqrt{5}) = x^2 - 5$

14. $(x - 2)^3 = x^3 - 3x^2(2) + 3x(2^2) - 2^3$

                $= x^3 - 6x^2 + 12x - 8$

15. $[(x + y) - z][(x + y) + z] = (x + y)^2 - z^2$
$$= x^2 + 2xy + y^2 - z^2$$

16. $2x^4 - 3x^3 - 2x^2 = x^2(2x^2 - 3x - 2)$
$$= x^2(2x + 1)(x - 2)$$

17. $x^3 + 2x^2 - 4x - 8 = x^2(x + 2) - 4(x + 2)$
$$= (x + 2)(x^2 - 4)$$
$$= (x + 2)(x + 2)(x - 2)$$
$$= (x + 2)^2(x - 2)$$

18. $8x^3 - 27 = (2x)^3 - 3^3$
$$= (2x - 3)((2x)^2 + (2x)(3) + 3^2)$$
$$= (2x - 3)(4x^2 + 6x + 9)$$

19. (a) $\dfrac{16}{\sqrt[3]{16}} = \dfrac{16}{\sqrt[3]{8 \cdot 2}} = \dfrac{16}{2} \dfrac{1}{2^{1/3}} \cdot \dfrac{2^{2/3}}{2^{2/3}} = 4 \cdot 2^{2/3} = 4\sqrt[3]{4}$

(b) $\dfrac{6}{1 - \sqrt{3}} = \dfrac{6}{1 - \sqrt{3}} \cdot \dfrac{1 + \sqrt{3}}{1 + \sqrt{3}} = \dfrac{6(1 + \sqrt{3})}{1 - 3} = -3(1 + \sqrt{3})$

(c) $\dfrac{1}{\sqrt{x + 2} - \sqrt{2}} \cdot \dfrac{\sqrt{x + 2} + \sqrt{2}}{\sqrt{x + 2} + \sqrt{2}} = \dfrac{\sqrt{x + 2} + \sqrt{2}}{(x + 2) - 2} = \dfrac{\sqrt{x + 2} + \sqrt{2}}{x}$

20. Shaded region = (area big triangle) − (area small triangle)
$$= \tfrac{1}{2}(3x)(\sqrt{3}x) - \tfrac{1}{2}(2x)(\tfrac{2}{3}\sqrt{3}x)$$
$$= \tfrac{1}{2} 3\sqrt{3}x^2 - \tfrac{2}{3}\sqrt{3}x^2 = \left(\tfrac{3}{2} - \tfrac{2}{3}\right)\sqrt{3}x^2 = \tfrac{5}{6}\sqrt{3}x^2$$

21. Midpoint $= \left(\dfrac{-2 + 6}{2}, \dfrac{5 + 0}{2}\right) = \left(2, \dfrac{5}{2}\right)$

Distance $= \sqrt{(-2 - 6)^2 + (5 - 0)^2}$
$$= \sqrt{64 + 25} = \sqrt{89} \approx 9.43$$

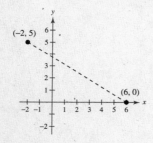

22.

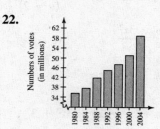

Chapter 1　Chapter Test Solutions

1.

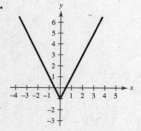

Intercepts: $(0, -1), \left(-\tfrac{1}{2}, 0\right), \left(\tfrac{1}{2}, 0\right)$

2.

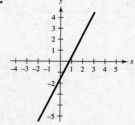

Intercepts: $\left(0, -\tfrac{8}{5}\right), \left(\tfrac{4}{5}, 0\right)$

3.

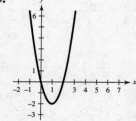

Intercepts: $(0, 0), (2, 0)$

4.

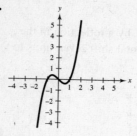

Intercepts: $(0, 0), (1, 0), (-1, 0)$

5.

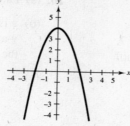

Intercepts: $(0, 4), (2, 0), (-2, 0)$

6.

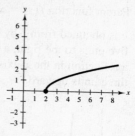

Intercept: $(2, 0)$

7. $5x + 2y = 3$

$$2y = -5x + 3$$

$$y = -\frac{5}{2}x + \frac{3}{2}$$

Slope $= -\frac{5}{2}$

(a) Parallel line

$$y - 4 = -\frac{5}{2}(x - 0)$$

$$y = -\frac{5}{2}x + 4$$

$$5x + 2y - 8 = 0$$

(b) Perpendicular line

$$y - 4 = \frac{2}{5}(x - 0)$$

$$y = \frac{2}{5}x + 4$$

$$2x - 5y + 20 = 0$$

8. Slope $= \dfrac{4 - (-1)}{-3 - 2} = \dfrac{5}{-5} = -1$

$$y + 1 = -1(x - 2)$$

$$y = -x + 1$$

9. No, for some x there corresponds more than one value of y. For instance, if $x = 1$, $y = \pm 1/\sqrt{3}$.

10. $f(x) = |x + 2| - 15$

(a) $f(-8) = |-8 + 2| - 15 = 6 - 15 = -9$

(b) $f(14) = |14 + 2| - 15 = 16 - 15 = 1$

(c) $f(t - 6) = |t - 6 + 2| - 15 = |t - 4| - 15$

11. $3 - x \geq 0 \implies$ domain is all $x \leq 3$.

12. $C = 5.60x + 24{,}000$

$P = R - C$

$= 99.50x - (5.60x + 24{,}000)$

$= 93.9x - 24{,}000$

13. $f(-x) = 2(-x)^3 - 3(-x)$

$= -2x^3 + 3x = -f(x)$

Odd

14. $f(-x) = 3(-x)^4 + 5(-x)^2$

$= 3x^4 + 5x^2 = f(x)$

Even

15. $h(x) = \frac{1}{4}x^4 - 2x^2 = \frac{1}{4}x^2(x^2 - 8)$

By graphing h, you see that the graph is increasing on $(-2, 0)$ and $(2, \infty)$ and decreasing on $(-\infty, -2)$ and $(0, 2)$.

16. $g(t) = |t + 2| - |t - 2|$

By graphing g, you see that the graph is increasing on $(-2, 2)$, and constant on $(-\infty, -2)$ and $(2, \infty)$.

17. Relative minimum: $(-3.33, -6.52)$

Relative maximum: $(0, 12)$

18. Relative minimum: $(0.77, 1.81)$

Relative maximum: $(-0.77, 2.19)$

19. (a) Parent function $f(x) = x^3$

(b) g is obtained from f by a horizontal shift five units to the right, a vertical stretch of 2, a reflection in the x-axis, and a vertical shift three units upward.

(c)

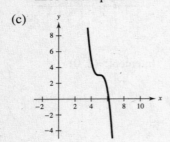

20. (a) Parent function $f(x) = \sqrt{x}$

(b) g is obtained from f by a reflection in the y-axis, and a horizontal shift seven units to the left.

(c)

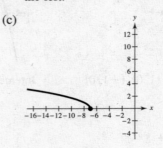

21. (a) Parent function $f(x) = |x|$

(b) $g(x) = 4|-x| - 7 = 4|x| - 7$ is obtained from f by a vertical stretch of 4 followed by a vertical shift seven units downward.

(c)

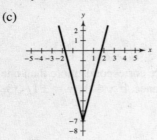

22. (a) $(f - g)(x) = x^2 - \sqrt{2 - x}$

Domain: $x \leq 2$

(b) $\left(\dfrac{f}{g}\right)(x) = \dfrac{x^2}{\sqrt{2 - x}}$

Domain: $x < 2$

(c) $(f \circ g)(x) = f\left(\sqrt{2 - x}\right) = 2 - x$

Domain: $x \leq 2$

(d) $(g \circ f)x = g(x^2) = \sqrt{2 - x^2}$

Domain: $-\sqrt{2} \leq x \leq \sqrt{2}$

23. $f(x) = x^3 + 8$

Yes, f is one-to-one and has an inverse function.

$$y = x^3 + 8$$
$$x = y^3 + 8$$
$$x - 8 = y^3$$
$$\sqrt[3]{x - 8} = y$$
$$f^{-1}(x) = \sqrt[3]{x - 8}$$

24. $f(x) = x^2 + 6$

No, f is not one-to-one, and does not have an inverse function.

25. $f(x) = \dfrac{3x\sqrt{x}}{8}$

Yes, f is one-to-one and has an inverse function.

$$y = \tfrac{3}{8}x^{3/2}, \ x \geq 0, y \geq 0$$
$$x = \tfrac{3}{8}y^{3/2}, \ y \geq 0, x \geq 0$$
$$\tfrac{8}{3}x = y^{3/2}$$
$$\left(\tfrac{8}{3}x\right)^{2/3} = y$$
$$f^{-1}(x) = \left(\tfrac{8}{3}x\right)^{2/3}, \quad x \geq 0$$

Chapter 2   Chapter Test Solutions

1. $\dfrac{12}{x} - 7 = -\dfrac{27}{x} + 6$

$\dfrac{39}{x} = 13$

$39 = 13x$

$3 = x \implies x = 3$

2. $\dfrac{4}{3x - 2} - \dfrac{9x}{3x + 2} = -3$

$4(3x + 2) - 9x(3x - 2) = -3(3x - 2)(3x + 2)$

$12x + 8 - 27x^2 + 18x = -3(9x^2 - 4)$

$-27x^2 + 30x + 8 = -27x^2 + 12$

$30x = 4$

$x = \dfrac{2}{15}$

3. $(-8 - 3i) + (-1 - 15i) = -9 - 18i$

4. $\left(10 + \sqrt{-20}\right) - \left(4 - \sqrt{-14}\right) = 6 + 2\sqrt{5}i + \sqrt{14}i = 6 + \left(2\sqrt{5} + \sqrt{14}\right)i$

5. $(2 + i)(6 - i) = 12 + 6i - 2i + 1 = 13 + 4i$

6. $(4 + 3i)^2 - (5 + i)^2 = (16 + 24i - 9) - (25 + 10i - 1) = -17 + 14i$

7. $\dfrac{8 + 5i}{6 - i} \cdot \dfrac{6 + i}{6 + i} = \dfrac{48 + 30i + 8i - 5}{36 + 1} = \dfrac{43}{37} + \dfrac{38}{37}i$

8. $\dfrac{5i}{2 + i} \cdot \dfrac{2 - i}{2 - i} = \dfrac{10i + 5}{4 + 1} = 1 + 2i$

9. $\dfrac{(2i - 1)}{(3i + 2)} \cdot \dfrac{2 - 3i}{2 - 3i} = \dfrac{6 - 2 + 4i + 3i}{4 + 9}$

$= \dfrac{4}{13} + \dfrac{7}{13}i$

10.

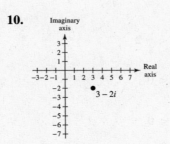

11. $f(x) = 3x^2 - 6 = 0$

$x \approx \pm 1.414$

12. $f(x) = 8x^2 - 2 = 0$

$x = \pm 0.5$

13. $f(x) = x^3 - 4x^2 + 5x$

$= 0$

$x = 0$

14. $f(x) = x - x^3$

$x = 0, \pm 1$

15. $x^2 - 10x + 9 = 0$

$(x - 1)(x - 9) = 0$

$x = 1, 9$

16. $x^2 + 12x - 2 = 0$

$x = \dfrac{-12 \pm \sqrt{12^2 - 4(-2)}}{2}$

$= -6 \pm \sqrt{38}$

17. $4x^2 - 81 = 0$

$4x^2 = 81$

$x^2 = \dfrac{81}{4}$

$x = \pm \dfrac{9}{2}$

18. $5x^2 + 14x - 3 = 0$

$(x + 3)(5x - 1) = 0$

$x = -3, \frac{1}{5}$

19. $3x^3 - 4x^2 - 12x + 16 = 0$

$x^2(3x - 4) - 4(3x - 4) = 0$

$(x^2 - 4)(3x - 4) = 0$

$x = 2, -2, \frac{4}{3}$

20. $x + \sqrt{22 - 3x} = 6$

$\sqrt{22 - 3x} = 6 - x$

$22 - 3x = (6 - x)^2$

$22 - 3x = 36 - 12x + x^2$

$x^2 - 9x + 14 = 0$

$(x - 2)(x - 7) = 0$

$x = 2$

($x = 7$ is extraneous.)

21. $(x^2 + 6)^{2/3} = 16$

$x^2 + 6 = 16^{3/2} = 64$

$x^2 = 58$

$x = \pm\sqrt{58} \approx \pm 7.616$

22. $|8x - 1| = 21$

$8x - 1 = 21$ or $-(8x - 1) = 21$

$8x = 22$ or $-8x = 20$

$x = \frac{11}{4}$ or $x = -\frac{5}{2}$

23. $8x - 1 > 3x - 10$

$5x > -9$

$x > -\frac{9}{5}$

$\left(-\frac{9}{5}, \infty\right)$

$-\frac{9}{5}$

-4 -3 -2 -1 0 1 2 3 4 5

24. $2|x - 8| < 10$

$|x - 8| < 5$

$-5 < x - 8 < 5$

$3 < x < 13$

3 4 5 6 7 8 9 10 11 12 13

25. $6x^2 + 5x + 1 \geq 0$

$(3x + 1)(2x + 1) \geq 0$

Critical numbers: $-\frac{1}{3}, -\frac{1}{2}$

$\left(-\infty, -\frac{1}{2}\right], \left[-\frac{1}{3}, \infty\right)$

$-\frac{1}{2}$ $-\frac{1}{3}$

-2   -1   0   1

26. $\dfrac{3 - 5x}{2 + 3x} < -2$

$\dfrac{3 - 5x}{2 + 3x} + 2 < 0$

$\dfrac{3 - 5x + 2(2 + 3x)}{2 + 3x} < 0$

$\dfrac{x + 7}{2 + 3x} < 0$

Critical numbers: $-7, -\dfrac{2}{3}$

Checking the three intervals, we obtain

$-7 < x < -\dfrac{2}{3}$.

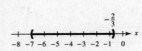

$-\frac{2}{3}$

-8 -7 -6 -5 -4 -3 -2 -1 0

27. $y = 18.30t - 76.2, \ r \approx 0.99622$

$y = 200$ for $t \approx 15$, or 2005

Chapters P–2   Cumulative Test Solutions

1. $\dfrac{14x^2y^{-3}}{32x^{-1}y^2} = \dfrac{7x^3}{16y^5}, x \neq 0$

2. $8\sqrt{60} - 2\sqrt{135} - \sqrt{15} = 16\sqrt{15} - 6\sqrt{15} - \sqrt{15} = 9\sqrt{15}$

3. $\sqrt{28x^4y^3} = 2x^2y\sqrt{7y}$

4. $4x - [2x + 5(2 - x)] = 4x - [-3x + 10]$
$$= -10 + 7x = 7x - 10$$

5. $(x - 2)(x^2 + x - 3) = x^3 + x^2 - 3x - 2x^2 - 2x + 6$
$$= x^3 - x^2 - 5x + 6$$

6. $\dfrac{2}{x + 3} - \dfrac{1}{x + 1} = \dfrac{2(x + 1) - (x + 3)}{(x + 3)(x + 1)}$

$$= \dfrac{x - 1}{(x + 3)(x + 1)}$$

7. $25 - (x - 2)^2 = [5 - (x - 2)][5 + (x - 2)]$
$$= (7 - x)(3 + x)$$

8. $x - 5x^2 - 6x^3 = -x(6x^2 + 5x - 1)$
$$= -x(6x - 1)(x + 1)$$
$$= x(x + 1)(1 - 6x)$$

9. $54 - 16x^3 = 2(27 - 8x^3)$
$$= 2(3 - 2x)(9 + 6x + 4x^2)$$

10. Midpoint $= \dfrac{((-7/2) + (5/2), 4 + (-8))}{2}$

$$= \left(-\dfrac{1}{2}, -2\right)$$

Distance $= \sqrt{\left(\dfrac{5}{2} - \left(-\dfrac{7}{2}\right)\right)^2 + (-8 - 4)^2}$

$$= \sqrt{36 + 144}$$
$$= \sqrt{180}$$
$$= 6\sqrt{5} \approx 13.42$$

11. $\left(x + \dfrac{1}{2}\right)^2 + (y + 8)^2 = 16$

12. $x - 3y + 12 = 0$
$$-3y = -x - 12$$
$$y = \dfrac{x}{3} + 4$$

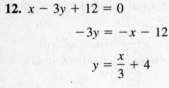

13. $y = x^2 - 9$

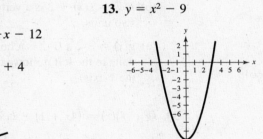

14. $y = \sqrt{4 - x}$

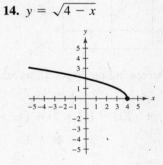

15. (a) Slope $= \dfrac{8 - 4}{-5 - (-1)} = -1$
$$y - 8 = -1(x + 5)$$
$$y = -x + 3$$
$$y + x = 3$$

(b) $(-1, 4), (0, 3), (1, 2)$

16. (a) $y - 1 = -2\left(x + \dfrac{1}{2}\right)$
$$y - 1 = -2x - 1$$
$$y = -2x$$
$$y + 2x = 0$$

(b) Three additional points: $(0, 0), (1, -2), (2, -4)$

17. (a) Vertical line: $x = -\frac{3}{7}$ or $x + \frac{3}{7} = 0$

(b) Three additional points:

$\left(-\frac{3}{7}, 0\right), \left(-\frac{3}{7}, 1\right), \left(-\frac{3}{7}, 2\right)$

18. $6x - y = 4$ has slope 6.

(a) $y - 3 = 6(x - 2)$

$y = 6x - 9$

(b) $y - 3 = -\frac{1}{6}(x - 2)$

$y = -\frac{1}{6}x + \frac{10}{3}$

19. $f(x) = \dfrac{x}{x - 2}$

(a) $f(5) = \dfrac{5}{5 - 2} = \dfrac{5}{3}$

(b) $f(2)$ is undefined.

(c) $f(5 + 4s) = \dfrac{5 + 4s}{(5 + 4s) - 2} = \dfrac{5 + 4s}{3 + 4s}$

20. $f(x) = \begin{cases} 3x - 8, & x < 0 \\ x^2 + 4, & x \geq 0 \end{cases}$

(a) $f(-8) = 3(-8) - 8 = -32$

(b) $f(0) = 0^2 + 4 = 4$

(c) $f(4) = 4^2 + 4 = 20$

21. $(-\infty, \infty)$

22. $5 + 7t \geq 0$

$7t \geq -5$

$t \geq -\frac{5}{7}$

$\left[-\frac{5}{7}, \infty\right)$

23. $9 - s^2 \geq 0$

$9 \geq s^2$

$[-3, 3]$

24. All $x \neq -\frac{2}{5}$

25. $g(-x) = 3(-x) - (-x)^3$

$= -3x + x^3 = -g(x)$

Odd

26. No, for some x there correspond two values of y.

27.

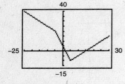

Decreasing on $(-\infty, 5)$, increasing on $(5, \infty)$

28. (a) $r(x) = \frac{1}{2}f(x)$ is a vertical shrink of f.

(b) $h(x) = f(x) + 2$ is a vertical shift upward of two units.

(c) $g(x) = -\sqrt[3]{x + 2}$ is a horizontal shift two units to the left, followed by a reflection in the x-axis.

29. $(f + g)(x) = (x^2 + 2) + (4x + 1)$

$= x^2 + 4x + 3$

30. $(g - f)(x) = (4x + 1) - (x^2 + 2)$

$= -x^2 + 4x - 1$

31. $(g \circ f)(x) = g(x^2 + 2)$

$= 4(x^2 + 2) + 1$

$= 4x^2 + 9$

32. $(fg)(x) = (x^2 + 2)(4x + 1)$

$= 4x^3 + x^2 + 8x + 2$

33. Yes, $h(x) = 5x - 2$ has an inverse function.

$$y = 5x - 2$$

$$x = 5y - 2$$

$$x + 2 = 5y$$

$$\frac{x + 2}{5} = y$$

$$h^{-1}(x) = \frac{x + 2}{5}$$

34.

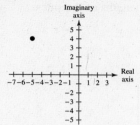

35. $f(x) = 4x^3 - 12x^2 + 8x = 0$

$$x = 0, 1, 2$$

36. $f(x) = \dfrac{5}{x} - \dfrac{10}{x - 3}$

$$x = -3$$

37. $f(x) = |3x + 4| - 2 = 0$

$$x = -2, -0.667$$

38. $f(x) = \sqrt{x^2 + 1} + x - 9 = 0$

$$x \approx 4.444$$

39. $\dfrac{x}{5} - 6 \le \dfrac{-x}{2} + 6$

$$\frac{x}{2} + \frac{x}{5} \le 12$$

$$\frac{7x}{10} \le 12$$

$$x \le \frac{120}{7} \approx 17.143$$

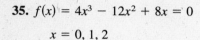

40. $2x^2 + x \ge 15$

$$2x^2 + x - 15 \ge 0$$

$$(2x - 5)(x + 3) \ge 0$$

Critical numbers: $-3, \dfrac{5}{2}$

Test intervals: $(-\infty, -3), \left(-3, \dfrac{5}{2}\right), \left(\dfrac{5}{2}, \infty\right)$

You obtain $(-\infty, -3] \cup \left[\dfrac{5}{2}, \infty\right)$.

41. $|7 + 8x| > 5$

$$7 + 8x > 5 \quad \text{or} \quad 7 + 8x < -5$$

$$8x > -2 \quad \text{or} \quad 8x < -12$$

$$x > -\frac{1}{4} \quad \text{or} \quad x < -\frac{3}{2}$$

42. $\dfrac{2(x - 2)}{x + 1} \le 0$

Critical numbers: $x = -1, x = 2$

Test intervals: $(-\infty, -1), (-1, 2), (2, \infty)$

Is $\dfrac{2(x - 2)}{x + 1} \le 0$?

$$-1 < x \le 2$$

43. $V = \dfrac{4}{3}\pi r^3 \implies r = \sqrt[3]{\dfrac{3}{4\pi}V}$

$$= \sqrt[3]{\dfrac{3}{4\pi}(370.7)}$$

$$\approx 4.456 \text{ inches}$$

44. (a) Let x and y be the lengths of the sides.

$$2x + 2y = 546 \implies y = 273 - x$$

$$A = xy = x(273 - x)$$

(b)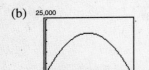

Domain: $0 < x < 273$

(c) If $A = 15000$, then $x = 76.23$ or 196.77.

Dimensions in feet:

76.23×196.77 or 196.77×76.23

45. (a) Linear: $y_1 = 770.3t + 1263$, $r \approx 0.98770$

(b)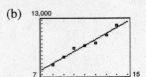

(c) For 2008, $t = 18$, $y_1 \approx \$15,128$ million.

For 2010, $t = 20$, $y_1 \approx \$16,669$ million.

(d) Answers will vary.

Chapter 3   Chapter Test Solutions

1. $y = x^2 + 4x + 3 = x^2 + 4x + 4 - 1 = (x + 2)^2 - 1$

Vertex: $(-2, -1)$

$x = 0 \implies y = 3$

$y = 0 \implies x^2 + 4x + 3 = 0 \implies (x + 3)(x + 1) = 0 \implies x = -1, -3$

Intercepts: $(0, 3)$, $(-1, 0)$, $(-3, 0)$

2. Let $y = a(x - h)^2 + k$. The vertex $(3, -6)$ implies that $y = a(x - 3)^2 - 6$. For $(0, 3)$ you obtain

$$3 = a(0 - 3)^2 - 6 = 9a - 6 \implies a = 1.$$

Thus, $y = (x - 3)^2 - 6 = x^2 - 6x + 3$.

3. (a) $y = -\frac{1}{20}x^2 + 3x + 5 = -\frac{1}{20}(x^2 - 60x + 900) + 5 + 45 = -\frac{1}{20}(x - 30)^2 + 50$

Maximum height: $y = 50$ feet

(b) The term 5 determines the height at which the ball was thrown. Changing the constant term results in a vertical shift of the graph and therefore changes the maximum height.

4. $f(x) = 4x^3 + 4x^2 + x = x(4x^2 + 4x + 1) = x(2x + 1)^2$

Zeros: 0 (multiplicity 1)

$-\frac{1}{2}$ (multiplicity 2)

5. $f(x) = -x^3 + 7x + 6$

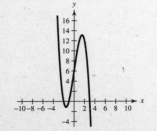

6. $x^2 + 1 \overline{\smash{\big)}\ 3x^3 + 0x^2 + 4x - 1}$

$$\phantom{x^2 + 1 \overline{\smash{\big)}\ }} \begin{array}{r} 3x \\ \hline 3x^3 + 0x^2 + 4x - 1 \\ 3x^3 + 3x \\ \hline x - 1 \end{array}$$

$$3x + \frac{x - 1}{x^2 + 1}$$

7.

$$
\begin{array}{r|rrrr}
2 & 2 & 0 & -5 & 0 & -3 \\
  &   & 4 & 8 & 6 & 12 \\
\hline
  & 2 & 4 & 3 & 6 & 9
\end{array}
$$

$$2x^3 + 4x^2 + 3x + 6 + \frac{9}{x-2}$$

8.

$$
\begin{array}{r|rrrr}
-2 & 3 & 0 & -6 & 5 & -1 \\
   &   & -6 & 12 & -12 & 14 \\
\hline
   & 3 & -6 & 6 & -7 & 13
\end{array}
$$

$$f(-2) = 13$$

9. Possible rational zeros:

$$\pm 24, \pm 12, \pm 8, \pm 6, \pm 4, \pm 3, \pm 2, \pm 1, \pm\tfrac{3}{2}, \pm\tfrac{1}{2}$$

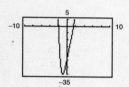

Rational zeros: $-2, \tfrac{3}{2}$

10. Possible rational zeros: $\pm 2, \pm 1, \pm\tfrac{2}{3}, \pm\tfrac{1}{3}$

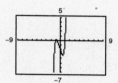

Rational zeros: $\pm 1, -\tfrac{2}{3}$

11. $f(x) = x^3 - 7x^2 + 11x + 19$

$= (x + 1)(x^2 - 8x + 19)$

For the quadratic,

$$x = \frac{8 \pm \sqrt{64 - 4(19)}}{2} = 4 \pm \sqrt{3}i.$$

Zeros: $-1, 4 \pm \sqrt{3}i$

$f(x) = (x + 1)\left(x - 4 + \sqrt{3}i\right)\left(x - 4 - \sqrt{3}i\right)$

12. $f(x) = x(x - 2)(x - 2 - i)(x - 2 + i)$

$= (x^2 - 2x)((x - 2)^2 + 1)$

$= (x^2 - 2x)(x^2 - 4x + 5)$

$= x^4 - 6x^3 + 13x^2 - 10x$

13. $f(x) = \left(x - (1 + \sqrt{3}i)\right)\left(x - (1 - \sqrt{3}i)\right)(x - 2)(x - 2)$

$= (x^2 - 2x + 4)(x^2 - 4x + 4)$

$= x^4 - 6x^3 + 16x^2 - 24x + 16$

14. $f(x) = x(x - 1 - i)(x - 1 + i)$

$= x((x - 1)^2 + 1)$

$= x(x^2 - 2x + 2)$

$= x^3 - 2x^2 + 2x$

15.

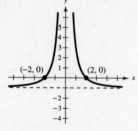

Vertical asymptote: $x = 0$

Intercepts: $(2, 0), (-2, 0)$

Symmetry: y-axis

Horizontal asymptote: $y = -1$

16. $g(x) = \dfrac{x^2 + 2}{x - 1} = x + 1 + \dfrac{3}{x - 1}$

Vertical asymptote: $x = 1$

Intercept: $(0, -2)$

Slant asymptote: $y = x + 1$

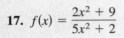

17. $f(x) = \dfrac{2x^2 + 9}{5x^2 + 2}$

Horizontal asymptote: $y = \dfrac{2}{5}$

y-axis symmetry

Intercept: $\left(0, \dfrac{9}{2}\right)$

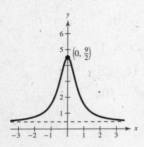

18. (a)

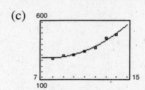

(b) $y = 5.582t^2 - 85.53t + 602.0$

(c)

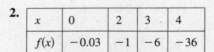

Yes, the model is a good fit.

(d) For 2005, $t = 15$ and $y \approx \$575$ billion.

For 2010, $t = 20$ and $y \approx \$1124$ billion.

(e) Answers will vary.

Chapter 4   Chapter Test Solutions

1.

x	-2	-1	0	1	2
$f(x)$	100	10	1	0.1	0.01

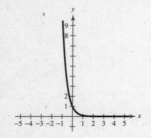

$f(x) = 10^{-x}$

Horizontal asymptote: $y = 0$

Intercept: $(0, 1)$

2.

x	0	2	3	4
$f(x)$	-0.03	-1	-6	-36

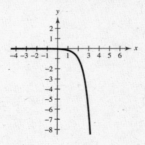

$f(x) = -6^{x-2}$

Horizontal asymptote: $y = 0$

Intercept: $\left(0, -\dfrac{1}{36}\right)$

3.

x	-2	-1	0	1	2
$f(x)$	0.9817	0.8647	0	-6.3891	-53.5982

$f(x) = 1 - e^{2x}$

Horizontal asymptote: $y = 1$

Intercept: $(0, 0)$

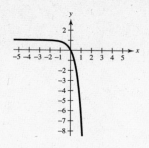

4. $\log_7 7^{-0.89} = -0.89 \log_7 7$

$\qquad\qquad = -0.89$

5. $4.6 \ln e^2 = 4.6(2) \ln e = 9.2$

6. $2 - \log_{10} 100 = 2 - 2 = 0$

7. $f(x) = -\log_{10} x - 6$

Domain: $x > 0$

Vertical asymptote: $x = 0$

x-intercept: $(10^{-6}, 0) \approx (0, 0)$

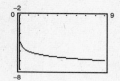

8. $f(x) = \ln(x - 4)$

Domain: $x > 4$

Vertical asymptote: $x = 4$

x-intercept: $(5, 0)$

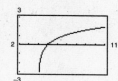

9. $f(x) = 1 + \ln(x + 6)$

Domain: $x > -6$

Vertical asymptote: $x = -6$

x-intercept: $(-5.632, 0)$

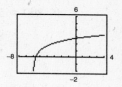

10. $\log_7 44 = \dfrac{\ln 44}{\ln 7} \approx 1.945$

11. $\log_{2/5}(0.9) = \dfrac{\ln(0.9)}{\ln(2/5)} \approx 0.115$

12. $\log_{24} 68 = \dfrac{\ln 68}{\ln 24} \approx 1.328$

13. $\log_2 3a^4 = \log_2 3 + \log_2 a^4 = \log_2 3 + 4 \log_2 a$

14. $\ln \dfrac{5\sqrt{x}}{6} = \ln 5 + \ln \sqrt{x} - \ln 6$

$\qquad\qquad = \ln 5 + \dfrac{1}{2} \ln x - \ln 6$

15. $\ln \dfrac{x\sqrt{x+1}}{2e^4} = \ln x + \ln \sqrt{x+1} - \ln(2) - \ln e^4$

$\qquad\qquad = \ln x + \dfrac{1}{2} \ln(x+1) - \ln 2 - 4$

16. $\log_3 13 + \log_3 y = \log_3(13y)$

17. $4 \ln x - 4 \ln y = \ln x^4 - \ln y^4 = \ln\left(\dfrac{x^4}{y^4}\right) = \ln\left(\dfrac{x}{y}\right)^4$

18. $\ln x - \ln(x+2) + \ln(2x-3) = \ln\left[\dfrac{x(2x-3)}{x+2}\right]$

19. $3^x = 81 = 3^4$

$\quad x = 4$

20. $\quad 5^{2x} = 2500$

$\quad 2x \ln 5 = \ln 2500$

$\qquad\quad x = \dfrac{1}{2}\dfrac{\ln 2500}{\ln 5} \approx 2.431$

21. $\log_7 x = 3$

$\qquad 7^3 = x$

$\qquad\, x = 343$

22. $\log_{10}(x - 4) = 5$

$\qquad\quad 10^5 = x - 4$

$\qquad\qquad x = 10^5 + 4$

$\qquad\qquad\, = 100{,}004$

23. $\dfrac{1025}{8 + e^{4x}} = 5$

$1025 = 40 + 5e^{4x}$

$985 = 5e^{4x}$

$e^{4x} = 197$

$4x = \ln(197)$

$x = \dfrac{1}{4}\ln(197) \approx 1.321$

24. $-xe^{-x} + e^{-x} = 0$

$e^{-x}(1 - x) = 0$

$x = 1$

25. $\log_{10} x - \log_{10}(8 - 5x) = 2$

$\log_{10}\!\left(\dfrac{x}{8 - 5x}\right) = 2$

$10^2 = \dfrac{x}{8 - 5x}$

$800 - 500x = x$

$800 = 501x$

$x = \dfrac{800}{501} \approx 1.597$

26. $2x \ln x - x = 0$

$2 \ln x = 1, \quad (x \neq 0)$

$\ln x = \dfrac{1}{2}$

$x = e^{1/2} \approx 1.649$

27. $\dfrac{1}{2} = 1e^{k(22)}$ (half-life is 22 years)

$\ln \dfrac{1}{2} = 22k$

$k = \dfrac{1}{22}\ln \dfrac{1}{2} = -\dfrac{1}{22}\ln 2 \approx -0.03151$

$A = e^{-0.03151(19)}$

≈ 0.54953 or   55% remains

28. (a) Quadratic model: $R = -0.092t^2 + 3.29t + 38.1$

Exponential model: $R = 46.99(1.026)^t$

Power model: $R = 36.00t^{0.233}$

(b)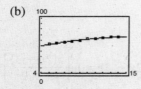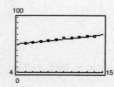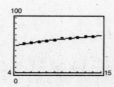

(c) The quadratic model is best for 2010, $t = 20$, and $R \approx 67.1$ billion dollars.

Chapter 5   Chapter Test Solutions

1. $x - y = 6 \implies y = x - 6$. Then $3x + 5(x - 6) = 2 \implies 8x = 32 \implies x = 4, y = 4 - 6 = -2$.

Answer: $(4, -2)$

2. $y = x - 1 = (x - 1)^3 \implies x = 1$ or $1 = (x - 1)^2 = x^2 - 2x + 1 \implies x^2 - 2x = 0$.

Thus, $x = 1$ or $x(x - 2) = 0 \implies x = 0, 1, 2$.

Answer: $(0, -1), (1, 0), (2, 1)$

3. $x - y = 3 \implies y = x - 3 \implies 4x - (x - 3)^2 = 7$

$$4x - (x^2 - 6x + 9) = 7$$

$$x^2 - 10x + 16 = 0$$

$$(x - 2)(x - 8) = 0$$

$x = 2, 8$

Answer: $(2, -1), (8, 5)$

4. $\begin{cases} 2x + 5y = -11 & \text{Equation 1} \\ 5x - y = 19 & \text{Equation 2} \end{cases}$

$-\frac{5}{2}$ times Eq. 1 added to Eq. 2 produces

$-\frac{27}{2}y = \frac{93}{2} \implies y = -\frac{31}{9}.$

Then $2x + 5\left(-\frac{31}{9}\right) = -11 \implies x = \frac{28}{9}.$

Answer: $\left(\frac{28}{9}, -\frac{31}{9}\right)$

5. $\begin{cases} 3x - 2y + z = 0 \\ 6x + 2y + 3z = -2 \\ 3x - 4y + 5z = 5 \end{cases}$

$\begin{cases} 3x - 2y + z = 0 \\ 6y + z = -2 \\ -2y + 4z = 5 \end{cases}$

$\begin{cases} 3x - 2y + z = 0 \\ y - 2z = -\frac{5}{2} \\ 13z = 13 \end{cases}$

$z = 1$

$y = 2(1) - \frac{5}{2} = -\frac{1}{2}$

$x = \frac{1}{3}\left(-1 + 2\left(-\frac{1}{2}\right)\right) = \frac{1}{3}(-2) = -\frac{2}{3}$

Answer: $\left(-\frac{2}{3}, -\frac{1}{2}, 1\right)$

6. $\begin{cases} x - 4y - z = 3 \\ 2x - 5y + z = 0 \\ 3x - 3y + 2z = -1 \end{cases}$

$\begin{cases} x - 4y - z = 3 \\ 3y + 3z = -6 \\ 9y + 5z = -10 \end{cases}$

$\begin{cases} x - 4y - z = 3 \\ y + z = -2 \\ -4z = 8 \end{cases}$

$z = -2$

$y = -2 - (-2) = 0$

$x = 3 + 4(0) + (-2) = 1$

Answer: $(1, 0, -2)$

7. $6 = a(0)^2 + b(0) + c \implies c = 6$

$2 = a(-2)^2 + b(-2) + c$

$\frac{9}{2} = a(3)^2 + b(3) + c$

Hence, $\begin{cases} 4a - 2b + 6 = 2 \text{ or } 2a - b = -2 \\ 9a + 3b + 6 = \frac{9}{2} \text{ or } 9a + 3b = -\frac{3}{2} \end{cases}.$

Solving this system for a and b, you obtain

$a = -\frac{1}{2}, b = 1$. Thus, $y = -\frac{1}{2}x^2 + x + 6.$

8. $\dfrac{5x - 2}{(x - 1)^2} = \dfrac{A}{x - 1} + \dfrac{B}{(x - 1)^2}$

$5x - 2 = A(x - 1) + B = Ax + (-A + B)$

$\begin{cases} A = 5 \\ -A + B = -2 \implies B = 3 \end{cases}$

$\dfrac{5x - 2}{(x - 1)^2} = \dfrac{5}{x - 1} + \dfrac{3}{(x - 1)^2}$

9. $\dfrac{x^3 + x^2 + x + 2}{x^4 + x^2} = \dfrac{A}{x} + \dfrac{B}{x^2} + \dfrac{Cx + D}{x^2 + 1}$

$x^3 + x^2 + x + 2 = Ax(x^2 + 1) + B(x^2 + 1) + (Cx + D)x^2$

$= (A + C)x^3 + (B + D)x^2 + Ax + B$

$\begin{cases} A + C = 1 \\ B + D = 1 \\ A = 1 \\ B = 2 \end{cases}$

$A = 1, C = 0, B = 2, D = -1$

$\dfrac{x^3 + x^2 + x + 2}{x^4 + x^2} = \dfrac{1}{x} + \dfrac{2}{x^2} - \dfrac{1}{x^2 + 1}$

10. $\begin{bmatrix} 2 & 1 & 2 & \vdots & 4 \\ 2 & 2 & 0 & \vdots & 5 \\ 2 & -1 & 6 & \vdots & 2 \end{bmatrix}$ row reduces to $\begin{bmatrix} 1 & 0 & 2 & \vdots & 1.5 \\ 0 & 1 & -2 & \vdots & 1 \\ 0 & 0 & 0 & \vdots & 0 \end{bmatrix}$.

Infinite number of solutions. Let $z = a$, $y = 2a + 1$, $x = 1.5 - 2a$.

Answer: $(1.5 - 2a, 1 + 2a, a)$, where a is any real number

11. $\begin{bmatrix} 2 & 3 & 1 & \vdots & 10 \\ 2 & -3 & -3 & \vdots & 22 \\ 4 & -2 & 3 & \vdots & -2 \end{bmatrix}$ row reduces to $\begin{bmatrix} 1 & 0 & 0 & \vdots & 5 \\ 0 & 1 & 0 & \vdots & 2 \\ 0 & 0 & 1 & \vdots & -6 \end{bmatrix}$.

Answer: $(5, 2, -6)$

12. (a) $A - B = \begin{bmatrix} 1 & 0 & 4 \\ -7 & -6 & -1 \\ 0 & 4 & 0 \end{bmatrix}$

(b) $3A = \begin{bmatrix} 15 & 12 & 12 \\ -12 & -12 & 0 \\ 3 & 6 & 0 \end{bmatrix}$

(c) $3A - 2B = \begin{bmatrix} 7 & 4 & 12 \\ -18 & -16 & -2 \\ 1 & 10 & 0 \end{bmatrix}$

(d) $AB = \begin{bmatrix} 36 & 20 & 4 \\ -28 & -24 & -4 \\ 10 & 8 & 2 \end{bmatrix}$

13. $\begin{bmatrix} -2 & 2 & 3 & \vdots & 1 & 0 & 0 \\ 1 & -1 & 0 & \vdots & 0 & 1 & 0 \\ 0 & 1 & 4 & \vdots & 0 & 0 & 1 \end{bmatrix}$

reduces to

$\begin{bmatrix} 1 & 0 & 0 & \vdots & -\frac{4}{3} & -\frac{5}{3} & 1 \\ 0 & 1 & 0 & \vdots & -\frac{4}{3} & -\frac{8}{3} & 1 \\ 0 & 0 & 1 & \vdots & \frac{1}{3} & \frac{2}{3} & 0 \end{bmatrix}$.

$A^{-1} = \begin{bmatrix} -\frac{4}{3} & -\frac{5}{3} & 1 \\ -\frac{4}{3} & -\frac{8}{3} & 1 \\ \frac{1}{3} & \frac{2}{3} & 0 \end{bmatrix}$

$A^{-1} \begin{bmatrix} 7 \\ -5 \\ -1 \end{bmatrix} = \begin{bmatrix} -2 \\ 3 \\ -1 \end{bmatrix}$

Answer: $(-2, 3, -1)$

14. $\begin{vmatrix} -25 & 18 \\ 6 & -7 \end{vmatrix} = (-25)(-7) - 6(18) = 67$

15. $\det(A) = \begin{vmatrix} 4 & 0 & 3 \\ 1 & -8 & 2 \\ 3 & 2 & 2 \end{vmatrix} = 4(-16 - 4) - 0 + 3(2 + 24)$

$= -80 + 78 = -2$

16. $\begin{vmatrix} x_1 & y_1 & 1 \\ x_2 & y_2 & 1 \\ x_3 & y_3 & 1 \end{vmatrix} = \begin{vmatrix} -1 & 1 & 1 \\ 4 & 11 & 1 \\ -1 & -5 & 1 \end{vmatrix}$

$= -1(16) - 1(4 + 1) + 1(-20 + 11)$

$= -16 - 5 - 9 = -30$

Area $= |-30| = 30$ square units

17. $x = \dfrac{\begin{vmatrix} 3 & -2 \\ -1 & 4 \end{vmatrix}}{\begin{vmatrix} 2 & -2 \\ 1 & 4 \end{vmatrix}} = \dfrac{10}{10} = 1$

$y = \dfrac{\begin{vmatrix} 2 & 3 \\ 1 & -1 \end{vmatrix}}{\begin{vmatrix} 2 & -2 \\ 1 & 4 \end{vmatrix}} = \dfrac{-5}{10} = -\dfrac{1}{2}$

Answer: $\left(1, -\dfrac{1}{2}\right)$

18. Upper left: $400 + x_2 = x_1$

Upper right: $x_1 + x_3 = x_4 + 600$

Lower left: $300 = x_2 + x_3 + x_5$

Lower right: $x_5 + x_4 = 100$

$$\begin{cases} x_1 - x_2 & = 400 \\ x_1 \quad + x_3 - x_4 & = 600 \\ x_2 + x_3 \quad + x_5 = 300 \\ x_4 + x_5 = 100 \end{cases}$$

Solving this system:

$$\begin{bmatrix} 1 & -1 & 0 & 0 & 0 & \vdots & 400 \\ 1 & 0 & 1 & -1 & 0 & \vdots & 600 \\ 0 & 1 & 1 & 0 & 1 & \vdots & 300 \\ 0 & 0 & 0 & 1 & 1 & \vdots & 100 \end{bmatrix} \rightarrow \begin{bmatrix} 1 & 0 & 1 & 0 & 1 & \vdots & 700 \\ 0 & 1 & 1 & 0 & 1 & \vdots & 300 \\ 0 & 0 & 0 & 1 & 1 & \vdots & 100 \\ 0 & 0 & 0 & 0 & 0 & \vdots & 0 \end{bmatrix}$$

Letting $x_3 = a$ and $x_5 = b$ be real numbers, we have:

$x_5 = b$

$x_4 = 100 - b$

$x_3 = a$

$x_2 = 300 - a - b$

$x_1 = 700 - b - a$

Chapters 3–5   Cumulative Test Solutions

1. $f(x) = -\frac{1}{2}(x^2 + 4x)$

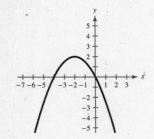

2. $f(x) = \frac{1}{4}x(x - 2)^2$

3. $x^3 + 2x^2 + 4x + 8 = (x + 2)(x^2 + 4)$

Zeros: $-2, \pm 2i$

4. Using a graphing utility, $x \approx 1.424$.

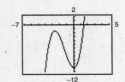

5. $\begin{array}{r|rrrr} 6 & 2 & -5 & 6 & -20 \\ & & 12 & 42 & 288 \\ \hline & 2 & 7 & 48 & 268 \end{array}$

$$\frac{2x^3 - 5x^2 + 6x - 20}{x - 6} = 2x^2 + 7x + 48 + \frac{268}{x - 6}$$

6. $f(x) = (x - 0)(x + 3)\left[x - \left(1 + \sqrt{5}i\right)\right]\left[x - \left(1 - \sqrt{5}i\right)\right]$

$\quad = x(x + 3)\left[(x - 1)^2 + 5\right]$

$\quad = (x^2 + 3x)(x^2 - 2x + 6)$

$\quad = x^4 + x^3 + 18x$

7. $f(x) = \dfrac{2x}{x - 3}$

Vertical asymptote: $x = 3$

Horizontal asymptote: $y = 2$

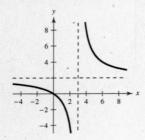

8. $f(x) = \dfrac{5x}{x^2 + x - 6}$

$\quad = \dfrac{5x}{(x + 3)(x - 2)}$

Vertical asymptotes: $x = -3, 2$

Horizontal asymptote: $y = 0$

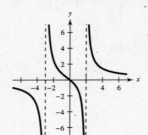

9. $f(x) = \dfrac{x^2 - 3x + 8}{x - 2}$

$\quad = x - 1 + \dfrac{6}{x - 2}$

Vertical asymptote: $x = 2$

Slant asymptote: $y = x - 1$

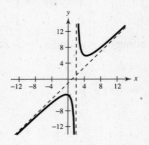

10. $g(x) = 3^{x+1} - 5$ is a horizontal shift one unit to the left, and a vertical shift five units downward.

11. $g(x) = -\log_{10}(x + 3)$ is a reflection in the x-axis and a horizontal shift three units to the left.

12. (a) All real numbers

(b) $8^{-x+1} > 0 \implies$ no x-intercepts

$\quad x = 0$: $y = 8 + 2 = 10$

Intercept: $(0, 10)$

(c) Asymptote: $y = 2$

13. (a) All real numbers

(b) $(0, 2)$

(c) Asymptote: $y = 0$

14. (a) All real numbers $x > 0$

(b) $(1, 0)$

(c) Asymptotes: $x = 0, y = 0$

15. (a) $2x - 1 > 0$

Domain: all $x > \frac{1}{2}$

(b) $\log_6(2x - 1) = -1$

$\quad\quad 6^{-1} = 2x - 1$

$\quad\quad\quad x = \frac{1}{2}\left(\frac{1}{6} + 1\right) = \frac{7}{12}$

Intercept: $\left(\frac{7}{12}, 0\right)$

(c) Asymptote: $x = \frac{1}{2}$

16. $\log_5 21 = \dfrac{\ln 21}{\ln 5} \approx 1.892$

17. $\log_9 6.8 = \dfrac{\ln 6.8}{\ln 9} \approx 0.872$

18. $\log_2\left(\dfrac{3}{2}\right) = \dfrac{\ln\left(\frac{3}{2}\right)}{\ln 2} \approx 0.585$

19. $\ln\!\left(\dfrac{x^2 - 4}{x^2 + 1}\right) = \ln[(x - 2)(x + 2)] - \ln(x^2 + 1)$

$\qquad\qquad = \ln(x - 2) + \ln(x + 2) - \ln(x^2 + 1)$

20. $2 \ln x - \ln(x - 1) + \ln(x + 1) = \ln\!\left[x^2\!\left(\dfrac{x + 1}{x - 1}\right)\right]$

21. $3^x + 5 = 32$

$\qquad 3^x = 27$

$\qquad\; x = 3$

22. $-2(6^{x+1} - 32) = 100$

$\qquad\quad 6^{x+1} - 32 = -50$

$\qquad\qquad\; 6^{x+1} = -18$

No solution

23. $3 \log_8(x + 1) = 2$

$\qquad \log_8(x + 1) = \tfrac{2}{3}$

$\qquad\qquad 8^{2/3} = x + 1$

$\qquad\qquad\quad 4 = x + 1$

$\qquad\qquad\quad x = 3$

24. $250e^{0.05x} = 500{,}000$

$\qquad\; e^{0.05x} = 2000$

$\qquad 0.05x = \ln 2000$

$\qquad\quad\; x = 20 \ln 2000$

$\qquad\qquad \approx 152.018$

25. $2x^2 e^{2x} - 2x e^{2x} = 0$

$\qquad (2x^2 - 2x)e^{2x} = 0$

$\qquad\qquad 2x^2 - 2x = 0$

$\qquad\qquad 2x(x - 1) = 0$

$\qquad\qquad\qquad x = 0, 1$

26. $\ln(2x - 5) - \ln x = 1$

$\qquad \ln \dfrac{2x - 5}{x} = 1$

$\qquad\qquad e = \dfrac{2x - 5}{x}$

$\qquad\qquad ex = 2x - 5$

$\qquad\; x(e - 2) = -5$

$\qquad\qquad x = \dfrac{-5}{e - 2} < 0$

No solution because

$\ln\!\left(\dfrac{-5}{e - 2}\right)$ does not exist.

27. $\begin{cases} 4x - 5y = 29 \\ -x + 9y = 16 \end{cases}$

From Equation 2, $x = 9y - 16$. Then

$4(9y - 16) - 5y = 29 \implies$

$\qquad\qquad 31y = 93 \implies y = 3,$

and $x = 9(3) - 16 = 11.$

Answer: $(11, 3)$

28. $\begin{cases} 2x - y^2 = 0 \\ x - y = 4 \implies x = y + 4 \end{cases}$

$\qquad 2(y + 4) - y^2 = 0$

$\qquad\qquad y^2 - 2y - 8 = 0$

$\qquad (y - 4)(y + 2) = 0$

$\qquad\qquad y = 4 \implies x = 8$

$\qquad\qquad y = -2 \implies x = 2$

$(2, -2), (8, 4)$

29. $4x - 3y = 0$

$2x - 9y = 5 \implies 4x - 18y = 10$

Subtracting, $15y = -10 \implies y = -\tfrac{2}{3} \implies x = -\tfrac{1}{2}.$

Answer: $\left(-\tfrac{1}{2}, -\tfrac{2}{3}\right)$

30. $\begin{cases} x - y + 3z = -1 \\ 2x + 4y + z = 2 \end{cases}$

$\begin{cases} x - y + 3z = -1 \\ \quad\; 6y - 5z = 4 \end{cases}$

Infinite number of solutions

Let $z = a.$

$\qquad 6y - 5a = 4 \implies y = \tfrac{1}{6}(4 + 5a)$

$x - \tfrac{1}{6}(4 + 5a) + 3a = -1 \implies x = \tfrac{1}{6}(-2 - 13a)$

Answer: $\left(\tfrac{1}{6}(-2 - 13a), \tfrac{1}{6}(4 + 5a), a\right)$

31. $\begin{bmatrix} 1 & 2 & -5 & \vdots & 1 \\ 3 & -2 & 1 & \vdots & 9 \\ 2 & 4 & 3 & \vdots & -11 \end{bmatrix}$ row reduces to

$\begin{bmatrix} 1 & 0 & 0 & \vdots & \frac{3}{2} \\ 0 & 1 & 0 & \vdots & -\frac{11}{4} \\ 0 & 0 & 1 & \vdots & -1 \end{bmatrix}$.

Answer: $\left(\frac{3}{2}, -\frac{11}{4}, -1\right)$

32. $\begin{bmatrix} 2 & 5 & -1 & \vdots & 4 \\ 1 & -1 & -3 & \vdots & 6 \\ 1 & 2 & 2 & \vdots & -6 \end{bmatrix}$ row reduces to

$\begin{bmatrix} 1 & 0 & 0 & \vdots & -2 \\ 0 & 1 & 0 & \vdots & 1 \\ 0 & 0 & 1 & \vdots & -3 \end{bmatrix}$.

Answer: $(-2, 1, -3)$

33. $3A - 2B = \begin{bmatrix} -7 & -10 & -16 \\ -6 & 18 & 9 \\ -12 & 16 & 7 \end{bmatrix}$

34. $5A + 3B = \begin{bmatrix} -18 & 15 & -14 \\ 28 & 11 & 34 \\ -20 & 52 & -1 \end{bmatrix}$

35. $AB = \begin{bmatrix} 3 & -31 & 2 \\ 22 & 18 & 6 \\ 52 & -40 & 14 \end{bmatrix}$

36. $BA = \begin{bmatrix} 5 & 36 & 31 \\ -36 & 12 & -36 \\ 16 & 0 & 18 \end{bmatrix}$

37. $\begin{vmatrix} 4 & -5 \\ 6 & -2 \end{vmatrix} = -8 + 30 = 22$

38. $\begin{vmatrix} 3 & 4 & \frac{1}{2} \\ 4 & -3 & 8 \\ -2 & 1 & 5 \end{vmatrix} = 3(-15 - 8) - 4(20 + 16) + \frac{1}{2}(4 - 6)$

$= 3(-23) - 4(36) - 1$

$= -214$

39. $\begin{vmatrix} 1 & 3 & 6 \\ 0 & -2 & 8 \\ 0 & 0 & 7 \end{vmatrix} = 1(-2)(7) = -14$

40. $\begin{vmatrix} -\frac{4}{5} & \frac{7}{5} & 1 \\ 0 & 3 & 1 \\ 2 & 0 & 1 \end{vmatrix} = -\frac{4}{5}(3) + 2\left(\frac{7}{5} - 3\right)$

$= -\frac{12}{5} - \frac{16}{5} = -\frac{28}{5}$

Area $= \frac{1}{2}\left(\frac{28}{5}\right) = \frac{14}{5}$

41. $\begin{vmatrix} -2 & -8 & 1 \\ 3 & 7 & 1 \\ 7 & 19 & 1 \end{vmatrix} = 0 \implies$ the points are collinear.

42. (a) $A = 200\left(1 + \dfrac{0.06}{12}\right)^{30(12)} \approx \$1204.52$

(b) $A = 200e^{0.06(30)} \approx \$1209.93$

43. $P = 228e^{kt}$

$166 = 228e^{k(-30)} \implies k \approx 0.01058$

For 2010, $P = 228e^{k(10)} \approx 253.4$ or 253,400 people.

44. (a) Quadratic model: $y = 0.0707x^2 - 0.183x + 1.45$, Coefficient of determination 0.99871

Exponential model: $y = 0.8915(1.2106)^x$, Coefficient of determination: 0.98862

Power model: $y = 0.7865x^{0.6762}$, Coefficient of determination: 0.93130

—CONTINUED—

44. —CONTINUED—

(b) Quadratic model: Exponential model: Power model:

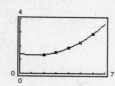   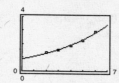   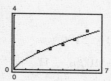

(c) The quadratic model is best because its coefficient of determination is closest to 1.

(d) For 2008, $x = 8$ and $y \approx \$4.51$. For 2010, $x = 10$ and $y \approx \$6.69$. Answers will vary.

45. (a) $C = 59.95x + 150,000$

$R = 200x$

(b)

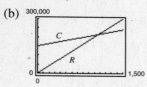

$59.95x + 150,000 = 200x$

$150,000 = 140.05x$

$x \approx 1071$ units

Must sell 1071 units to break even.

46. $\begin{cases} 2l + 2w = 76 \\ lw = 352 \implies l = \dfrac{352}{w} \end{cases}$

$2\left(\dfrac{352}{w}\right) + 2w = 76$

$704 + 2w^2 = 76w$

$w^2 - 38w + 352 = 0$

$(w - 22)(w - 16) = 0$

$w = 16, 22$

Dimensions: 16×22 meters

Chapter 6   Chapter Test Solutions

1. $a_n = \left(-\dfrac{2}{3}\right)^{n-1}$

$a_1 = \left(-\dfrac{2}{3}\right)^{1-1} = \left(-\dfrac{2}{3}\right)^0 = 1$

$a_2 = -\dfrac{2}{3}$

$a_3 = \left(-\dfrac{2}{3}\right)^2 = \dfrac{4}{9}$

$a_4 = \left(-\dfrac{2}{3}\right)^3 = -\dfrac{8}{27}$

$a_5 = \left(-\dfrac{2}{3}\right)^4 = \dfrac{16}{81}$

2. $a_1 = 12, \ a_{k+1} = a_k + 4$

$a_2 = 12 + 4 = 16$

$a_3 = 16 + 4 = 20$

$a_4 = 20 + 4 = 24$

$a_5 = 24 + 4 = 28$

3. $b_1 = -x$

$b_2 = \dfrac{x^2}{2}$

$b_3 = -\dfrac{x^3}{3}$

$b_4 = \dfrac{x^4}{4}$

$b_5 = -\dfrac{x^5}{5}$

4. $b_1 = -\dfrac{x^3}{3!} = -\dfrac{x^3}{6}$

$b_2 = -\dfrac{x^5}{5!} = -\dfrac{x^5}{120}$

$b_3 = -\dfrac{x^7}{7!}$

$b_4 = -\dfrac{x^9}{9!}$

$b_5 = -\dfrac{x^{11}}{11!}$

5. $\dfrac{11!4!}{4!7!} = \dfrac{11!}{7!}$

$= \dfrac{11 \cdot 10 \cdot 9 \cdot 8 \cdot 7!}{7!}$

$= 11 \cdot 10 \cdot 9 \cdot 8$

$= 7920$

6. $\dfrac{n!}{(n+1)!} = \dfrac{n!}{(n+1)n!} = \dfrac{1}{n+1}$ **7.** $\dfrac{2n!}{(n-1)!} = \dfrac{2n(n-1)!}{(n-1)!} = 2n$ **8.** $a_n = n^2 + 1,\ n = 1, 2, 3, \ldots$

9. $a_n = dn + c$

$c = a_1 - d = 5000 - (-100) = 5100 \implies$

$a_n = -100n + 5100 = 5000 - 100(n-1)$

10. $a_n = a_1 r^{n-1},\ a_1 = 4,\ r = \frac{1}{2} \implies a_n = 4\left(\frac{1}{2}\right)^{n-1}$

11. $\displaystyle\sum_{n=1}^{12} \dfrac{2}{3n+1}$

12. $2 + \dfrac{1}{2} + \dfrac{1}{8} + \dfrac{1}{32} + \cdots = \displaystyle\sum_{n=1}^{\infty} 2\left(\dfrac{1}{4}\right)^{n-1}$

$= \displaystyle\sum_{n=0}^{\infty} \dfrac{1}{2^{2n-1}}$

13. $\displaystyle\sum_{n=1}^{7} (8n - 5) = 8\left(\dfrac{7(8)}{2}\right) - 5(7) = 224 - 35 = 189$

14. $\displaystyle\sum_{n=1}^{8} 24\left(\dfrac{1}{6}\right)^{n-1} = 24\left(\dfrac{1 - (1/6)^8}{1 - (1/6)}\right)$

$= 24\left(\dfrac{6}{5}\right)\left(1 - \left(\dfrac{1}{6}\right)^8\right)$

$\approx 28.79998 \approx 28.80$

15. $\displaystyle\sum_{n=0}^{\infty} \dfrac{(-1)^n 2^n}{5^{n-1}} = \displaystyle\sum_{n=0}^{\infty} 5\left(\dfrac{(-1)2}{5}\right)^n$

$= 5\left(\dfrac{1}{1 - (-2/5)}\right)$

$= 5\left(\dfrac{5}{7}\right)$

$= \dfrac{25}{7}$

16. (1) For $n = 1$, $3 = \dfrac{3(1)(1+1)}{2}$.

(2) Assume $S_k = 3 + 6 + \cdots + 3k = \dfrac{3k(k+1)}{2}$.

Then $S_{k+1} = 3 + 6 + \cdots + 3k + 3(k+1)$

$= S_k + 3(k+1)$

$= \dfrac{3k(k+1)}{2} + 3(k+1)$

$= \dfrac{k+1}{2}[3k + 6]$

$= \dfrac{3(k+1)(k+2)}{2}$.

Therefore, the formula is true for all positive integers n.

17. $(2a - 5b)^4 = (2a)^4 - 4(2a)^3(5b) + 6(2a)^2(5b)^2 - 4(2a)(5b)^3 + (5b)^4$

$= 16a^4 - 160a^3b + 600a^2b^2 - 1000ab^3 + 625b^4$

18. $_9C_3 = 84$ **19.** $_{20}C_3 = 1140$ **20.** $_9P_2 = \dfrac{9!}{7!}$ **21.** $_{70}P_3 = \dfrac{70!}{67!}$

$= 9 \cdot 8 = 72$

$= 70 \cdot 69 \cdot 68$

$= 328{,}440$

22. $4 \cdot {}_nP_3 = {}_{n+1}P_4$

$$4\frac{n!}{(n-3)!} = \frac{(n+1)!}{(n-3)!}$$

$$4n! = (n+1)!$$

$$n = 3$$

23. $26 \cdot 10 \cdot 10 \cdot 10 = 26{,}000$ ways

24. ${}_{25}C_4 = \dfrac{25!}{21! \, 4!} = \dfrac{25 \cdot 24 \cdot 23 \cdot 22}{24} = 12{,}650$ ways

25. There are 6 red face cards $\Rightarrow$ probability $= \frac{6}{52} = \frac{3}{26}$.

26. $\dfrac{1}{{}_{11}C_6} = \dfrac{1}{462}$

27. (a) $\left(\dfrac{30}{60}\right)\left(\dfrac{30}{60}\right) = \dfrac{1}{2} \cdot \dfrac{1}{2} = \dfrac{1}{4}$

(b) $\dfrac{11}{60} \cdot \dfrac{11}{60} = \dfrac{121}{3600} \approx 0.0336$

(c) $\dfrac{1}{60} \approx 0.0167$

28. $1 - 0.75 = 0.25 = 25\%$

Chapter 7   Chapter Test Solutions

1. $y^2 = 8x = 4(2)x$

Vertex: $(0, 0)$

Focus: $(2, 0)$

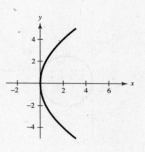

2. $4x^2 + y^2 = 4$

$$x^2 + \frac{y^2}{4} = 1$$

Vertices: $(0, \pm 2)$

$$c^2 = a^2 - b^2$$

$$= 4 - 1 = 3$$

Foci: $\left(0, \pm\sqrt{3}\right)$

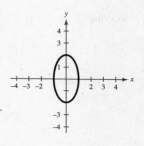

3. $x^2 - 4x - 4y^2 = 0$

$$x^2 - 4x + 4 - 4y^2 = 4$$

$$(x - 2)^2 - 4y^2 = 4$$

$$\frac{(x-2)^2}{4} - y^2 = 1$$

Hyperbola

Center: $(2, 0)$

Vertices: $(4, 0), (0, 0)$

$c^2 = a^2 + b^2 = 2^2 + 1^2 = 5$

Foci: $\left(2 \pm \sqrt{5}, 0\right)$

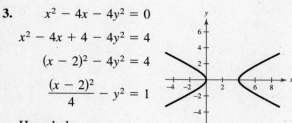

4. Center: $(0, 0)$

Radius $= \dfrac{1}{2}(16) = 8$

$x^2 + y^2 = 64$

5. Center: $\left(\dfrac{2-8}{2}, \dfrac{-8+2}{2}\right) = (-3, -3)$

Radius $= \dfrac{1}{2}\sqrt{(2-(-8))^2 + (-8-2)^2} = 5\sqrt{2}$

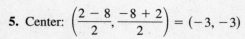

$(x + 3)^2 + (y + 3)^2 = 50$

6. Vertex: $(0, 7)$

$$(y - 7)^2 = 4p(x - 0)$$

$$(0 - 7)^2 = 4p(-8 - 0)$$

$$49 = -32p \implies p = -\frac{49}{32}$$

$$(y - 7)^2 = -\frac{49}{8}x$$

$$(y - 7)^2 = 4\left(-\frac{49}{32}\right)x$$

7. Vertex: $(6, -2), p = 2$

$$(y + 2)^2 = 4(2)(x - 6)$$

$$(y + 2)^2 = 8(x - 6)$$

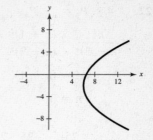

8. Center: $(4, 2), a = 4, b = 2$

$$\frac{(x - 4)^2}{16} + \frac{(y - 2)^2}{4} = 1$$

9. Center: $(-6, 3), a = 7, b = 4$

$$\frac{(x + 6)^2}{16} + \frac{(y - 3)^2}{49} = 1$$

10. Vertical transverse axis, $a = 3$

Asymptotes:

$$y = \pm\frac{a}{b}x = \pm\frac{3}{2}x \implies b = 2$$

$$\frac{y^2}{9} - \frac{x^2}{4} = 1$$

11. Center: $(-1, 6)$

$$a = 6, b = \frac{1}{4}$$

Asymptotes:

$$y = 6 \pm \frac{6}{1/4}(x + 1)$$

$$y = 6 \pm 24(x + 1)$$

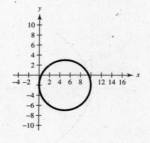

12. $(x^2 - 10x + 25) + (y^2 + 4y + 4) = -4 + 25 + 4$

$$(x - 5)^2 + (y + 2)^2 = 25, \text{ circle}$$

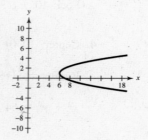

13. $y^2 - 2y + 1 = x - 7 + 1$

$$(y - 1)^2 = x - 6, \text{ parabola}$$

14. $(x^2 + 20x + 100) + 4(y^2 - 10y + 25) = 300 + 100 + 100$

$$(x + 10)^2 + 4(y - 5)^2 = 500$$

$$\frac{(x + 10)^2}{500} + \frac{(y - 5)^2}{125} = 1, \text{ ellipse}$$

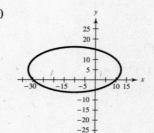

15. $x = t^2 - 6, y = \frac{1}{2}t - 1$

$t = 2(y + 1) \implies x = [2(y + 1)]^2 - 6$

$\qquad\qquad\qquad = 4(y + 1)^2 - 6 \implies$

$(y + 1)^2 = \frac{1}{4}(x + 6), \quad$ parabola

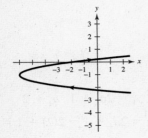

16. $x = \sqrt{t^2 + 2}, y = \frac{t}{4}$

$t = 4y \implies x = \sqrt{16y^2 + 2} \implies$

$\qquad\qquad\qquad x^2 = 16y^2 + 2 \implies$

$\qquad\qquad x^2 - 16y^2 = 2$

Right portion of hyperbola, $\left(x \geq \sqrt{2}\right)$

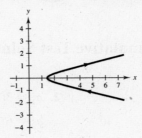

17. $x = 4t, y = |t - 6|$

$t = \frac{x}{4} \implies y = \left|\frac{x}{4} - 6\right| = \frac{1}{4}|x - 24|$

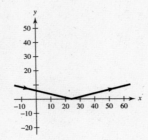

18.

19. $x = t, y = 7t + 6$

$\quad x = -t, y = -7t + 6$

(other answers possible)

20. $x = t, y = t^2 + 10$

$\quad x = t + 1, y = (t + 1)^2 + 10$

(other answers possible)

21. Sample answers:

$x = 4 - t^2$

$y = t$

$x = 4 - 4t^2$

$y = 2t$

22. Sample answers:

$x = \pm\sqrt{16 - 4t^2}$

$y = t$

$x = \pm\sqrt{16 - t^2}$

$y = \frac{1}{2}t$

23. Vertex: $(0, 16) \implies x^2 = 4p(y - 16)$

$(6, 14)$ on parabola $\implies 36 = 4p(14 - 16) = -8p \implies p = -\frac{9}{2}$

$x^2 = 4\left(-\frac{9}{2}\right)(y - 16)$

$x^2 = -18(y - 16)$

Let $y = 0 \implies x^2 = -18(0 - 16) = 288 \implies x = \pm12\sqrt{2}.$

Width $= 2\left(12\sqrt{2}\right) = 24\sqrt{2} \approx 33.94$ meters

24. $\dfrac{x^2}{92.5^2} + \dfrac{y^2}{109.5^2} = 1$

$\dfrac{x^2}{8556.25} + \dfrac{y^2}{11{,}990.25} = 1$

Ellipse

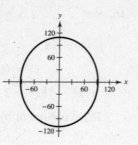

Chapters 6–7   Cumulative Test Solutions

1. $a_1 = \dfrac{(-1)^{1+1}}{2(1)+3} = \dfrac{1}{5}$

$a_2 = \dfrac{-1}{7}$

$a_3 = \dfrac{1}{9}$

$a_4 = \dfrac{-1}{11}$

$a_5 = \dfrac{1}{13}$

2. $a_1 = 3(2)^{1-1} = 3$

$a_2 = 6$

$a_3 = 12$

$a_4 = 24$

$a_5 = 48$

3. $a_{n+1} = 2a_n + 1,\; a_1 = 2$

$a_2 = 2a_1 + 1 = 2(2) + 1 = 5$

$a_3 = 2(5) + 1 = 11$

$a_4 = 23$

$a_5 = 47$

4. $a_n = \dfrac{(-1)^n x^n}{n!}$

$a_1 = \dfrac{(-1)^1 x^1}{1!} = -x$

$a_2 = \dfrac{(-1)^2 x^2}{2!} = \dfrac{x^2}{2}$

$a_3 = -\dfrac{x^3}{6}$

$a_4 = \dfrac{x^4}{24}$

$a_5 = -\dfrac{x^5}{120}$

5. $\dfrac{49!}{46!} = \dfrac{49 \cdot 48 \cdot 47 \cdot 46!}{46!}$

$\quad = 49 \cdot 48 \cdot 47$

$\quad = 110{,}544$

6. $\dfrac{10!}{5!2!} = \dfrac{10 \cdot 9 \cdot 8 \cdot 7 \cdot 6 \cdot 5!}{5!2!}$

$\quad = 10 \cdot 9 \cdot 8 \cdot 7 \cdot 3$

$\quad = 15{,}120$

7. $\dfrac{(n+1)!}{2n!} = \dfrac{(n+1)n!}{2n!}$

$\quad = \dfrac{n+1}{2}$

8. $\dfrac{(2n)!}{(2n+1)!} = \dfrac{(2n)!}{(2n+1)(2n)!}$

$\quad = \dfrac{1}{2n+1}$

9. $\displaystyle\sum_{k=1}^{6} (7k-2) = 7\dfrac{6(7)}{2} - 2(6)$

$\quad = 7(21) - 12$

$\quad = 135$

10. $\displaystyle\sum_{i=1}^{250} (4i^3 + 5) = 4\sum_{i=1}^{250} i^3 + \sum_{i=1}^{250} 5$

$\quad = \dfrac{4(250)^2(251)^2}{4} + 5(250)$

$\quad = 3{,}937{,}563{,}750$

11. $\displaystyle\sum_{i=1}^{100}(6i^2 + 4i - 3) = 6\sum_{i=1}^{100}i^2 + 4\sum_{i=1}^{100}i - \sum_{i=1}^{100}3$

$$= 6\frac{100(101)(201)}{6} + 4\frac{100(101)}{2} - 300$$

$$= 2{,}050{,}000$$

12. $\displaystyle\sum_{k=3}^{6}(k-1)(k+2) = \sum_{k=3}^{6}(k^2 + k - 2)$

$$= 10 + 18 + 28 + 40 = 96$$

13. $\displaystyle\sum_{n=0}^{10}9\left(\frac{3}{4}\right)^n = \sum_{n=1}^{11}9\left(\frac{3}{4}\right)^{n-1}$

$$= 9\frac{1 - (3/4)^{11}}{1 - (3/4)}$$

$$= 36\left(1 - \left(\frac{3}{4}\right)^{11}\right)$$

$$\approx 34.4795$$

14. $\displaystyle\sum_{n=1}^{\infty}8(0.9)^{n-1} = \sum_{n=0}^{\infty}8(0.9)^n = \frac{8}{1 - 0.9} = \frac{8}{0.1} = 80$

15. For $n = 1$, $3 = 1(2(1) + 1) = 3$. Assume true for $n = k$. Then,

$$3 + 7 + \cdots + (4k - 1) + [4(k + 1) - 1] = k(2k + 1) + [4k + 3]$$

$$= 2k^2 + 5k + 3 = (2k + 3)(k + 1)$$

$$= (k + 1)[2(k + 1) + 1].$$

16. $(x + 5)^4 = x^4 + 20x^3 + 150x^2 + 500x + 625$

17. $(2x + y^2)^5 = 32x^5 + 80x^4y^2 + 80x^3y^4 + 40x^2y^6 + 10xy^8 + y^{10}$

18. $(x - 2y)^6 = x^6 - 12x^5y + 60x^4y^2 - 160x^3y^3 + 240x^2y^4 - 192xy^5 + 64y^6$

19. $(2x - 1)^8 = 256x^8 - 1024x^7 + 1792x^6 - 1792x^5 + 1120x^4 - 448x^3 + 112x^2 - 16x + 1$

20. $\dfrac{5!}{2!2!1!} = \dfrac{120}{4} = 30$ **21.** $\dfrac{6!}{3!} = 6 \cdot 5 \cdot 4 = 120$ **22.** $\dfrac{10!}{2!2!2!} = 453{,}600$ **23.** $\dfrac{10!}{3!2!2!} = 151{,}200$

24. Hyperbola

Center: $(-5, -3)$

Vertical transverse axis

Vertices: $(-5, 3), (-5, -9)$

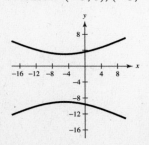

25. Ellipse with center $(2, -1)$

Vertices: $(2, 2), (2, -4)$

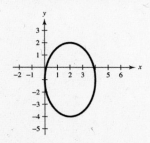

26. Hyperbola: $\dfrac{y^2}{16} - \dfrac{x^2}{16} = 1$

Center: $(0, 0)$

Vertices: $(0, \pm 4)$

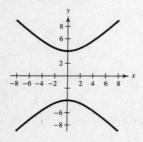

27. $(x^2 - 2x + 1) + (y^2 - 4y + 4) = -5 + 1 + 4 = 0$

$$(x - 1)^2 + (y - 2)^2 = 0$$

Point: $(1, 2)$

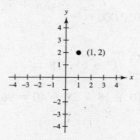

28. $y^2 - 6y + 9 = 6x - 42 + 9$

$(y - 3)^2 = 6x - 33$

Parabola

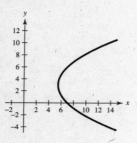

29. $9(x^2 + 6x + 9) + 25y^2 = 144 + 81$

$9(x + 3)^2 + 25y^2 = 225$

$$\dfrac{(x + 3)^2}{25} + \dfrac{y^2}{9} = 1$$

Ellipse

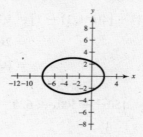

30. Radius: $\sqrt{(6 - (-2))^2 + (1 - 4)^2} = \sqrt{73}$

Center: $(-2, 4)$

$(x + 2)^2 + (y - 4)^2 = 73,$ circle

31. Vertex: $(2, 3) \implies (x - 2)^2 = 4p(y - 3)$

$(0, 0)$ on parabola

$(0 - 2)^2 = 4p(0 - 3)$

$4 = 4p(-3) \implies p = -\frac{1}{3}$

$(x - 2)^2 = 4\left(-\frac{1}{3}\right)(y - 3)$

$(x - 2)^2 = -\frac{4}{3}(y - 3)$

32. Center: $(1, 4), a = 5, b = 2$

$$\dfrac{(x - 1)^2}{25} + \dfrac{(y - 4)^2}{4} = 1$$

33. Center: $(0, -4), a = 2$

$$\dfrac{(y + 4)^2}{4} - \dfrac{x^2}{b^2} = 1$$

$(4, 0)$ on hyperbola

$$\dfrac{(0 + 4)^2}{4} - \dfrac{4^2}{b^2} = 1 \implies$$

$$\dfrac{16}{b^2} = 4 - 1 = 3 \implies b^2 = \dfrac{16}{3}$$

$$\dfrac{(y + 4)^2}{4} - \dfrac{x^2}{16/3} = 1$$

34. The center is $(0, 2)$ and $c = 2$. The transverse axis is vertical and $\dfrac{a}{b} = \dfrac{1}{2}$.

$$b^2 = c^2 - a^2 = 4 - \left(\frac{b}{2}\right)^2 \implies \frac{5b^2}{4} = 4 \implies b^2 = \frac{16}{5} \text{ and } a^2 = \frac{4}{5}$$

Thus, $\dfrac{(y-2)^2}{4/5} - \dfrac{x^2}{16/5} = 1$.

35. $a = \dfrac{[6 - (-2)]}{2} = 4$

$b = 1$

Center: $\left(\dfrac{6-2}{2}, 3\right) = (2, 3)$

$\dfrac{(x-2)^2}{16} + \dfrac{(y-3)^2}{1} = 1$

36. $(y - k)^2 = 4p(x - h)$

$p = 2$

$(y - 2)^2 - 4(2)(x + 6)$

$(y - 2)^2 = 8(x + 6)$

37. $x = 2t + 1, \ y = t^2$

(a), (b)

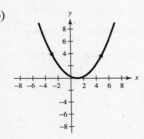

(c) $t = \frac{1}{2}(x - 1) \implies y = \left(\frac{1}{2}(x - 1)\right)^2$, parabola

38. $x = 8 + 3t, \ y = 4 - t$

(a), (b)

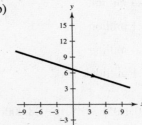

(c) $t = \frac{1}{3}(x - 8) \implies y = 4 - \frac{1}{3}(x - 8)$

$\qquad\qquad\qquad\quad = -\frac{1}{3}x + \frac{20}{3}, \ \text{line}$

39. $x = 4 \ln t, \ y = \frac{1}{2}t^2$

(a), (b)

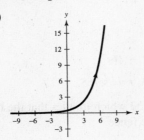

(c) $t = e^{x/4} \implies y = \frac{1}{2}e^{x/2}$

40. $y = 3x - 2$

Sample answers:

$x = t$

$y = 3t - 2$

$x = -t$

$y = -3t - 2$

41. Sample answers:

$x = \pm\sqrt{t^2 + 16}$

$y = t$

$x = t$

$y = \pm\sqrt{t^2 - 16}$

42. $y = \dfrac{2}{x}$

Sample answers:

$x = t$

$y = \dfrac{2}{t}$

$x = -t$

$y = -\dfrac{2}{t}$

43. Sample answers:

$x = t$

$y = \dfrac{e^{2t}}{e^{2t} + 1}$

$x = \dfrac{1}{2}t$

$y = \dfrac{e^{t}}{e^{t} + 1}$

44. $x = 2 + (6 - 2)t \implies x = 4t + 2$

$y = -3 + (4 + 3)t \qquad y = 7t - 3$

45. (a) $a_n = 288.25n^2 - 1038.9$

$a_2 = 288.25(2)^2 - 1038.9 \approx \$114.1$ million (2002)

$a_3 \approx \$1555.4$ million (2003)

$a_4 \approx \$3573.1$ million (2004)

$a_5 \approx \$6167.4$ million (2005)

$a_6 \approx \$9338.1$ million (2006)

(b) Adding, total revenue $\approx \$20,748$ million.

(c) $a_7 \approx \$13,085$ million (2007)

$a_8 \approx \$17,409$ million (2008)

$a_9 \approx \$22,309$ million (2009)

$a_{10} \approx \$27,786$ million (2010)

46. $28,000 + 28,000(1.05) + \cdots + 28,000(1.05)^{14} = 28,000 \displaystyle\sum_{n=1}^{15} (1.05)^{n-1}$

$$= 28,000 \dfrac{1 - 1.05^{15}}{1 - 1.05} \approx \$604,199.78$$

47. There are two choices for the first digit (4 or 5). Then there remain two choices for the second digit, and one for the third. Thus, probability $= \frac{1}{4}$.

48. $a = 8, b = 7$

$\dfrac{x^2}{49} + \dfrac{y^2}{64} = 1$

49. $x = 99.6t$

$y = 3 + 57.5t - 16t^2$

When $x = 375$, $t \approx 3.765$.

Then $y \approx -7.32$ which implies that the ball does not go over the fence.